Foundations of ELECTRIC POWER

J.R. COGDELL

University of Texas at Austin
ECE Department
Austin, Texas

PRENTICE HALL

Upper Saddle River, New Jersey 07458

Library of Congress Cataloging-in Publication Data

Cogdell, J. R.
Foundation of electrical power / J. R. Cogdell.
p. cm.
Includes bibliographical references and index.
ISBN 0-13-907767-7
1. Electric motors 2. Power electronics I. Title.
TK2511.C64 1999
621.46—dc21 98—44923
CIP

Acquisitions editor: **Alice Dworkin**
Editorial/production supervision: **Sharyn Vitrano**
Copy editor: **Barbara Danziger**
Managing editor: **Eileen Clark**
Editor-in-chief: **Marcia Horton**
Director of production and manufacturing: **David W. Riccardi**
Manufacturing buyer: **Pat Brown**
Editorial assistant: **Dan De Pasquale**

A Pearson Education Company
Upper Saddle River, NJ 07458

Printed in the United States of America
10 9 8 7 6 5 4 3 2 1

ISBN 0-13-907767-7

Prentice-Hall International (UK) Limited,London
Prentice-Hall of Australia Pty. Limited, Sydney
Prentice-Hall Canada Inc., Toronto
Prentice-Hall Hispanoamericana, S.A., Mexico
Prentice-Hall of India Private Limited, New Delhi
Prentice-Hall of Japan, Inc., Tokyo
Pearson Education Asia Pte. Ltd., Singapore
Editora Prentice-Hall do Brasil, Ltda., Rio de Janeiro

Contents

Preface

The need for this book.

About 10 years ago, I started teaching electrical engineering to nonmajors. We used a well-known text, but I found that my students had trouble with the book. The encyclopedic scope of this text of necessity forced it to be superficial. I wanted a text that presented the important ideas in depth and left many of the details for future learning in further study or professional practice. So I wrote *Foundations of Electrical Engineering,* choosing "Foundation" for the title to point to the few important principles that upheld the entire superstructure of electrical engineering. *Foundations of Electrical Engineering* was well received and went into a much-changed second edition in 1996.

The Foundations series. The *Foundations of Electrical Engineering* has now been divided into three stand-alone books: *Foundations of Electric Circuits, Foundations of Electronics,* and *Foundations of Electric Power.* The purpose was to reduce the student's cost and to target specific one-semester (one- or two-quarter) courses as follows:

One-semester course	Foundation books required
Circuits for majors or nonmajors	*Foundations of Electric Circuits*
Circuits and electronics for nonmajors	*Foundations of Electric Circuits* (skipping Chaps. 5 and 7) and *Foundations of Electronics*
Electronics for nonmajors	*Foundations of Electronics*
Power for majors or nonmajors	*Foundations of Electric Power*

For a full-year survey of electrical engineering, including topics in circuits, electronics, and power, *Foundations of Electrical Engineering* is the best choice.

The present volume is comprised of Chaps. 6, 13–18 of *Foundations of Electrical Engineering*, 2d Edition, specifically the chapters on topics related to electric power. The only addition is Appendix A on Linear Systems. These materials were designed for a course at the University of Texas for electrical and mechanical engineers that emphasizes electric motors. The existence of such a course reflects the present-day reality that very few electrical engineers study electric power. Electric power is certainly not losing importance; the knowledge simply is moving from electrical engineers to the end user such as the mechanical engineer who needs to choose a motor to drive a pump.

This book, therefore, is written to all engineers. For this reason we have included a chapter on electrical physics, since most likely the student has thought little about electric and magnetic fields since the freshman physics course. Similarly, Appendix A on Linear Systems is included to serve the student who has not had a course on this subject but needs to consider the dynamic response of a motor in a system. Without going into great depth or mathematical sophistication, this appendix introduces the principal ideas and common notation in system theory and supports sections in the text that discuss motors as system components.

Prerequisites

We address this book to students who have completed one year of college calculus and physics. We work with linear differential equations and the algebra of complex numbers, but the techniques of solution are explained at an appropriate level. A few simple ideas from mechanics are introduced in discussions of dynamics and steady-state operations of motors, primarily Newton's second law. Most important is knowledge of electric circuit theory, including AC circuits and power and energy relationships in AC circuits. The necessary background is found in our *Foundations of Electric Circuits*.

Pedagogy of the book.

One way to view the structure of an engineering subject is shown in the following diagram:

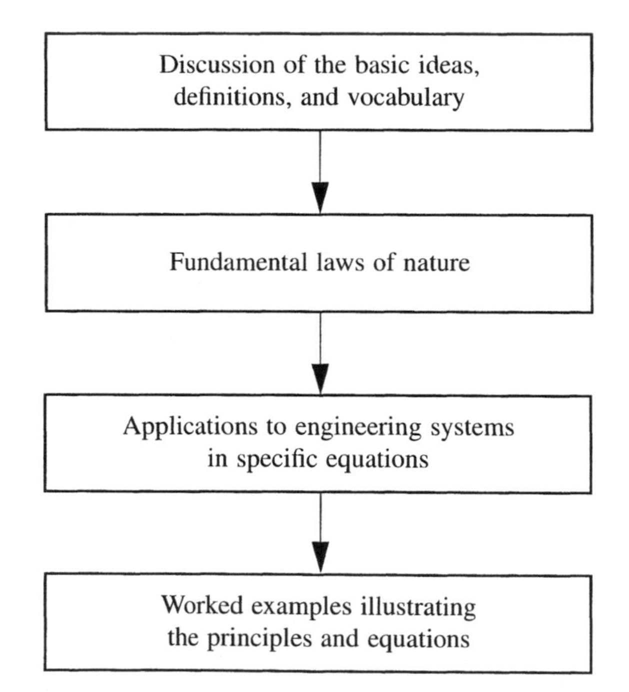

The instructor's emphasis goes from top to bottom. The structure in the instructor's (and author's) mind is primarily ideas and vocabulary, followed by laws, equations, then examples—from the general to the specific. But most students seem to learn in the opposite order: first examples, then equations, then laws, and finally ideas. One suspects some students never go beyond studying the examples, and certainly many believe the equations are what's important. When the instructor probes their understanding of the more general principles by giving a quiz problem that differs from previous examples, many students serve up memorized "solutions" or protest that "the quiz wasn't anything like the lectures and homework." The goals and needs of both instructors and students were paramount in the writing and design of this book.

Aids to learning

This text addresses the viewpoints of both students and instructors in a number of ways

Conservation of Energy

- **Ideas.** We have identified eight basic ideas of electrical engineering and identified them with a lightbulb icon in the margin, as shown at left. All but two of these ideas are used in the present volume. These icons hopefully will convince students that electrical engineering is based upon a few foundational ideas that come up again and again.

vocabulary for important words

- **Key terms** are italicized and emphasized by a note in the margin, as shown, when they are first introduced and defined. *Vocabulary* cannot be overemphasized because we communicate with others and, indeed, do our own thinking with words. Each chapter ends with a Glossary that defines many key terms and refers to the context where the words first appear in the text.
- **Causality diagrams**[1] that show cause/effect relationships are given in complicated situations. One reason students have trouble solving problems is that they look at equations and don't see what the important variables are. Understanding consists largely in knowing the causal connections between various factors in a problem so that equations are written with a purpose. Causality diagrams picture these cause/effect relationships.

LEARNING OBJECTIVE

This alerts students to important material

- **Objectives.** Chapters begin with stated objectives and end with summaries that review how those objectives have been met. In between, we place marginal pointers that alert the student to where the material relates directly to one of the stated objectives. Our intention is to give road signs along the way to keep our travelers from losing their way.

EXAMPLE P.1

This is the title of an example

The numerous examples are boxed, titled, and numbered.

SOLUTION:
Solutions are differentiated from problem statements, as shown. Students are lured beyond passively studying the examples by a WHAT IF? challenge at the end.

WHAT IF?

What if the student were asked to rework the example with a slight change?[2] The answer to the WHAT IF? challenges appears in a footnote for easy checking of results.

---

[1] See page 168 for an example.

[2] That would get them involved, wouldn't it?

- **Check your understanding.** We have "Check Your Understanding" questions and problems, with answers, after major sections.

Problems

Three types of problems have different levels of difficulty and appear at three places in the development

1. WHAT IF? challenges follow most examples. These problems present a slight variation on the examples and are intended to involve the student actively in the principles illustrated by the examples. The answers are given in a footnote.
2. Check Your Understanding problems follow most sections. These are intended as occasions for review and quick self-testing of the material in each section. The answers follow the problems.
3. The numerous end-of-chapter problems range from straightforward applications, similar to the examples, to quite challenging problems requiring insight and refined problem-solving skills. Answers are given to the odd-numbered problems. We are convinced that the only path to becoming a good problem solver passes through a forest of nontrivial problems.

Acknowledgments

I gratefully acknowledge the assistance of many colleagues at the University of Texas at Austin: Lee Baker, David Bourell, David Brown, John Davis, Mircea Driga, "Dusty" Duesterhoeft, Bill Hamilton, Om Mandhana, Charles Roth, Irwin Sandberg, Ben Streetman, Jon Valvano, Bill Weldon, Paul Wildi, Quanghan Xu, and no doubt others. My warmest thanks go to my friend, Jian-Dong Zhu, who worked side by side with me in checking the answers to the end-of-chapter problems. Special thanks go to my son-in-law, David Brydon, who introduced me to Mathematica and to my son, Thomas Cogdell, who produced most of the figures in the text. My heartfelt thanks for numerous corrections, improvements, and wise advice go to the reviewers of *Foundations of Electrical Engineering*: William E. Bennett, U.S. Naval Academy; Richard S. Marleau, University of Wisconsin-Madison; Phil Noe, Texas A&M University; Ed O'Hair, Texas Tech University; and Terry Sculley and Carl Wells, Washington State University and to Dolon Williams for numerous corrections and suggestions.

I wish to thank my wife for her support and encouragement. Finally, I thank the Giver of all good gifts for the joy I have in teaching and writing about electrical engineering.

John R. Cogdell

CHAPTER

1 Electric Power Systems

Three-Phase Power

Power Distribution Systems

Introduction to Electric Motors

Chapter Summary

Glossary

Problems

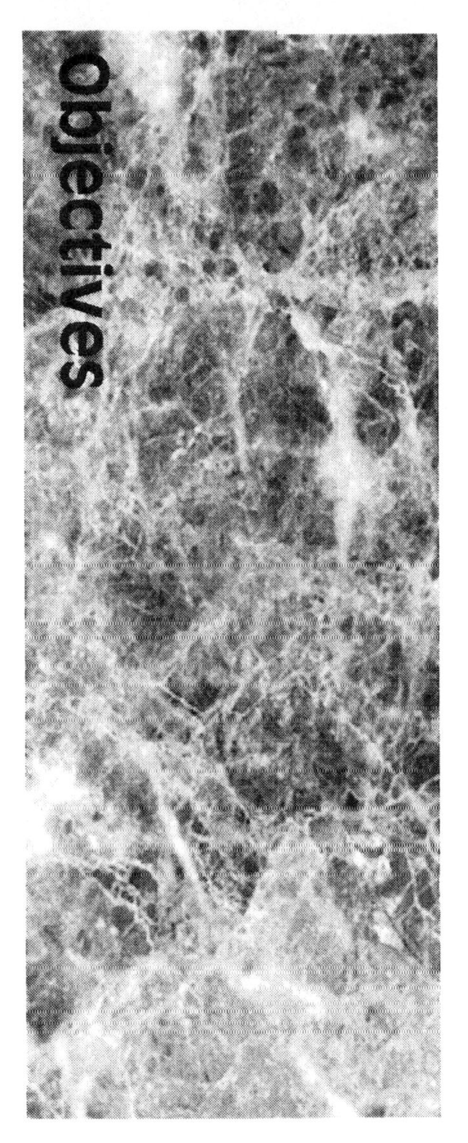

1. To understand the nature and advantages of three-phase power
2. To understand how to connect a generator in wye or delta
3. To understand how to calculate voltage, current, and power in balanced wye- and delta-connected loads
4. To understand how to derive and use the per-phase equivalent circuit of a balanced three-phase load
5. To understand how to connect single-phase transformers for three-phase transformation
6. To understand how transmission-line impedance affects power and voltage loss on the line
7. To understand how motor and load interact to establish steady-state operation
8. To understand how to analyze and interpret nameplate information of induction motors

This chapter applies the concepts of ac circuit to power distribution systems and electric motors. Three-phase circuits are introduced, including three-phase transformers. The final section on electric motors examines motor–load interaction and shows how to analyze the nameplate information on the two most common types of electric motors.

1.1 THREE-PHASE POWER

Importance of Electric Power Systems

The generation and distribution of electric power is the business of large electric utility companies. Various fuels and sources are used, such as coal, natural gas, fuel oil, nuclear, and water power, not to mention some quaint and futuristic schemes such as wind power, tidal and wave power, solar power, and burning household garbage.

The normal way to generate[1] electrical power is to burn the primary fuel and produce high-pressure steam in a boiler. The steam drives a turbine that turns an electrical generator. The voltage from the generator is transformed to high voltage for transmission over long distances. For reliability, all the generators in a geographic region are synchronized and are linked together with transmission lines to exchange real and reactive power.

Near the location of the industrial or residential customers, the voltage is lowered from transmission levels, typically, 120–500 kV, to distribution levels, typically 4.8–34 kV. For the residential consumer the voltage is lowered to 120/240 V, single-phase.

The power is generated, transmitted, and distributed in three-phase form. Only very near to the customer is the power changed from three-phase to single-phase power. This chapter begins with the introduction and study of three-phase electric systems. We then discuss three-phase transmission and distribution systems. Finally, we give a brief introduction to electric motors.

Introduction to Three-Phase Power Systems

Importance of three-phase systems. If you looked out your window at this moment, you would probably see some power lines. Count the wires and you will likely find there to be four. Go examine a pole closely and you will see that at each pole, one of the four wires is connected to a conductor that comes down the pole and enters the ground.

When you are driving cross country and see a large electrical transmission line, you will again see four wires. One of them, running along the top of the towers, will be noticeably smaller than the other three. If you look closely, you will again see that the small wire is grounded at every tower.

The three ungrounded wires in these transmission systems are driven by three ac generators. The grounded wire increases safety and protection from lightning. Power is conveyed by the three larger wires in the form of three-phase electric power. The overwhelming majority of the world's electric power is generated and distributed as three-phase power. For example, if you were to examine a catalog of industrial-grade motors, you will discover that all the larger electric motors, bigger than a few horsepower, would be three-phase motors.

[1] "Convert" would be a better word, because the energy is normally converted from chemical energy to heat, and finally to electrical energy.

LEARNING OBJECTIVE 1.

To understand the nature and advantages of three-phase power

What is three-phase power? Physically, there are three wires that carry the power, and often a fourth wire, called the *neutral*, which is grounded. In enclosed cables, the active wires are normally colored red, black, and blue; and the neutral, if present, is white or gray. The phases are traditionally designated *A*, *B*, and *C*, and the time-domain voltages between them are as shown in Fig. 1.1. The voltages are expressed mathematically as

$$\begin{aligned} v_{AB}(t) &= V_p\cos(\omega t) \\ v_{BC}(t) &= V_p\cos(\omega t - 120^\circ) \\ v_{CA}(t) &= V_p\cos(\omega t - 240^\circ) \end{aligned} \tag{1.1}$$

IDEA 7 The Frequency Domain

The frequency-domain picture for a three-phase system is shown in Fig. 1.2. We have used $\underline{\mathbf{V}}_{AB}$ as our phase angle[2] reference and shown $\underline{\mathbf{V}}_{BC}$ following by 120°, then $\underline{\mathbf{V}}_{CA}$. This is known as an *ABC* phase sequence and corresponds to the time-domain representation in Fig. 1.1 and Eqs. (1.1).

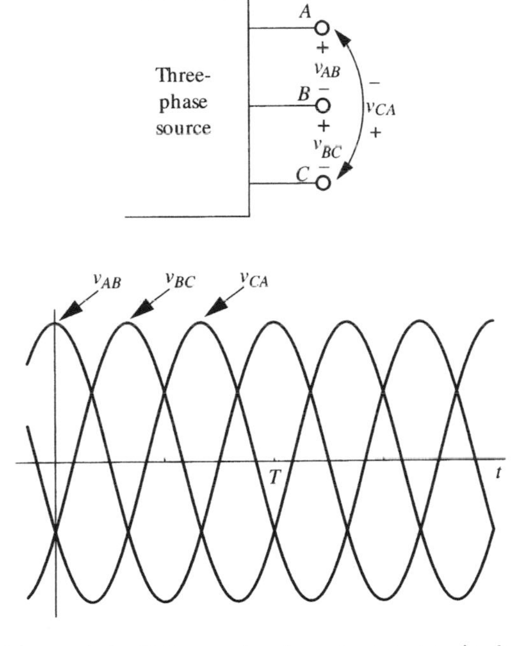

Figure 1.1 Voltages of a three-phase system in the time domain.

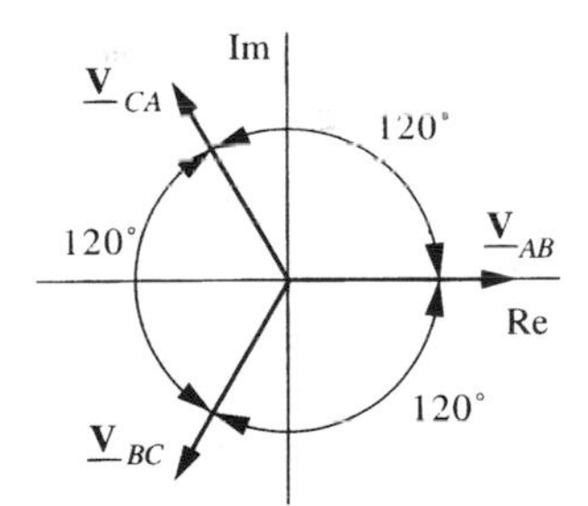

Figure 1.2 Voltages of a three-phase system in the frequency domain.

Advantages of three-phase systems. It is easily shown that single-phase power produces a pulsating flow of energy. A *smooth* flow of energy from source to load is achieved by a balanced three-phase system. If we have identical resistive (*R*) loads con-

[2] We apologize for using "phase" to mean two separate concepts in this section, but this vocabulary is standard. The three voltages have different phases for their sources, and from this property, the sources themselves came to be called "phases." So "phase" means the phase angle of a sinusoidal quantity as well as the source of that sinusoid. In this section, we use "phase angle" to refer to the phase of a sinusoid. The magnitude of the phasor is the rms value of the sinusoid.

nected between the three phases, the instantaneous flow of power would be given by Eq. (1.2). We use the trigonometric identity $\cos^2\alpha = [1 + \cos(2\alpha)]/2$ to derive the second and third forms

$$\begin{aligned} p(t) &= \frac{v_{AB}^2(t)}{R} + \frac{v_{BC}^2(t)}{R} + \frac{v_{CA}^2(t)}{R} \\ &= \frac{V_p^2}{2R}[1 + \cos 2(\omega t) + 1 + \cos 2(\omega t - 120°) + 1 + \cos 2(\omega t - 240°)] \\ &= \frac{3V_p^2}{2R} + \frac{V_p^2}{2R}\underbrace{[\cos(2\omega t) + \cos(2\omega t - 240°) + \cos(2\omega t - 480°)]}_{\text{add to zero at all times}} \end{aligned} \tag{1.2}$$

We see a constant term and a term that appears to be time-varying at twice the source frequency. Actually, *the second term adds to zero at all times.* This is easily shown by a phasor diagram; indeed, the phasors representing these terms give the same phasor diagram as Fig. 1.2.[3] Clearly, the phasor sum of the three symmetrical phasors is zero, and hence the fluctuating power term is also zero at all times. Thus, Eq. (1.2) reduces to

$$p(t) = \frac{3V_p^2}{2R} \quad \text{(a constant)} \tag{1.3}$$

This constant flow of energy effects general smoothness of operation in three-phase electrical equipment. A rough analogy is suggested by comparing an engine having one cylinder with an engine having many cylinders—clearly, the multicylinder engine runs smoother.

Compared with a single-phase system, distribution losses are proportionally less for a three-phase system. Additionally, three-phase motors offer advantages over single-phase motors in both startup and run characteristics. In short, three-phase systems are supremely important for the generation, distribution, and use of electrical power, particularly in industrial settings.

Three-Phase Power Sources

LEARNING OBJECTIVE 2.

To understand how to connect a generator in wye or delta

Three single-phase sources. Three-phase generators produce three single-phase voltages with the required 120° phase-angle shifts, which are internally connected to produce a three-phase source. In this section, we pretend that the three single-phase voltages are brought out of the generator to a terminal board,[4] and our job is to connect the resulting six terminals together to produce a three-wire, three-phase source. The terminal board is shown in Fig. 1.3, and the phasor diagram of the available voltages is shown in Fig. 1.4.

terminal

Delta (Δ) and wye (Y) connections. The symmetry of the desired phasor voltages suggests that we require some sort of symmetrical connection for the three volt-

[3] A phase angle of −480° is the same as −120°.

[4] A *terminal* is the end of a wire from an electrical device. A *terminal board* is a place where terminals are made available for connection.

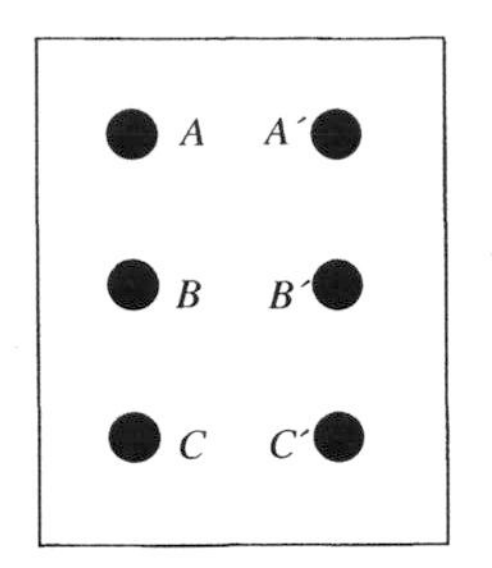

Figure 1.3 External terminals for the three separate phases. These must be connected to produce a three-phase system.

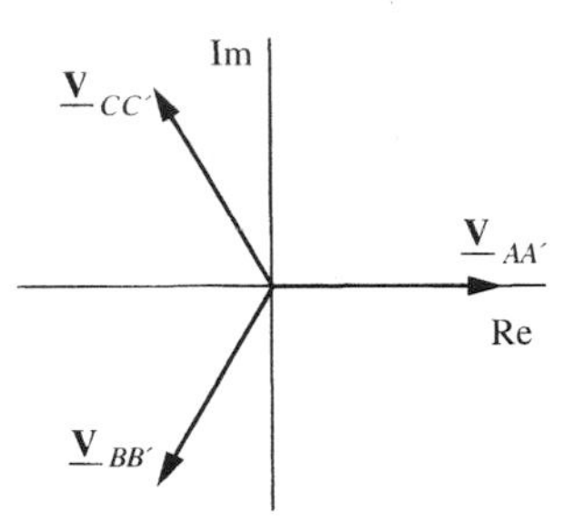

Figure 1.4 Phasor voltages in the three coils. These must be connected externally to make a true three-phase system.

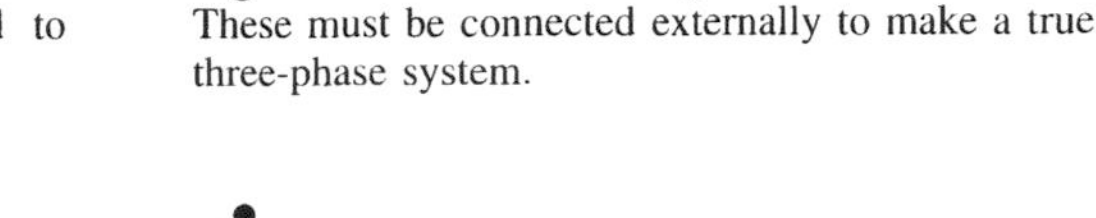

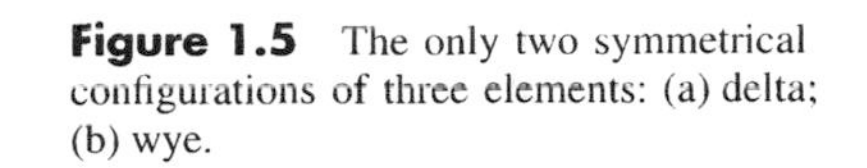

Figure 1.5 The only two symmetrical configurations of three elements: (a) delta; (b) wye.

ages represented in Fig. 1.4. Only two symmetrical configurations exist for three elements connected end to end, and these are shown in Fig. 1.5. The closed ring is usually called a *delta* (for the Greek letter Δ), even when it is drawn upside down or on its side. The configuration with a common point is sometimes called a star configuration, but in three-phase terminology, it is more often called a *wye configuration* (a phonetic spelling of the letter Y), regardless of orientation.

These symmetric configurations suggest two solutions of the question posed earlier. We require three terminals for a source of three-phase power. The delta has three terminals, and hence the three-phase outputs connect these terminals. The wye, on the other hand, has four terminals, counting the common connection in the center. This gives us a place to connect the fourth wire mentioned earlier, the neutral wire that is grounded. We have to connect the terminals to have the geometric symmetry of the delta or wye configurations in Fig. 1.5, but we must also retain the electrical symmetry in Fig. 1.2.

delta connection

Delta connection. The delta ties the three generators in a closed ring. Figures 1.6(a) and 1.6(b) show one possible connection for the delta. We must be careful, however, when we close the ring, for the voltage must be small to avoid a large circulating current. Closing the delta is like jump starting a car having a weak battery, as shown in Fig. 1.6(c). The circuit can be closed if the polarities are correct. With the connection marked "Yes," there will be at most a small voltage across the gap and only small currents will flow through the batteries. But if the polarities are wrong, there will be approximately 24 V across the gap and a huge current will flow if the connection is made.

Similarly, if we are to close the ring of generators in Fig. 1.6(b), we require the voltage across the gap to be small. This requires that

$$\underline{\mathbf{V}}_{AC'} = \underline{\mathbf{V}}_{AA'} + \underline{\mathbf{V}}_{BB'} + \underline{\mathbf{V}}_{CC'} = 0 \tag{1.4}$$

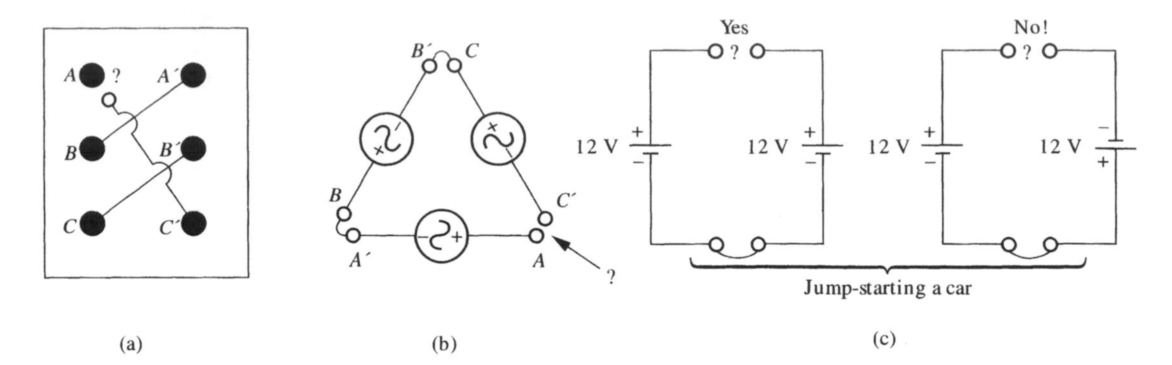

Figure 1.6 (a) Potential delta connection; (b) the voltage across A and C' must be zero if the ring is to be closed; (c) the polarity must correct before closing the circuit.

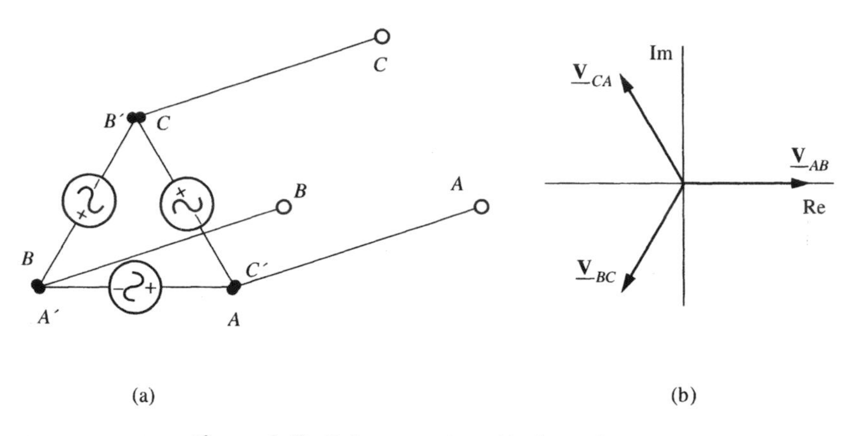

Figure 1.7 Delta connection with phasor diagram.

Notice that Eq. (1.4) follows from the rule for adding subscripted voltages because A' is connected to B and B' is connected to C. The general rule is

$$V_{AB} = V_{AX} + V_{XB} \tag{1.5}$$

We can extend this to $V_{AB} = V_{AX} + V_{YB}$ if X and Y are connected. The phasor diagram for the sum in Eq. (1.4) is easily derived from Fig. 1.4; clearly, the sum is small, ideally zero. Thus, it is safe to close the ring of generators and bring out the connected terminals as a three-phase source. Figure 1.7(b) shows the phasor diagram of the final connection in Fig. 1.7(a). Another possible delta connection would result with A connected to B', B connected to C', and C connected to A'. We leave the investigation of this possibility to the reader.

wye connection

Wye connection. A wye connection results from connecting A', B', and C', as shown in Fig. 1.8(a), with three wires for the three-phase power (A, B, C) and a common point ($A'B'C'$) for a neutral. With this connection, the magnitudes of $\underline{\mathbf{V}}_{AC}$, $\underline{\mathbf{V}}_{BA}$, and $\underline{\mathbf{V}}_{CB}$ are $\sqrt{3}$ greater than those of the component voltages, $\underline{\mathbf{V}}_{AA'}$, $\underline{\mathbf{V}}_{BB'}$, and $\underline{\mathbf{V}}_{CC'}$, and the phase angle of $\underline{\mathbf{V}}_{AC}$ lies at $-30°$ relative to $\underline{\mathbf{V}}_{AA'}$. These combinations are illus-

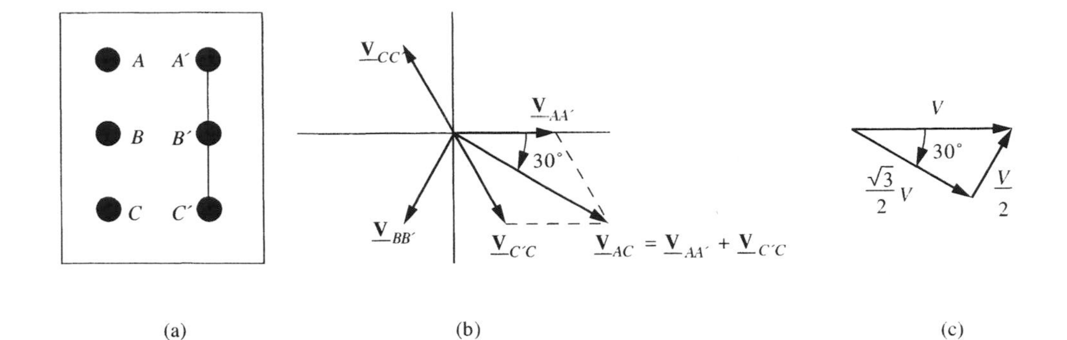

Figure 1.8 Possible wye connection.

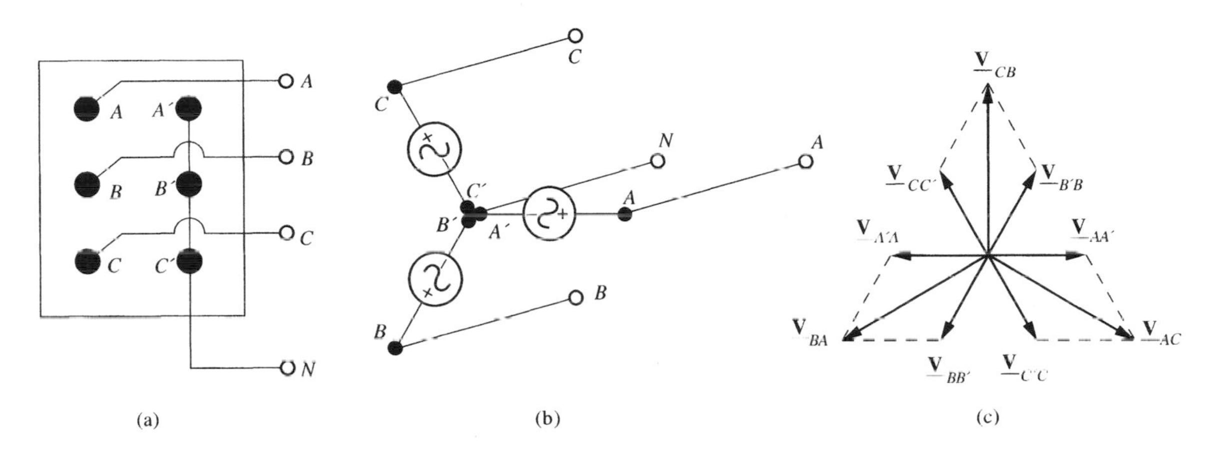

Figure 1.9 Wye connection. The line voltages are $\sqrt{3}$ larger than the phase voltages.

trated in Fig. 1.8(b) by the forming of $_{AC}$ through addition of $_{AA'}$ and $_{C'C}$, which is the negative of $\underline{\mathbf{V}}_{CC'}$. The $\sqrt{3}$ comes from the 30° right triangle, as shown in Fig. 1.8(c). In Figs. 1.9(a) and 1.9(b), we label the three lines A, B, and C, and the neutral N. Figure 1.9(c) shows the final phasor diagram of the wye connection. The wye connection leads to the four-wire system that we described at the beginning of this section.

three-phase voltage, three-phase current, line voltage, line current

Three-phase voltage and current. The *three-phase voltage,* or *line voltage,* is the voltage between the lines carrying the power. The *three-phase currents,* or *line currents* are the currents in the three lines. For a balanced system, these quantities are simply the voltage and current measured by a voltmeter between any two lines and an ammeter in any of the lines. Hence, a 460-V three-phase system has 460 V (rms) between any two of the three lines carrying the power.

Line and phase voltage and current for delta and wye connections. A three-phase generator consists of three interconnected single-phase generators,[5] as

[5] Actually, there is only one generator, but it has three single-phase *windings* that are interconnected. Here we call these "generators" to indicate single-phase voltage sources.

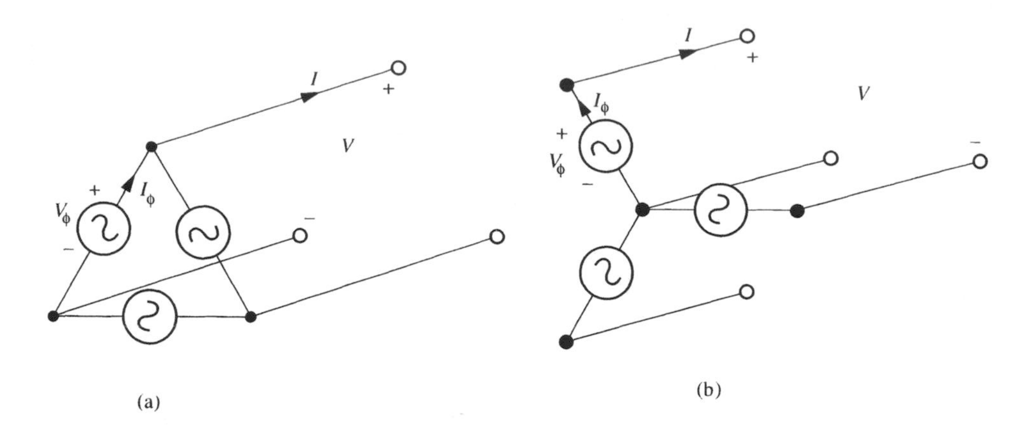

Figure 1.10 Three-phase (line) and phase voltage and current for a (a) delta- and (b) wye-connected three-phase system.

phase voltage, phase current

shown in Figs. 1.7 and 1.9. The *phase voltage*, V_ϕ, and *current*, I_ϕ, are the voltage across an individual generator and the current through that generator. For a delta connection, the phase voltage is the three-phase, or line voltage; but for a wye connection, the phase voltage is the three-phase voltage divided by $\sqrt{3}$, as shown in Fig. 1.9(c). The phase current is the same as the three-phase, or line current for a wye connection, but for the delta connection, the phase current is the three-phase current divided by $\sqrt{3}$. The phase and line voltage and current are shown in Fig. 1.10 and summarized in Table 1.1.

TABLE 1.1 Relationships between Phase and Three-Phase Voltages for Wye and Delta Connections

	Three-Phase Voltage, V	*Three-Phase Current, I*
Wye	$V_\phi = V/\sqrt{3}$	$I_\phi = I$
Delta	$V_\phi = V$	$I_\phi = I/\sqrt{3}$

EXAMPLE 1.1 Three-phase sources

Three single-phase sources are wye-connected as a three-phase source that measures 208 V for the three-phase voltage. What would be the three-phase voltage if the generators were delta-connected?

SOLUTION:
The phase, or line-to-neutral, voltage of the wye-connected system is $208/\sqrt{3} = 120$ V. Thus, the delta connection would give a three-phase voltage of 120 V.

WHAT IF? What if one of the three phases in the delta is reversed, but the ring is not closed? What would be the voltages between the four sets of terminals?[6]

Phase rotation. The phase rotation came out *ABC* in both systems we developed, but *ACB* is also a possibility. The physical phase rotation is very important; for example, the rotational direction of a three-phase motor depends on the phase rotation of the input voltages. In practice, the three wires are arbitrarily labeled *A*, *B*, and *C*, and one of several techniques is used to determine whether the phase rotation is *ABC* or *ACB*.

Other possible connections. We have now shown the two ways for connecting the voltages in Fig. 1.3 to give a three-wire, symmetrical power system. Actually, we have shown one version of each way, for there is an alternative delta or wye. For example, we can make *A*, *B*, and *C* the neutral for a wye.

Three-Phase Loads

LEARNING OBJECTIVE 3.

To understand how to calculate voltage, current, and power in balanced wye- and delta-connected loads

Delta-connected resistors. Like three-phase generators, three-phase loads can be connected in delta or wye. Figure 1.11 shows a balanced three-phase resistive load connected in delta. The source of the three-phase power is not shown; we assume *ABC* phase rotation and a three-phase voltage $V = |\underline{\mathbf{V}}_{AB}| = |\underline{\mathbf{V}}_{BC}| = |\underline{\mathbf{V}}_{CA}|$. We show no neutral because the load offers no place for connecting a neutral. In practice, however, a delta-connected load, such as a three-phase motor, is housed in a physical structure that is normally grounded directly to earth ground and through a neutral. However, the motor circuit would be floating.

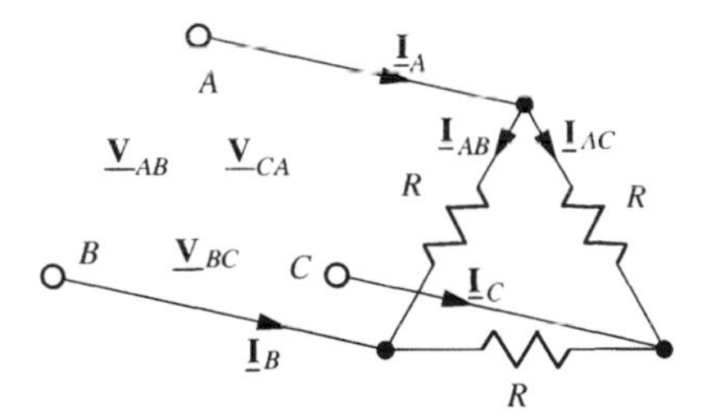

Figure 1.11 Delta-connected load.

IDEA 1 Conservation of Charge

With a load connected in delta, we must distinguish between the line currents and the phase currents flowing in the resistors. Figure 1.12 shows the phase currents, $\underline{\mathbf{I}}_{AB}$, $\underline{\mathbf{I}}_{BC}$, and $\underline{\mathbf{I}}_{CA}$. These currents have the same phase angle as their corresponding line voltages. Any line current, say, $\underline{\mathbf{I}}_A$, can be determined by phasor addition of the phase currents. Kirchhoff's current law at the top node is

$$\underline{\mathbf{I}}_A = \underline{\mathbf{I}}_{AB} + \underline{\mathbf{I}}_{AC} \qquad (1.6)$$

[6] 120, 120, 120, and 240 V.

but

$$\underline{\mathbf{I}}_{AC} = -\underline{\mathbf{I}}_{CA} \tag{1.7}$$

and hence the currents add as in Fig. 1.12. The other line currents could be determined similarly; indeed, the picture develops like that of the wye generator connection in Fig. 1.9(c). We see that the line currents are $\sqrt{3}$ greater than the phase currents. Thus, the second row of Table 1.1 is valid for delta-connected loads as well as sources. For the resistive load, the line current in A has the same phase angle as the average of the phase angles of $\underline{\mathbf{V}}_{AB}$ and $\underline{\mathbf{V}}_{AC}$. The phase-angle relationships for a resistive load are shown in Fig. 1.13.

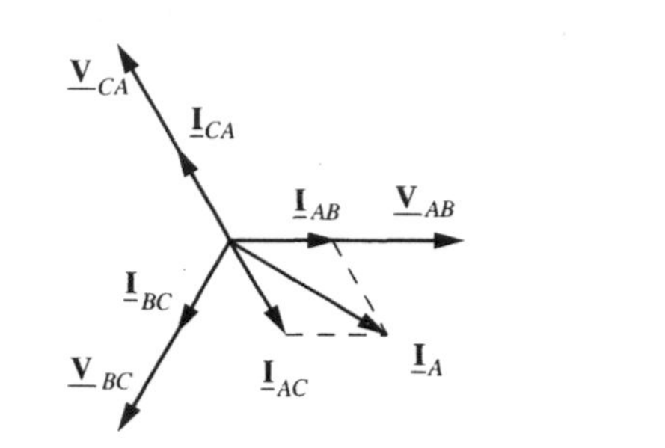

Figure 1.12 Phase current addition to yield line current.

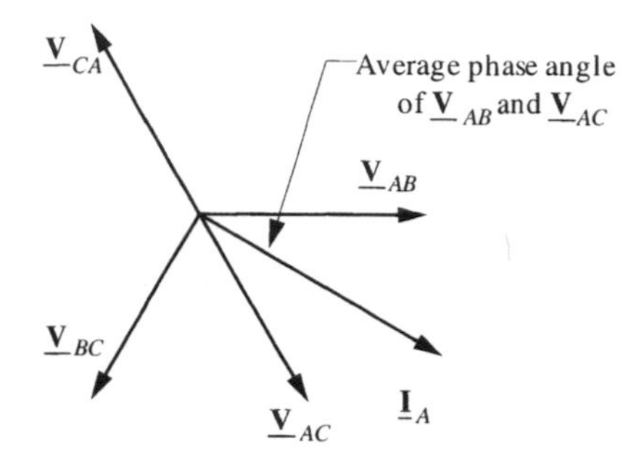

Figure 1.13 For a resistive load, the line current is in phase angle with the average phase angle of the voltages to the other two lines.

EXAMPLE 1.2

Delta-connected load

A delta-connected resistive load has a line voltage of 460 V and a line current of 32 A. Find the resistance values.

SOLUTION:

The phase voltage is $V_\phi = 460$ V and the phase current is $I_\phi = 32/\sqrt{3} = 18.5$ A. Thus, the phase impedance is

$$|\underline{\mathbf{Z}}_\phi| = \frac{V_\phi}{I_\phi} = \frac{460}{18.5} = 24.9\ \Omega \tag{1.8}$$

WHAT IF?

What if one of the resistors were missing? What would be the line currents?[7]

[7] 32, 18.5, and 18.5 A.

General loads in delta. With balanced loads that are not resistive, the phasor diagram shown in Fig. 1.12 changes only slightly. The magnitude of the phase currents is computed by dividing the line voltages, which are also the phase voltages, by the magnitude of the phase impedance. The phase currents will lead or lag the line voltages according to the angle of the phase impedance. Hence, the line currents will also be shifted in phase angle by the angle of the impedance. We give an example later.

Power in delta connections. The total power to a load, $P_{3\phi}$, is the sum of the powers delivered to the three phase impedances; and this would be

$$P_{3\phi} = 3\,P_\phi = 3\,V_\phi I_\phi \times PF \tag{1.9}$$

where *PF* is the power factor of the phase impedance. In Eq. (1.9), V_ϕ and I_ϕ represent the rms values of the phase voltage and current. We desire, however, to express the total power in terms of line voltage and current. The phase currents are often inaccessible for measurements, but the line voltage and current always can be measured. Consequently, we introduce the line voltage and current

$$P = 3V\frac{I}{\sqrt{3}} \times PF = \sqrt{3}\,VI \times PF \tag{1.10}$$

where V and I are the rms line voltage and current, respectively. In the application of Eq. (1.10), the power factor is the cosine of the angle of the phase impedance and is *not* the phase angle between line current and line voltage. The power-factor angle is, however, the angle between the phase angle of the line current and the average phase angle of the voltages to the two other lines, as illustrated in Fig. 1.13.

EXAMPLE 1.3 Delta-connected *RL* impedances

A 230-V three phase power system supplies 2000 W to a delta-connected balanced load with a power factor of 0.9 lagging. Determine the line currents, the phase currents, and the phase impedance. Draw a phasor diagram.

SOLUTION:
First, we calculate the magnitude of the line currents from Eq. (1.10).

$$P = \sqrt{3}\,VI \times PF \Rightarrow I = \frac{2000}{\sqrt{3}(230)(0.9)} = 5.58 \text{ A (rms)} \tag{1.11}$$

The phase currents are smaller by $\sqrt{3}$, so

$$I_\phi = \frac{I}{\sqrt{3}} = \frac{5.58}{\sqrt{3}} = 3.22 \text{ A} \tag{1.12}$$

This allows us to calculate the impedance in each phase of the delta. The angle of the impedance is implied by the power factor: $\theta = \cos^{-1}(0.9) = +25.8^\circ$, + because the current is lagging (inductive).

$$Z_\phi = \frac{V_\phi}{I_\phi} \angle \cos^{-1}(PF) = \frac{230}{3.22} \angle \cos^{-1}(0.9) = 71.4 \angle +25.8^\circ\ \Omega \tag{1.13}$$

We can now draw the phasor diagram, Fig. 1.14. We use $\underline{\mathbf{V}}_{AB}$ for the phase-angle reference, with the other line voltages placed symmetrically in *ABC* sequence. The phase currents lag by 25.8°, as shown. The line currents can be computed by phasor addition of the phase currents, as we did in Fig. 1.12, but another approach is to use our earlier results to place the line currents behind the phase currents by 30° and greater by $\sqrt{3}$. Whichever way is chosen, only one line current need be determined, $\underline{\mathbf{I}}_A$ for example, and the other two can be constructed by symmetry.

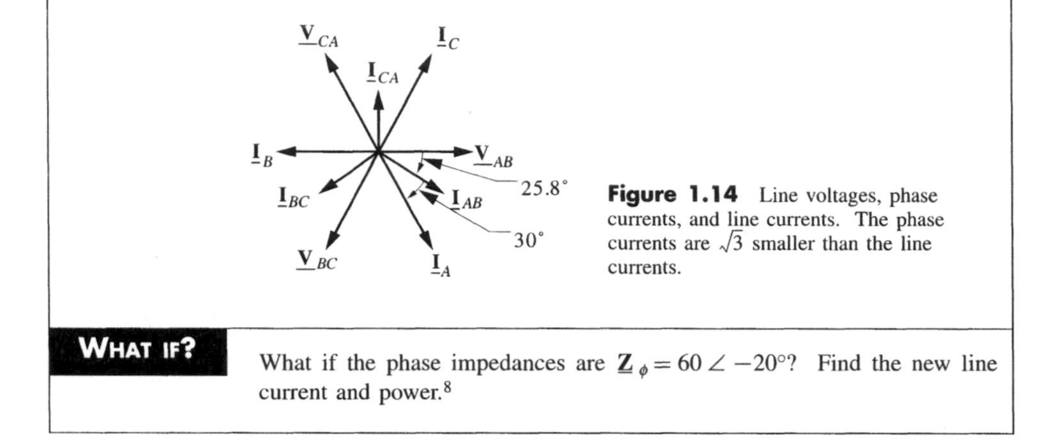

Figure 1.14 Line voltages, phase currents, and line currents. The phase currents are $\sqrt{3}$ smaller than the line currents.

WHAT IF?

What if the phase impedances are $\underline{\mathbf{Z}}_\phi = 60 \angle -20^\circ$? Find the new line current and power.[8]

Wye-connected loads. Figure 1.15 shows a load connected in wye.[9] We have shown no connection to the neutral, labeled *N*, but there would often be a connection between the load neutral and the source neutral, if such existed, and the neutral is often grounded. For a perfectly balanced load, no current would flow in the neutral connection because the three line currents add to zero. We now calculate the line-to-neutral voltages, $\underline{\mathbf{V}}_{AN}$, $\underline{\mathbf{V}}_{BN}$, and $\underline{\mathbf{V}}_{CN}$, and the line currents, $\underline{\mathbf{I}}_A$, $\underline{\mathbf{I}}_B$, and $\underline{\mathbf{I}}_C$.

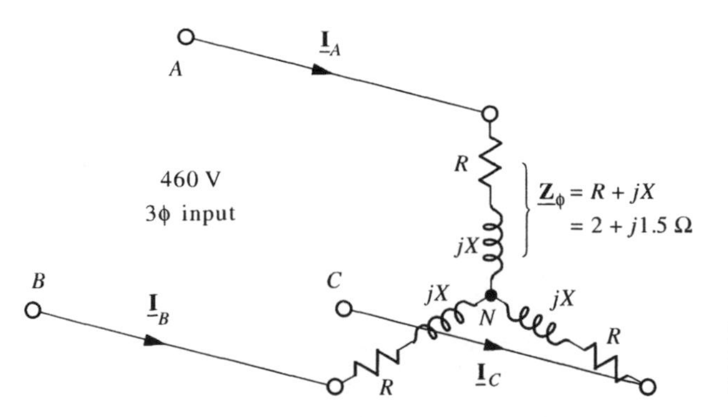

Figure 1.15 Wye-connected load. To find the line currents, we must determine the line-to-neutral voltages.

[8] 6.64 A and 2485 W.

[9] On the input voltage, "3ϕ" is a common abbreviation for "three phase."

Line-to-neutral voltages. The phase impedances are given, $\underline{\mathbf{Z}}_\phi$, hence, it is clear that the line currents can be determined once the line-to-neutral voltages are known; for example,

$$\underline{\mathbf{I}}_A = \frac{\underline{\mathbf{V}}_{AN}}{\underline{\mathbf{Z}}_\phi} \tag{1.14}$$

The line-to-neutral voltages can be determined from consideration of the symmetry of the circuit. It is convenient to pretend that we know the line-to-neutral voltages (which we do not) and determine from them the line-to-line voltages. The relationship between these two sets of three-phase voltages then becomes known and we henceforth can deduce either set of voltages from the other. We assume the *ABC* phase sequence; hence, the line-to-neutral voltages must appear as in Fig. 1.16, assuming that we make $\underline{\mathbf{V}}_{AN}$ the phase reference.

First, we determine $\underline{\mathbf{V}}_{AB}$. We can express $\underline{\mathbf{V}}_{AB}$ in terms of $\underline{\mathbf{V}}_{AN}$ and $\underline{\mathbf{V}}_{NB}$:

$$\underline{\mathbf{V}}_{AB} = \underline{\mathbf{V}}_{AN} + \underline{\mathbf{V}}_{NB} \tag{1.15}$$

Equation (1.15) becomes more useful when we reverse the subscripts on $\underline{\mathbf{V}}_{NB}$ and change the sign:

$$\underline{\mathbf{V}}_{BN} = -\underline{\mathbf{V}}_{NB} \Rightarrow \underline{\mathbf{V}}_{AB} = \underline{\mathbf{V}}_{AN} - \underline{\mathbf{V}}_{BN} \tag{1.16}$$

Equation (1.16) is represented in Fig. 1.16, with the negative of $\underline{\mathbf{V}}_{BN}$ drawn and added to $\underline{\mathbf{V}}_{AN}$. We note that $\underline{\mathbf{V}}_{AB}$ leads $\underline{\mathbf{V}}_{AN}$ by 30° and is somewhat greater in magnitude. The phasor addition is identical to that shown in Fig. 1.8(b) and the magnitudes of phase and line voltages have the ratio $\sqrt{3}$, just as the currents do in the delta-connected load. The remaining line voltages, $\underline{\mathbf{V}}_{BC}$ and $\underline{\mathbf{V}}_{CA}$, may be determined by similar reasoning, or more directly by arranging them in *ABC* sequence, each 120° from $\underline{\mathbf{V}}_{AB}$.

Figure 1.17 shows the results. The line-to-neutral voltages lag the corresponding line voltages by 30°, when we consider the two voltages with, say, *A* written first, like $\underline{\mathbf{V}}_{AB}$ and $\underline{\mathbf{V}}_{AN}$. But a better way to think about the phase angle is to realize that the phase angle of the corresponding line-to-neutral voltage lies between the phase angles of the two line-to-line voltages that connect to the same point. Thus, $\underline{\mathbf{V}}_{AN}$ will lie halfway between $\underline{\mathbf{V}}_{AB}$ and $\underline{\mathbf{V}}_{AC}$. This phase-angle relation, together with the magnitude ra-

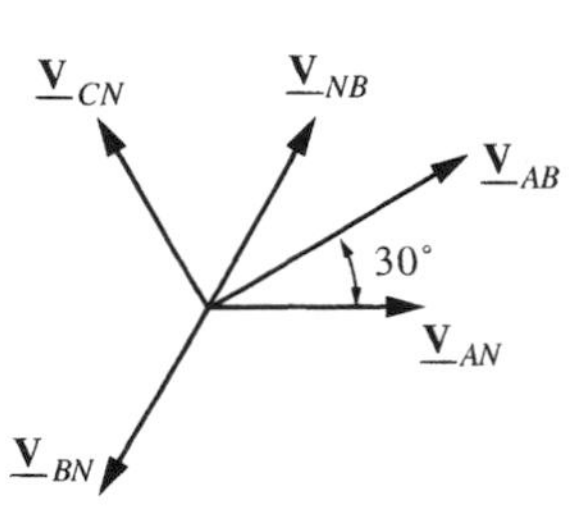

Figure 1.16 Determining line-to-line voltages from line-to-neutral voltages.

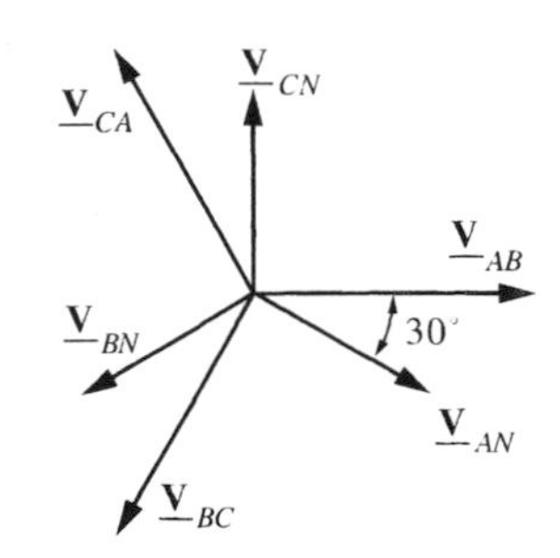

Figure 1.17 The line-to-neutral voltages are smaller by $\sqrt{3}$ and lag line-to-line voltages by 30°.

tio of $1/\sqrt{3}$, allows us to determine easily the line-to-neutral voltages from the set of line-to-line voltages, or vice versa, as shown in Fig. 1.17.

EXAMPLE 1.4

Wye-connected load

Find the line and phase current and line and phase voltage for the wye-connected load shown in Fig. 1.15.

SOLUTION:
We make $\underline{\mathbf{V}}_{AB}$ the phase-angle reference, that is, $\underline{\mathbf{V}}_{AB} = 460 \angle 0^\circ$ V. Hence, the line voltage is 460 V (rms). From Fig. 1.17, we see that $\underline{\mathbf{V}}_{AN}$ will lie at -30° and be smaller by $\sqrt{3}$, or $\underline{\mathbf{V}}_{AN} = 266 \angle -30^\circ$ V; so the phase voltage is 266 V (rms). Hence, the current in line A is

$$\underline{\mathbf{I}}_A = \frac{\underline{\mathbf{V}}_{AN}}{\underline{\mathbf{Z}}_\phi} = \frac{266 \angle -30^\circ}{2 + j1.5} = \frac{266 \angle -30^\circ}{2.50 \angle 36.9^\circ} = 106.2 \angle -66.9^\circ \text{ A} \tag{1.17}$$

The other line currents can be determined similarly or by symmetry from $\underline{\mathbf{I}}_A$. Thus, the phase and line currents, which are the same for the wye-connected load, are 106.2 A.

Power in wye-connected loads. The total power to the wye-connected load is three times the power to each phase of the load, P_ϕ. Thus, we can compute the total power with

$$P_{3\phi} = 3\,P_\phi = 3\,V_\phi I_\phi \times PF \tag{1.18}$$

where $P_{3\phi}$ is the power in the load, V_ϕ is the phase rms voltage, the line-to-neutral voltage in this instance, I_ϕ is the phase rms current, also the line current in this instance, and PF is the power factor of the phase impedance.

EXAMPLE 1.5

Power in wye-connected load

Find the total power to the three-phase load in the previous example.

SOLUTION:
The phase voltage and current were determined to be 266 V and 106.2 A, respectively. The power factor is the angle of the phase impedance, $2 + j1.5 = 2.50 \angle 36.9^\circ$; so the power factor is $\cos 36.9^\circ$, lagging. Using Eq. (1.18), we find the power in the load to be

$$P_{3\phi} = 3(266)\,(106.2)\,(\cos 36.9^\circ) = 67.7 \text{ kW} \tag{1.19}$$

WHAT IF?

What if one phase impedance were missing? What would be the power?[10]

[10] 33.9 kW.

Using line voltage and current. The neutral of the wye-connected load might not be accessible for voltage measurement; hence, it is desirable to express the total power in terms of the line voltage and current. Using the $\sqrt{3}$ ratio between phase voltage and line voltage, we may convert Eq. (1.18) to

$$P_{3\phi} = \left(\frac{V}{\sqrt{3}}\right)(I) \times PF = \sqrt{3}VI \times PF \tag{1.20}$$

where V is the line voltage, I is the line current, and PF is the power factor of the load. In Eq. (1.20), the power factor is the cosine of the angle of the phase impedance.

Equation (1.20) is identical to Eq. (1.10), which was developed for the delta-connected load. Thus, the formula is general and applies to all balanced three-phase loads. Clearly, the two load configurations cannot be distinguished by external measurement but can be identified in practice only by examining the internal connections in the three-phase load.

IDEA 6 **Equivalent Circuits**

Delta–wye conversions. This does not mean, however, that the *same* set of phase impedances are equivalent when connected first in delta and then in wye. Indeed, the appearance of the circuits suggests that the delta gives parallel paths, whereas the wye gives series paths. This appearance suggests that the line current for the delta would be larger than for the wye if the same phase impedance were used for each connection. It can be shown that the ratio is 3:1; that is, three identical impedances will draw three times the current (and three times the power) when connected in delta, as compared to when they are connected in wye.

IDEA 5 **Impedance Level**

Thus, the delta and wye are equivalent if the phase impedances differ by a factor of 3, with the delta connection having the higher impedance level. This equivalence, shown in Fig. 1.18, is often useful in solving three-phase problems. We leave proof of this equivalence for a problem at the end of this chapter.

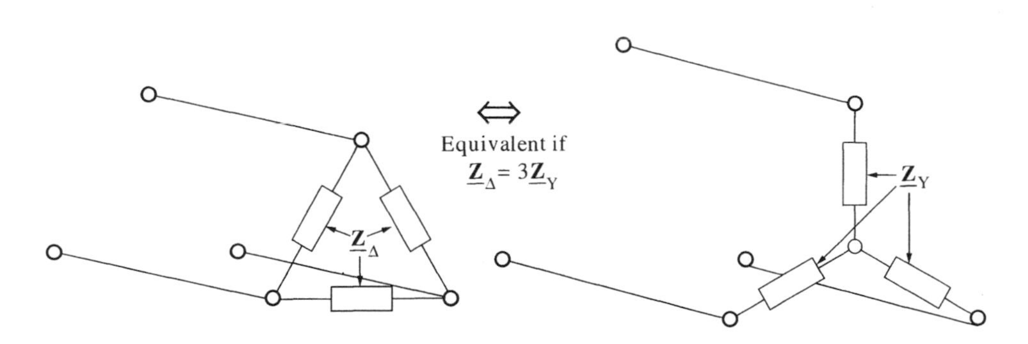

Figure 1.18 Delta and wye equivalence.

EXAMPLE 1.6 Delta load with line losses

Determine the power to the delta-connected load in Fig. 1.19.

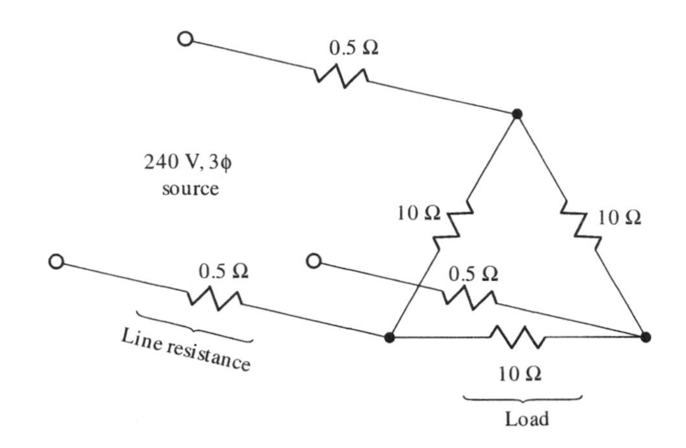

Figure 1.19 Find the power to the delta-connected load.

SOLUTION:

The 10-Ω resistors represent the load, but we must consider the resistance of the wires leading to the load, which is represented by the 0.5-Ω resistors. The presence of the wire resistance undermines our previous approach for solving delta-connected loads, but if we convert the delta to an equivalent wye load, we can solve the problem.

Figure 1.20 shows the circuit after conversion to wye. The wire resistance can now be combined with the load resistance to yield a phase resistance of 3.83 Ω, and the rms line-to-neutral voltage is $240/\sqrt{3} = 139$ V. The line current thus is $139/3.83 = 36.1$ A, and the total power to the wires plus load is $\sqrt{3}(240)(36.1) = 15.0$ kW. The wire losses are $3(36.1)^2(0.5) = 1960$ W, the rest of the power going to the delta-connected load.

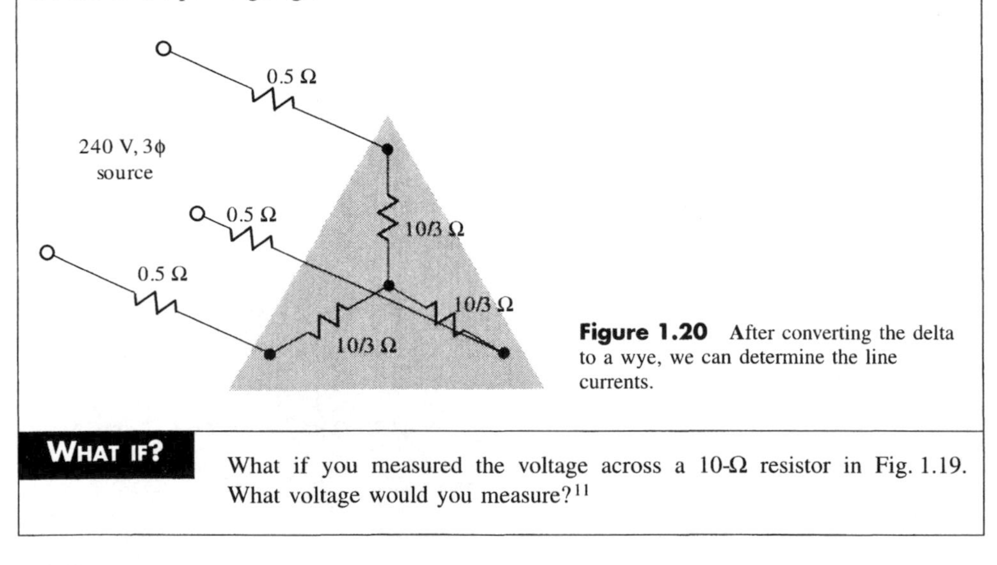

Figure 1.20 After converting the delta to a wye, we can determine the line currents.

WHAT IF?

What if you measured the voltage across a 10-Ω resistor in Fig. 1.19. What voltage would you measure?[11]

[11] 209 V.

LEARNING OBJECTIVE 4.

To understand how to derive and use the per-phase equivalent circuit of a balanced three-phase load

Equivalent Circuits

Per-Phase Equivalent Circuits

Need for a simpler model. Three-phase circuits are awkward to draw, and much of the drawing is unnecessary for a balanced circuit because each phase is identical. A single-phase model can represent the voltage, current, and power relationships in the three-phase circuit without this redundancy.

The *per-phase equivalent circuit* is a single-phase circuit that represents a balanced three-phase circuit. The per-phase model can represent any three-phase system, whether source or load is connected in delta or wye; thus, the system is that shown in Fig. 1.21. The power source establishes the voltage, V, and the load determines the current and the power factor. The voltage, current, and power factor on the line indicate the power delivered to the load.

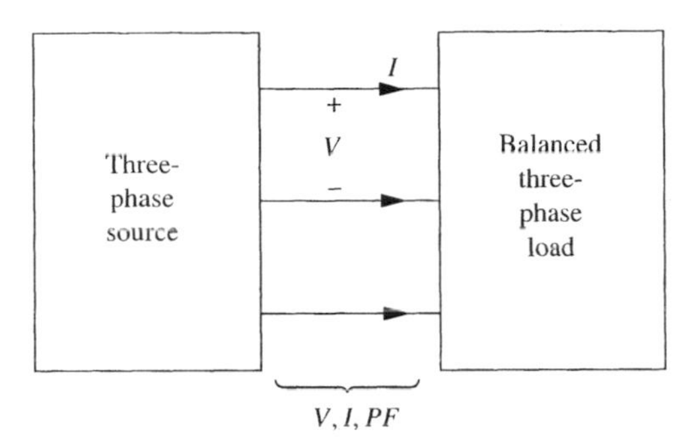

Figure 1.21 The per-phase circuit is based on the line voltage, current, and power factor and not on the internal connections of source and load.

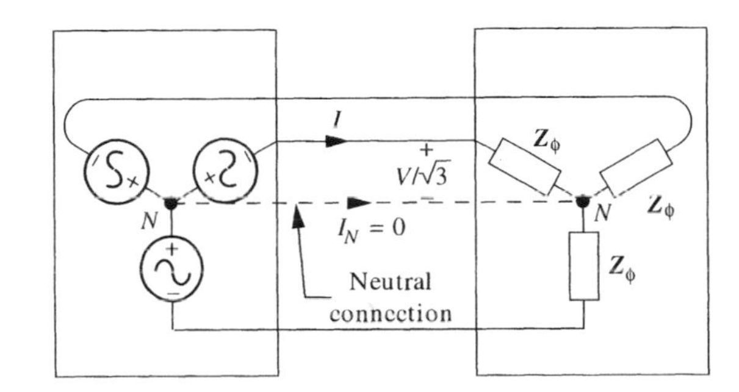

Figure 1.22 The wye–wye circuit with neutral in place can be considered three single-phase circuits sharing a wire.

Wye–wye basis. The per-phase circuit is derived *as if* the three-phase source and load were wye-connected, as shown in Fig. 1.22. We have drawn the neutral connection dashed because it actually may be missing. Even if present, the current in the neutral will be zero for a balanced load, as indicated; thus, the neutrals of the source and the load are at the same voltage, as if connected by a short circuit. This equivalent short circuit connecting the neutrals divides the three-phase circuit into three, identical single-phase circuits sharing a common neutral. We may take the inner circuit, where we marked the voltage and current, as representing one-third of the three-phase system. The per-phase equivalent circuit, shown in Fig. 1.23, introduces the per-phase voltage, current, and impedance, $\underline{\mathbf{V}}_{pp}$, $\underline{\mathbf{I}}_{pp}$, and $\underline{\mathbf{Z}}_{pp}$, respectively.

Figure 1.23 Per-phase equivalent circuit.

Relationship between three-phase and per-phase quantities. The relationship between the three-phase quantities and the per-phase quantities are shown in Table 1.2. The per-phase voltage is $V/\sqrt{3}$ because, as shown in Fig. 1.22, the single-phase circuit consists of one line of the three-phase system plus the neutral. The currents and the power factor are the same. The per-phase impedance is the same as the load phase impedance if the load is wye-connected, but is one-third the load phase impedance if delta-connected. The power in the per-phase circuit is one-third the full power in the three-phase circuit, as the name "per phase" suggests. The same would be true for apparent, reactive, and complex power.

TABLE 1.2 Relationships between Three-Phase and Per-Phase Quantities

	Three-Phase Circuit	*Per-Phase Circuit*
Voltage	V	$V_{pp} = V/\sqrt{3}$
Current	I	$I_{pp} = I$
Impedance	Z_{Δ} or Z_Y	$Z_{pp} = Z_Y$ or $Z_{\Delta}/3$
Power factor	PF	$PF =$ same
Power	$P_{3\phi} = \sqrt{3}VI \times PF$	$V_{pp}I_{pp} \times PF = P_{3\phi}/3$

EXAMPLE 1.7 Motor per-phase equivalent circuit

A three-phase induction motor operates with the following conditions: 3 hp, 3515 rpm, 230 V, 8.8 A, efficiency = 80.0%. Derive a per-phase equivalent circuit for the motor.

SOLUTION:

The per-phase voltage and current come directly from the three-phase voltage and current:

$$V_{pp} = \frac{230}{\sqrt{3}} = 133 \text{ V} \qquad \text{and} \qquad I_{pp} = 8.8 \text{ A} \tag{1.21}$$

and thus the magnitude of the per-phase impedance is established

$$|Z_{pp}| = \frac{V_{pp}}{I_{pp}} = \frac{133}{8.8} = 15.1 \ \Omega \tag{1.22}$$

To determine the angle of the impedance, we need the phase angle of the current, which is not given, or the power factor, which is not given either but may be determined from the power quantities. The output power of the motor is 3 hp × 746 W/hp = 2240 W, and the input power follows from this and the efficiency

$$\text{Efficiency} = \eta = \frac{P_{out}}{P_{in}} \Rightarrow P_{in} = \frac{P_{out}}{\eta} = \frac{2240}{0.800} = 2800 \text{ W} \tag{1.23}$$

Thus, the power factor of both the three-phase and the per-phase circuits is

$$PF = \frac{P_{pp}}{V_{pp}I_{pp}} = \frac{2800/3}{133 \times 8.8} = 0.798 \tag{1.24}$$

and hence the angle of the per-phase impedance is $\cos^{-1}(0.798) = 37.1°$ and the per-phase impedance is $\underline{\mathbf{Z}}_{pp} = 15.1 \angle 37.1° = 12.0 + j9.09\ \Omega$. *Note:* (1) we divided the total electrical input power by 3 to convert to power per phase, and (2) we assumed a positive angle for the impedance (inductive) because an induction motor draws lagging current. The per-phase equivalent circuit is that shown in Fig. 1.23 with $\underline{\mathbf{V}}_{pp} = 133 \angle 0°$ V, $\underline{\mathbf{Z}}_{pp} = 15.1 \angle 37.1°\ \Omega$, and $\underline{\mathbf{I}}_{pp} = 8.8 \angle -37.1°$ A.

What if the motor were connected for 460-V operation? In this case, the input current would be 4.4 A, but the power, speed, and efficiency would be unchanged. Find the new per-phase quantities.[12]

unbalanced three-phase systems

Unsymmetric loads. A per-phase equivalent circuit is possible only for a balanced system. A three-phase load becomes *unbalanced* when the phase impedances are not identical. This is an undesirable situation and is avoided in practice if possible. When unbalanced loads are connected in delta, calculation of the phase and line currents becomes tedious, though straightforward. All phase and line currents must be calculated individually because symmetry has been lost.

When unbalanced loads are connected in wye, the analysis is straightforward only when the neutral of the load is connected to the neutral of the three-phase source. With the neutral connected, the three loads operate in effect as single-phase loads that share the neutral connection. Current will flow in the neutral wire for an unbalanced load.

When there exists no neutral wire in the unbalanced wye connection, complications arise in the calculation of the line-to-neutral voltages and the line currents. Because the neutral of the load is no longer at the same voltage as the neutral of the source, the first step in solving the problem is to calculate the voltage of the neutral of the load. Then one can proceed to solve for the line currents. Such calculations are routinely performed with computers.

Check Your Understanding

1. A three-phase circuit measures 762 V (rms) between earth ground and line *A*. What would be the voltage between lines *B* and *C*?
2. For a delta-connected load, the phase and line currents are the same. True or false?
3. A three-phase load uses 50 kW at 480 V. The current is 64 A. Find the reactive power required by the inductive load.
4. Three 120-Ω resistors connected in wye are equivalent to what resistances connected in delta?

Answers. **(1)** 1320 V; **(2)** false; **(3)** +18.2 kVAR; **(4)** 360 Ω.

1.2 POWER DISTRIBUTION SYSTEMS

Introduction. The power that is indispensable to our civilization is generated in large central power plants, transformed to high voltages for transmission over long distances, lowered to moderate voltages for distribution throughout a geographic region such as a small city, and finally reduced to the voltage levels required by industrial and residential consumers. Only at the last stage is the power available in single-phase form; the generation, transmission, and distribution use three-phase.

[12] $V_{pp} = 266$ V, $I_{pp} = 4.4$ A, and $\mathbf{Z}_{pp} = 60.4 \angle 37.1^\circ\ \Omega$.

In this section, we deal with the transmission and distribution aspects of the power system. In the next section, we deal with ac motors, which are a major consumer of electric power and have special importance to engineers. The operation of ac generators is described in Chap. 4.

Voltage levels. For reasons intrinsic to good design, large three-phase generators produce voltage in the range of 11–25 kV. As discussed before, the voltage is raised to much higher values, 120–500 kV, for transmission. The voltage is lowered, perhaps by degrees, to smaller values, 7.2–23 kV, for distribution over a geographic area. As near to the customer as possible, within the building for a commercial customer or on a pole outside for a residential customer, the voltage is transformed finally to the standard utilization voltages, 120 and 240 V for residential, perhaps higher voltages for industrial customers. All these voltage transformations, with the exception of the last one, are performed by three-phase transformers.

LEARNING OBJECTIVE 5.
To understand how to connect single-phase transformers for three-phase transformation

Three-Phase Transformers

Three-phase power may be transformed by three single-phase transformers or by a single three-phase transformer. Economics favors the latter for larger transmission transformers and the former for smaller distribution transformers. The principles are the same in both cases. We speak in the following as if we are using three single-phase transformers.

In three-phase transformation, primaries and secondaries can be connected in delta (Δ) or wye (Y). This gives four possible combinations: Y–Δ, Δ–Y, Δ–Δ, and Y–Y. We now analyze the Y–Δ connection and summarize the other three connections in a table.

The Y–Δ connection. Figure 1.24 shows a Y–Δ connection in two circuit representations. We assume three single-phase transformers with voltages of $V_p : V_s$ and currents

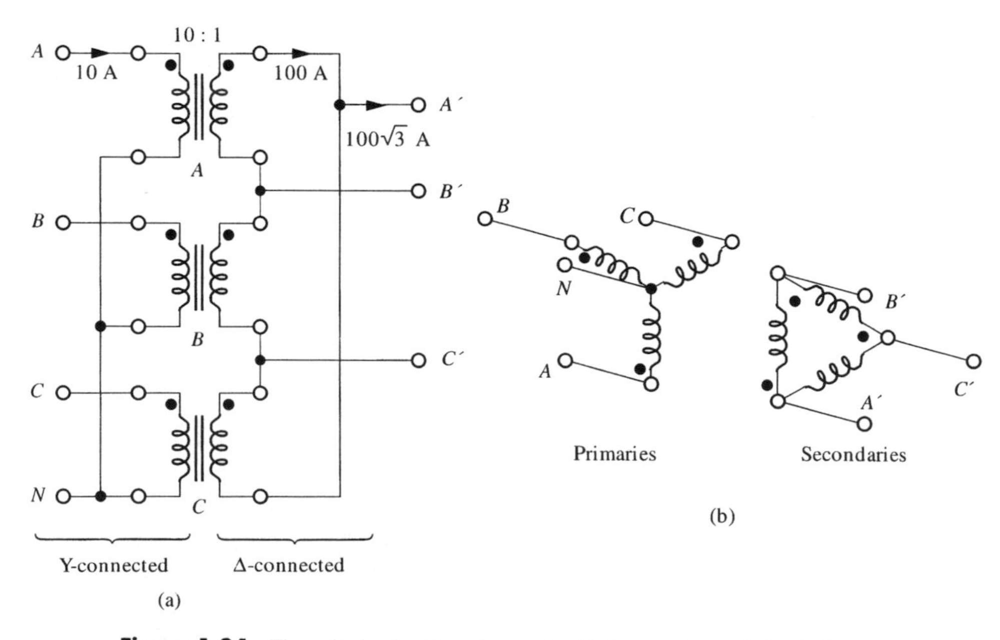

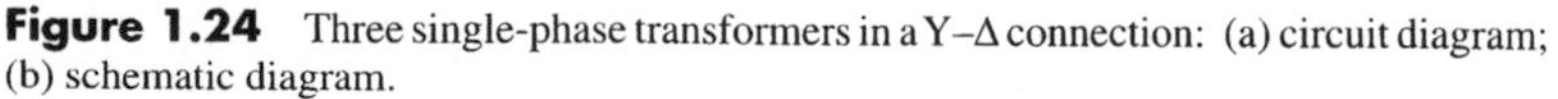

Figure 1.24 Three single-phase transformers in a Y–Δ connection: (a) circuit diagram; (b) schematic diagram.

of $I_p{:}I_s$ with apparent power $S = V_p I_p = V_s I_s$. Operated at rated voltage and current, the primary line voltage and current are $V = \sqrt{3}\,V_p$ and $I = I_p$, for an apparent power of $\sqrt{3}\,VI = 3V_pI_p = 3S$. Thus, the apparent power rating of the three single-phase transformers connected for three-phase transformation is three times the apparent power rating of each component transformer. The secondary line voltage is V_s and the rated current is $\sqrt{3}\,I_s$; hence, the apparent power rating again is $3S$. In both cases, the rating of the three-phase connection is three times the rating of the single-phase transformers. This is true for all four possible connections.

EXAMPLE 1.8 Y–Δ transformer connection

Three 2400/240-V, 24-kVA (each) single-phase transformers are connected with the 2400 side in wye and the 240 side in delta. Find the rated three-phase voltage, current, and apparent power ratings on both sides of the transformer.

SOLUTION:
The allowed currents in the primary and secondary windings are 10 A and 100 A, respectively. Operated at rated voltage and current, the primary line voltage and current are $\sqrt{3} \times 2400 = 4160$ V and 10 A, respectively, for an apparent power of $\sqrt{3}\,VI = 72$ kVA. The secondary line voltage is 240 V and the rated current is $\sqrt{3} \times 100 = 173$ A. Hence, the apparent power in the secondary is also 72 kVA.

WHAT IF? What if the transformers were connected Δ–Δ?[13]

The phasor diagram for the Y–Δ connection is shown in Fig. 1.25. The secondary voltages are one-tenth the primary line-to-neutral voltage and 30° out-of-phase angle. This suggests that care should be used in designing and installing such systems because if multiple paths exist, and they do in most power systems, phase angles must be correct.

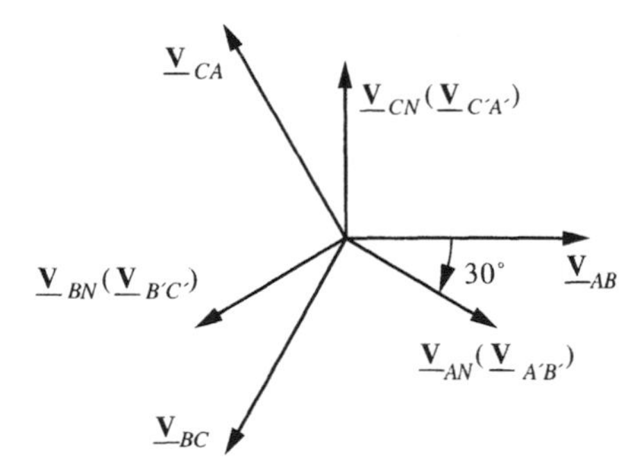

Figure 1.25 Phasor diagram for a Y-Δ transformer connection. Voltage magnitudes are not to scale.

[13] Primary: 2400 V, 17.3 A, 72 kVA; secondary: 240 V, 173 A, 72 kVA.

TABLE 1.3 Voltage and Currents Resulting from Three-Phase Transformer Connections*

Connections	*Primary V*	*Primary I*	*Secondary V*	*Secondary I*
Y–Y	$\sqrt{3}V_p$	I_p	$\sqrt{3}V_s$	I_s
Y–Δ	$\sqrt{3}V_p$	I_p	V_s	$\sqrt{3}I_s$
Δ–Y	V_p	$\sqrt{3}I_p$	$\sqrt{3}V_s$	I_s
Δ–Δ	V_p	$\sqrt{3}I_p$	V_s	$\sqrt{3}I_s$

*The winding (single-phase) voltage and current are V_p and I_p on the primary and V_s and I_s on the secondary. The three-phase voltage and current are V and I, respectively.

Applications of transformer connections. The four possible transformer connections are presented in Table 1.3. The winding (single-phase) voltage and current are V_p and I_p on the primary and V_s and I_s on the secondary, and the three-phase voltage and current are given in the table. Each of the connections find use:

- **Wye–wye.** This connection is occasionally used in high-voltage transmission lines. In addition to providing a neutral, the wye connection has the virtue of having lower voltage across the windings and is desirable for high-voltage connections. When used, Y–Y transformers often have a third set of windings, a *tertiary winding,* that is connected in delta to reduce harmonics in the power system.
- **Delta–wye.** This connection is used to step up the voltage from generation levels to the high voltages required for long-distance transmission. Also can be used to reduce from distribution levels to consumer voltage levels.
- **Wye–delta.** This connection is used to step down the voltage from transmission levels to distribution levels.
- **Delta–delta.** This connection is used at distribution and consumer levels. This connection has the unique property that one leg of the delta can be omitted in the so-called *V* or open-delta connection.[14] This configuration is useful because it can operate at 58% of the rating that the full delta would have and hence the open-delta connection is often installed in a growing system for later expansion.

open-delta connection

Per-phase equivalent circuits for transmission and distribution systems. All the connections just listed, and indeed the entire transmission and distribution system, can be represented by an equivalent per-phase (single-phase) circuit. After presenting per-unit calculations, we proceed to describe the properties of power transmission and distribution systems.

[14] The open-delta connection is drawn in the figure for Problem 1.21.

Per-Unit Calculations

per-unit calculations

Introduction. The *per-unit* system of calculation is a method of *normalizing* electrical circuit calculations so that voltage transformations in transformers become inconsequential. Other benefits of the per-unit system are (1) many of the $\sqrt{3}$'s inherent in three-phase calculations are eliminated, and (2) on a per-unit basis all generators, transmission lines, transformers, motors, etc., look more or less alike.

Base values. In power system calculations, we are concerned with calculating voltages, currents, impedances, and the three types of power, apparent power (S), real power (P), and reactive power (Q). We need base values of these quantities to normalize actual circuit quantities; but voltage, current, impedance, and power are interrelated such that only two may be chosen as base values and the other two base values may be calculated. Normally, voltage (V or kV) and apparent power (VA, kVA, and MVA) are used as primary base values, and current and impedance base values are secondary.

Converting to per unit (pu). To convert a circuit quantity to per unit (pu), divide by the base value. Thus

$$V_{pu} = \frac{V}{V_{base}} \quad \text{and} \quad S_{pu} = \frac{S}{S_{base}} \tag{1.25}$$

where the unsubscripted quantities are the circuit values. For example, we have a 240/120-V, 12-kVA transformer that is operated at 10 kVA with 220 V on the primary. If we use the rated values as base, the per-unit voltage and power are

$$V_{pu} = \frac{220\,\text{V}}{240\,\text{V}} = 0.917 \quad \text{and} \quad S_{pu} = \frac{10\,\text{kVA}}{12\,\text{kVA}} = 0.833 \tag{1.26}$$

The per-unit quantities are unitless. We can normalize current and impedance also, but we must first establish consistent bases from the base voltage and apparent power.

Base values for current and impedance. Because the apparent power is voltage times current, we can derive the current base as

$$I_{base} = \frac{S_{base}}{V_{base}} = \frac{12\,\text{kVA}}{240\ \text{V}} = 50\ \text{A} \tag{1.27}$$

and similarly for impedance

$$Z_{base} = \frac{V_{base}}{I_{base}} = \frac{V_{base}^2}{S_{base}} = \frac{(240\ \text{V})^2}{12\,\text{kVA}} = 4.8\ \Omega \tag{1.28}$$

EXAMPLE 1.9 Per-unit circuit calculations

A 240/120-V, 12-kVA transformer that is operated at 220 on the primary has a 10-Ω load on the secondary. Derive a per-unit circuit and calculate the per-unit current in both primary and secondary and the per-unit power. Figure 1.26 shows the circuit.

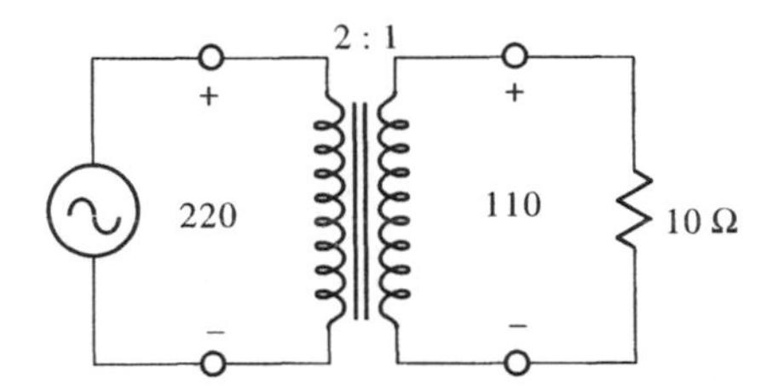

Figure 1.26 The transformer will disappear in the per-unit circuit.

SOLUTION:

We have the base values for the primary in Eq. (1.26). We choose the rated voltage and the kVA for the secondary base values, which makes the transformer disappear, as we shall see. Thus, the base values for the secondary are: from Eqs. (1.27) and (1.28)

$$I_{base} = \frac{12\,\text{kVA}}{120\,\text{V}} = 100\ \text{A} \quad \text{and} \quad Z_{base} = \frac{(120\,\text{V})^2}{12\,\text{kVA}} = 1.2\ \Omega \tag{1.29}$$

The per-unit voltage on both primary and secondary is 0.917, and the per-unit load impedance is

$$Z_{pu} = \frac{10\ \Omega}{1.2\ \Omega} = 8.33; \quad \text{hence} \quad I_{pu} = \frac{V_{pu}}{Z_{pu}} = \frac{0.917}{8.33} = 0.110 \tag{1.30}$$

Thus, the current in the secondary is 0.110× 100 A = 11 A. The per-unit current in the primary is the same, 0.110, because both actual and base currents are reduced by the turns ratio. Thus, the current in the primary is 0.110× 50 A = 5.5 A. Note the per-unit voltage and current are the same on both sides of the transformer; it becomes in effect a 1:1 transformer, as shown in Fig. 1.27. Hence, the per-unit power is the same on both sides of the ideal transformer

$$P_{pu} = V_{pu} I_{pu} = 0.917 \times 0.110 = 0.101 \tag{1.31}$$

so the actual power is 0.101 × S_{base} = 0.101 × 12,000 VA = 1210 W.

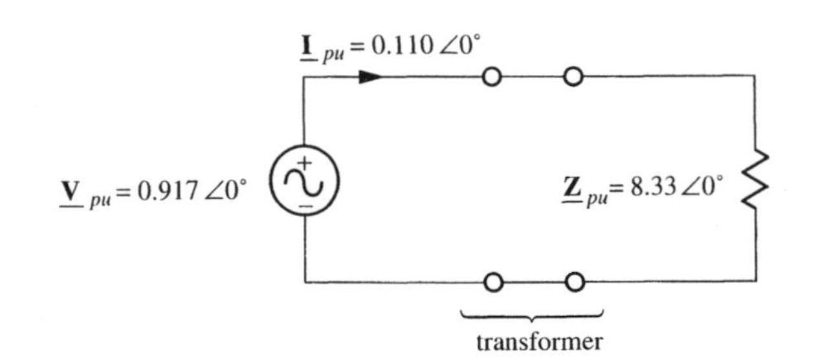

Figure 1.27 The transformer is gone.

WHAT IF?

What if we used 10 kVA as the base apparent power? What would be the new bases? Would the per-unit quantities change? Would the actual quantities change?[15]

Transmission Properties

LEARNING OBJECTIVE 6.

To understand how transmission-line impedance affects power and voltage loss on the line

Introduction. In this section, we investigate how the performance of a transmission system depends upon the real and reactive power passing through that system. Figure 1.28 shows a simple per-phase circuit model of a transmission system composed of a generator, a line represented by resistance and inductance, and a load. The transformers of the system are not shown because we can eliminate them either through reflecting all impedances to the transmission line part of the circuit or by using per-unit calculations. The resistance of the system will be ignored except when power loss calculations are made, and the reactance of the system consists of the inductive reactance of the transmission line and the transformers. The generator is represented by a voltage magnitude, $\underline{\mathbf{V}}_g = V_g \angle \delta$, at an angle, δ, with respect to the load voltage.

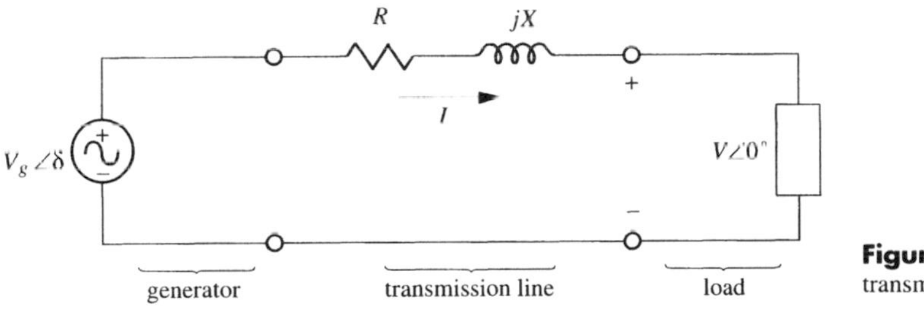

Figure 1.28 Per-phase model of power transmission system.

Analysis of the system. In the following analysis, we assume that the load voltage, $\underline{\mathbf{V}} = V \angle 0°$, is fixed both in magnitude and angle. Thus, the current, and consequently the real and reactive power in the system, depend only on the load. We analyze the circuit to show how the rms voltage loss in the line, $\Delta V = V_g - V$, the phase angle shift, δ, and the losses depend on the power flow in the system. Of course, we could perform an exact analysis, including the resistance, but we will gain adequate insight by ignoring the resistance.

Figure 1.29 shows the phasor diagram of the system with all quantities on a per-phase basis. We have assumed the voltage across the impedance of the transmission system, $jX\underline{\mathbf{I}}$, to be small relative to the voltages of source and load because this is typical. If we ignore the small difference between $\underline{\mathbf{V}}_g$ and its projection on the real axis, we have

[15] On the primary, the new base current would be 41.7 A and base impedance 5.76 Ω. On the secondary, the new base current would be 83.3 A and base impedance 1.44 Ω. The per-unit quantities change to $I_{pu} = 0.132$ and $S_{pu} = 0.121$, but the actual currents and power are unchanged.

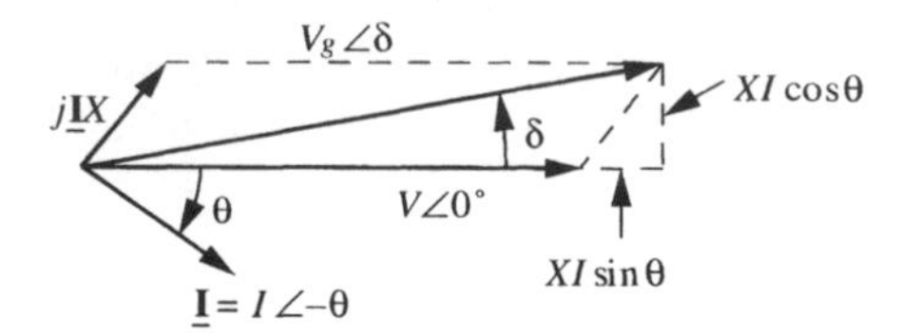

Figure 1.29 Phasor diagram for the transmission system.

$$V_g = V + XI\sin\theta = V + \frac{XQ}{V} \tag{1.32}$$

because $Q = VI\sin\theta$. Thus, the voltage loss in the line, for fixed system reactance and load voltage, depends approximately on the reactive power flow

$$\Delta V = \frac{X}{V} \times Q \tag{1.33}$$

angle of transmission

The *angle of transmission*, δ, can be derived geometrically

$$\delta = \sin^{-1}\left(\frac{XI\cos\theta}{V_g}\right) \approx \frac{XI\cos\theta}{V} = \frac{X}{V} \times P \tag{1.34}$$

because $P = VI\cos\theta$. Thus, the angle of transmission depends on the real power on the line. To find the line losses, we must include the line resistance, R

$$P_{loss} = I^2R = \frac{S^2}{V^2} \times R = \frac{R}{V^2} \times P^2 + \frac{R}{V^2} \times Q^2 \tag{1.35}$$

Thus, both real and reactive power flow contribute to the loss in the transmission system.

EXAMPLE 1.10 Transmission line

A 345-kV (receiving-end voltage) transmission line is 160 km long. The impedance/km is $0.034 + j0.32\ \Omega$/km per phase. The load requires 500 MW and 100 MVAR. At each end are 1000-MVA transformers with 0.1 per-unit reactance per phase based on the transformer voltage and apparent power rating. Find the voltage loss, ΔV, the angle of transmission, and the line losses on a per-unit basis.

SOLUTION:

We use the line voltage and the transformer apparent power rating as bases. We need to calculate the base impedance to normalize the transmission line impedance.

$$Z_{base} = \frac{V_{base}^2}{S_{base}} = \frac{(345\,\text{kV})^2}{1000\,\text{MVA}} = 119\ \Omega \tag{1.36}$$

so the per-unit line impedance is

$$\mathbf{Z}_{pu} = \frac{\mathbf{Z}_{line}}{Z_{base}} = \frac{(0.034 + j0.32) \times 160}{119} = 0.0457 + j0.430 \tag{1.37}$$

and hence $R_{pu} = 0.0457$ and $X_{pu} = 0.430 + 0.2$ after we add the reactance of the transformers. The per-unit powers are $P_{pu} = 500\text{ MW}/1000\text{ MVA} = 0.5$, and similarly $Q_{pu} = 0.1$. Thus, Eq. (1.33) gives the voltage drop to be

$$\Delta V_{pu} = \frac{X_{pu}}{V_{pu}} \times Q_{pu} = \frac{0.630}{1} \times 0.1 = 0.063 \tag{1.38}$$

Hence, there is a 6.3% voltage drop over the transmission line. The angle of transmission is given by Eq. (1.34)

$$\delta \approx \frac{X_{pu}}{V_{pu}} \times P_{pu} = \frac{0.630}{1} \times 0.5 = 0.315 \text{ radians} = 18.1^\circ \tag{1.39}$$

Thus, there exists an 18.1° phase-angle shift between generator and load. Finally, the power loss is given by Eq. (1.35)

$$\begin{aligned} P_{loss(pu)} &= \frac{R_{pu}}{V_{pu}^2} \times P_{pu}^2 + \frac{R_{pu}}{V_{pu}^2} \times Q_{pu}^2 \\ &= \frac{0.0457}{1^2} \times (0.5)^2 + \frac{0.0457}{1^2} \times (0.1)^2 = 0.0119 \end{aligned} \tag{1.40}$$

so the per-unit line loss is 1.19%, which on a base of 1000 MVA would be 11.9 MW. Thus the transmission efficiency is $(500 - 11.9)/500 = 97.6\%$.

WHAT IF?

What if the load were 800 MVA at a lagging power factor of 0.9? Find voltage and power loss in percent.[16]

Power-factor correction in a transmission system. Because the voltage and power loss in the transmission system depend on the reactive power flow, the power company has an interest in minimizing Q in the system. This it does by forcing large consumers to control their power factor and by using capacitors in transmission and distribution systems. Figure 1.30(a) models a source, load, and distribution line, which includes resistance and inductance, and Fig. 1.30(b) shows capacitors added to the distribution system by the power company to improve performance. With leading current, the load voltage can be equal to or greater than the source voltage. Thus, adding capacitors to the system can reduce line power and voltage losses.

Figure 1.31(a) shows the phasor diagram for the system with 0.9 power factor, lagging current. The lagging current, combined with the inductive impedance of the line, results in a large voltage loss in the line. Figure 1.31(b) shows the phasor diagram with

[16] $\Delta V = 22.0\%$ and loss $= 4.06\%$.

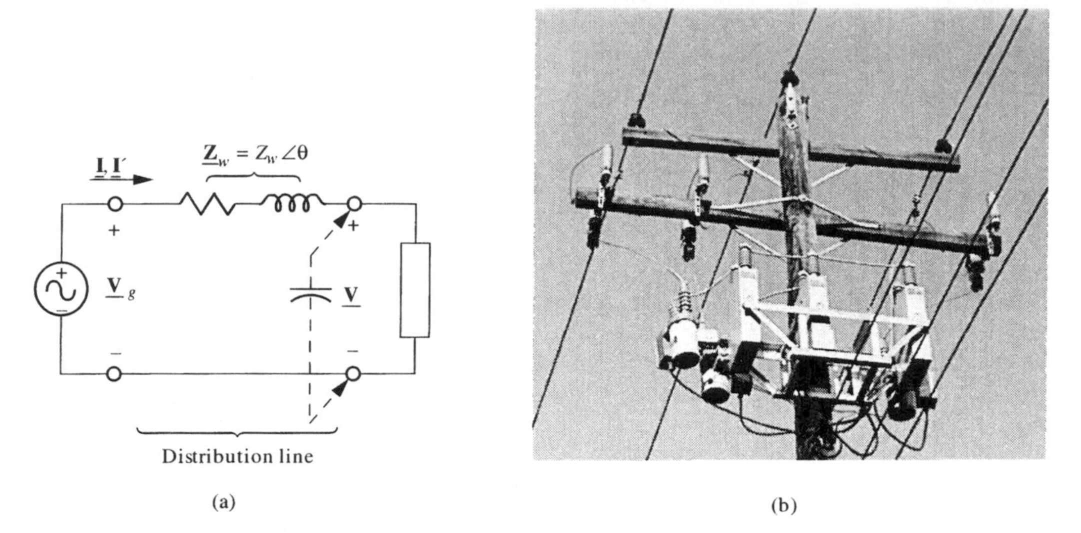

Figure 1.30 (a) Capacitors placed near the load improve efficiency and voltage regulation because the transmission system is inductive; (b) a 12.5-kV three-phase distribution power line with six capacitors between the lines and neutral.

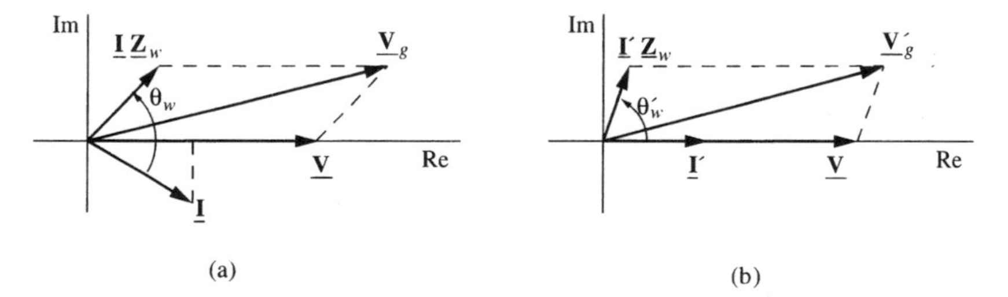

Figure 1.31 (a) Phasor diagram for the system with 0.9 power factor and no capacitors; (b) phasor diagram for the systems with the same load and unity power factor (with capacitors). Note that line current is smaller and $\underline{V}$ and $\underline{V}_g$ are more nearly equal in (b).

unity power factor and the same load. Here line current is smaller, and load and source voltages are more nearly equal.

Check Your Understanding

1. If a line voltage is 0.9 per unit, what is the per-phase voltage (line-to-neutral) in per unit?
2. The base voltage is 13,200 V and the base power is 100 kVA. Find the base current and impedance.
3. If the *PF* is 0.9 and the line loss is 1000 W, what would be the line loss if the *PF* were corrected to unity?

Answers. **(1)** 0.9, the $\sqrt{3}$'s go away; **(2)** 7.58 A, 1742 Ω; **(3)** 810 W.

1.3 INTRODUCTION TO ELECTRIC MOTORS

Introduction. A large part of the load on an electric power system consists of electric motors. At home (air conditioners, hair dryer, blender), in the shop (tools, fans), in the office (printers, cooling fans for electronic devices), and of course in factories, electric motors are everywhere. Although there are many types of electric motors, at least 90% are induction motors: single-phase induction motors for small jobs and three-phase for big.

Chapters 4 to 6 present physical principles, electrical equivalent circuits, and applications of synchronous, induction, and dc motors. In this section, we give information relevant to all motors and discuss the steady-state and dynamic responses of motors. We give detailed analysis of nameplate information on three-phase and single-phase induction motors. Much practical information about a motor can be deduced from its nameplate by applying basic knowledge of ac circuits and simple mechanics.

Terminology

rotor, stator

Rotor and stator. Figure 1.32 suggests an electrical motor. It has an electrical input of voltage and current, a mechanical output in the form of torque and rotation, and losses represented as heat. The motor consists mechanically of a *stator*, which does not rotate, a *rotor*, which can rotate, and an air gap to permit motion.

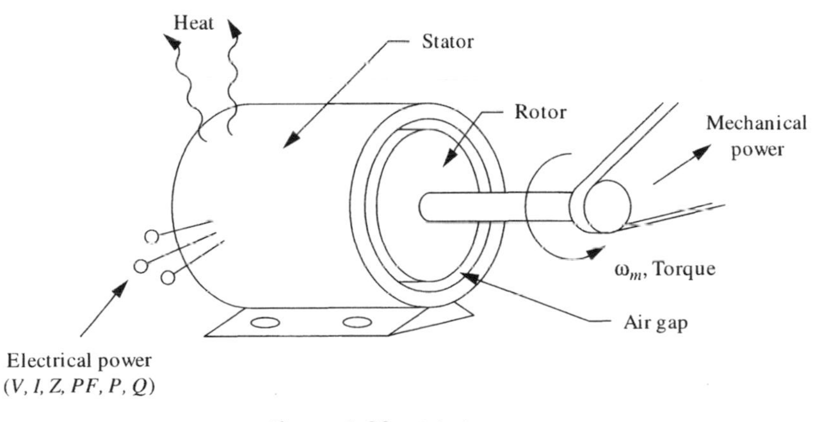

Figure 1.32 A basic motor.

armature, field

Armature and field. Electrical force (torque) is produced by currents interacting with magnetic flux. We distinguish electrically between the *field* circuit, which produces a magnetic flux, and the *armature* circuit, which carries a current. The field usually has many turns of wire carrying relatively small currents, and the armature normally has few turns of larger wire carrying relatively large currents. Depending on the type of motor, the field might be on the rotor or stator, and the armature is always the opposite. For induction motors, the field is on the stator and the armature on the rotor.

Types of electric motors. Although there are many types of electric motors, we deal here with two types: the three-phase induction motor and the single-phase induction motor. After discussing certain general principles that relate to all motors, we discuss and analyze the nameplate information for both these types of motors.

Motor Characterization in Steady-State Operation

Electrical input. We consider first steady-state operation, such as a fan motor turning at constant speed. The electrical quantities of interest in steady state are the input voltage, current, and the power factor. In many cases, analysis of the motor suggests an equivalent circuit that accounts for losses, energy storage, and the conversion of electrical power into mechanical power. The electrical operation depends in part on the mechanical load, as will be discussed presently.

Mechanical output. The output characteristic of a motor is the output torque, $T_M(\omega_m)$, as a function of rotation speed, ω_m. The operating speed of the motor is jointly determined by the output torque characteristic of the motor and the torque requirement of the load.

Basic equations for three-phase motor. The input electrical power is

$$P_{in} = \sqrt{3}VI \times PF \tag{1.41}$$

and the output mechanical power is

$$P_{out} = \omega_m \times T_M \tag{1.42}$$

where ω_m is the mechanical speed in radians/second and T_M is the output torque in newton-meters. Two other quantities of interest are the losses, P_{loss}, and the efficiency, η

$$P_{loss} = P_{in} - P_{out} \text{ and } \eta = \frac{P_{out}}{P_{in}} \tag{1.43}$$

Basic equations for single-phase motors. Equations (1.41) through (1.43) are also valid for single-phase motors except that the input power equation lacks the $\sqrt{3}$ for single phase.

EXAMPLE 1.11 Single-phase motor

Consider a 60-Hz, single-phase induction motor with the following nameplate information:[17] 1 hp, 1725 rpm, 115 V, 14.4 A, efficiency = 68%. Find the output torque, the power factor, and the cost of operating the motor if electric power costs 9.6 cents/kWh.

SOLUTION:

The output power is 1 hp × 746 W/hp = 746 W, so the input power is, from Eq. (1.43)

$$P_{in} = \frac{P_{out}}{\eta} = \frac{746}{0.68} = 1097 \text{ watts} \tag{1.44}$$

[17] We will discuss nameplate information presently. For now, assume that the actual operation of the motor is given by the nameplate information.

The power factor follows from Eq. (1.41) without the $\sqrt{3}$

$$PF = \frac{P_{in}}{VI} = \frac{1097}{115 \times 14.4} = 0.662 \tag{1.45}$$

and the current is lagging for this type of motor. The output torque is

$$T_{out} = \frac{P_{out}}{\omega_m} = \frac{746}{1725(2\pi/60)} = 4.13 \text{ N-m} \tag{1.46}$$

where the mechanical speed in rpm has been converted to radians/second. For continuous operation and nameplate conditions, the cost would be

$$\frac{1097 \text{ W}}{1000 \text{ W/kW}} \times \frac{24 \text{ hours}}{1 \text{ day}} \times \frac{9.6 \text{ cents}}{1 \text{ kWh}} \times \frac{1 \text{ dollar}}{100 \text{ cents}} = \$\,2.53/\text{day} \tag{1.47}$$

WHAT IF? What if the motor is running without a load?[18]

The Motor with a Load

LEARNING OBJECTIVE 7. To understand how motor and load interact to establish steady-state operation

Typical motor and load characteristics. Figure 1.33 shows the diverse torque characteristics that can be achieved for several types of electrical motors. Even within a specific motor type, the designer can tailor the motor characteristics within broad limits. Figure 1.34 shows representative torque requirements for various mechanical loads.

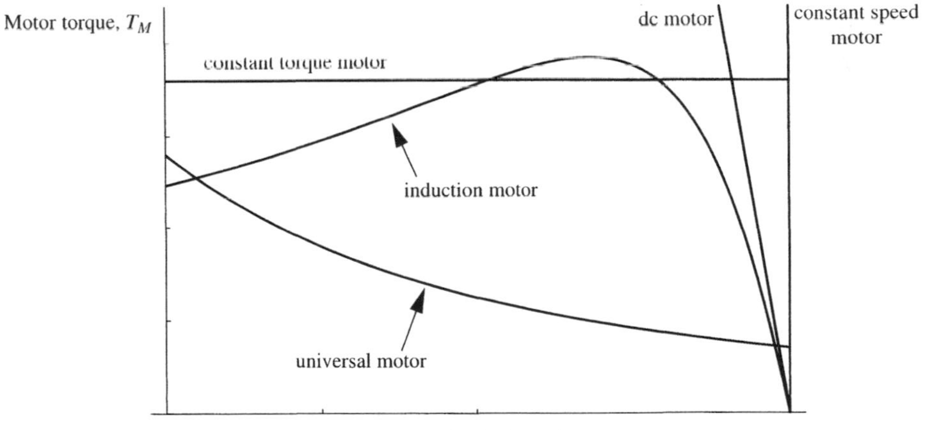

Figure 1.33 Motor torque characteristics.

[18] All we know in that case is that $P_{out} = 0$ and $T_M = 0$. But the cost would not be zero due to losses of the motor.

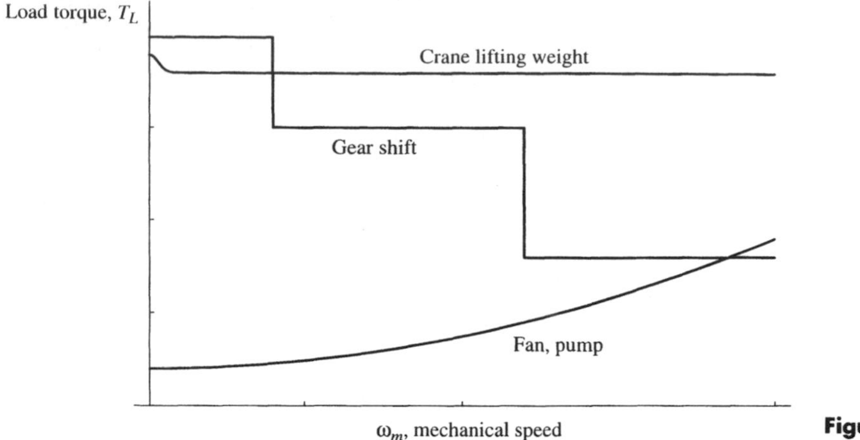

Figure 1.34 Load characteristics.

blocked-rotor torque, breakover torque

Motor–load interaction. System operation is jointly determined by load requirements, $T_L(\omega_m)$, and motor characteristics, $T_M(\omega_m)$. Consider, for example, connecting a three-phase induction motor to a fan starting from rest, as shown in Fig. 1.35. We have identified the motor-starting torque, often called the *blocked-* or *locked-rotor torque*, the maximum torque, often called the *breakover torque*, and the no-load speed. For rotation to occur, the starting torque of the motor must exceed the starting-torque requirement of the load. The excess torque, $\Delta T(\omega_m) = T_M - T_L$, will accelerate the system to the speed where the two characteristics cross, which would be the steady-state speed of the motor–load system. This condition, $T_M = T_L$, determines the steady-state speed of the system. Thus, the speed, output torque, and output power of the motor depend in part on the load requirements. If, for example, the load required less torque, the motor would run slightly faster and supply less power.

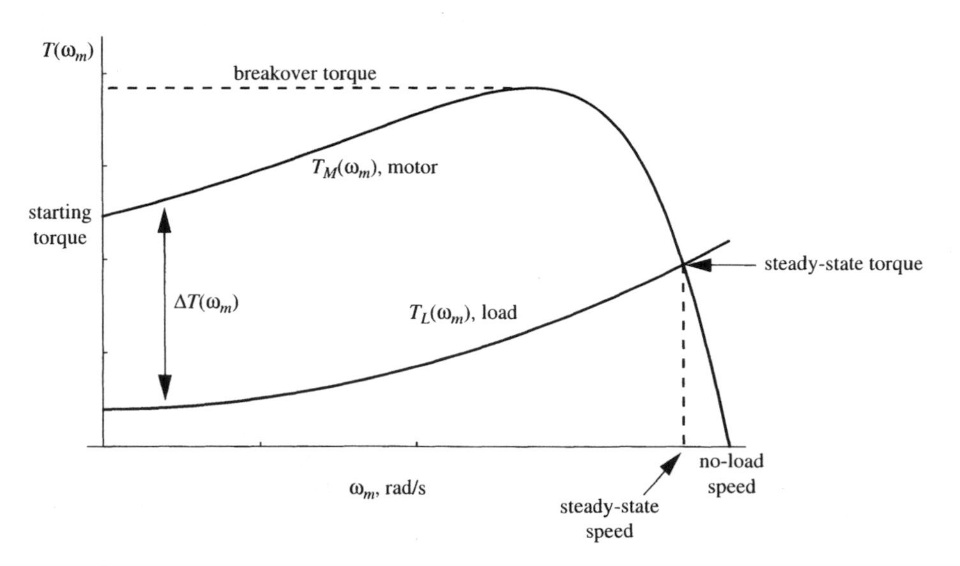

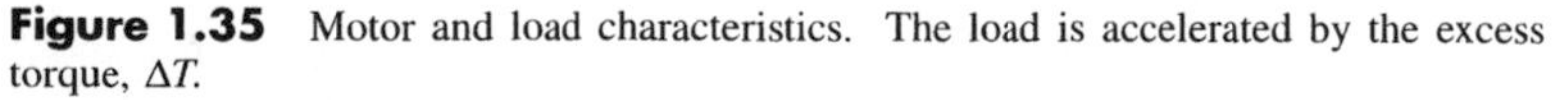
Figure 1.35 Motor and load characteristics. The load is accelerated by the excess torque, ΔT.

Dynamic Operation

run-up time

In addition to steady-state operation, we are also interested in transient, or dynamic, operation. For example, we may need to predict the starting current and the time required to reach the steady-state speed, which is called the *run-up time*. In dynamic operation, the rotor and load moments of inertia become factors, and energy processes are more complicated than for steady-state operation.

Run-up time. We can calculate the time required to reach steady-state speed by integrating the equation of motion. In general, for a rotational system

$$J\frac{d\omega_m}{dt} = T_M - T_L = \Delta T(\omega_m) \quad (1.48)$$

where ω_m represents mechanical rotation speed in radians/second, J the combined moment of inertia of motor and load, T_M the motor output torque, and T_L the load torque requirement. We may integrate Eq. (1.48) from zero time, when the motor is stopped, to the run-up time, t_{ru}, when the motor reaches the steady-state speed

$$t_{ru} = \int_0^{\omega_{ss}} \frac{J d\omega_m}{\Delta T(\omega_m)} \approx \sum \frac{J\Delta\omega_m}{\Delta T(\omega_i)} \quad (1.49)$$

where ω_{ss} is the steady-state speed. We may approximate the run-up time as indicated by the second form of Eq. (1.49) and illustrated in Fig. 1.36.

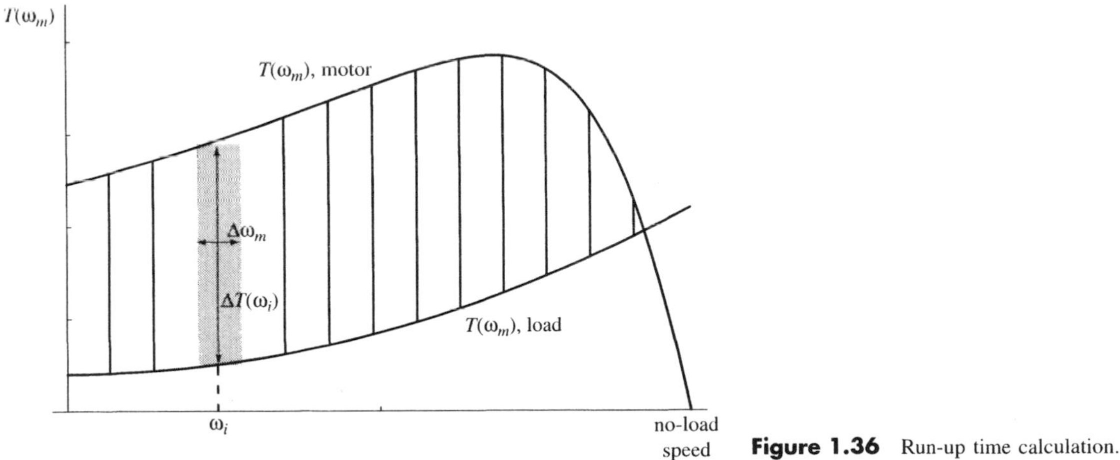

Figure 1.36 Run-up time calculation.

EXAMPLE 1.12 Run-up calculation

Consider a motor with a constant output torque of 5 N-m, driving a load requiring a torque proportional to speed. The steady-state speed is 800 rpm, and the combined moment of inertia of motor and load is 0.02 kg-m^2. Find the time required to reach equilibrium speed, starting from standstill.

SOLUTION:

The load requirement would be $T = C\omega_m$, where C is a constant to be determined. The equilibrium speed would be $\omega_{ss} = 800 \times 2\pi/60 = 83.8$ rad/s; hence, the constant is

$$T_M - T_L = 0 \Rightarrow 5 = C \times 83.8 \Rightarrow C = 0.0597 \text{ N-m/(rad/s)} \tag{1.50}$$

By Eq. (1.49), the run-up time is

$$\begin{aligned} t_{ru} &= 0.02 \int_0^{83.8} \frac{d\omega_m}{5 - 0.0597\omega_m} \\ &= \frac{0.02}{-0.0597} \ln(5 - 0.0597\omega_m)\Big|_0^{83.8} \end{aligned} \tag{1.51}$$

Equation (1.51) leads to an infinite result at the upper limit. The difficulty is that the system speed approaches steady state asymptotically, so theoretically it never reaches equilibrium.[19] One way to approximate the run-up speed is to assume a "final" speed slightly below the steady-state speed. For Eq. (1.51), changing the upper limit to 98% of the steady-state speed gives a run-up time of 1.311 s.

WHAT IF?

What if we define the final speed as 99.5% of final speed? What would be the run-up time?[20]

Summary. In this section, we discussed some fundamentals that apply to all motors. We presented equations relating to the input electrical variables and the output mechanical variables. We discussed how the output-torque characteristic of the motor interacts with the torque requirement of the load to establish the steady-state speed, torque, and power in the motor–load system.

In the next sections, we present and interpret the nameplate information for a three-phase induction motor and a single-phase induction motor, with most of the emphasis on the former. These two types of motors account for the vast majority of motor applications. Our purpose is to give sufficient information to analyze nameplate information. The details of motor models and operational characteristics for these and other electric motors are given in Chapters 4 to 6.

LEARNING OBJECTIVE 8.

To understand how to analyze and interpret nameplate information of induction motors

Three-Phase Induction Motor Nameplate Interpretation

Induction motor principle. In an induction motor, the field windings are on the stator. The currents in the field windings set up a rotating magnetic flux, which in turn induces voltage in the rotor. The voltage in the rotor drives currents that interact with the flux to produce torque. Because the relative motion between the rotating flux and the rotor conductors produces the voltage, the rotor will always turn slower than the flux rotates. The induction motor works like a fluid clutch or automatic transmission. Instead of fluid, the working medium is magnetic flux.

[19] Of course, a real motor does reach equilibrium. Our linear *model* leads here to a mathematical problem.

[20] 1.78 seconds.

Nameplate information. Three-phase induction motors are available from one-third horsepower to thousands of horsepower. The nameplate of a specific 60-Hz three-phase induction motor includes the following information:

- 50 horsepower
- Three phases
- 1765 rpm
- NEMA 326T frame
- 208–230/460 V
- 140–122/61 A
- Time rating: continuous
- Insulation class: F
- 1.15 service factor
- Maximum ambient temperature: 40°C
- NEMA code: G
- NEMA design: B
- NEMA nominal efficiency: 92.4%
- NEMA minimum efficiency: 91.0%

The motor nameplate also might give the motor type: drip-proof, totally enclosed fan cooled (TEFC), explosion-proof, etc.

service factor

Power rating. The ***power rating***[21] of this motor is 50 hp. This is a nominal rating and does not mean that the motor will put out 50 hp on all, or possibly on any, occasions. The actual power out of the motor depends on the load demands, as explained earlier—nor is 50 hp the maximum power of the motor. The maximum power that the motor can put out on a continuous basis is the nameplate power times the service factor, 50 hp × 1.15 = 57.5 hp in this case. Thus, the ***service factor*** is something of a "safety factor"; the 50-hp rating is, as we said before, the nominal power rating of the motor.

The nameplate power is significant in the following sense: *If* the motor is supplied with rated voltage, 460 V, and *if* the load demands exactly 50 hp, *then* the motor speed will be 1765 rpm *and* the input current will be 61.0 A.

Motor speed. The nameplate ***speed*** is 1765 rpm, and the significance of this speed is given in the preceding paragraph. With a power frequency of 60 Hz, induction motors run slightly below the standard speeds of 3600, 1800, and 1200 rpm, for the reason given above. If this motor is unloaded, the motor speed will be approximately 1800 rpm.

Motor output torque. The nameplate torque, T_{NP}, is not given but may be deduced from the output power and speed. Equation (1.42) yields

$$T_{NP} = \frac{P_{out}}{\omega_m} = \frac{50 \times 746}{1765 \times (2\pi/60)} = 202 \text{ N-m} \tag{1.52}$$

where we changed horsepower to watts and rpm to radians/second to make the units consistent. Again we stress that this is the nameplate torque; we do not know the actual torque unless we know what the load demands.

Torque versus speed. In the normal operating range, the output torque of a three-phase induction motor is approximately a straight line between zero torque at the no-load speed, ω_{NL}, and nameplate torque at nameplate speed, as shown in Fig. 1.37. Thus, the torque equation is of the form

$$T_M(\omega_m) = \text{slope} \times (\omega_m - \omega_{NL}) \tag{1.53}$$

[21] The nameplate terms will be in bold italics in the following discussion, for easy reference.

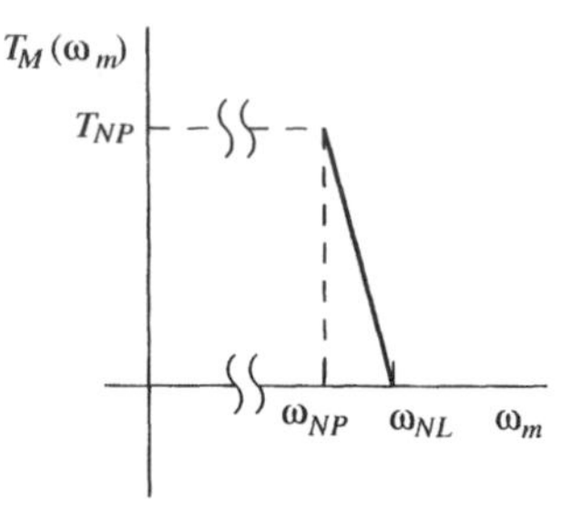

Figure 1.37 The torque characteristic is a straight line between nameplate and no-load conditions.

where the slope is $T_{NP}/(\omega_{NP} - \omega_{NL})$. Thus, the torque as a function of speed in the normal operating region is

$$T_M(\omega_m \text{ or } n) = T_{NP}\left(\frac{\omega_{NL} - \omega_m}{\omega_{NL} - \omega_{NP}}\right) = 202\left(\frac{1800 - n}{1800 - 1765}\right) \quad (1.54)$$

where we expressed speed in both rpm and rad/s.

EXAMPLE 1.13 Pump load

The 50-hp motor with the torque given by Eq. (1.54) drives a pump with a torque requirement of

$$T_L = 1.2 + 100\left(\frac{n}{1800}\right)^2 \text{ Nm} \quad (1.55)$$

Find the speed at which the motor–pump operates and the power required by the pump.

SOLUTION:

The motor–pump will operate where the motor output matches the pump demand

$$1.2 + 100\left(\frac{n}{1800}\right)^2 = 202\left(\frac{1800 - n}{1800 - 1765}\right) \quad (1.56)$$

which is a quadratic equation yielding $n = 1783$ rpm and $-188{,}598$ rpm. Clearly, the second answer represents the analytic extension of both motor and pump characteristics into regions where their models are invalid; the correct answer is 1783 rpm. The power is the speed in rad/s times the torque, which we can get from either motor or pump:

$$P_{out} = \omega_m T_M (\text{or } \omega_m T_L) \quad (1.57)$$

$$= 1783(2\pi/60)\left[1.2 + 100\left(\frac{1783}{1800}\right)^2\right] = 18{,}540 \text{ watts}$$

which is slightly less than 25 hp.

WHAT IF? What if the pump torque required $T_L = 100 + n/18$ N-m?[22]

[22] $n = 1766$ rpm and power $= 49$ hp.

NEMA frame 326T. NEMA is an acronym for the National Electrical Manufacturers Association, an industry association that has standardized many aspects of electrical power equipment. In this case, the mechanical dimensions of motors are standardized; for example, the motor output shaft diameter is $2\frac{1}{8}$ inches and is centered 8 inches above the base of all motors built on a ***NEMA 326T frame***.

Voltage and current ratings. The motor ***voltage rating*** is 208–230/460 V and the ***current rating*** is 140–122/61 A. We set aside the 208-V/140-A rating for a moment and concentrate on the 230/460-V rating. This reveals that each phase of the motor has multiple windings that can be connected in series or parallel. Figure 1.38(a) shows one phase of the high-voltage (460 V)/low-current (61 A) series connection, and Fig. 1.38(b) shows the low-voltage (230 V)/high-current (122 A) parallel connection. As you can see, the individual windings get the same voltage and current for both connections and the motor performance would be unchanged if indeed the voltage were exactly 230 or 460 volts. In practice, the series connection is preferred if 460-V three-phase power is available because the lower current requires smaller wire to supply the motor. Put another way, the impedance level of the motor is four times higher and the output impedance requirements of the power system are less demanding with the high-voltage/low-current connection.

IDEA 5 **Impedance Level**

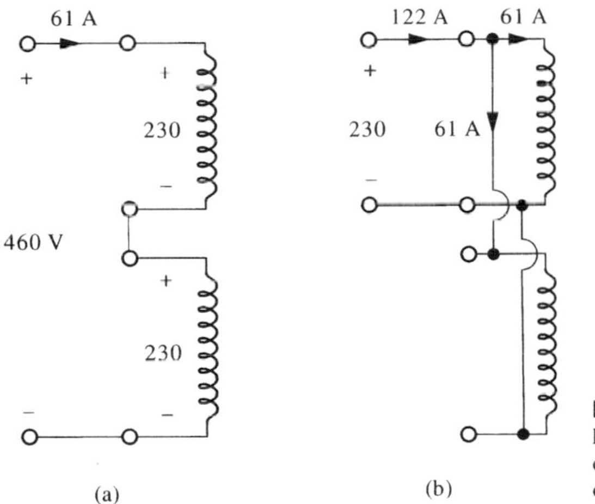

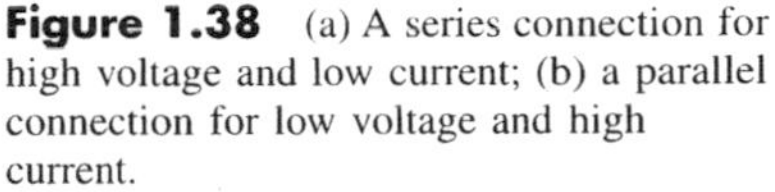
Figure 1.38 (a) A series connection for high voltage and low current; (b) a parallel connection for low voltage and high current.

The 208-V rating. The 208-V rating tells us the lowest voltage the motor should be supplied, and the current for 50 hp would be 140 A. First of all, $208 = 120\sqrt{3}$, so 208-V systems exist as 120-V three-phase connected in wye. But more common is low voltage due to older standards or long connecting wire, say, with pumps scattered around an oil field. Low voltage is dangerous to the motor because motors tend to overheat when the voltage is low, as explained in what follows, and heat shortens the motor lifetime.

Constant impedance or constant power? Up to this point, we have spoken of impedance as if it were constant, and this is true for many circuit elements. But a motor tends to take constant power from the electrical system and hence does not offer constant impedance. To see the difference, examine the contrast in Eq. (1.58)

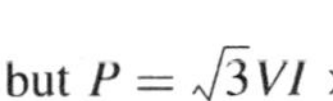

$$Z = \frac{V}{I} = \frac{(\downarrow)}{(\downarrow)} \text{ but } P = \sqrt{3}VI \times PF = \sqrt{3}(\downarrow)(\uparrow) \times PF \tag{1.58}$$

For constant impedance, the current goes down when the voltage goes down. This is the characteristic of resistors, inductors, capacitors, and many other devices as well. But for constant power, the current goes *up* when the voltage goes down provided the power factor does not change. This is how a motor behaves because the power of the motor depends largely on load demands.

Impedance Level

Thus, low voltage forces a high current, even if the voltage source is ideal. In practice, the high current lowers the voltage further due to the output impedance of the power system. Thus, when you use a portable electric tool such as a chain saw or lawn mower, the manufacturer will warn you to use only 100 feet of extension cord: Too long a cord produces low voltage, which produces higher current, which produces even lower voltage, etc. Because the resistive losses of the motor increase with the square of the current, the motor can rapidly overheat if the extension cord is too long.

EXAMPLE 1.14 Long extension cord

A saws-all electric saw is rated at 120 V, 8 A single phase. The saw is used with 100 ft of 16/3[23] extension cord. How much do the motor losses increase as a result of the extension cord?

SOLUTION:

The resistance for No. 16 wire is 4.02 Ω/1000 ft, and we have 200 feet of wire considering both conductors. Thus, the nominal 8 A of current causes a voltage drop of

$$\Delta V \approx \frac{4.02\ \Omega}{1000\ \text{ft}} \times 200\ \text{ft} \times 8\ \text{A} = 6.4\ \text{V} \tag{1.59}$$

and hence the voltage at the saw is about 113.6 V, assuming 120 V for the supply. Because the saw requires constant power, this loss in voltage causes a further increase in current to

$$\Delta I \approx 8\ \text{A} \times \frac{120}{113.6} = 8.45\ \text{A} \tag{1.60}$$

which will cause a further decrease in voltage, but we will ignore further effects. The increase in motor loss will be

$$\Delta\text{loss} \approx \left(\frac{8.45}{8}\right)^2 = 1.116, \text{ or a } 11.6\% \text{ increase} \tag{1.61}$$

WHAT IF? What if 150 ft of extension cord is used?[24]

Thermal specifications. The ***time rating*** is continuous operation. The ***insulation class*** of F tells us that the maximum temperature of the insulation should not exceed 155°C. The importance of the ***ambient temperature*** of 40°C is self-evident. These

[23] 16/3 = No. 16 size power wires with a ground.

[24] The motor losses increase by 18.3%.

specifications remind us again that the motor lifetime depends on the thermal environment affecting the temperature of its windings.

NEMA code G. The ***NEMA code*** tells us the starting-current requirements of the motor. NEMA classifies motors according to locked rotor (starting) kVA, and code G means 5.60 to 6.29 kVA/hp. Hence, for this motor, the starting current lies between 609 and 684 A with the 460-V connection, and twice these values for the 230-V connection. Because of these large currents, large induction motors are frequently started with reduced voltage and the voltage is increased as the motor speed increases.

NEMA design B. All three-phase induction motors have similar characteristics, but motors may be tailored somewhat for specific properties:

- Design A has high run efficiency and high breakover torque, with moderate starting torque and high starting current.
- Design B has moderate run efficiency, moderate starting torque with low starting current, and moderate breakover torque.
- Design C is similar to design B but has higher starting torque and is more expensive.
- Design D has relatively low run efficiency, extremely high starting torque which is also the breakover torque, and low starting current.

Our motor is ***design B***, which is the general-purpose motor.

Efficiency. The nameplate gives a nominal (92.4%) and minimum (91.0%) ***efficiency***. The nominal is what one would expect on an average, and we assume this value to calculate the nameplate input power from Eq. (1.43)

$$P_{in} = \frac{P_{out}}{\eta} = \frac{50 \times 746}{0.924} = 40,370 \text{ W} \tag{1.62}$$

and the input electrical power allows us to calculate the nameplate power factor from Eq. (1.41)

$$PF = \frac{P_{in}}{S} = \frac{40,370 \text{ W}}{48,600 \text{ VA}} = 0.831 \tag{1.63}$$

where $S = \sqrt{3}\,VI = \sqrt{3} \times 460 \times 61 = 48{,}600$ VA is the nameplate VA rating of the machine. We also calculate the nameplate reactive power required by the motor:

$$Q_{in} = \pm\sqrt{S^2 - P_{in}^2} = +\sqrt{(48.6)^2 - (40.4)^2} = +27.0 \text{ kVAR} \tag{1.64}$$

where the positive sign is used because an induction motor has lagging current.

The motor is designed to have maximum efficiency at nameplate conditions and hence the efficiency will be fairly constant in the vicinity of nameplate operation. However, the efficiency falls as the motor is lightly loaded.

Summary. We discussed the nameplate information for a three-phase induction motor. From the nameplate information, we deduced nameplate torque, the torque–speed characteristic, power factor, and reactive power. We stress again that the actual motor conditions depend on the load demands.

Single-Phase Induction Motor

Nameplate information. Single-phase induction motors come in many varieties and sizes; we now discuss the type of single-phase motor used to power farm machinery, stationary shop tools, compressors, and the like.[25] Such induction motors are made in sizes from one-fourth horsepower to 10 horsepower and are used where three-phase power is unavailable. The nameplate information on one motor is as follows:

- Phase 1
- Frame L56
- Hz 60
- Volts 115
- 40°C
- SF 1.35
- Type CS
- HP 1/3
- RPM 1725
- Amps 5.8
- Insulation class A

We now discuss the nameplate information that is not self-evident.

Type CS. Single-phase induction motors come in three types, CS, SP, and CR:

- CS: A capacitor-start/induction-run motor has high starting torque and moderate run efficiency.
- SP: A split-phase motor has low starting torque and moderate run efficiency.
- CR: A capacitor-start/capacitor-run motor has high starting torque and high run efficiency, and is more expensive.

RPM 1725. The meaning is self-evident. However, we should mention that the no-load speed of this motor would not be 1800, as for a three-phase motor, but would be somewhere around 1780–1790 rpm. Thus, we cannot derive the torque–speed characteristics without measuring the no-load speed. Once we know the no-load speed, we can derive the torque–speed characteristic as for a three-phase motor.

Other specifications. The remainder of the specifications are similar to those for three-phase motors. Torque can be calculated from power and speed. Efficiency is not normally given, so losses and power factor cannot be calculated from the nameplate information.

Check Your Understanding

1. In a motor the field current is usually larger than the armature current. True or false?
2. The output torque of a motor in operation depends in part on the load requirements. True or false?
3. The product of the power factor and the efficiency of an ac motor is always equal to the output mechanical power divided by the input electrical apparent power. True or false?

Answers: **(1)** False; **(2)** true; **(3)** true.

[25] Small fans, hand-held tools, and many household appliances use other types of single-phase motors.

CHAPTER SUMMARY

Chapter 1 introduces electric power distribution and utilization in electrical motors. Three-phase circuits are investigated. The per-phase model is developed and per-unit normalization is defined and illustrated. System aspects of electric motors are discussed, and the analysis of the nameplate information for three- and single-phase electric motors is illustrated.

Objective 1: To understand the nature and advantages of three-phase power. The three-phase system is described in the time and frequency domains. The advantages include efficiency of distribution, smoothness of power flow, and characteristics in starting and operating induction motors.

Objectives 2 and 3: To understand how to analyze a generator or load in wye or delta connection. We derive the phase and amplitude relationships between line voltage, line current, and power and phase voltage, phase current, and power for both wye and delta connections. We show that three-phase power depends only on line voltage, current, and power factor and not on the load or source connections.

Objective 4: To understand how to derive and use the per-phase equivalent circuit of a balanced three-phase load. The per-phase equivalent circuit is a single-phase circuit that models the state of a balanced three-phase circuit. The per-phase current and power factor are the same as the three-phase line current and power factor, but the per-phase voltage is equal to the line-to-neutral voltage of the three-phase system. The per-phase power is one-third the three-phase power. The per-phase equivalent circuit is independent of the actual connection of the three-phase load or source.

Objective 5: To understand how to connect single-phase transformers for three-phase transformation. Three-phase power may be transformed either by one three-phase transformer or by three single-phase transformers. Connecting primaries and secondaries in wye or delta gives four possible connections, each having different voltage and current ratios and different advantages and disadvantages.

Objective 6: To understand how transmission-line impedance affects power and voltage loss on the line. Transmission of electrical power over large distances involves significant effects of line impedance. An approximate analysis shows the effect of line inductance on power angle and power magnitude. We introduce the definitions and advantages of the per-unit system of normalization.

Objective 7: To understand how motor and load interact to establish steady-state operation. We examine dynamic and steady-state interactions of motors with their loads. Motor power and speed depend on load demands.

Objective 8: To understand how to analyze and interpret nameplate information of induction motors. Nameplate information gives a benchmark for comparing motors. The nameplate directly or indirectly gives power, speed, torque, voltage, current, efficiency, and other information. The dangers of operating a motor at low voltage are illustrated.

GLOSSARY

Angle of transmission, p. 28, the difference in phase angle between the voltage at the sending and receiving ends of a power transmission system; depends on real power flow.

Armature, p. 31, the windings of an electric motor that carry large currents.

Base values, p. 25, voltage, current, and power quantities to normalize circuit voltage, current, and power in per-unit calculations.

Delta (Δ) connection, p. 7, a three-phase connection that ties three generator windings or loads in a closed ring, with no neutral connection.

Field, p. 31, the part of an electric motor that creates a large magnetic flux.

Nameplate, p. 37, the information attached to a device that gives information about its inputs and outputs.

NEMA, p. 39, an acronym for the National Electric Manufacturer's Association, an industrial association that has standardized many aspects of electric power equipment.

Open-delta connection or V connection, p. 48, a three-phase transformer connection that uses two single-phase transformers.

Per-phase equivalent circuit, p. 19, a single-phase circuit that represents a balanced three-phase circuit.

Per-phase impedance, p. 19, an impedance in a per-phase equivalent circuit, which represents some aspect of a corresponding impedance in a three-phase circuit.

Per-unit calculations, p. 25, a method for normalizing electrical circuit calculations.

Phase current, p. 9, the current in a generator winding supplying a three-phase system or in one leg of a three-phase load.

Phase impedance, p. 12, the phasor phase voltage divided by the phasor phase current.

Phase sequence, p. 11, the time-sequence order by which the voltages or currents in a three-phase system reach their maximum.

Phase voltage, p. 9, the voltage across a generator winding supplying a three-phase system, or the voltage across one leg of a three-phase load.

Rotor, p. 31, the part of a motor that rotates.

Run-up time, p. 35, the time required for a motor–load to reach steady-state speed with a given load condition.

Service factor, p. 37, the ratio of maximum power to nameplate power on a motor, something of a safety factor.

Stator, p. 31, the part of a motor that does not rotate.

Terminal, p. 6, the end of a wire connected to an electrical device.

Terminal board, p. 6, a place where electrical terminals are made available for connection.

Three-phase current, p. 9, the current in the lines carrying the power in a three-phase system.

Three-phase transformer, p. 22, a transformer in which all phases share a common magnetic system.

Three-phase voltage, p. 9, the voltage between the lines carrying the power in a three-phase system.

Unbalanced load, p. 21, a load with unequal phase impedances, resulting in unequal line currents.

Wye (Y) connection, p. 8, a three-phase connection that ties three generator windings or three loads to a common point, which is the neutral connection.

PROBLEMS

Section 1.1: Three-Phase Power

1.1. Pick two values of ωt in Eq. (1.2) and show by direct calculation that the time-varying terms cancel.

1.2. For Fig. 1.6(a), develop a delta connection by first connecting A to B'. Draw a phasor diagram of the resulting system. Is the phase rotation ABC or ACB?

1.3. Three 230-V (rms) generators are connected in a three-phase wye configuration to generate three-phase power. The load consists of three balanced impedances, $\underline{\mathbf{Z}}_L = 2.6 + j1.8\ \Omega$, connected in delta.

- **(a)** Find the line current an ammeter would measure.
- **(b)** Find the apparent power.
- **(c)** Find the real power to the load.
- **(d)** What is the phase angle between $\underline{\mathbf{I}}_A$ and $\underline{\mathbf{V}}_{AB}$, assuming ABC rotation?

1.4. For the three-phase power system shown in Fig. P1.4, a voltmeter measures 146 V between line A and the neutral N. Find the following:

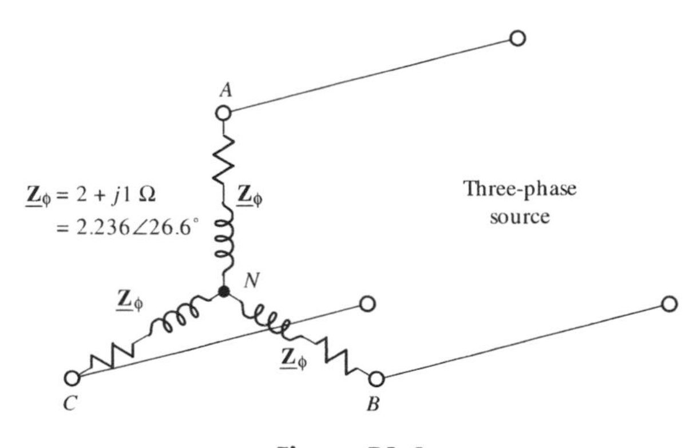

Figure P1.4

- **(a)** Line voltage.
- **(b)** Line current.
- **(c)** Load power factor.
- **(d)** Apparent power.
- **(e)** Real power.

1.5. In Fig. P1.5, the voltage between A and N is 120 V as measured by a standard meter. Let this voltage be the phase reference. The phase impedance is $\underline{\mathbf{Z}}_\phi = 6.2 + j2.7 = 6.76 \angle 23.5°\ \Omega$.

- **(a)** What is $\underline{\mathbf{V}}_{AB}$ as a phasor?
- **(b)** What would an ammeter measure as the line current?
- **(c)** What is the apparent power?
- **(d)** What is the real power?

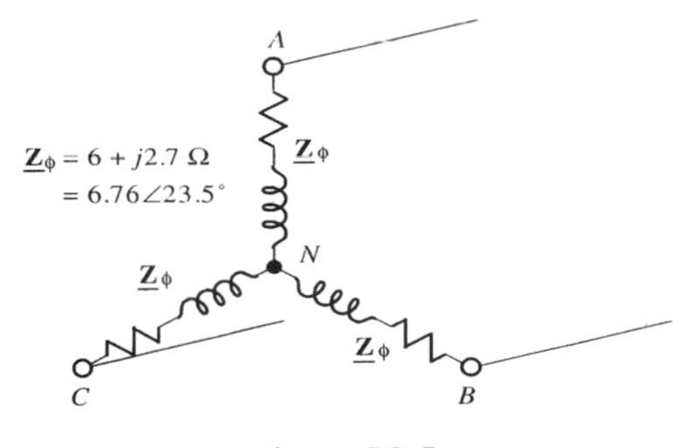

Figure P1.5

1.6. Figure P1.6 shows a three-phase source and load.

- **(a)** Draw appropriate connections.
- **(b)** Find the phase voltage.
- **(c)** Find the line current.
- **(d)** Determine the power factor.

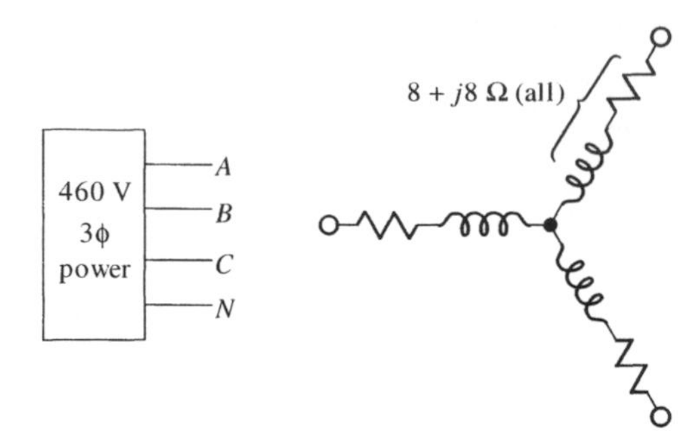

Figure P1.6

(e) What is the power in the load?

(f) Assuming $\underline{\mathbf{V}}_{AB} = 460 \angle 0°$, find $\underline{\mathbf{I}}_A$ as a phasor.

1.7. For the three-phase delta-connected load in Fig. P1.7, we have the line voltage and current to be $\underline{\mathbf{V}}_{AB} = 480 \angle 0°$ V and $\underline{\mathbf{I}}_A = 10 \angle -30$ A.

(a) What is $\underline{\mathbf{V}}_{CA}$?

(b) What is the phase current in the load, rms value?

(c) What is the time-average power into the load?

(d) What is the phase impedance?

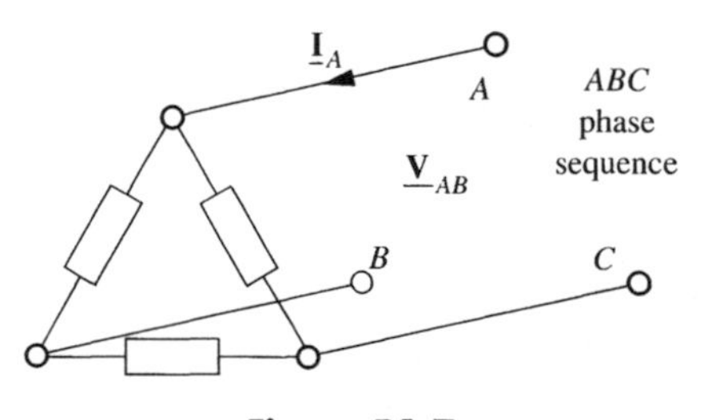

Figure P1.7

1.8. Three 230-V rms generators are connected in delta to form a three-phase source. A balanced wye-connected load has phase impedances of $12 + j7\ \Omega$. Find the rms values of the following:

(a) The phase voltage of the source.

(b) The line voltage of the system.

(c) The phase voltage of the load.

(d) The phase current of the load.

(e) The line current of the system.

(f) The phase current of the source.

(g) What is the power factor of the load?

(h) What is the real power to the load?

1.9. In the three-phase circuit shown in Fig. P1.9, find the following:

(a) The line current that would be measured by an ammeter.

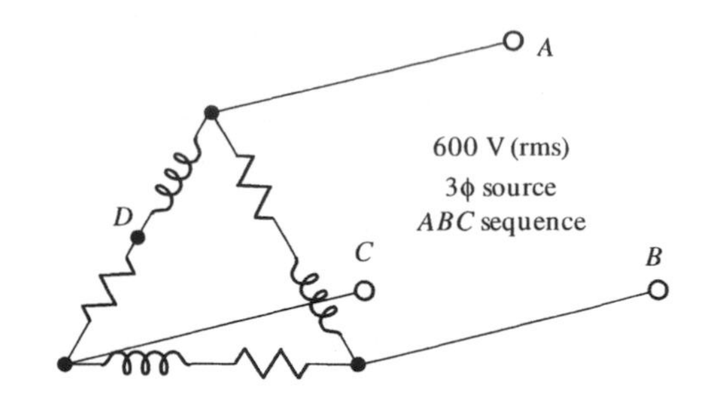

Figure P1.9

(b) The power factor of the three-phase load.

(c) The voltage that would be measured between B and D by a voltmeter.

1.10. In the three-phase circuit shown in Fig. P1.10, find the following:

(a) The line current that would be measured by an ammeter.

(b) The power factor of the three-phase load.

(c) The voltage that would be measured between B and D by a voltmeter.

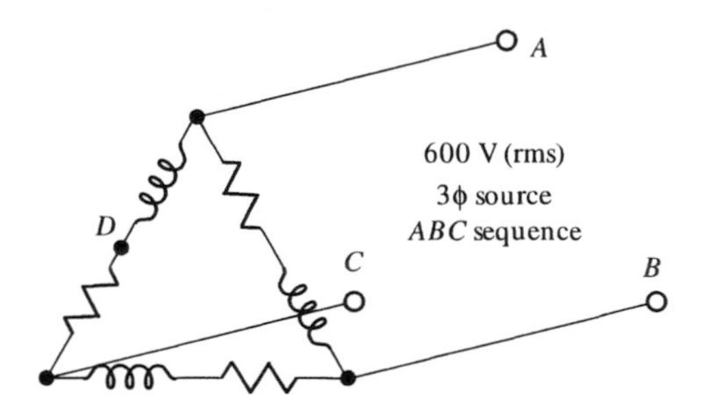

Figure P1.10

1.11. For the three-phase circuit connected in delta shown in Fig. P1.11, find the following:

(a) The load power factor. Assume lagging.

(b) The line current, rms.

(c) The magnitude of the phase impedance.

(d) The reactive power to each phase impedance.

1.12. For the three-phase circuit connected in delta shown in Fig. P1.12, find the following:

(a) The load power factor. Assume lagging.

(b) The line current, rms.

(c) The magnitude of the phase impedance.

(d) The reactive power to each phase impedance.

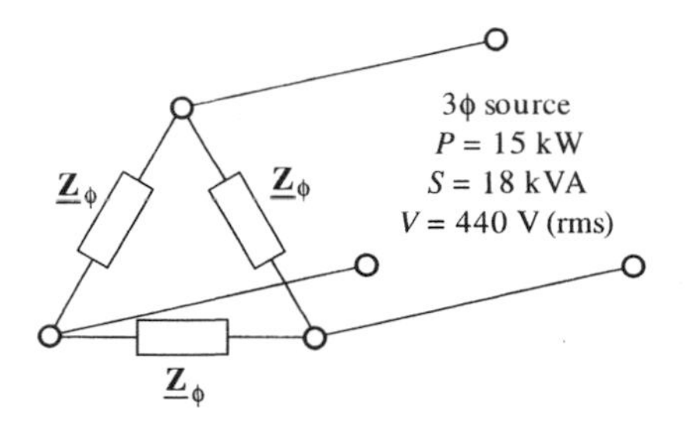

Figure P1.11

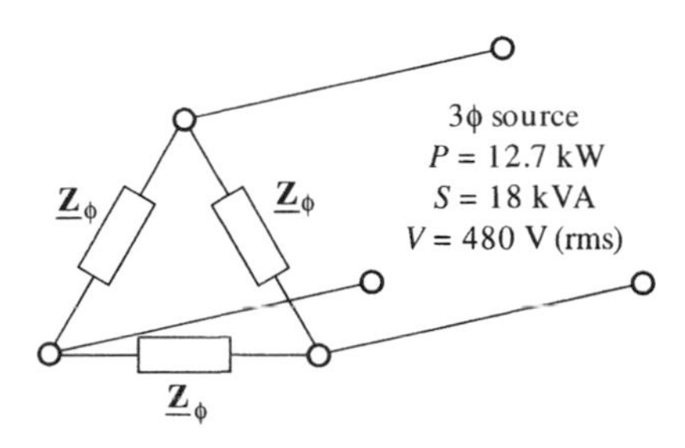

Figure P1.12

1.13. For the three-phase system shown in Fig. P1.13, find the following:

- **(a)** Phase voltage.
- **(b)** Line voltage.
- **(c)** Phase current.
- **(d)** Line current.
- **(e)** Phase impedance.
- **(f)** Apparent power.
- **(g)** Power factor.
- **(h)** Real power to the load.

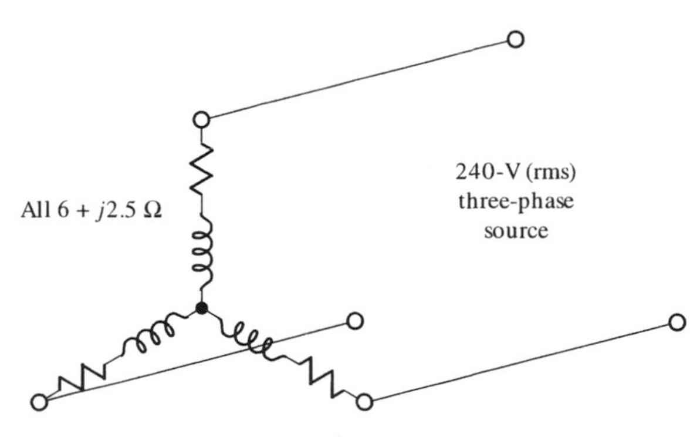

Figure P1.13

1.14. A balanced, wye-connected three-phase load is shown in Fig. P1.14. The current in line A is $\underline{I}_A = 8\angle 0°$ A. The voltage from B to the neutral point is $\underline{V}_{BN} = 120\angle -90°$ V.

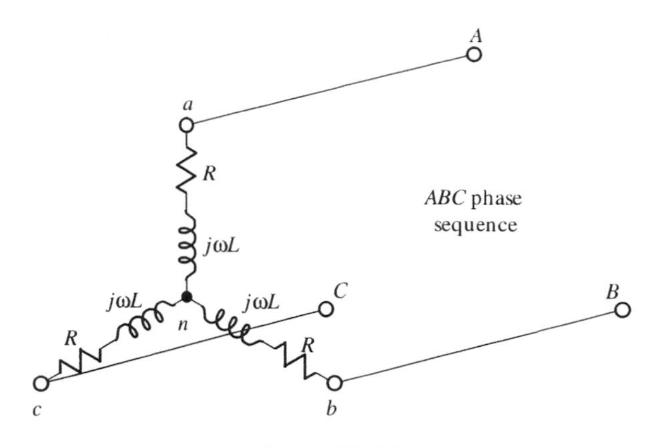

Figure P1.14

- **(a)** Find the three-phase voltage that a voltmeter would read.
- **(b)** Find the real and reactive power into the entire three-phase load.
- **(c)** Determine R and L, assuming 60-Hz operation.

1.15. Demonstrate the equivalence of the delta and wye circuits in Fig. 1.18. Assume an impedance $\mathbf{Z}_\Delta \angle \theta$ for the delta and an impedance $\mathbf{Z}_Y \angle \theta$ for the wye and compute the complex power for each. Equate these powers and confirm the 3:1 ratio shown in Fig. 1.18.

1.16. Three identical resistors are placed in a wye configuration and draw a total of 150 W from a three-phase source. What power would the same resistors draw if placed in delta?

1.17. For the three-phase circuit shown in Fig. P1.17, the 0.1-Ω resistors represent the resistance of the distribution system. Find the following:

- **(a)** Total power out of the source, including line and load.
- **(b)** Line losses.
- **(c)** Distribution system efficiency.

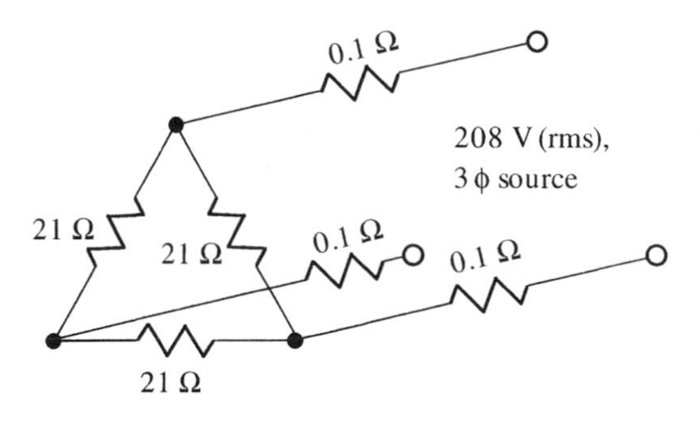

Figure P1.17

1.18. Using delta–wye transformations, determine the total power given to the delta and wye loads in Fig. P1.18, not counting the losses in the 0.1-Ω resistors that represent losses in the connecting wires. *Hint:* The neutrals of two balanced wye loads have the same voltage and hence may be considered as connected.

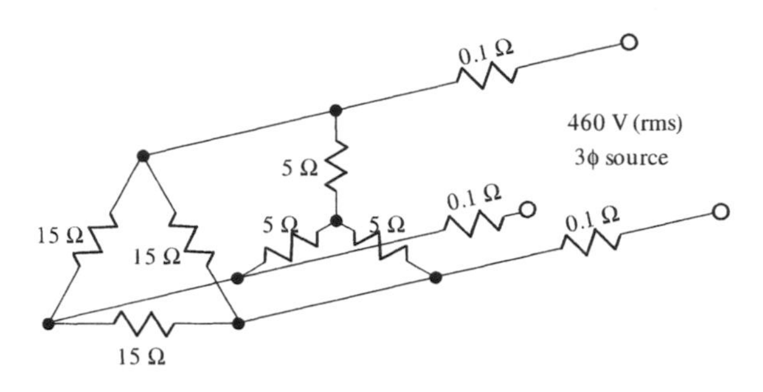

Figure P1.18

1.19. The city of Austin, Texas, distributes power with a three-phase system with 12.5 kV between the power-carrying wires. But each group of houses is served from one phase and ground, and transformed to 240/120 V by a pole transformer, as shown in Fig. P1.19.

(a) What is the turns ratio (primary/secondary turns) of the pole transformer to give 240 V, center-tapped?

(b) When a 1500-W hair dryer is turned on, how much does the current increase in the high-voltage wire? Assume the power factor is unity and the transformer is 100% efficient.

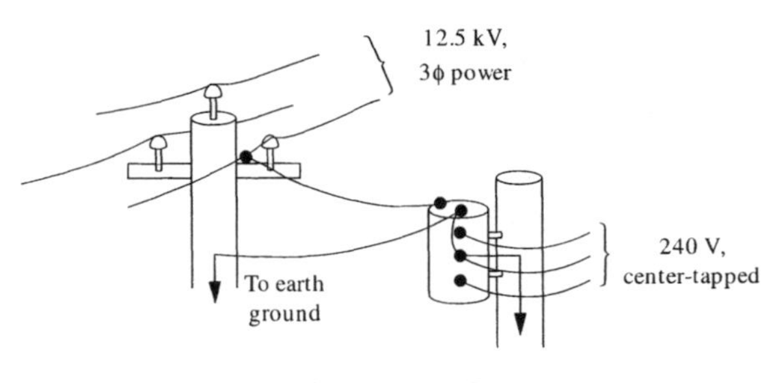

Figure P1.19

1.20. A workman finds a three-phase cable with four wires. He labels the wires 1, 2, 3, and 4 and measures the following voltages: $V_{12} = 150$ V and $V_{23} = 260$ V. What are V_{13} and V_{34}?

In the text, we showed how to transform three-phase power with the use of three single-phase transformers. There are two ways to transform three-phase power with *two* single-phase transformers. The next two problems investigate these methods. In them, we will transform 460 V three phase to 230 V three phase; hence, the transformers have a turns ratio of 2:1. *Hint*: In both figures, the geometric orientation hints of the phasor relationships.

1.21. The configuration shown in Fig. P1.21 is called the "open-delta" or V connection, for obvious reasons. Identical 2:1 transformers are used.

(a) Show that if *ABC* is 460-V balanced three-phase, *abc* is 230-V balanced three-phase. Consider the *ABC* voltages to be a three-phase set and prove the *abc* set is three-phase.

(b) If the load is 30 kVA, find the required kVA rating of the transformers to avoid overload. [You can solve this independent of part (a).]

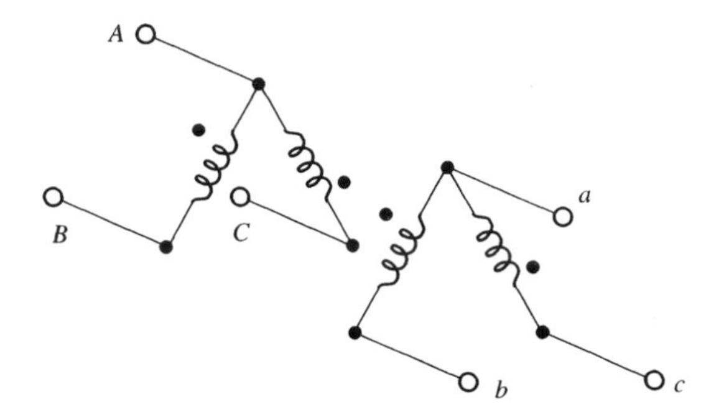

Figure P1.21 The *V* or open-delta transformer connection.

1.22. The circuit shown in Fig. P1.22 is called the T connection. For this connection, the 2:1 transformers are not identical but have different voltage and kVA ratings. The bottom transformer is center-tapped so as to have equal, in-phase voltages for each half.

(a) Find the voltages V_1 and V_2 to make this transform 460-V to 230-V three-phase.

(b) If the load is 30 kVA, find the required rating of each transformer to avoid overload.

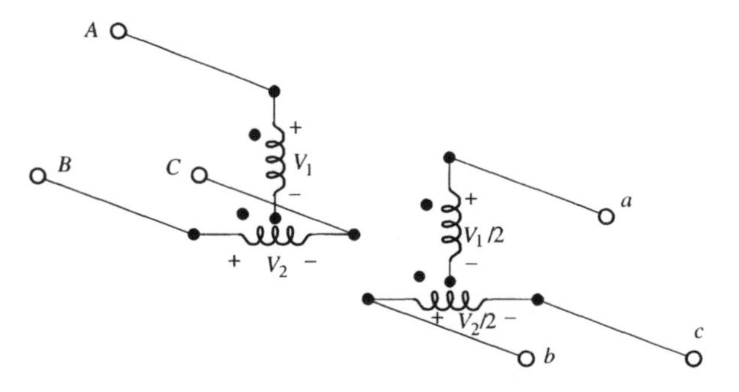

Figure P1.22 The *T* transformer connection.

Section 1.2: Power Distribution Systems

1.23. A wye-connected, balanced, three-phase load has a three-phase voltage of 480 V and requires a kVA of 20 kV at a *PF* of 0.94, lagging.

(a) What is the per-phase, per-unit equivalent of the load, using the three-phase voltage and 42 kVA for bases?

(b) What if the load is in delta?

1.24. A 500-MVA, 22-kV ac generator is represented on a per-phase equivalent circuit as an ideal ac voltage source in series with an inductive reactance. The internal reactance is Y-connected, 1.1 per unit. Find the actual reactances that are connected in Y.

1.25. A 12/138-kV, 50-MVA three-phase transformer has a per-phase inductive reactance of $j0.005\ \Omega$, referred to the primary side. Find the per-unit reactance for the transformer and give a per-unit, per-phase equivalent circuit for the transformer.

1.26. A transmission line has an impedance of $100 + j500\ \Omega$, including both wires. The 3ϕ voltage at the generator output is 22 kV. The voltage is stepped up to 345 kV in a Δ–Y transformer connection, sent over a transmission line, and then stepped down to a nominal 23 kV with a Y–Δ transformer connection. The load is 30 MW at a lagging *PF* of 0.92.

(a) Give a per-phase equivalent circuit showing the per-phase voltage and current on the line.

(b) Find the approximate per-unit voltage loss. Ignore the transformer inductance.

(c) Find the approximate per-unit power loss.

1.27. The circuit shown in Fig. P1.27 represents a 60-Hz power distribution system. The distribution voltage is 8 kV and the load is 20 houses, each requiring on the average 12 kW at 0.95 *PF*, lagging. The line impedance is shown. Assume an ideal transformer.

(a) What would be the magnitude of the current in the primary of the transformer?

(b) What are the line power losses?

(c) What value of capacitor across the transformer primary minimizes line losses? Consider that the load voltage is constant.

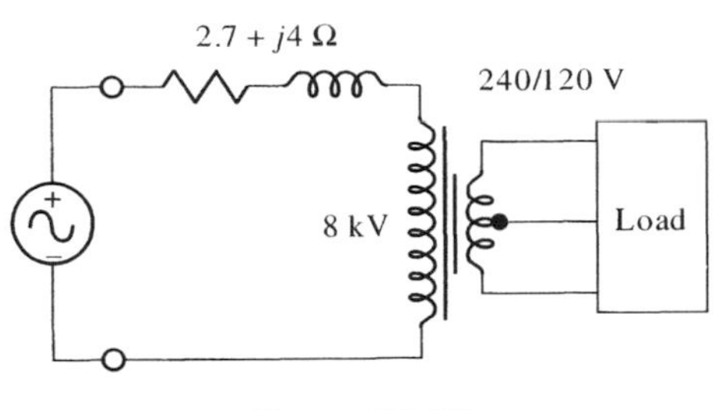

Figure P1.27

(d) What are the line power losses with the corrected *PF*?

1.28. The circuit in Fig. P1.28 shows a load that requires 500 kVA at a lagging *PF* of 0.84, a voltage of 13 kV (maintained constant), and frequency of 60 Hz. The $2 + j10\ \Omega$ represents the impedance of the distribution line. Making no approximations, you are to calculate the line losses and voltage difference between load and source voltage under two conditions:

(a) The circuit as shown.

(b) The circuit with a capacitor at the load to correct the power factor to unity with the same real power. Also determine the value of the required capacitor.

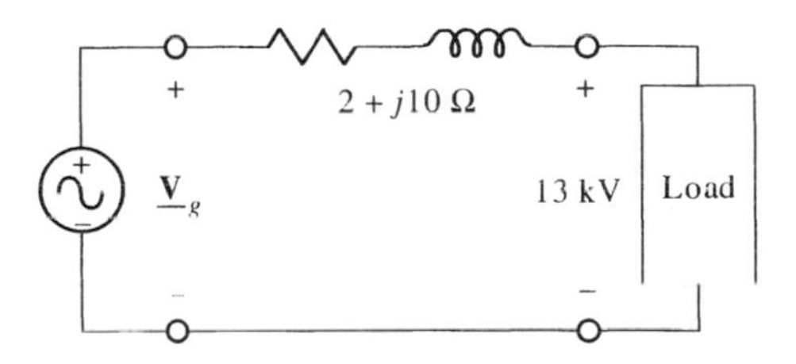

Figure P1.28

1.29. A 138-kV overhead line has a series resistance and inductive reactance of 0.16 and 0.41 Ω/km, respectively. Find the magnitude of the voltage required at the sending end of the line for 138 kV at the receiving end if the line is 80 km long. The apparent power on the line is 12.5 MVA at 0.95 *PF*, lagging.

Section 1.3: Introduction to Electric Motors

1.30. A motor has the output torque characteristic shown in Fig. P1.30. The load torque characteristic is

$$T_L(n) = 10\left(\frac{n}{1200}\right)^2 \text{ N-m}$$

(a) Determine the operating speed.

(b) Determine the output power.

(c) What is the maximum possible output power from the motor?

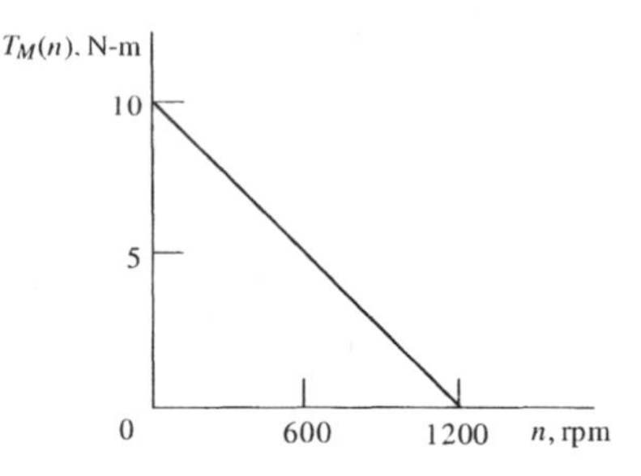

Figure P1.30

1.31. A motor has the parabolic output torque characteristic shown in Fig. P1.31. The load torque characteristic is

$$T_L(n) = 3 + 4\left(\frac{n}{1800}\right) \text{ N-m}$$

(a) Determine the operating speed.
(b) Determine the output power.
(c) What is the maximum possible output power from the motor?

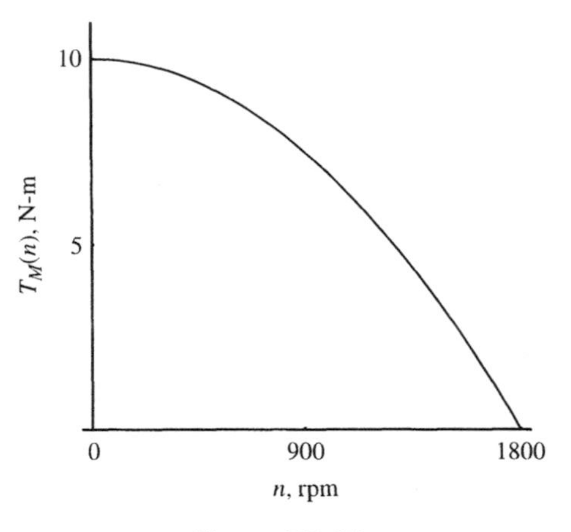

Figure P1.31

1.32. A motor has an output power given by the formula

$$P(\omega_m) = 12\omega_m - \frac{10\omega_m^2}{120\pi} \text{ watts}$$

(a) What is the starting torque of the motor?
(b) Find the no-load speed in rpm.
(c) Find the higher speed for 1-hp output power.
(d) Find the maximum output power.

1.33. A motor has an output torque characteristic given by the equation

$$T_M(\omega_m) = 50 - \frac{(\omega_m - 40)^2}{80} \text{ N-m}$$

where ω_m is the mechanical speed of the motor in radians/second.

(a) Find the no-load speed of the motor.
(b) What is the maximum torque of the motor?
(c) What is the maximum power of the motor?

1.34. The torque characteristics of two motors are shown in Fig. P1.34. The load requires a constant torque of 3 N-m.

(a) Which motor would have the longer run-up time if starting the load from standstill?
(b) Determine the power required for the load in steady state if driven by motor #2.

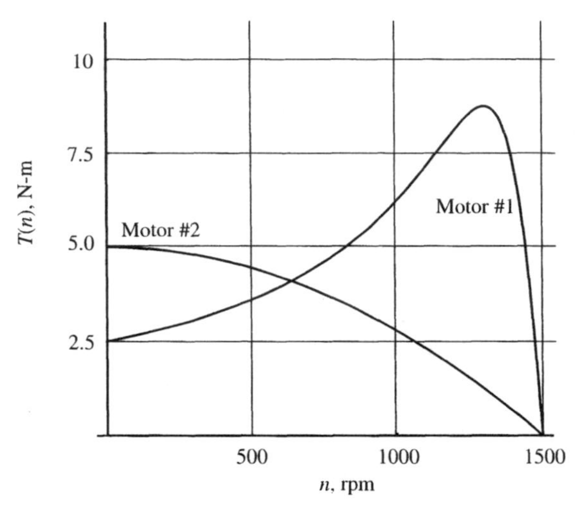

Figure P1.34

1.35. An electric motor has a torque–speed characteristic given by the equation

$$T_M(\omega_m) = \frac{100}{(5 + K\omega_m)^2} \text{ N-m}$$

where K is a constant. The motor puts out 1 hp at 12,000 rpm.

(a) Find K.
(b) What is the starting torque?
(c) What is the maximum power out of the motor?

1.36. The output torque of a motor is given in Fig. P1.36.

(a) Find the blocked-rotor torque.
(b) Find the no-load speed in rpm.
(c) At what speed does the motor put out 12 hp?
(d) If a load requires 20 N-m, what speed would the motor–load run in rpm?

1.37. A motor has the torque given by the equation

$$T_M(\omega_m) = 50 + 0.1\omega_m - \frac{(\omega_m - 40)^2}{80} \text{ N-m}$$

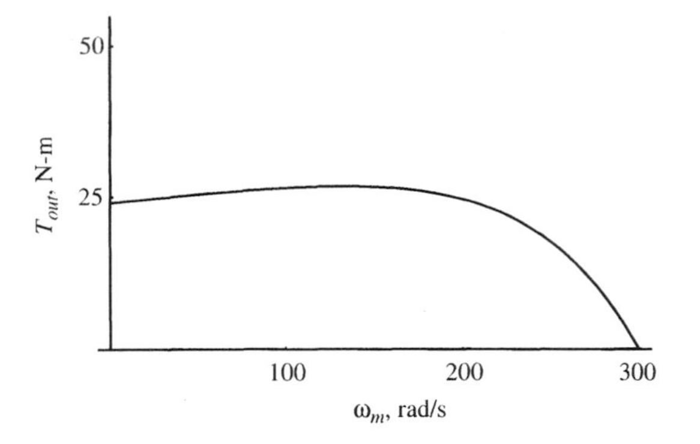

Figure P1.36

(a) Find the starting torque.
(b) Find the no-load speed of the motor.
(c) Find the maximum torque of the motor.
(d) Find the maximum power of the motor.

1.38. A motor has an output torque characteristic that is well approximated by the function,

$$T(n) = 12 \cos\left(\frac{\pi}{2} \times \frac{n}{1200}\right) \text{ N-m,}$$

where $0 < n < 1200$ is the speed in rpm.
(a) Find the starting torque.
(b) Find the no-load speed.
(c) Find the maximum power out of the motor in hp.

1.39. A motor with a moment of inertia $J = 0.5$ kg-m^2 is turning a load at a constant speed of 1160 rpm. When the load is suddenly disconnected, the instantaneous angular acceleration is +12 rad/s^2. Find the load torque at 1160 rpm.

1.40. Write the differential equation for a motor–load system where $T_M(\omega_m) = T_M$, a constant, and $T_L(\omega_m) = C\omega_m$, similar to the example on page 35. Use the notation J = moment of inertia, K = torque constant, and ω_{ss} = steady-state speed. This system fits the conditions of first-order transients; hence the initial value, final value, and time constant establish the response. Determine the time constant. Calculate the run-up time for 98% of equilibrium speed based on the differential equation solution in terms of T_M, J, and ω_{ss}.

1.41. A fan requires a driving torque of the form $T_L(\omega_m) = K\omega_m^2$. The fan requires $\frac{1}{2}$ hp of drive power on the shaft to turn 1800 rpm. The fan is driven by an electrical motor with the output characteristic given in Fig. P1.41.

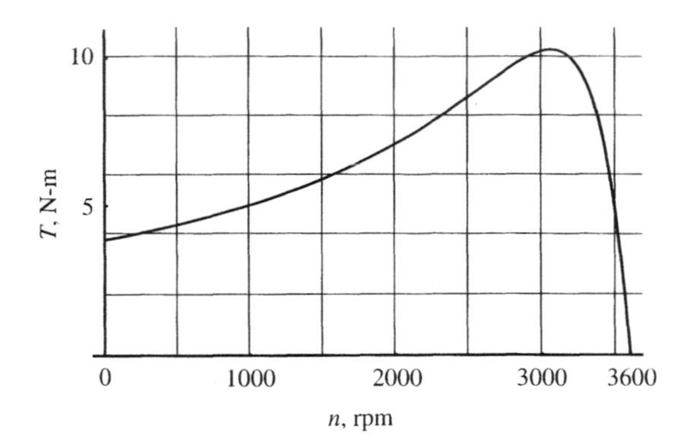

Figure P1.41

(a) Find the constant K, with torque and speed expressed in mks units, N-m and rad/s.
(b) Find the speed in rpm at which the fan will operate.
(c) Find the approximate time it takes the fan to reach its final speed. The moment of inertia of the motor–load is $J = 0.06$ kg-m^2.

1.42. Calculate the steady-state speed and run-up time for the motor and load used in the example on page 35 if the load torque requirement is changed to

$$T_L(\omega_m) = 0.001\omega_m^2 \text{ N-m.}$$

1.43. A motor–load system has a run-up characteristic given by

$$n(t) = 1180(1 - e^{-t/0.6s}) \text{ rpm.}$$

The motor–load moment is $J = 0.2$ kg-m^2, and the load torque requirement is constant at 8 N-m. Find the following:
(a) Motor-starting torque.
(b) Motor torque at 1180 rpm.
(c) Time required to reach 96% of final speed.

1.44. A motor–load has a combined moment of inertia of 2 kg-m^2. The output torque of the motor is

$$T_M(n) = 10\left[1 - \left(\frac{n}{1800}\right)\right] \text{ N-m}$$

where n is the speed in rpm. The required load torque is a constant 6 Nm.
(a) What is the equilibrium speed in rpm?
(b) How long would it take the system to reach 99% of the final speed starting from a standstill?

1.45. Figure P1.45 shows the input current, efficiency, and power factor of a 230-V three-phase motor from no load to full load.

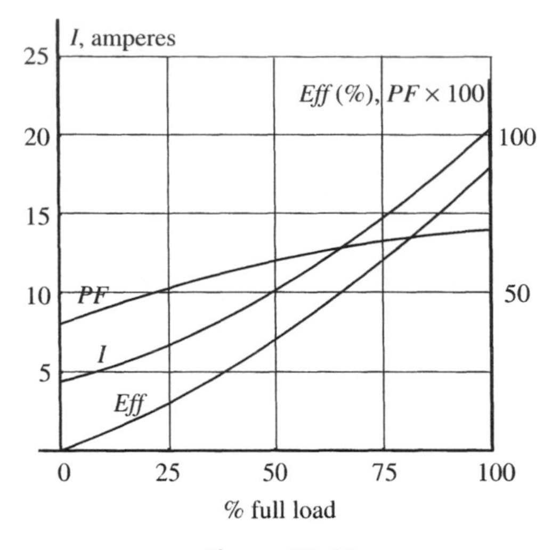

Figure P1.45

(a) What is the rated output power of the motor in hp?
(b) What is the reactive power used by the motor at 50% load? The current is lagging.
(c) What are the losses of the motor at no load?

1.46. A 208-V three-phase motor runs 1720 rpm at a load requiring 9.0 N-m torque. The motor draws 5.5 A of current and the power factor is 0.92, lagging.
(a) Find the input power.
(b) Find the losses of the motor.
(c) Find the efficiency of the motor.

1.47. A 60-Hz, three-phase induction motor has the following nameplate information: 25 hp, 1755 rpm, 91.7% eff., 230/460 V, 61.8/30.9 A, service factor 1.15.
(a) Find the torque output at nameplate conditions.
(b) What is the apparent power into the machine at nameplate conditions?
(c) Find the power factor at nameplate conditions.
(d) Determine the reactive power into the machine at nameplate conditions, assuming lagging current.

1.48. An induction motor has the following nameplate information: 60 Hz, three phase, 7.5 hp, 1750 rpm, 230/460 V, 22.0/11.0 A, 86.0% eff., NEMA frame 213T, 1.15 SF, \$376 list, \$250.79 wholesale, 129.0 lb shipping weight. For this type of motor, the blocked rotor torque is 135% and the breakover torque is 185% of the nameplate torque.
(a) Find the input power factor at nameplate conditions.
(b) Find the nameplate starting torque of the motor.
(c) Find the losses of the motor at nameplate conditions.
(d) What range of power in hp can the motor sustain constantly without failure under normal thermal conditions?

1.49. An engineer is comparing two 60-Hz, 1.5-hp, three-phase motors. Motor A is a high-efficiency motor with the following nameplate information: 1.5 hp, 230/460 V, 1740 rpm, 85.5% eff., 4.4/2.2 A, 1.15 SF, \$146.41. Motor B has the following nameplate information: 1.5 hp, 230/460 V, 1740 rpm, 82.5% eff, 5.2/2.6 A, 1.15 SF, \$130.07. Based upon 2000 hours of operation a year at nameplate rating and 8¢/kWh, how much money would motor A save per year, not including the motor investment cost?

1.50. An induction motor has the following nameplate information: single-phase, $\frac{1}{2}$ hp, 50 Hz, 1425 rpm, 110/220 V, 9.2/4.6 A, 1.25 SF, Class B insulation, 40°C.
(a) Find the output torque under nameplate conditions.
(b) What is the maximum output power that can be produced by the motor in sustained operation without overheating? (Assume 40°C ambient temperature.)
(c) Assuming that the motor power factor and efficiency are numerically equal, what are the motor losses at nameplate condition?

1.51. A three-phase, 60-Hz induction motor has the following nameplate information: 7.5 hp, 1755 rpm, 230/460 V, 19.4/9.7 A, 88.5% eff., 1.15 SF. We may assume the no-load speed of the motor to be 1800 rpm.
(a) Find the power factor at nameplate conditions.
(b) Find the output torque at nameplate conditions.
(c) Find the speed at which the motor and load will operate if the load requires a torque of

$$T_L(n) = 10 + 12.8(n/1800)^2 \text{ N-m}.$$

1.52. A three-phase, 60-Hz induction motor has the following nameplate information: 30 hp, 39 A, 1760 rpm, 460 V, 87.5% eff, 1.15 SF. The no-load speed of the motor is 1799.4 rpm. The load torque is given by the equation

$$T_L(n) = 10 + 12\left(\frac{n}{1800}\right) + 15\left(\frac{n}{1800}\right)^2 \text{ N-m}$$

where n is the speed in rpm. Find the operating speed of the motor–load system.

1.53. A three-phase, 60-Hz induction motor has the following nameplate information: 25 hp, 1755 rpm, 230/460 V, 64.0/32.0 A, 91.0% efficiency, 1.15 SF.

(a) Find the reactive power requirement at nameplate conditions.
(b) Find the motor losses at nameplate conditions.
(c) Find the nameplate output torque.
(d) At what speed will the motor produce 10 hp?

1.54. The torque–speed characteristic shown in Fig. P1.54 is the normal operating region (no load to nameplate load) for a three-phase induction motor. At full load the efficiency is 67.7% and the current and voltage are 1.1 A and 230 V, respectively.

(a) What is the hp rating of the motor?
(b) Find the motor losses at nameplate conditions.
(c) If the voltage dropped to 220 V at nameplate power, what would be the current, approximately?

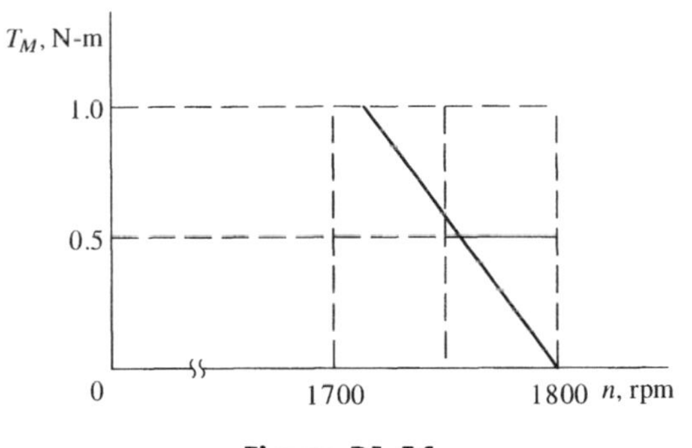

Figure P1.54

1.55. A 60-Hz, single-phase induction motor has the following nameplate information: 1 hp, 1740 rpm, 115/230 V, 13.6/6.8 A, service factor 1.0, power factor 0.73, starting current 50 A at 115 V, starting torque 9.0 lb-ft.

(a) What is the output torque under nameplate conditions?
(b) Determine the efficiency at nameplate conditions.
(c) Determine the reactive power into the machine under nameplate conditions, assuming lagging current.

1.56. A motor catalog lists the following motor: 5 hp, 230 V, 12.8 A. 1740 rpm, 60 Hz, 85.5% efficiency, 1.15 SF. Assume operation at nameplate conditions and assume losses are equally divided between rotor and stator.

(a) Is this a three-phase or single-phase motor? Explain how you know.
(b) What is the power factor of the motor at nameplate conditions?
(c) Find the motor losses at nameplate conditions.

1.57. A 3-hp, 120-V, 1160-rpm, 60-Hz, single-phase motor has input current of 30 A and a power factor of 0.76. At nameplate conditions, find the motor losses and the output torque.

1.58. A single-phase induction motor has the following nameplate information: $\frac{1}{2}$ hp, 1140 rpm, 115/230 V, 10.0/5.0 A, SF 1.25. Assume the power factor at nameplate is 0.78, lagging. The torque–speed curve is shown in Fig. P1.58.

(a) Find the efficiency of the motor at nameplate conditions.
(b) Find the ratio of the starting torque to full-load torque.
(c) Determine the torque as a function of speed in rpm in the region from zero to nameplate power.

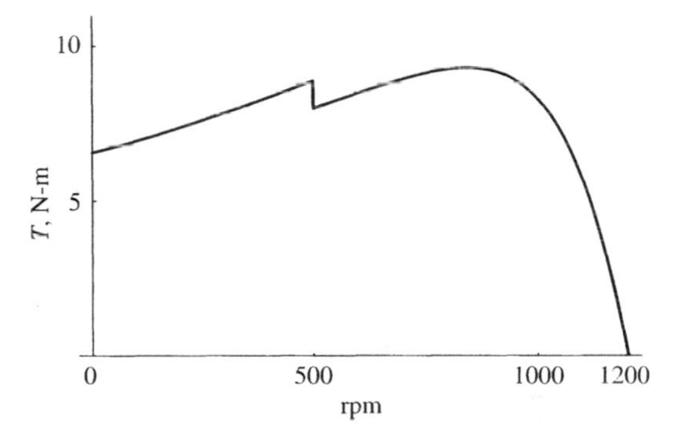

Figure P1.58

1.59. A motor catalog gives the following nameplate information for a 60-Hz, single-phase induction motor: 3 hp, 1740 rpm, 115/230 V, 38.0/19.0 A, SF 1.15. Losses are known to be 17% of the output power at nameplate condition.

(a) Find the output torque of the motor at nameplate conditions.
(b) Find the power factor of the motor at nameplate conditions.
(c) Find the maximum power that the motor can sustain for a long period.
(d) What is the maximum power that the motor can produce for a short period without stalling?

1.60. A motor catalog lists the following information about 60-Hz, 1-hp motors: **Motor A:** single-phase, 1725 rpm, 115/230 V, 15.0/7.5 A, no efficiency given, 30 lb; **Motor B:** three-phase, 1750 rpm, 230/460 V, 3.5/1.75 A, 77.2% efficiency, 32 lb.

(a) Calculate the efficiency of the single-phase motor, assuming that the motors have the same power factor.

(b) Which motor has the greater output torque at nameplate conditions and what is it?

(c) Which motor has the greater apparent power requirement at nameplate conditions and what is it?

1.61. Two single-phase induction motors are being considered for an application: **Motor A:** 1 hp, 1725 rpm, 115/230 V, 9.2/4.6 A, 1.25 SF, $225.27, 37 lb. **Motor B:** 1 hp, 1725 rpm, 115/230 V, 14.8/7.4 A, 1.0 SF, $182.99, 30 lb.

(a) Assume that for each motor, the power factor and the efficiency are equal. Find the efficiencies of both motors.

(b) If electric energy cost 3.6 ¢/kWh and the motors were run continuously at nameplate conditions, how soon (days) would Motor A justify its increased cost through saving of electric energy?

General Problems

1.62. Two pump jacks in the oil field require three-phase induction motors to drive the pumps. The wells are 325 yards apart and are connected by three No. 10 wires (1 Ω/1000 ft). One of the wells is shallow and requires a 10-hp motor; the other well is deep and requires a 50-hp motor. For the 10-hp motor the nameplate information is: 60 Hz, 1750 rpm, 230/460 V, 26.2/13.1 A, 88.5% eff., 1.35 SF. For the 50-hp motor, the nameplate information is: 60 Hz, 1775 rpm, 230/460 V, 90.2% eff., 121.4/60.7 A, 1.35 SF. The motor is connected for 460-V operation, and must be operated between 440 and 460 V for the warranties to be valid. Assume the motors are operated at their nameplate power. Power is brought to the motors from a 13-kV feeder line, and three single-phase transformers, connected in Y on the primaries and Δ on the secondaries, supply 460-V power to the line that serves the motors.

(a) Determine a suitable location for the transformers. That is, at what point between the motors should the transformers be located in order to minimize losses and to provide the acceptable voltages to the motors?

(b) Give the voltage and current in primary and secondary windings of the single-phase transformers.

(c) Find the overall efficiency of the system with both motors running at nameplate power.

1.63. A junior engineer needs 12 hp to drive a load, so he takes a 10-hp motor and a 2-hp motor, both three-phase, and puts them on the same shaft. The motor nameplates read: 10 hp, 1750 rpm, 60 Hz, 230/460 V, 26.2/13.1 A, 88.5% eff., 1.1 service factor; 2 hp, 1725 rpm, 60 Hz, 230/460 V, 6.6/3.3 A, 80.0% eff., 1.35 service factor. Assume operation in the small-slip region and negligible rotational loss.

(a) What speed does the system run, assuming exactly 12 hp for the load?

(b) Is either motor overloaded? To answer, you must calculate the motor power for both motors and explain your conclusion.

Answers to odd-numbered problems

1.1. $\cos(0^\circ) + \cos(-240^\circ) + \cos(-480^\circ) = 0$ and $\cos(20^\circ) + \cos(-220^\circ) + \cos(-460^\circ) = 0$.

1.3. **(a)** $218 \angle -34.7^\circ$ A; **(b)** 151 kVA; **(c)** 124 kW; **(d)** -64.7°.

1.5. **(a)** $120\sqrt{3} \angle +30^\circ$ V; **(b)** 17.7 A;**(c)** 6390 W; **(d)** 5860 W.

1.7. **(a)** $480 \angle -240^\circ$ V; **(b)** 5.77 A rms; **(c)** 8.31 kW; **(d)** $83.1 \angle 0^\circ$ Ω.

1.9. **(a)** 45.5 A rms; **(b)** 0.876; **(c)** 311 V, rms.

1.11. **(a)** 0.833; **(b)** 23.6 A rms; **(c)** 32.3 Ω; **(d)** 3.32 kVAR.

1.13. **(a)** 139 V rms; **(b)** 240 V rms; **(c)** 21.3 A rms; **(d)** 21.3 A rms; **(e)** $6 + j2.5$ Ω; **(f)** 8860 W; **(g)** 0.923; **(h)** 8180 W.

1.15. Proof.

1.17. **(a)** 6090 W; **(b)** 85.8 W; **(c)** 98.6 %.

1.19. **(a)** 30.1:1; **(b)** 0.208 A, rms.

1.21. **(a)** Calculate V_{bc} from the other two and it comes out right; **(b)** 34.6 kVA for the two transformers.

1.23. **(a)** $2.10 \angle -19.9^\circ$; **(b)** same.

1.25. **(a)** 0.00174 pu; **(b)** transformer is 1:1 with $j0.00174$ in primary or secondary.

1.27. **(a)** 31.6 A, rms; **(b)** 2690 W; **(c)** 3.27 μF; **(d)** 2430 W.

1.29. 140,050 ∠ 1.01° V.

1.31. **(a)** 1188 rpm; **(b)** 702 W; **(c)** 726 W at 1039 rpm.

1.33. **(a)** 103.2 rad/s; **(b)** 50 N-m; **(c)** 2740 W at 65.5 rad/s.

1.35. **(a)** 6.35×10^{-3}; **(b)** 4 N-m; **(c)** 787 W at 7520 rpm.

1.37. **(a)** 30 N-m; **(b)** 110 rad/s; **(c)** 54.2 N-m at 44.0 rad/s; **(d)** 3200 W at 70.1 rad/s.

1.39. 6 N-m at 1160 rpm.

1.41. **(a)** 2.23×10^{-4}; **(b)** about 3500 rpm; **(c)** 3.6 s.

1.43. **(a)** 49.2 N-m; **(b)** 8 N-m; **(c)** 1.93 s.

1.45. **(a)** 7.1 hp; **(b)** 3030 VAR; **(c)** 771 W.

1.47. **(a)** 101 N-m; **(b)** 24.6 kVA; **(c)** 0.826; **(d)** +13.9 kVAR.

1.49. $7.61/year.

1.51. **(a)** 0.818; **(b)** 30.4 N-m; **(c)** 1767 rpm.

1.53. **(a)** 15.2 kVAR; **(b)** 1845 W; **(c)** 101.5 N-m; **(d)** 1782 rpm.

1.55. **(a)** 4.09 N-m; **(b)** 65.3%; **(c)** 1070 VAR.

1.57. 498 W and 18.4 N-m.

1.59. **(a)** 12.3 N-m; **(b)** 0.599; **(c)** 3.45 hp; **(d)** no way to determine.

1.61. **(a)** 84.0% for A and 66.2% for B; **(b)** 205 days.

1.63. **(a)** 183.0 rad/s; **(b)** 10.6 hp for 10-hp motor and 1.4 hp for 2-hp motor. Neither is overloaded.

CHAPTER

The Physical Basis of Electromechanics

1. To understand the definition of the electric field and its relationship to voltage
2. To understand the definition and nature of magnetic fields and their relationship to current and magnetic materials
3. To understand how to calculate voltage in stationary and moving wires with Faraday's law
4. To understand how to calculate stored energy in linear and nonlinear magnetic systems
5. To understand how to analyze and model energy conversion in a linear transducer

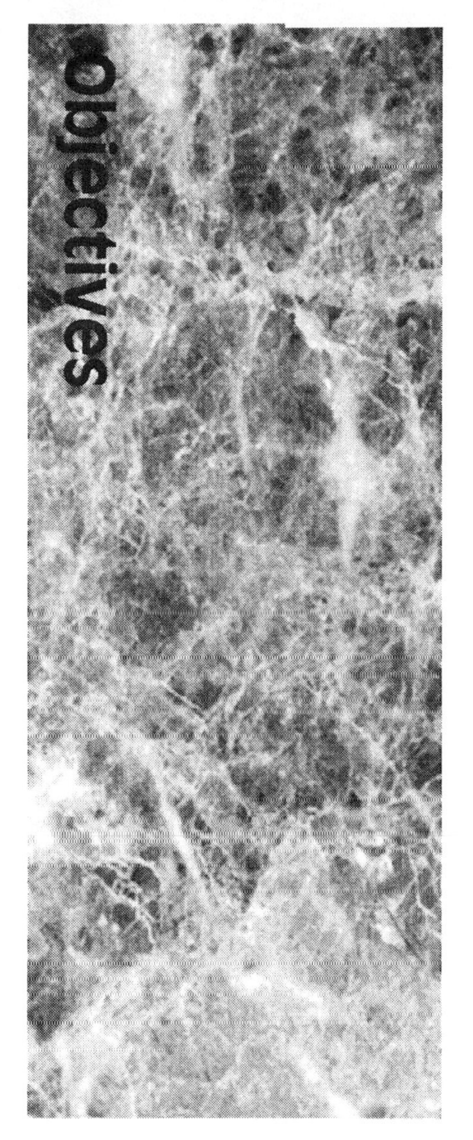

This review of electrical physics is useful because principles of electromechanical devices are based on electric and magnetic fields. This review will help you understand the basic laws that govern important electric power devices such as motors and transformers.

Introduction to Electromechanics

Importance of electrical energy conversion. Electric motors are such an important part of modern civilization that we hardly notice them. But if you look, you see them everywhere: in your car (windshield wipers, starter motor, window lifts); in the office (floor buffers, hard disk drives on computers); in the home (hair dryers, clocks, large appliances); and of course in the factory. Electromechanical devices are also required in the generation of electric power from basic energy sources and in the distribution of such power to customers. Also important are the electrical transformer, which operates on principles similar to those of electromechanical devices, and other magnetic devices such as electrical relays, bells (doorbells, alarm bells), solenoids (electric locks, starter solenoids on cars), magnets, and inductors used in electronics and fluorescent lights.

Motor principles and applications. Few engineers outside the major manufacturers design a new electrical motor. Thus, we do not emphasize the design of electromechanical equipment; rather, we aim at imparting a physical understanding of electromechanics and the practical knowledge that would guide one to choose the right device for a specific need.

Contents of this chapter. Here we review the electrical physics underlying electromechanics. We define electric and magnetic fields, and we state the basic laws that operate in electromechanical devices. Magnetic field and force concepts are emphasized because virtually all electromechanical devices utilize magnetic phenomena.

2.1 ELECTRIC FORCES AND ELECTRIC FIELDS

Forces between Charges

electrostatic force, electric energy

Two experimental results. Figure 2.1(a) shows an experiment that one can perform with a battery and two sheets of metal. The battery forces opposite charges to the sheets, and the sheets exhibit an small attractive force tending to pull them together. This force can be explained as an attraction between the excess positive and negative charges on the sheets. We call the force an *electrostatic force* because it is associated with the relative location of the charges; and energy associated with this type of force is called electrostatic energy, or more simply, *electric energy*. Electrostatic forces are responsible for lightning, because charges become separated through the breakup of water

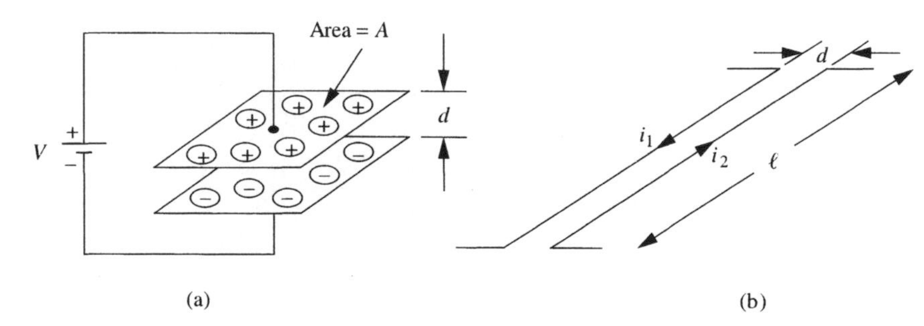

Figure 2.1 Two simple experiments.

droplets in clouds, and electrostatic forces are used in photocopy machines to form images on a selenium surface with charged bits of dry ink. Electrostatic forces indicate the presence of an electric field.

magnetic force, magnetic energy

Figure 2.1(b) shows another simple experiment with currents in long, parallel wires, which for the current directions indicated exhibit a small force of repulsion. This type of force requires moving charges; thus, such a force exists between currents. This we call *magnetic force*, and energy associated with this type of force is *magnetic energy*. Such magnetic forces operate in electric motors and magnets and also can be used to deflect beams of electrons in TV tubes. Magnetic forces indicate the presence of a magnetic field.

Forces and fields. It is convenient to describe these forces in terms of electric and magnetic fields. Rather than thinking of the forces as produced directly between stationary and moving charges, we assert that one set of charges, due to its presence and motion, sets up electric and magnetic fields, and that other charges experience force through interacting with, or responding to, these fields.

Electric Fields

electric field

Definition of an electric field. An *electric field* is defined as the vector force on a stationary charge divided by that charge; thus

$$\vec{E} = \frac{\vec{f}}{q} \text{ N/C} \tag{2.1}$$

LEARNING OBJECTIVE 1.

To understand the definition of electric field and its relationship to voltage

where $\vec{E}$ is the vector electric field in newtons/coulomb and $\vec{f}$ is the vector force in newtons on the charge q in coulombs. The direction of the electric field is the direction of force on a positive charge. A single positive charge produces an electric field directed outward from the charge. Although the electric field is a vector field, in this book we deal only with cases where the direction of the electric field is evident so as to eliminate the need for vector mathematics.

dipole

Electric-field patterns. Figure 2.2(a) shows the electric field pattern from an isolated positive charge; Fig. 2.2(b) shows the electric field pattern of a *dipole*, an associ-

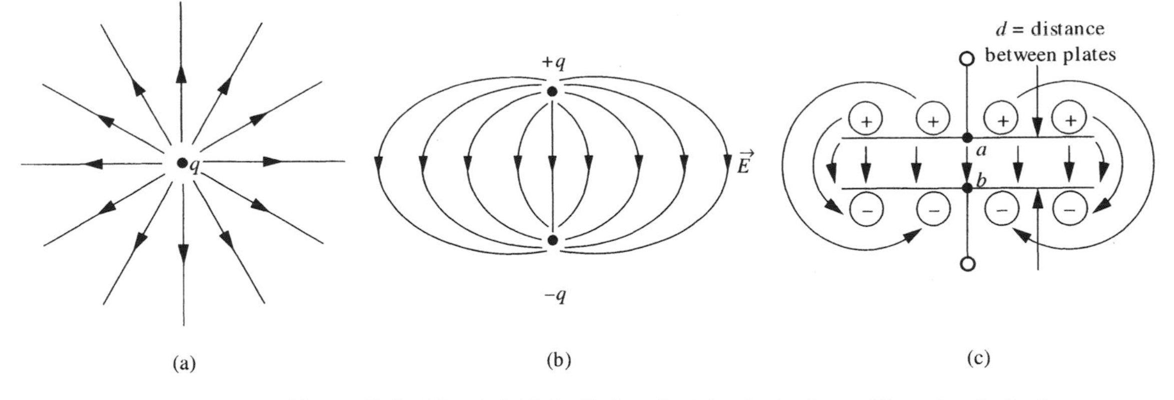

Figure 2.2 Electric-field distributions for (a) a single charge, (b) an electric dipole, and (c) a capacitor.

ated pair of equal but opposite charges; and Fig. 2.2(c) shows the electric field for a capacitor. In the second two cases, the electric-field lines start on positive charges and end on negative charges.

EXAMPLE 2.1 **Force on a water droplet**

A water droplet having an excess charge of $5e$ is located in a vertical electric field of +10,000 N/C. What is the force on the droplet?

SOLUTION:
Using Eq. (2.1) and $e = -1.60 \times 10^{-19}$ C, we find the force to be

$$f = 5 \times (-1.60 \times 10^{-19}) \times 10^{4} = -8.01 \times 10^{-15}\ \text{N} \tag{2.2}$$

WHAT IF? What if you want the direction?[1]

Dimensions of an electric field. The units of an electric field are newton/coulombs, but this is equivalent to volt/meter, as shown by the following

$$\frac{\text{N}}{\text{C}} = \frac{\text{N} \times \text{meter}}{\text{C} \times \text{meter}} = \frac{\text{joule}}{\text{C}} \times \frac{1}{\text{meter}} = \frac{\text{volt}}{\text{meter}} \tag{2.3}$$

because a joule/coulomb is a volt. Thus, we can interpret an electric field as a voltage per unit distance, or alternately, we can interpret a voltage as the integral of an electric field. We now investigate this second interpretation.

voltage, fringing

Voltage and electric fields. Recall that the *voltage* from point a to point b is defined as the work done by the electrical system in moving a charge from a to b, W_{ab}, divided by the magnitude of the charge

$$V_{ab} = W_{ab}/q \tag{2.4}$$

Let us consider the charged capacitor of Fig. 2.1(a), shown in cross-section in Fig. 2.2(c). Opposite excess charges spread over the top and bottom plates of the capacitor. The electric field is directed from the positive to the negative charges because that is the direction a positive charge would tend to move. Between the plates of the capacitor, the electric-field is essentially constant in magnitude and direction, but the electric field lines outside the plates spread out, which is called *fringing*.

Conservation of Energy

Voltage calculation. The work done by the electrical system in moving a test charge q from a to b is defined in Eq. (2.4) as $W_{ab} = qV_{ab}$. We can express this work mechanically in terms of force and displacement, with the integral of the vector dot product of the force and displacement

[1] Downward, opposite the field direction.

$$W_{ab} = \int_a^b \vec{f} \cdot \vec{d\ell} = q \int_a^b \vec{E} \cdot \vec{d\ell} \text{ J} \tag{2.5}$$

where $\vec{f} = q\vec{E}$ is the force on the test charge and $\vec{d\ell}$ is the vector increment of distance along the path of the line integral. Two comments are required before we continue to investigate the relationship between voltage and the electric field.

1. The relationship between the electric field and voltage that follows from Eq. (2.5) results from defining voltage as the voltage (or potential) *drop* from point a to point b.
2. We use the notation of vector calculus in writing the equations describing electric (and magnetic) fields because these are required to express the relationships mathematically. We apply the equations only in situations where, due to symmetry, we do not require application of vector calculus. In the present instance, for example, we perform the line integral only in the region where the electric field is important, and we follow a path aligned with the electric field. On this path, the vector line integral becomes an ordinary integral.

The integral in Eq. (2.5) describes the summation of the electric-field component times the increment in distance in the direction of the path followed. For the capacitor, we perform the integral near the middle of the capacitor, where the electric field goes directly from the top to the bottom plate, and we move the charge in a straight line from top to bottom. Thus, the line integral reduces to the ordinary integral

$$W_{ab} = \int_a^b f\ dx = \int_a^b qE\ dx \Rightarrow V_{ab} = \frac{W_{ab}}{q} = \int_a^b E\ dx \text{ V} \tag{2.6}$$

where E is the magnitude of the electric field and dx is the incremental displacement of q in moving from a to b. Electric fields associated with charge distributions, like gravitational fields, are conservative fields, which means that the work done, and hence the voltage, is independent of the path taken from a to b. Equation (2.6) shows that the voltage between two points can be expressed as the integral of the electric field or, alternatively, that a voltage indicates the presence of an electric field.

Electric field in the capacitor. We may apply Eq. (2.6) to the capacitor shown in Fig. 2.2(c) to determine the electric field between the plates of the capacitor, given the voltage. In this region, E is constant:

$$V_{ab} = \int_a^b E\ dx = E \int_a^b dx = Ed \Rightarrow E = \frac{V_{ab}}{d} \text{ V/m} \tag{2.7}$$

where d is the distance between the plates. Equation (2.7) suggests the interpretation mentioned earlier, that the electric field is a voltage per unit distance in space.

Summary. Electric fields are produced by charges and exert force on other charges. An electric field can be considered a voltage stretched over space.

EXAMPLE 2.2 Electric field in a capacitor

A cylindrical capacitor is constructed by rolling a 0.0005-inch-thick plastic material between metal conductors. The maximum electric field the plastic material can tolerate is 50×10^6 V/m. What should be the voltage rating of the resulting capacitor?

SOLUTION:

From Eq. (2.7) we calculate the maximum voltage to be

$$V_{max} = E_{max} \times d = 50 \times 10^6 \text{ V/m} \times 0.0005 \text{ in.} \times 0.0254 \text{ m/in.} = 635 \text{ V} \tag{2.8}$$

Check Your Understanding

1. Give the units for an electric field.
2. An electron accelerates downward under the influence of an electric field. What is the direction of the field?
3. Estimate the peak electric field at an appliance outlet.

Answers. **(1)** Newtons/coulomb or volts/meter; **(2)** upward; **(3)** $\approx$ 12 kV/m.

2.2 MAGNETIC FORCES AND MAGNETIC FIELDS

Currents and Magnetic Forces

Force equation. Magnetic force is produced when one set of moving charges interacts with another set of moving charges; or more simply, magnetic forces exist between currents. In Fig. 2.1(b), we show two long parallel wires of length ℓ separated by a distance d. The total force of repulsion, F, acting on the wires is described by Ampère's force law

$$F = \mu_0 \frac{i_1 i_2 \ell}{2\pi d} \text{ N} \tag{2.9}$$

where μ_0 is the permeability of free space and has a numerical value of $4\pi \times 10^{-7}$ henry/meter in the mks system. In Eq. (2.9), i_1 and i_2 are referenced in opposite directions.

EXAMPLE 2.3 Force in a lamp cord

Estimate the force/meter between the two wires carrying current to a 60-W, 120-V lamp.

SOLUTION:

The current is 60 W/120 V = 0.5 A (rms). Thus, from Eq. (2.9)

$$\frac{F}{\ell} = 4\pi \times 10^{-7} \frac{(0.5)(0.5)}{2\pi(2 \times 10^{-3})} = 2.50 \times 10^{-5} \text{ N/m} \tag{2.10}$$

where we have used 2 mm as the distance between the two conductors.

WHAT IF? What if you want the force between two conductors in a power plant with 8 kA separated by 6 inches?[2]

---

[2] 84.0 newtons/meter.

Our approach here, as for electrostatics, is to assert that one current produces a magnetic field in its vicinity, and the other current interacts with that field to experience a force. Here we consider the force on the wire carrying i_2 due to the magnetic field produced by i_1. Thus, we divide the problem into two parts: first, the relationship between i_1 and its magnetic field, H_1; and, second, the interaction of i_2 with H_1 to produce a force on the wire carrying i_2.

Currents and their magnetic fields. Currents produce magnetic fields just as stationary charges produce electric fields. Magnetic fields are vector fields, but we can gain a sufficient understanding of the nature of magnetic fields and of many important devices using magnetic fields without vector mathematics. Because the current produces a magnetic field, we anticipate a theory for determining the field, given the current. Such a theory exists, but is mathematically difficult to apply. There is a relatively simple relationship, Ampère's circuital law, that allows us to determine the magnetic field in several important geometries.

Ampère's circuital law. Unlike electric fields, which start on positive charges and end on negative charges, magnetic field lines encircle their source currents. Figures 2.3(a) and 2.3(b) show magnetic field configurations from a long straight wire and a coil of wire, respectively. Ampère's circuital law is a conservation principle that relates the integral of the magnetic field around a closed contour to the current passing through the area enclosed by the contour.

LEARNING OBJECTIVE 2.

To understand the definition and nature of magnetic fields and their relationship to current and magnetic materials

$$\oint_C \vec{H} \cdot \vec{d\ell} = i \tag{2.11}$$

where $\vec{H}$ is the magnetic field in amperes/meter, $\vec{d\ell}$ is a vector element of distance, C is the contour of integration, and i is the current passing through the area bounded by C. This gives the picture of magnetic-field lines circling currents, but always in such a way that, if you follow a magnetic-field line, the spatial integral of the magnetic-field strength around the path equals the current encircled. If the path is farther away from the current, the field will be weaker because the integration path length is longer.

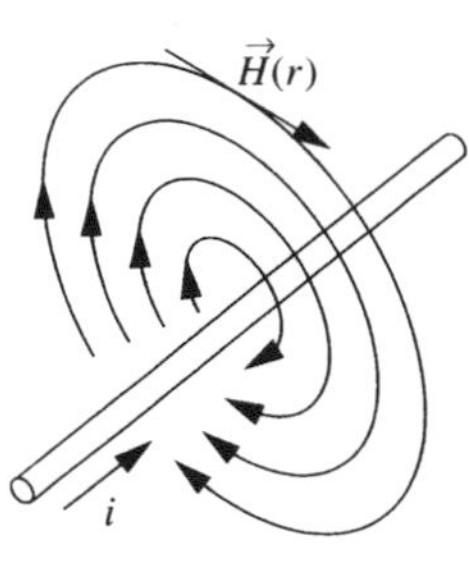

(a)

(b)

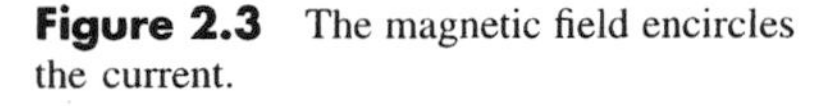

Figure 2.3 The magnetic field encircles the current.

Magnetic field of a long straight wire. We use Ampère's circuital law to calculate the magnetic field due to an infinitely long wire, suggested in Fig. 2.3(a). Here the magnetic-field lines must by symmetry encompass the current in circular paths, as shown, and hence the magnitude of the magnetic field $|\vec{H}(r)|$ at a radius r from the wire must be constant in magnitude. Thus, the magnetic field is given by

$$\oint \vec{H}\cdot\vec{d\ell} = |\vec{H}(r)|\oint |\vec{d\ell}| = |\vec{H}(r)| \times 2\pi r = i \tag{2.12}$$

or

$$H(r) = |\vec{H}(r)| = \frac{i}{2\pi r}\ \frac{\text{A}}{\text{m}} \tag{2.13}$$

where $H(r)$ is the magnitude of the magnetic field at radius r. Thus, the magnetic field is proportional to the current and, in this case, inversely proportional to the distance from the wire.

Magnetic forces. We return to Fig. 2.1(b) and the force equation in Eq. (2.9), which we rearrange into the form

$$f_2 = \frac{F}{\ell} = \mu_0\left(\frac{i_1}{2\pi d}\right) i_2 = \mu_0 H_1 i_2 \ \text{N/m} \tag{2.14}$$

magnetic field

where f_2 is the force in newtons/meter on i_2, and H_1 is the magnetic field due to i_1 at the location of i_2.[3] Although derived for a specific geometry, Eq. (2.14) is valid generally and, indeed, can serve as a definition of *magnetic field.* The force per meter on the wire carrying i_2 depends on the magnetic field due to i_1 at the wire and the current in the wire. Later, we reformulate Eq. (2.14) in vector form and elaborate on the importance of the quantity $\mu_0 H_1$. For now, we accomplished our goal of separating the force into two terms: the magnetic field produced by i_1 and the force between that field and the wire carrying i_2.

EXAMPLE 2.4 Wire force

A wire carrying 1000 A experiences a force of 2 N/m. What is the magnetic field at the wire?

SOLUTION:
From Eq. (2.14)

$$H = \frac{F/\ell}{\mu_0 i} = \frac{2}{4\pi \times 10^{-7} \times 1000} = 1590 \ \text{A/m} \tag{2.15}$$

Actually, this is the component of $\vec{H}$ perpendicular to i; thus, 1590 A/m is the minimum field at the wire.

[3] Equation (2.14) is valid if $\vec{H}_1$ is perpendicular to the direction of i_2.

Direction of the Magnetic Field

Right-hand rule. Ampère's circuital law, as stated, gives only the magnitude of the magnetic field. To determine the direction of the field, we use the right-hand rule. Actually, there are several forms of the right-hand rule that relate the directions of current and magnetic field. In Fig. 2.3(a), the convenient form is: Take the *wire* in your right hand with your thumb in the direction of the current; your fingers then encircle the wire in the direction of the magnetic field.

For a coil, as in Fig. 2.3(b), the following is convenient: Take the *coil* in your right hand with the fingers pointing in the direction of the conventional current. Then your thumb points in the direction of the magnetic field *inside* the coil.

magnetic-field direction, north and south magnetic poles

Magnetic poles. The direction of the magnetic field as given by the right-hand rule is consistent with the definition based on the magnetic-pole concept, which came first historically. The early theory of magnetism was developed along the same lines as electrostatics through the concept of magnetic poles. A compass was thought to consist of a north and a south magnetic pole, with the north pole of the compass indicating the direction of north. The *direction of the magnetic field* is defined as the direction a compass would indicate as northerly. Because opposites attract, it follows that the north *geographic* pole contains a south *magnetic* pole, as indicated in Fig. 2.4(a). A *north magnetic pole* is defined to be a region where the magnetic-field lines emerge from an object, the Earth in this case, and a *south magnetic pole* is a region where the magnetic-field lines enter an object. Thus, Fig. 2.4(a) shows the magnetic-field lines emerging from the north *magnetic* pole at the south *geographic* pole and disappearing into the south *magnetic* pole located at the north *geographic* pole.

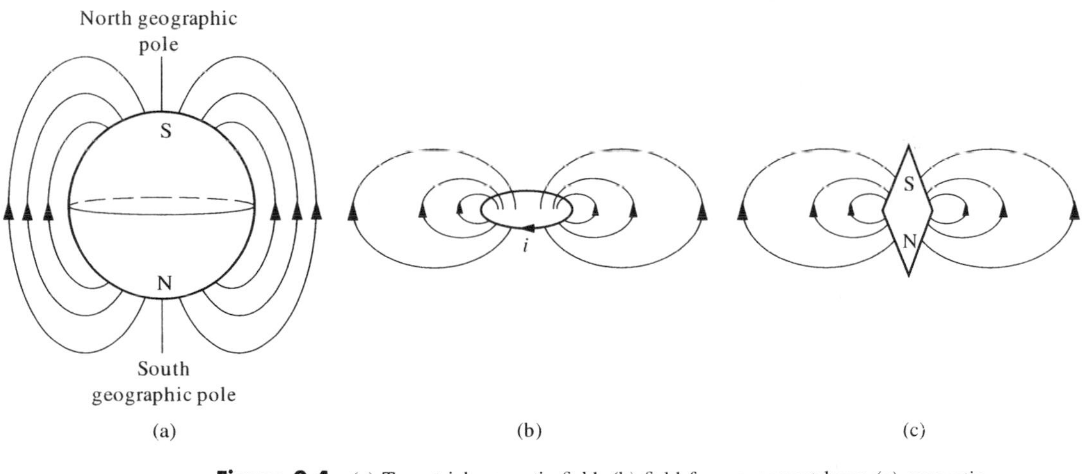

Figure 2.4 (a) Terrestrial magnetic field; (b) field from a current loop; (c) magnetic dipole.

magnetic dipole, magnetic moment

A compass is analogous to an electric dipole and is called a *magnetic dipole* or *magnetic moment.* A small loop of current also has a dipole field as shown in Fig. 2.4(b), and Fig. 2.4(c) shows a dipole created by a pair of magnetic poles. Although isolated magnetic poles, magnetic charges, have not been detected by physicists, some basic atomic particles possess a magnetic moment.

Magnetic Effects in Matter

ferromagnetism, magnetic domain

Ferromagnetism. Magnetic fields interact with matter through several effects. The important effect from an engineering viewpoint is ferromagnetism, which arises out of the inherent magnetic moment associated with the spin of the orbiting electrons, Fig. 2.5(a). In certain materials, notably iron, the magnetic moments of the orbital electrons interact with each other to produce a region of magnetic coherence over an extended region of the material, a *magnetic domain*. A magnetic domain can be thought of as a microscopic magnet and is capable of interacting significantly with an external field. Under normal conditions, the magnetic domains are randomly oriented, as suggested by Fig. 2.5(b), and produce no net effect. However, under the influence of an external magnetic field, the boundaries of the domains move such that the magnetic domains in the direction of the applied field grow in size at the expense of those in other directions, and the net effect is a considerable enhancement of the magnetic flux. This effect, called *ferromagnetism*, can establish large magnetic fluxes as required for electromechanical devices.

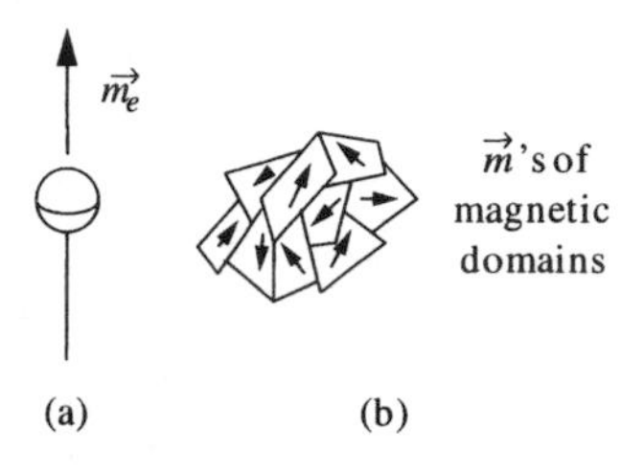

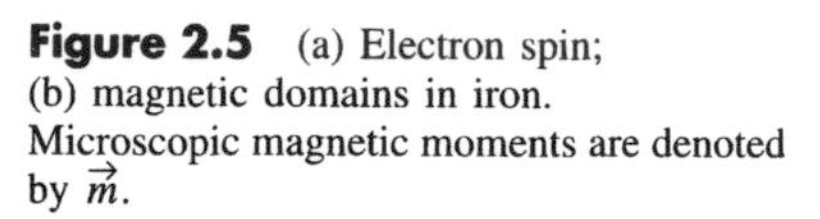
Figure 2.5 (a) Electron spin; (b) magnetic domains in iron. Microscopic magnetic moments are denoted by $\vec{m}$.

magnetic flux density

Magnetic flux density. Consider the state of matter that is magnetized by an external field, such that the magnetic domains produce a net magnetic moment. We define the magnetic dipole density, $\vec{M}$, as the vector sum of all the magnetic moments over a volume of space, divided by that volume. The magnetic dipole density characterizes the effect of the magnetic field on the magnetic state of the matter. The *magnetic flux density*, $\vec{B}$, combines the applied and the induced magnetism, as defined in Eq. (2.16)

$$\vec{B} = \mu_0(\vec{H} + \vec{M}) \tag{2.16}$$

The units of magnetic flux density are webers/meters^2, which has been given the honorary unit tesla (T) after Nikola Tesla (1856–1943).

Causal relationships. The causal relationships in producing magnetic flux are summarized in Fig. 2.6. The magnetic field, $\vec{H}$, can be thought of as a magnetic stress on space, and the magnetic flux density, $\vec{B}$, a magnetic strain. In other words, the magnetic field, $\vec{H}$, tends to magnetize space, and the magnetic flux density, $\vec{B}$, is the total magnetic effect that results.

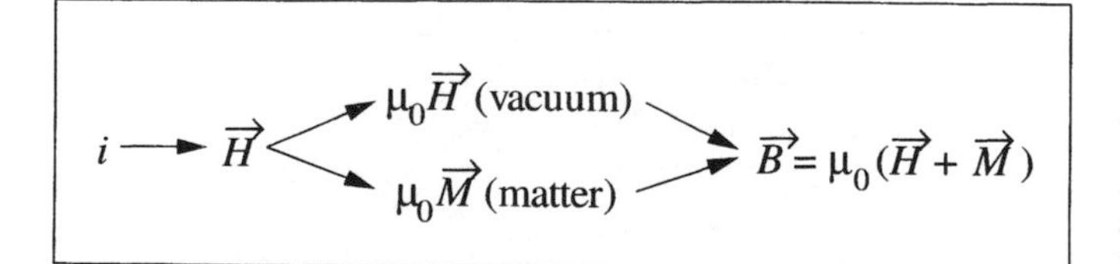

Figure 2.6 Causality diagram for the magnetization of space. The current creates the magnetic field that magnetizes space and any matter in space. The net magnetization combines both effects.

Magnetic properties of iron. Because iron is important to electromechanics, we concentrate on the ferromagnetic properties of iron. For moderate applied fields, the induced magnetic dipole density is proportional to the applied field, and the magnetic flux density is proportional to the magnetic field. Equation (2.17) expresses this proportionality

$$\vec{M} \propto \vec{H} \Rightarrow \vec{B} = \mu_i \vec{H} = \mu_0 \mu_r \vec{H} \tag{2.17}$$

permeability, relative permeability

where μ is called the *permeability* of the iron. If we think of $\vec{H}$ as the cause of the magnetization and $\vec{B}$ as the effect, then the constant μ in Eq. (2.17) indicates the magnitude of the induced effect. The constant μ_r is called the *relative permeability* of the material and is essentially unity for all materials except ferromagnetic materials, notably iron, for which the value of the relative permeability lies typically in the range 1000 to 10,000. Thus for iron, a small magnetic cause, $\vec{H}$, creates a large magnetic effect, $\vec{B}$.

hysteresis curve

Magnetic saturation and hysteresis. A more complete picture of magnetic effects in iron must show the effects of magnetic saturation and hysteresis, as indicated in Fig. 2.7. Here we show the effect of applying an external magnetic field to unmagnetized iron. The magnetism curve starts at the origin (a) and increases linearly as the magnetic field magnetizes the iron. The slope of the curve in this region is $\mu = \mu_0 \mu_r$, Eq. (2.17). Eventually, however, all the magnetic domains align with the applied magnetic field, and the curve flattens out as the iron becomes magnetically saturated (b). If the applied field is then reduced to zero, the flux density follows a different curve because the iron tends to retain its previous magnetized state. The thermodynamic forces that disorient the magnetic domains in the iron do not fully overcome the order imposed by the external field, and the iron retains a residual magnetism (c). We thus have produced a permanent magnet. As we reverse the applied magnetic field, the iron eventually becomes magnetized in the reverse direction until it again saturates (d). If we continue to cycle the magnetic state of the iron by applying ac current to the coil creating the applied magnetic field, the curve continues to follow an S-shaped curve, which is called a *hysteresis curve*. The area enclosed by the hysteresis curve is the energy loss per unit volume per cycle. This loss heats the iron and is one reason why electric motors and transformers become hot.

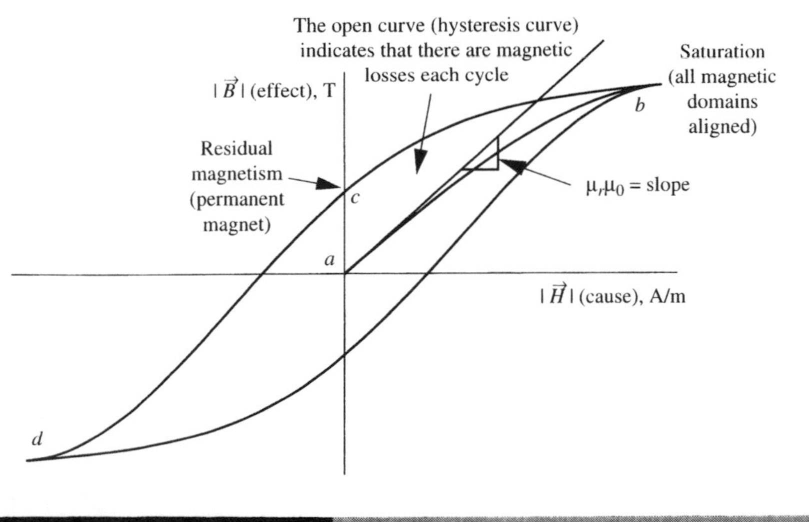

Figure 2.7 A magnetic hysteresis curve for iron.

Air and iron paths for magnetic flux. An iron path has a strong effect on the magnetic flux density, as illustrated by the hypothetical experiment shown in Fig. 2.8. Figure 2.8(a) shows a coil energized by a current. The resulting magnetic field, $\vec{H}$, has the shape shown and conforms to Ampère's circuital law, Eq. (2.11). Because $\vec{M}$ is zero for air, the magnetic flux density is given by $\vec{B} = \mu_0 \vec{H}$ and is relatively small. An identical coil wound around an iron core, Fig. 2.8(b), produces a rather different flux pattern. The magnetic field, $\vec{H}$, is distorted by the presence of the iron. The magnetic fields in the region around the coil have roughly the same magnitude as before, because Ampère's circuital law must still be satisfied for the same coil and current. Within the iron, however, the magnetic flux density, $\vec{B}$, is much larger than before by the magnitude of μ_r, say, 5000, resulting in a large magnetic flux that follows the iron path.

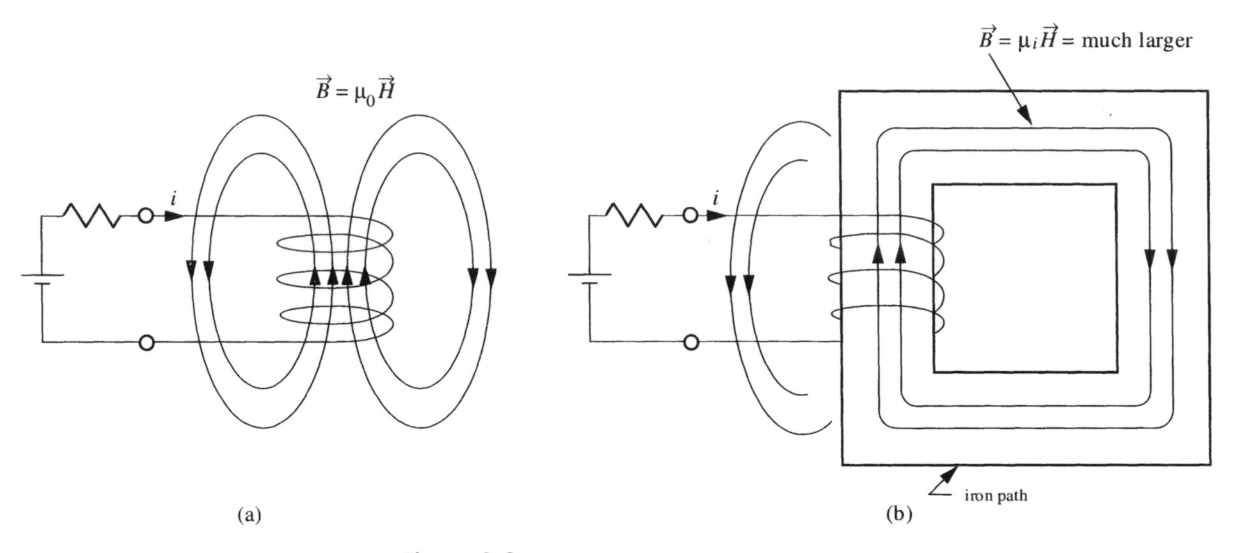

Figure 2.8 (a) A coil in air; (b) the iron path for flux of the same coil.

Interaction with current. The force given in Eq. (2.14) has a direction in space. A general form of Ampère's force law that gives the direction is the vector cross product of current and flux density

$$\vec{f}_2 = \vec{i}_2 \times \vec{B}_1 \text{ N/m} \tag{2.18}$$

where $\vec{f}_2$ is the vector force per unit length on the wire carrying $\vec{i}_2$ and $\vec{B}_1$ is the vector flux density due to $\vec{i}_1$. Equation (2.18) is a form of Ampère's force law.

Right-hand rule. We also have a right-hand rule relating current, magnetic flux density, and force from Eq. (2.18). In this case, point the fingers of your right hand in the direction of the current; then bend them in the direction of the magnetic flux. Your thumb indicates the direction of the force, as shown in Fig. 2.9, consistent with the definition of a vector product in a right-hand system, as Eq. (2.18) requires.

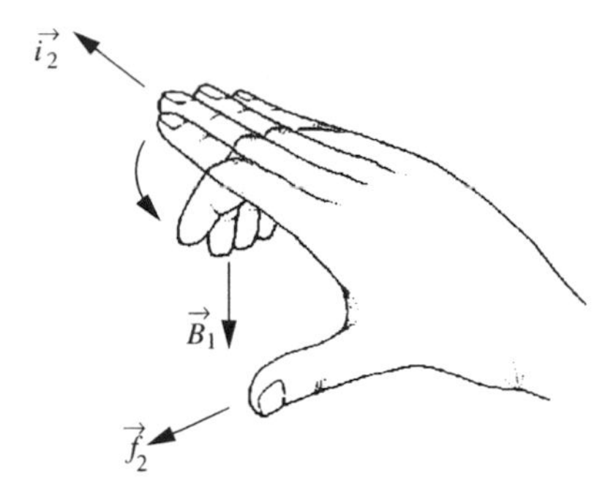

Figure 2.9 For the direction of $\vec{i}_2 \times \vec{B}_1$, put the fingers of the right hand in the direction of $\vec{i}_2$, and then swing to the direction of $\vec{B}_1$. The thumb points in the direction of the force.

EXAMPLE 2.5 **Right-hand rule**

Work out the magnitude and directions of the flux and force for the case shown in Fig. 2.1(b) using the appropriate right-hand rules.

SOLUTION:
Putting the thumb of the right hand in the direction of i_1, we determine from the right-hand rule that B_1 is upward at the position of the wire carrying i_2 and has a magnitude, from Eq. (2.13), of $\mu_0 i_1/2\pi d$. Crossing $\vec{i}_2$ into $\vec{B}_1$, we see that the force on the wire carrying i_2 is directed away from the wire carrying i_1.

WHAT IF? What if the currents are in the same direction?[4]

Summary. Iron enhances greatly the magnetic flux and controls its distribution in space. Large magnetic flux densities produce correspondingly large forces on conductors carrying currents. Thus, electromechanical devices normally use magnetic structures made of iron.

Magnetic Flux (Φ) and Magnetic Flux Linkage (λ)

Magnetic flux density is an important quantity because magnetic flux, of which $\vec{B}$ gives the spatial distribution, plays an important role in electromechanical energy conversion. In this section, we discuss the properties of magnetic flux and then begin our examination of the role played by magnetic flux in electromechanical energy conversion.

Properties of magnetic flux. Magnetic flux lines encircle the currents that generate them. Thus, magnetic flux lines exist only in closed loops. This means that magnetic flux is conserved in any closed spatial region: What goes in must come out. This is expressed by

$$\oint\!\!\oint_S B_\perp \, da = 0 \tag{2.19}$$

[4]The wires are attracted.

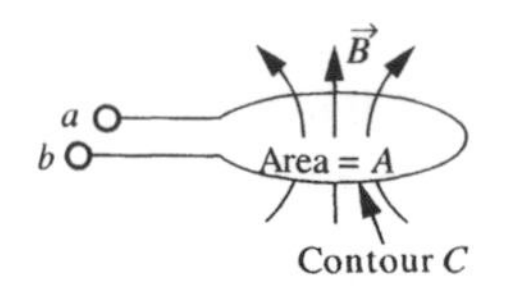

Figure 2.10 A one-turn coil.

where S is a closed surface, da is a differential element of area on the surface, and $B_\perp$ is the component of magnetic flux density perpendicular to da. Magnetic flux lines, therefore, behave like the flow lines of an incompressible fluid. It follows that the amount of magnetic flux passing through a loop of wire having an area A enclosed by the contour of the wire C, as shown in Fig. 2.10, is a uniquely defined quantity

$$\Phi = \int_A B_\perp \, da \tag{2.20}$$

magnetic flux

Equation (2.20) gives a mathematical expression for calculating the *magnetic flux*, Φ, passing through such an area. The magnetic flux is a scalar quantity with units of webers (Wb), after Wilhelm Weber (1804–1891).

EXAMPLE 2.6 Flux in a coil

A circular coil has a diameter of 2 cm and a flux of 5×10^{-8} Wb passing through it. Find the average flux density in the coil.

SOLUTION:

The average flux density would be the flux divided by the area

$$B_{avg} = \frac{\Phi}{A} = \frac{5 \times 10^{-8}}{\pi(0.01)^2} = 1.59 \times 10^{-4} \text{ tesla} \tag{2.21}$$

WHAT IF? What if the coil is in air and you want the average magnetic field?[5]

magnetic flux linkage

Flux linkage. For a coil of wire, the number of turns in the coil becomes important. The flux passing through the coil is the product of the number of turns, n, and the flux passing through a single turn, Φ. This product is called the *magnetic flux linkage* of the coil, λ

$$\lambda = n\Phi \text{ weber-turns} \tag{2.22}$$

Fig. 2.11 shows the causality relationships between i, $\vec{H}$, $\vec{B}$, Φ and λ.

The definition of flux linkage in Eq. (2.22) assumes that every turn of the coil has the same flux passing through it. Faraday's law, discussed in the next section, shows that the concept of flux linkage, broadly construed, is not founded on this assumption.

$$i \longrightarrow \vec{H} \longrightarrow \vec{B} = \mu \vec{H} \longrightarrow \int B_\perp \, da \longrightarrow \lambda = n\Phi$$

Figure 2.11 Extension of causal diagram in Fig. 2.6 to include magnetic flux and flux linkage.

[5] For air, $\mu = \mu_0$, so the average magnetic field is $1.59 \times 10^{-4}/\mu_0 = 127$ A/m.

Check Your Understanding

1. A wire is stretched around the equator, closed into a loop, and excited by a current traveling westward. Does the magnetic field from the current aid or oppose the Earth's magnetic field above the surface of the Earth?
2. At the south geographic pole, which direction (N, E, S, W, up, or down) does a magnetic compass indicate as "north"?
3. What is the value of the integral of the magnetic field around the electric cord of a lighted 120-V, 60-W electric light?
4. A cosmic ray with a positive charge is deflected in what direction by the Earth's magnetic field?

Answers. **(1)** aid; **(2)** up; **(3)** zero because the currents in the two wires add to zero; **(4)** eastward.

2.3 DYNAMIC MAGNETIC SYSTEMS

Induced Voltage

LEARNING OBJECTIVE 3.

To understand how to calculate voltage in stationary and moving wires with Faraday's law

Faraday's law. In 1831, Michael Faraday published his law of electromagnetic induction relating voltage in a coil to time-varying magnetic flux

$$v_t = \frac{d\Phi}{dt} \text{ V} \tag{2.23}$$

where v_t is the voltage per turn induced by a time-varying magnetic flux, Φ, through a coil. A coil having n turns constitutes a series connection of the n turns; hence, the total induced voltage is

$$v = nv_t = n\frac{d\Phi}{dt} = \frac{d\lambda}{dt} \text{ V} \tag{2.24}$$

EXAMPLE 2.7 **Induced voltage**

An inductor has 120-V, 60-Hz voltage applied. Calculate the flux linkage.

SOLUTION:
Letting the ac line voltage have a phase of zero, we have $v(t) = 120\sqrt{2}\cos(120\pi t)$ V. We may integrate Eq. (2.24) to give

$$\lambda = \int 120\sqrt{2}\cos(120\pi t)\,dt = \frac{\sqrt{2}}{\pi}\sin(120\pi t) + C \tag{2.25}$$

where the constant of integration, C, represents a possible dc flux. The magnitude of the ac flux linkage is thus $\sqrt{2}/\pi = 0.450$ weber-turn.

The importance of flux linkage. Equation (2.24) shows that the induced voltage is the time derivative of the flux linkage of a coil. The last form of the equation is valid, and indeed serves as a definition of the flux linkage when all the magnetic flux does not pass through every turn of a coil. For example, if we crumpled a length of wire into a random tangle, we would not have a coil with well-defined turns, but the flux linkage would still be a meaningful quantity and would still be related through Eq. (2.24) to the induced voltage. In other words, the tangle of wire would still have an inductance. We investigate the relationship between flux linkage and inductance after we consider the sign of the induced voltage.

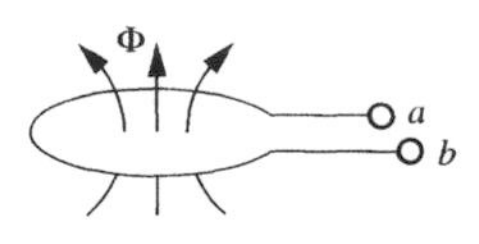

Figure 2.12 Lenz's law expresses an action-reaction principle determining the polarity of the induced voltage.

Lenz's law. The polarity of the induced voltage may be determined through Lenz's law. Consider the situation shown in Fig. 2.12 and assume that the magnetic flux, Φ, is changing with time. According to Faraday's law, a voltage v_{ab} is induced. Lenz's law states that if the loop were closed, *a* connected to *b*, the current would flow in the direction to produce a flux *inside the coil* opposing the original flux *change*. For example, let us say that the flux is increasing in the coil in Fig. 2.12. If we connected a resistor between *a* and *b*, then current would flow to generate a *reaction* magnetic flux opposing the increase; in this case, the reaction flux would be downward. Using the right-hand rule, we point our right-hand thumb downward and determine that the current flows clockwise as seen from the top. The current flows through the external connection from *a* to *b*; hence, the induced voltage v_{ab} must be positive. The external circuit will see *a* as + and *b* as − because current flows out of the + and into the − *from the viewpoint of the external connection*. Thus, if Φ were increasing, v_{ab} would be numerically positive.

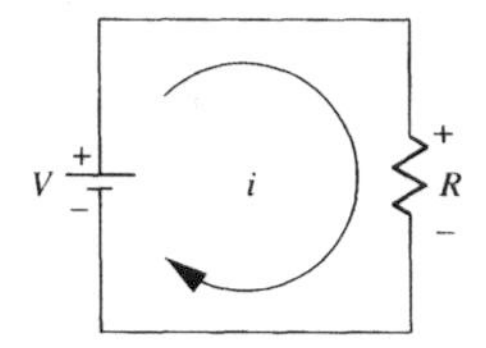

Figure 2.13 The physical current flows *out of* the + on the source (battery) and *into* the + on the load.

This reasoning is based on the fact that *for a source*, the physical current flows out of the + terminal, as shown in Fig. 2.13. The voltage described by Faraday's law acts as a source for the external circuit; thus, the + polarity marking should be put at the terminal where the current flows out. Of course, if the coil is open-circuited, no current will flow, but the + voltage is produced nevertheless.

Inductance. The circuit-theory concept of inductance is related to the flux linkage through Faraday's law. We have but to compare Faraday's law in Eq. (2.24) with the circuit-theory definition of inductance

$$v = \frac{d}{dt}\lambda \quad \text{and} \quad v = \frac{d}{dt}Li \tag{2.26}$$

inductance

to define *inductance* as

$$\lambda = Li \Rightarrow L = \frac{\lambda}{i} \tag{2.27}$$

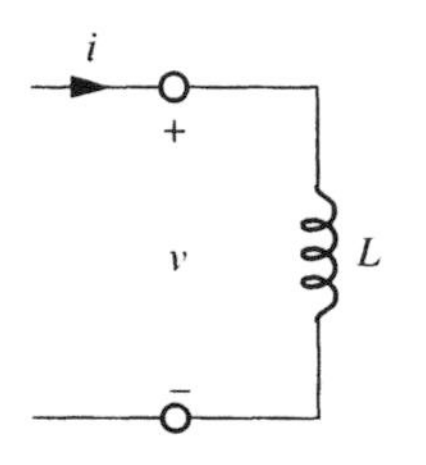

Figure 2.14 The load set of voltage and current variables expresses Lenz's law for the circuit variables.

Four comments:

1. Equation (2.27) defines inductance. To calculate the inductance of a coil, assume a current, determine the flux linkage, and use Eq. (2.27).
2. Although we normally write the definition of inductance with the inductance treated as a constant, and hence bring L outside the derivative in Eq. (2.26), the form shown is required when the inductance can change with time, as would be the case, for example, if a coil were part of a moving system.
3. The polarity of the voltage in Eq. (2.24), which may be resolved through application of Lenz's law, is established in the circuit-theory case through a standard sign convention. We use a load-set convention for inductors, Fig. 2.14, which fixes the

relationship between the voltage polarity and the current reference direction (and hence the direction of the magnetic flux).

4. In a *linear* magnetic system, the flux linkage is proportional to the current, and hence the inductance defined in Eq. (2.27) is constant and depends only on the geometry of the coil and the magnetic properties of the surroundings. However, if iron is involved in the inductor, then the inductance can be a function of the current because the *B–H* curve, Fig. 2.7, is not linear.

EXAMPLE 2.8 **Inductance calculation**

In Example 2.7, the current in the inductor is 200 mA(rms). Calculate the inductance.

SOLUTION:
The peak current is $200\sqrt{2}$ mA and the peak flux linkage is $\sqrt{2}/\pi$; hence, from Eq. (2.27)

$$L = \frac{\lambda_{peak}}{i_{peak}} = \frac{\sqrt{2}/\pi}{0.2\sqrt{2}} = 1.59 \text{ H} \tag{2.28}$$

WHAT IF? What if you use impedance instead of flux linkage to calculate the inductance?[6]

Coupled coils. When a magnetic system has multiple coils, the flux linkage in a coil can have contributions from all other coils. Consider, for example, the two-coil system shown in Fig. 2.15. The flux linkage in coil (1) has contributions from coils (1) and (2)

$$\lambda_1 = n_1(\Phi_{11} \pm \Phi_{12}) = \lambda_{11} \pm \lambda_{12} \tag{2.29}$$

where Φ_{11} is the flux in coil (1) due to i_1 and Φ_{12} is the flux in coil (1) due to i_2. The sign depends on the relative polarity of the coils. Faraday's law for the voltage in coil (1) is

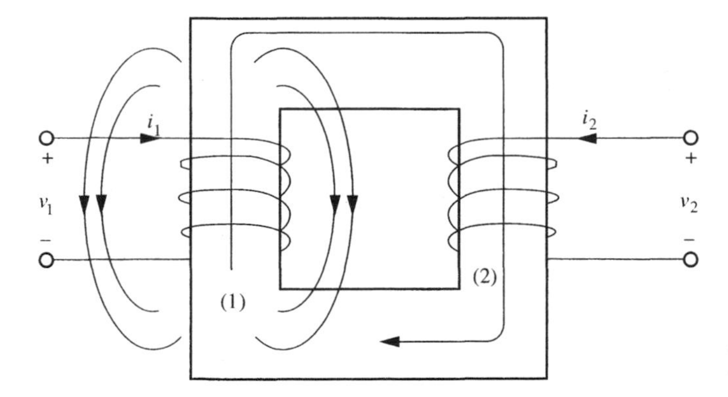

Figure 2.15 Magnetic system with two coupled coils. The flux in coil (2) has contributions from i_1 and i_2.

[6] The impedance magnitude is 600 Ω; $L = 600/(2\pi \times 60) = 1.59$ H, as before.

$$v_1 = \pm\left(\frac{d\lambda_{11}}{dt} \pm \frac{d\lambda_{12}}{dt}\right) \tag{2.30}$$

where the signs are established by Lenz's law.

Mutual inductance. The circuit model for the coils involves circuit elements such as the self-inductance of coil (1), L_{11}, and the mutual inductance, M_{12}, between coils (1) and (2), where

$$L_{11} = \frac{\lambda_{11}}{i_1} = \frac{n_1\Phi_{11}}{i_1} \qquad \text{and} \qquad M_{12} = \frac{\lambda_{12}}{i_2} = \frac{n_1\Phi_{12}}{i_2} \tag{2.31}$$

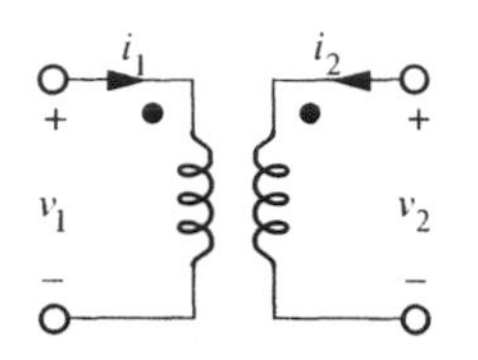

Figure 2.16 The standard dot convention with polarities defined in the customary manner. When physical currents are both into the dots, the fluxes from the coils add. This convention leads to the positive signs in Eqs. (2.33).

In these terms Eq. (2.30) becomes

$$v_1 = L_{11}\frac{di_1}{dt} + M_{12}\frac{di_2}{dt} \tag{2.32}$$

The signs in Eq. (2.32) are all positive because we adopted the standard dot convention for polarities shown in Fig. 2.16. According to this convention, positive currents into the dots produce fluxes that add. That convention, and the load sets relating voltage and currents, yields positive signs in the coupled equations of the system

$$\begin{aligned} v_1 &= L_{11}\frac{di_1}{dt} + M_{12}\frac{di_2}{dt} \\ v_2 &= M_{21}\frac{di_1}{dt} + L_{22}\frac{di_2}{dt} \end{aligned} \tag{2.33}$$

About self- and mutual inductance. What we have hitherto called "inductance" is simply the self-inductance of an isolated coil. The mutual inductance has two important properties:

- **Reciprocity.** It can be shown from Kirchhoff's laws that, for linear media, $M_{12} = M_{21}$. Thus it is customary to drop the subscripts and use L_1, L_2, and M for the self-inductances and mutual inductance for two coupled coils. With more than two coils, the subscripts must be retained for the mutual inductances.
- **Coefficient of coupling.** From conservation of energy and the nonnegative property of energy, it can be shown that the mutual inductance is equal to or smaller than the geometric mean of the self-inductances

$$M = k\sqrt{L_1 L_2}\,, \qquad \text{where } 0 < k < 1 \tag{2.34}$$

where k is the coefficient of coupling between the coils. When all of the flux in coil (1) also passes through coil (2), $k = 1$, but for weakly coupled coils, k is near zero.

EXAMPLE 2.9 Coupled coils

Two coils have self-inductances of $L_1 = 100$ mH and $L_2 = 1$ mH, and a coupling coefficient of $k = 0.5$. Find $v_1(t)$ and $v_2(t)$ if $i_1 = 10\cos(4000\pi t)$ mA and $i_2 = 0$, open circuit. Assume the polarity markings of Fig. 2.16.

SOLUTION:
From Eq. (2.34)

$$M = 0.5 \times \sqrt{100 \times 1} = 5 \text{ mH} \tag{2.35}$$

Thus, from Eqs. (2.33)

$$\begin{aligned} v_1(t) &= 0.100 \times \frac{d}{dt}[0.01 \cos(4000\pi t)] = -12.6 \sin(4000\pi t) \text{ V} \\ v_2(t) &= 0.005 \times \frac{d}{dt}[0.01 \cos(4000\pi t)] = -0.628 \sin(4000\pi t) \text{ V} \end{aligned} \tag{2.36}$$

WHAT IF? What if $i_1 = 0$ and $i_2 = 10 \cos(4000\pi t)$ mA?[7]

Stored Energy in a Magnetic System

Conservation of Energy

We apply conservation of energy to a lossless magnetic system represented by the inductor in Fig. 2.14. We may compute the energy stored in the system by integrating the input power, $p = vi$. The incremental energy input is then

$$dW_m = p\,dt = vi\,dt = \frac{d\lambda}{dt}\,i\,dt = i\,d\lambda \tag{2.37}$$

LEARNING OBJECTIVE 4.

To understand how to calculate stored energy in linear and nonlinear magnetic systems

where dW_m is the incremental increase in the stored magnetic energy in time dt. The total stored energy is thus

$$W_m = \int_0^{\lambda} i(\lambda')\,d\lambda' \tag{2.38}$$

where λ' is a dummy variable of integration. The integral in Eq. (2.38) corresponds to the shaded area in Fig. 2.17.

EXAMPLE 2.10 Maximum energy

Find the maximum energy that can be stored in a magnetic system in which $\lambda(i) = 10(1 - e^{-2i})$ Wb-turns.

SOLUTION:
The maximum energy is stored when the current goes to infinity and the flux linkage goes to 10 webers. From Eq. (2.38)

$$i(\lambda) = -\frac{1}{2}\ln\left(1 - \frac{\lambda}{10}\right) \Rightarrow W_m = \int_0^{10} -\frac{1}{2}\ln\left(1 - \frac{\lambda}{10}\right) d\lambda = 5 \text{ J} \tag{2.39}$$

[7] $v_1(t) = -0.628 \sin(4000\pi t)$ V and $v_2(t) = -0.126 \sin(4000\pi t)$ V.

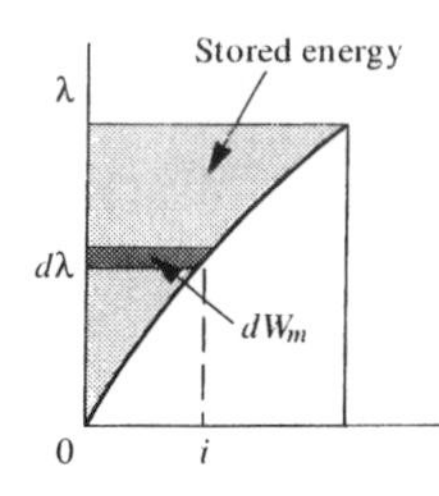

Figure 2.17 The stored energy corresponds to the shaded area.

WHAT IF? What if $\lambda = 5\sqrt{i}$? What is the maximum energy for this system?[8]

Linear magnetic systems. When a magnetic system is linear, the inductance is constant and Eq. (2.38) can be evaluated as shown in Eq. (2.40)

$$W_m = \int_0^{\lambda} \frac{\lambda'}{L}\, d\lambda' = \frac{\lambda^2}{2L} = \frac{1}{2} Li^2 \tag{2.40}$$

where i is the final value of the current. Thus, for a linear inductor, the stored energy depends on the final state of the current, or flux linkage, and the inductance, which characterizes the geometry of the system.

Relating λ and i to $\vec{B}$ and $\vec{H}$ Because the flux linkage, λ, is proportional to the magnetic flux density, $\vec{B}$, and the magnetic field, $\vec{H}$, is proportional to the current, it follows that the λ–i graph, as shown in Fig. 2.17, and the B–H plot, as exemplified by Fig. 2.7, would be identical except for scaling factors. This proportionality has several implications:

- Because the shaded area in Fig. 2.17 represents the stored energy, the corresponding area in the B–H curve must relate to energy. Indeed, the area in the B–H diagram corresponding to the shaded area in Fig. 2.17 represents the stored energy per unit volume in the magnetic system

  $$w_m = \int_0^{B} H\, dB' = \frac{B^2}{2\mu_0} = \frac{\mu_0 H^2}{2} \quad \text{J/m}^3 \tag{2.41}$$

  where w_m is the energy density, and H and B are the magnitudes of the magnetic field and flux density, respectively. The second and third forms in Eq. (2.41) assume a linear medium. The total energy in the system may be computed from the integral of the energy density over the entire volume of the magnetic system.

  $$W_m = \int_V w_m\, dv \text{ joules} \tag{2.42}$$

  when dv is a differential volume element of the volume, V.
- If the magnetic system contains iron, the λ–i curve will possibly be nonlinear and exhibit hysteresis, as shown in Fig. 2.7. When there is hysteresis, the magnetic system returns less energy to the external system than it received when energized. Thus, there is magnetic loss resulting physically from the reorienta-

[8] No limit exists.

tion of the magnetic domains by the external source. The loss per unit volume is the area inside the hysteresis curve in the B–H diagram. The energy lost to the electrical system through hysteresis loss appears as thermal energy in the iron.

Conductors Moving in Magnetic Flux

Faraday's law. Faraday's law in Eq. (2.24) is valid also for coils in which the wire is moving through the magnetic flux, provided the rate of flux change includes the effect of the motion. An alternate approach is to separate the voltage induced by a time-varying flux from the voltage induced in a moving conductor.

LEARNING OBJECTIVE 3.

To understand how to calculate voltage in stationary and moving wires with Faraday's law

Faraday's law for moving conductors. The calculation of the voltage induced in a moving conductor can be complicated, but the case of a straight wire moving through a uniform magnetic flux field is

$$v = \vec{\ell} \cdot \vec{u} \times \vec{B} = \ell u B_{\perp} \sin\theta \tag{2.43}$$

where $\vec{\ell}$ is the vector length of the wire with the arrowhead at the + end for v, $\vec{u}$ is its velocity, $\vec{B}_{\perp}$ is the flux density perpendicular to the wire, and θ is the angle between the motion and the perpendicular flux direction, all shown in Fig. 2.18. If the two ends of the bar were connected through a resistor, current would flow in the direction of $\vec{u} \times \vec{B}$, out of the end marked + and into the end marked − in Fig. 2.18. Thus, the force on a positive carrier is positive in the direction indicated by the vector product $\vec{u} \times \vec{B}$. This force can be combined with the electrostatic force in the Lorentz force law for the force on a moving charge

$$\vec{f} = q(\vec{E} + \vec{u} \times \vec{B}) \tag{2.44}$$

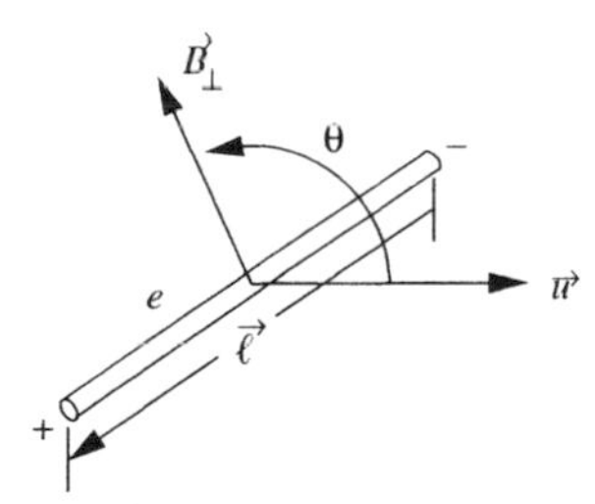

Figure 2.18 A conductor moving through magnetic flux generates a voltage called an electromotive force (emf).

Electromotive force (emf). The generation of a voltage by a magnetic flux, whether through flux change or through motion, indicates the presence of an electric field in the wire. This electric field integrated with Eq. (2.6) through the conductor path produces the voltage. The voltage described by Faraday's law is called an *electromotive force,* or *emf,* because such voltages can deliver energy to an electric circuit. The electric field due to a charge distribution is always conservative; KVL expresses conservation of energy for the resulting voltages. But the voltages induced by magnetic flux are nonconservative and, like a dc battery, can keep a current going indefinitely. Thus, an emf indicates an energy–conversion process.

electromotive force, emf

EXAMPLE 2.11 Rotating loop

Find the voltage induced in the rectangular loop of wire shown in Fig. 2.19. The loop has a length of ℓ, a width W, and is turning with an angular velocity ω, as shown.

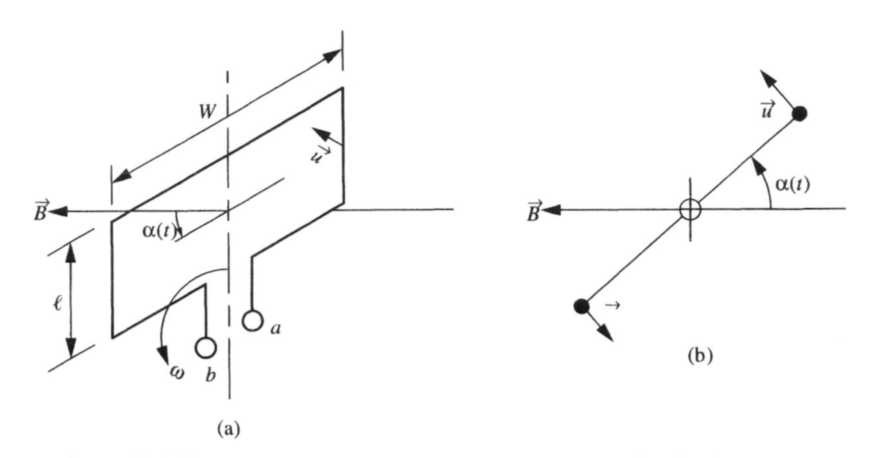

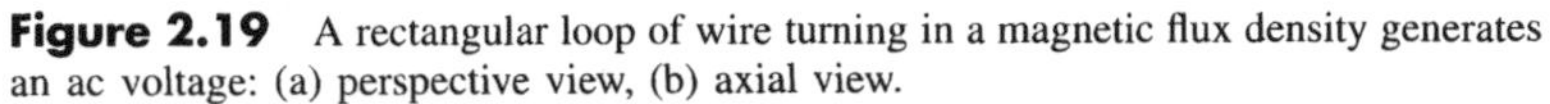

Figure 2.19 A rectangular loop of wire turning in a magnetic flux density generates an ac voltage: (a) perspective view, (b) axial view.

SOLUTION:

The plane of the coil makes an angle $\alpha(t)$ relative to the direction of the magnetic flux density, as shown in Fig. 2.19(b). The emf is produced in the parts of the coil that are parallel to the axis of rotation; the linear velocity of this wire is $u = \omega W/2$. The angle between $\vec{u}$ and $\vec{B}$ is $90^\circ - \alpha(t)$; hence, the emf from one wire is

$$v = \ell u\, B_\perp \sin[90^\circ - \alpha(t)] = \ell\omega \frac{W}{2} B_\perp \sin[90^\circ - \alpha(t)] \tag{2.45}$$
$$= \frac{\omega\Phi_m}{2}\cos[\alpha(t)]$$

where $\Phi_m = B_\perp W\ell$ is the maximum flux in the coil, when $\alpha(t) = 90^\circ$. An equal voltage is induced in the other wire. The voltages add in series, so the total motion-induced voltage is

$$v_{ab} = \omega\Phi_m \cos[\alpha(t)] = \omega\Phi_m \cos(\omega t) \tag{2.46}$$

because $\alpha(t) = \omega t$. Note that $\vec{u} \times \vec{B}$ would force current out of terminal a, which gives the polarity.

WHAT IF? What if you use Eq. (2.23)?[9]

[9]The flux through the coil is $\Phi(t) = \Phi_m \sin[\alpha(t)] = \Phi_m \sin(\omega t)$, and the emf by Faraday's law is $v_{ab} = d\Phi(t)/dt = \omega\Phi_m \cos(\omega t)$. The sign determined from Lenz's law is the same for both methods.

Electromechanical Energy Conversion

In this section, we show how Faraday's law and Ampère's force law describe the transformation of energy from electrical form, voltage and current, to mechanical form, force and velocity, and vice versa.

LEARNING OBJECTIVE 5.

To understand how to analyze and model energy conversion in a linear transducer

Electromechanical transducer. We will investigate the energy conversion process for the transducer shown in Fig. 2.20(a). The magnetic flux is furnished by an external magnet. The electrical circuit consists physically of a battery (V), a resistor (R), two stationary rails, and a movable bar that can roll or slide along the rails with electrical contact. We assume that the bar has an active length ℓ between the rails and is initially stationary. We close the switch and observe the sequence of effects that follow:

1. Current does not start immediately due to the inductance of the circuit.[10] The time constant of L/R is very small, however.
2. The current quickly reaches the value V/R.
3. A force is exerted on the bar due to the interaction between the current and magnetic flux. The magnitude of this force, by Eq. (2.18), is $F = iB\ell$ to the right, and the bar begins to move with a velocity u. The instantaneous mechanical power *out of* the bar, p_m, is

$$p_m = Fu = iB\ell u \quad \text{W} \tag{2.47}$$

This power comes from the electrical circuit, ultimately from the battery, and accelerates the bar, does work against an external force, or both.

Equivalent Circuits

4. The motion of the bar produces an electromotive force. This is the condition shown in Fig. 2.18 with $\theta = 90°$; thus, by Eq. (2.43), $v = uB\ell$. The polarity of the emf is positive where the current enters the moving bar. The equivalent circuit, shown in

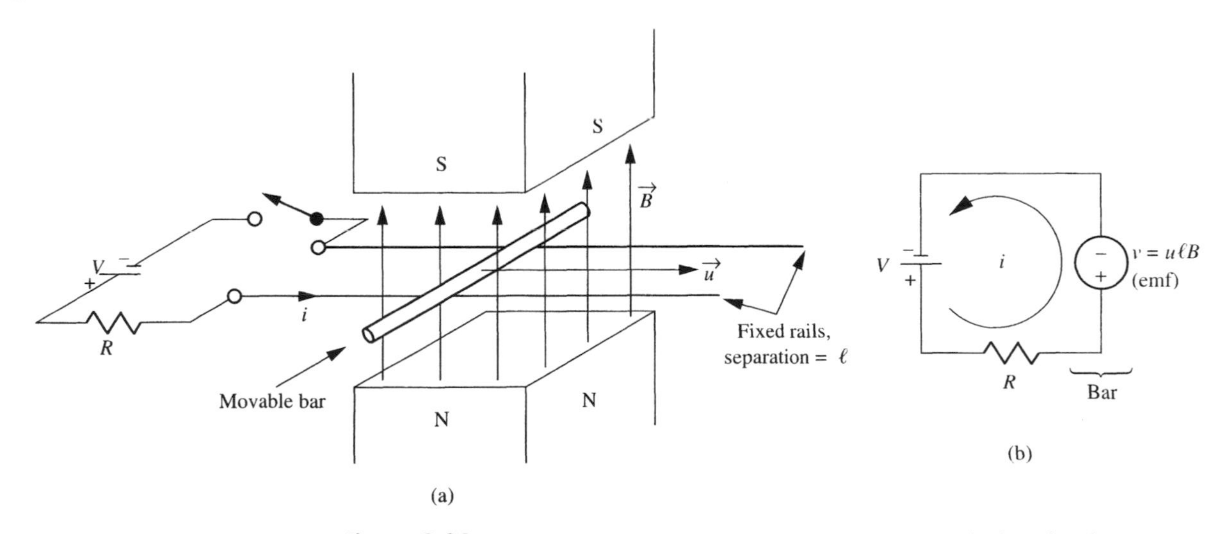

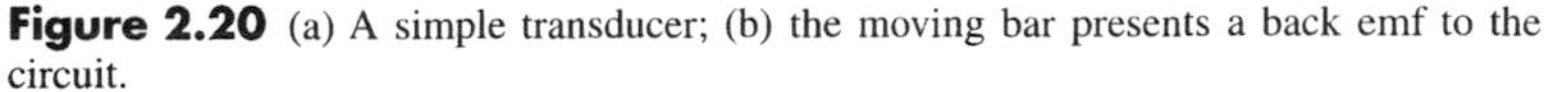
Figure 2.20 (a) A simple transducer; (b) the moving bar presents a back emf to the circuit.

[10] It is a one-turn coil.

Fig. 2.20(b), models the electrical aspects of the system. The moving bar generates a "back" emf that opposes the current.

Conservation of Energy

5. The instantaneous electrical power *into* the bar, p_e, is

$$p_e = vi = uB\ell i \tag{2.48}$$

Comparison of Eqs. (2.48) and (2.47) shows that the electrical input and the mechanical output powers are equal.

6. The dynamics of the electromechanical system involve the electrical system

$$i = \frac{V - v}{R} = \frac{V - B\ell u}{R} \tag{2.49}$$

and the mechanical system

$$F = B\ell i = B\ell\left(\frac{V - B\ell u}{R}\right) = \text{mechanical force balance} \tag{2.50}$$

7. The bar has an equilibrium speed of $u = V/\ell B$ at which the current, and hence the generated force, is zero. If moved faster by an external mechanical force, the bar becomes a generator, reverses the current direction, and supplies electrical power to the resistor and battery. In this circumstance, we could easily show that the input mechanical power is equal to the electrical output power supplied by the bar to the resistor and battery.

Causality diagram. Figure 2.21 shows causality for the electromechanical transducer. The mechanical system is unspecified except that the bar is free to move.

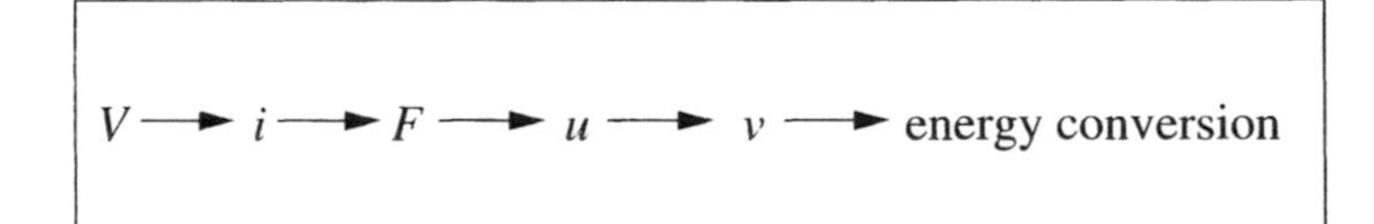

Figure 2.21 Causality diagram for the electromechanical transducer. The mechanical system is unspecified, but it is assumed that the electrical force causes motion.

We draw four conclusions:

1. The electrical input power to the bar (ignoring resistive losses) is transformed to mechanical output power with 100% efficiency.
2. The energy lost to the electrical circuit via the emf appears as mechanical energy. The emf voltage source represents electrically the coupling between electrical and mechanical systems.
3. This structure is a two-way transducer between the electrical and mechanical systems. The bar may represent an electrical load or source, depending on its velocity.
4. The magnetic flux, though an agent in the electromechanical coupling, does not participate directly in the energy exchange.

Whether motor or generator, the moving bar couples energy between electrical and mechanical systems with 100% efficiency. We have ignored mechanical losses such as friction and electrical losses associated with the resistance of the bar, but these are modeled easily in their respective systems.

EXAMPLE 2.12 Accelerating bar

Assume the bar has a mass M but no friction. Determine its motion.

SOLUTION:
From Eq. (2.50)

$$M\frac{du}{dt} = \frac{B\ell}{R}(V - B\ell u) \Rightarrow \frac{MR}{(B\ell)^2}\frac{du}{dt} + u = \frac{V}{B\ell} \tag{2.51}$$

The bar accelerates to a final velocity $u_\infty = V/B\ell$ with a time constant $\tau = MR/(B\ell)^2$.

WHAT IF? What if there is a frictional force Du?[11]

Summary. Through an analysis of a linear transducer, we established several important concepts. We showed that currents in wires moving through a magnetic flux effect electromechanical energy conversion. The moving conductor generates an electromotive force. When the current flows against the electromotive force, electrical energy is converted to mechanical energy; hence, the system acts as an electric motor. When the current is driven by the electromotive force, mechanical energy is converted to electrical energy, and the system acts as an electric generator. In both cases, an emf voltage source in the equivalent circuit represents the exchange of energy between the electrical and mechanical systems.

Check Your Understanding

1. Figure 2.22(a) shows a one-turn coil. The magnetic flux is decreasing. Mark the physical polarity (+ and −) on the coil terminals according to Lenz's law.

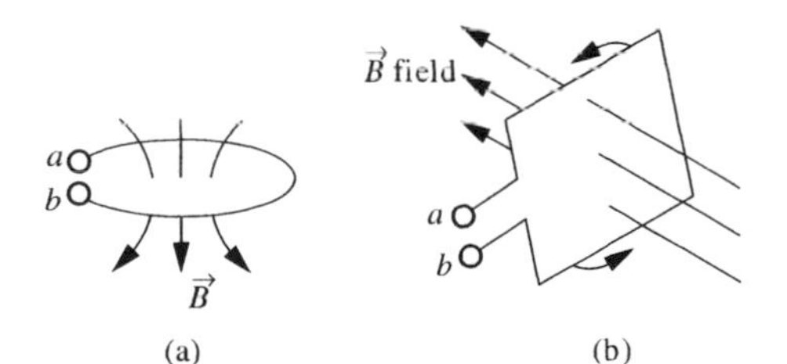

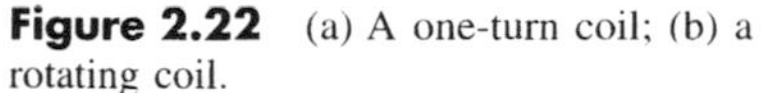

Figure 2.22 (a) A one-turn coil; (b) a rotating coil.

2. For the rotating coil shown in Fig. 2.22(b), determine the numerical sign of v_{ab} at the moment and angle shown.
3. What are the units of $\vec{u} \times \vec{B}$ in terms of volts, amperes, meters, and/or seconds? All are not required.
4. The deflection of the electron beam in a TV tube is accomplished with magnetic flux. If the TV set is facing south and the beam is deflected west, what is the direction of the magnetic flux established by the deflection coils?

Answers. **(1)** b is +; **(2)** v_{ab} is negative; **(3)** volts/meter, same as for electric fields; **(4)** down.

[11] Then $u_\infty = VB\ell/[RD + (B\ell)^2]$ and $\tau = MR/[RD + (B\ell)^2]$.

CHAPTER SUMMARY

This chapter reviews the basic laws underlying electromechanical devices:

1. Ampère's circuital law, Eq. (2.11), which relates a current to the magnetic field it creates.
2. Ampère's force law, Eq. (2.18), which describes the force/meter on a current-carrying conductor in magnetic flux.
3. Faraday's law, Eq. (2.24), which describes the emf produced by changing flux and/or moving conductors.

The presentation and exploration of these laws require definition of a number of quantities and explanation of practical matters such as the magnetic properties of iron.

The section on dynamic magnetic systems is important because Faraday's law, stored magnetic energy, and electromotive forces generated by moving conductors are the heart of electromechanics. The analysis of the linear transducer, which can act as either motor or generator, introduces the main ideas that are explored in the next four chapters.

Objective 1: To understand the definition of electric field and its relationship to voltage. An electric field is voltage stretched over space. The electric field is important to our purpose because electric fields are present in energy conversion.

Objective 2: To understand the definition and nature of magnetic fields and their relationship to current and magnetic materials. Magnetic fields support the efficient transformation of energy between electrical and mechanical forms. Ampère's circuital law relates a current to its magnetic field. Ferromagnetic materials, notably iron, increase the magnitude of the magnetic field and control its path through space.

Objective 3: To understand how to calculate voltage in stationary and moving wires with Faraday's law. A voltage is generated by a changing magnetic flux or by motion of a conductor through a magnetic flux. In a generator such voltage represents energy entering the electrical circuit, and in a motor such voltage represents energy leaving the electrical circuit as mechanical work.

Objective 4: To understand how to calculate stored energy in linear and nonlinear magnetic systems. We derive the equation for stored magnetic energy from the principle of conservation of energy. Stored energy provides the basis for calculating magnetic force in a magnetic system.

Objective 5: To understand how to analyze and model energy conversion in a linear transducer. To introduce energy conversion, we analyze the coupling between a dc circuit and a mechanical system represented by a movable bar carrying current. We see that the magnetic field supports the energy transformation but does not exchange energy. The energy-conversion efficiency is 100% between electrical and mechanical forms, although electrical and mechanical losses lower conversion efficiency in a real transducer.

In adopting a modest mathematical level, we presented the physics in a piecemeal manner that obscures the unity and elegance of the subject. In some cases, we have given only a special case of a physical law or treated a vector equation as if it were a scalar equation. Many details and subtleties were ignored. Our goal was to introduce the important concepts and equations that are used in the following chapters, not to give a full exposition of electrical physics.

Chapter 3 builds on the physical principles and laws reviewed in this chapter in the analysis of magnetic structures, specifically inductors, transformers, and transducers.

GLOSSARY

Ampère's circuital law, p. 63, a conservation principle relating current to the magnetic field it produces.

Electric energy, p. 58, the energy associated with motion against an electric force.

Electric field, p. 59, the vector force on a stationary charge divided by that charge; can be considered a voltage stretched over distance.

Electromotive force, p. 77, a voltage generated by a magnetic flux, whether through flux change or through motion.

Electrostatic force, p. 58, the electric force associated with the relative location of two or more charges.

Faraday's law, p. 71, the law relating voltage and changing magnetic flux linkage to a conductor.

Ferromagnetism, p. 66, the effect by which the magnetic domains in iron or another magnetic material align with an external magnetic field to produce a large magnetic flux.

Hysteresis curve, p. 67, the magnetic flux density versus magnetic field for a material, showing the nonlinearity and magnetic memory of the material.

Inductance, p. 72, a measure of a system's ability to store magnetic energy due to self-coupling of its magnetic field.

Lenz's law, p. 72, the rule that voltages produced by changing magnetic flux will produce currents that oppose the flux change if a path for current exists.

Magnetic energy, p. 59, the energy associated with motion against a magnetic force.

Magnetic field, p. 64, a magnetic stress on space produced by current. Magnetic field lines encircle the currents that produce them.

Magnetic flux, p. 69, a measure of the magnetization of a region of space, or the magnetic coupling to a region of space.

Magnetic flux density, p. 66, the magnetization of space that combines the effects of currents and atomic magnetism.

Magnetic flux linkage, p. 70, a measure of the magnetic coupling to a conductor, often the flux through a single turn times the number of turns in a coil.

Magnetic force, p. 62, the force between two currents or between a current and the atomic magnetism of a magnetic material.

Magnetic pole, p. 65, a way of identifying sources of magnetic field, particularly at the surface of a magnetic material.

Mutual inductance, p. 74, a measure of the magnetic coupling between two sets of conductors.

North magnetic pole, p. 65, a region where the magnetic field lines emerge from a surface.

Permeability, p. 67, a measure of the ability of a material to become magnetized.

Relative permeability, p. 67, the permeability divided by the permeability of space.

South magnetic pole, p. 65, a region where the magnetic field lines enter a surface.

Transducer, p. 79, a device for converting one physical quantity into another, such as physical motion to voltage, often to produce a signal to monitor the physical quantity.

Voltage, p. 60, the work done by an electrical system in moving a charge between two points, divided by the magnitude of the charge.

PROBLEMS

Section 2.2: Magnetic Forces and Magnetic Fields

2.1. Two infinite wires carry currents of I in opposite directions, as shown in Fig. P2.1. At what value(s) of x outside the wires does the magnetic field have the same magnitude as it has at $x = 0$?

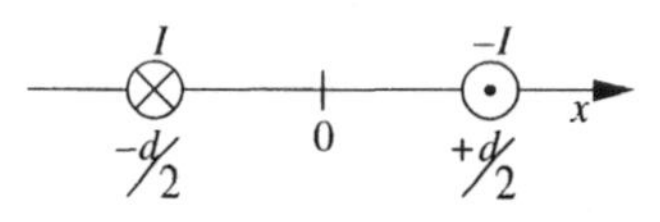

Figure P2.1

2.2. An infinite straight wire carries 10 A dc, as shown in Fig. P2.2. A nearby rectangular loop carries 5 A dc.

(a) Find the flux passing through the rectangular loop due to the 10-A current. Count downward as positive.

(b) Find the total force on the rectangular loop. Give the direction of the force relative to the infinite wire. *Hint:* The forces on the radial sides of the loop cancel by symmetry.

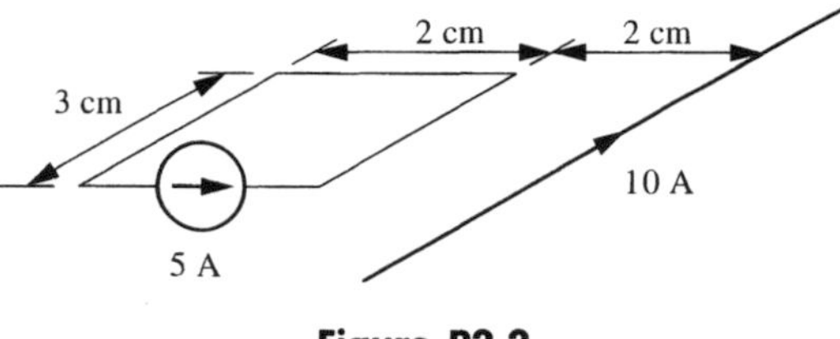

Figure P2.2

2.3. An auto battery is being charged with a current of 20 A.

(a) Estimate the magnetic flux density at the top of the battery, neglecting any effect due to the iron in the vicinity.

(b) Draw a picture showing the battery and the direction of the flux.

2.4. Two parallel infinite wires, a and b, are 4 meters apart, as shown in Fig. P2.4. In the plane of the wires, the flux is upward for $x < 0$ and for $1 < x < 4$, the cross-hatched regions shown. The current in a is 2 A, into the paper, as shown. Determine the current in b in magnitude and direction.

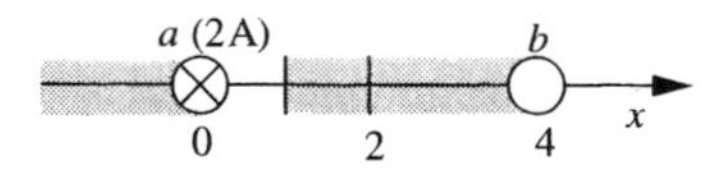

Figure P2.4

2.5. Two wires, infinite in the y-direction, are located at $x = 0$ and $x = 1$ cm, as shown in Fig. P2.5. The wires carry dc currents of 6 A and 3 A, as shown.

(a) Find the magnetic field at $z = 0$, $x = 0.3$ cm. Give direction and magnitude.

(b) What is the force per meter on the wire labeled b?

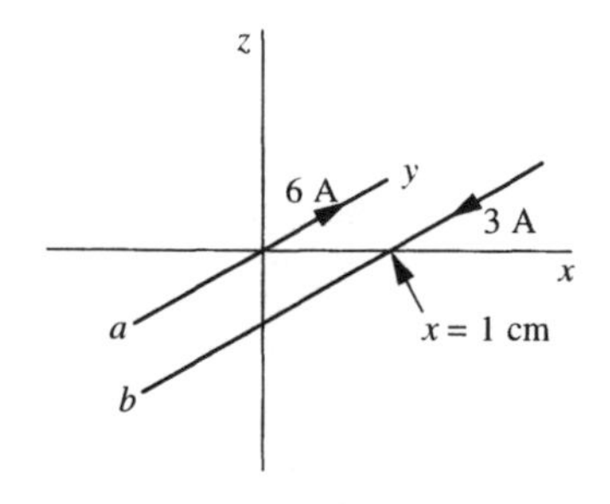

Figure P2.5

(c) If the magnetic field were integrated in a circular path at radius $r = 10$ meters from the y-axis, what would be the magnitude of the result?

2.6. An infinite wire with 0.1-inch diameter carries 2 A in the $-z$-direction, as shown in Fig. P2.6. The wire is on the axis of an iron pipe with an ID of 0.75 inch and an OD of 1.0 inch. The permeability of the iron is $1200\,\mu_0$; assume μ_0 elsewhere.

(a) Find the magnitude of the magnetic field and the magnetic flux density at the surface of the wire.

(b) Find the approximate magnitude of the magnetic field and the magnetic flux density in the iron.

(c) Find the magnetic flux/meter between the wire and the iron.

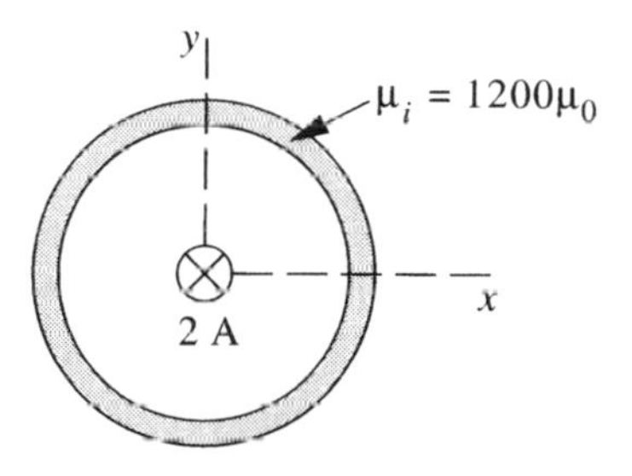

Figure P2.6

2.7. In a power plant, a large current is divided between three symmetrically placed conductors, as shown in Fig. P2.7. The centers of the conductors are a distance 20 cm apart, and all conductors carry the same current $i(t) = 1000\cos(\omega t)$ A.

(a) Find the peak force/meter exerted on any of the conductors by the other two conductors.

(b) Find the peak value of the magnetic flux density at a point b halfway between two of the conductors.

(c) Approximate the peak magnetic field at a distance of 10 meters from the wires, assuming no iron in the vicinity.

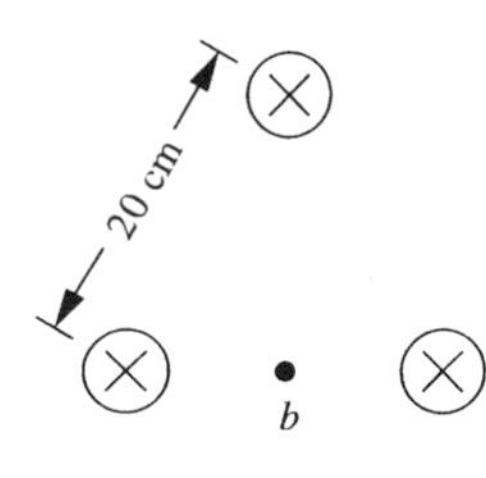

Figure P2.7

2.8. Three infinite, parallel wires of infinitesimal cross-section carry currents of +1 A, −3 A, and +2 A (all dc currents), respectively, as shown in Fig. P2.8. Note that d_1 and d_2 are positive numbers that sum to 1 meter. Assume that d_1 and d_2 are adjusted such that the net force on the center conductor is zero.

(a) Find the force/meter on the 1-A conductor counting force to the right as positive.

(b) Find the force/meter on the 2-A conductor.

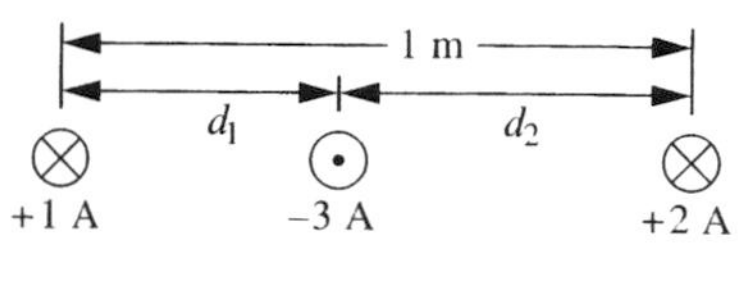

Figure P2.8

2.9. For the two wires shown in Fig. P2.5; Find the following:

(a) At what point on the x-axis between the wires is the flux density a minimum?

(b) At what finite point on the x-axis is the flux density zero?

2.10. An infinite wire along the y-axis has 10 A in the $+y$ direction, as shown in Fig. P2.10. Locate a second wire on the x–y plane such that there is zero magnetic field at $x = 1$ cm and the force between the wires is 0.002 newton/meter. Give the direction of the force (attraction or repulsion?).

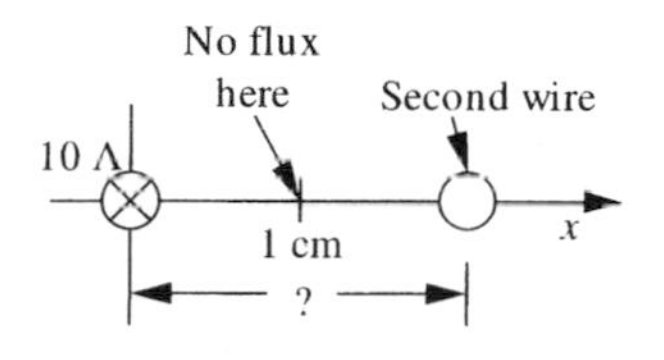

Figure P2.10

2.11. Figure P2.11 shows two infinite wires aligned with the x- and y-axes. The wires carry 1 A dc in the $+y$ and 1.5 A dc in the $-x$ direction, as shown.

(a) Where in the x–y plane is the magnetic field zero?

(b) In what regions of the x–y plane is the magnetic flux density out of the paper (in the positive z-direction)?

(c) Calculate the torque on the conductor carrying 1 A due to the field from the conductor carrying 1.5 A. Calculate either total torque or torque per meter, whichever is appropriate.

2.12. Two infinite wires are 2 cm apart and have diameters of 3 mm each, as shown in Fig. P2.12.

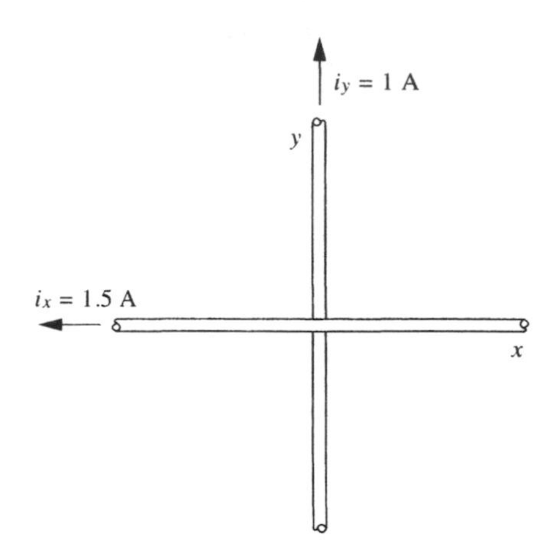

Figure P2.11

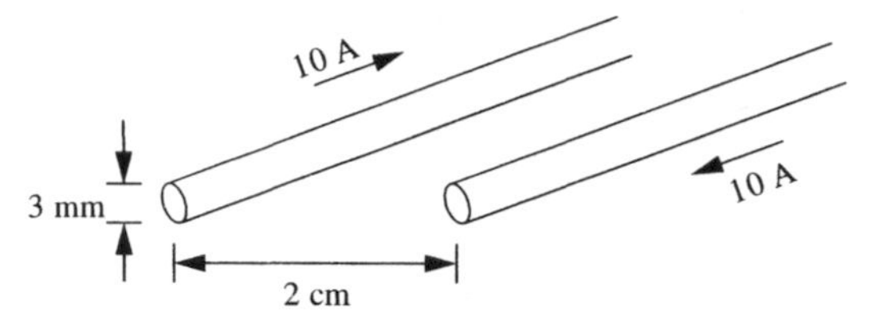

Figure P2.12

Each carries a current of 10 A, in opposite directions. Find the flux/meter passing between the wires, counting upward as positive.

2.13. Figure P2.13 shows a strip conductor carrying a current I in the $+y$-direction (into the paper.) The conductor is very thin and has a width w. The current is evenly distributed along the width of the conductor. The strip conductor can be analyzed as many wires, side by side, each having a current $di = (I/w)\,dx$ A.

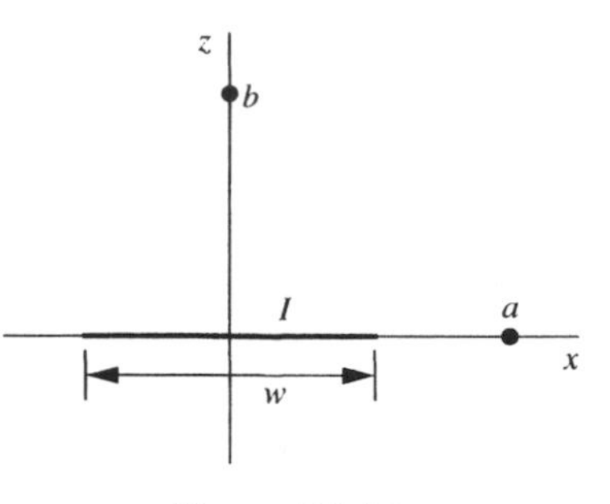

Figure P2.13

(a) What is the direction of the magnetic field at $x = a$?

(b) What is the magnitude of the magnetic field at $x = a$?

(c) What is the direction of the magnetic field at $y = b$?

(d) What is the magnitude of the magnetic field at $y = b$?

2.14. Assuming that lightning consists of electrons moving from earth to cloud, what direction (north, east, south, or west?) would a lightning bolt tend to move due to magnetic forces? Explain your answer.

2.15. The Lone Ranger is pursuing an outlaw who is fleeing toward the setting sun. He fires a silver bullet, which had a slight positive charge on it, toward the fugitive. Which way is the bullet deflected by the Earth's magnetic field (up, down, north, south, east, or west)?

Section 2.3: Dynamic Magnetic Systems

2.16. For the circuit and coil in Fig. P2.16, find the following:

(a) Draw the magnetic flux pattern, including flux direction.

(b) Would a compass needle inside the coil indicate north to be up or down?

(c) If the inductance of the coil were $L = 0.05$ H, find its magnetic flux linkage. Note that the wire in the coil and the rest of the circuit has a resistance of 1 Ω, in addition to the 2 Ω in the lumped resistance.

(d) As the battery weakens and the current decreases, an induced voltage appears across the coil. Would the + of the induced voltage be at the top or bottom of the coil?

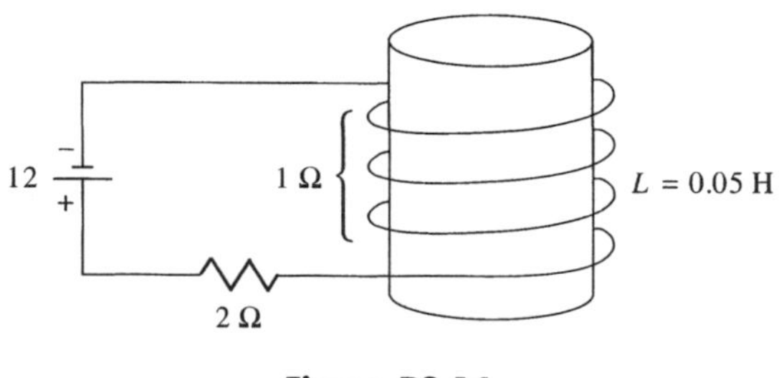

Figure P2.16

2.17. The Earth's magnetic field at the equator has a strength of about 30 μT. Consider an aircraft flying east over the equator at 600 mph. The wing span is 100 ft and the fuselage is 15 ft in diameter.

(a) To what part of the aircraft do the conduction electrons tend to move?

(b) What is the maximum emf created by the aircraft's motion?

2.18. Figure P2.18 shows a lossless, nonlinear inductor that is energized with a battery. After the switch is closed, the current is observed to increase as $i(t) = 0.05t^{1.2}$ A. Determine the flux linkage as a function of the current, $\lambda(i)$.

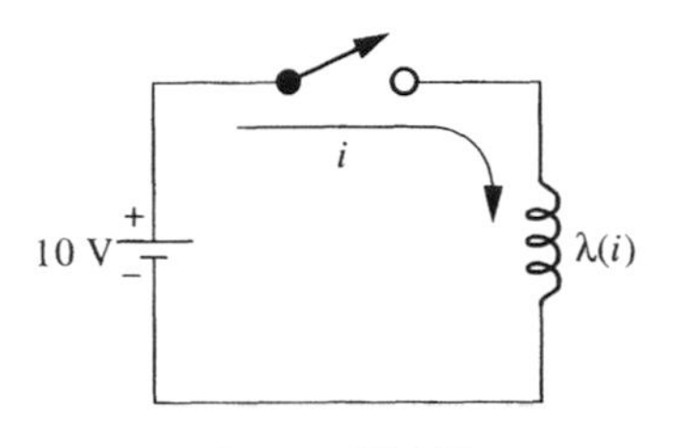

Figure P2.18

2.19. An inductor with a nonlinear magnetization curve described by $i = 0.2\lambda^2$ is connected to a 12-V battery by a switch that closes at $t = 0$, as shown in Fig. P2.19. Determine the stored energy as a function of time after the switch closure.

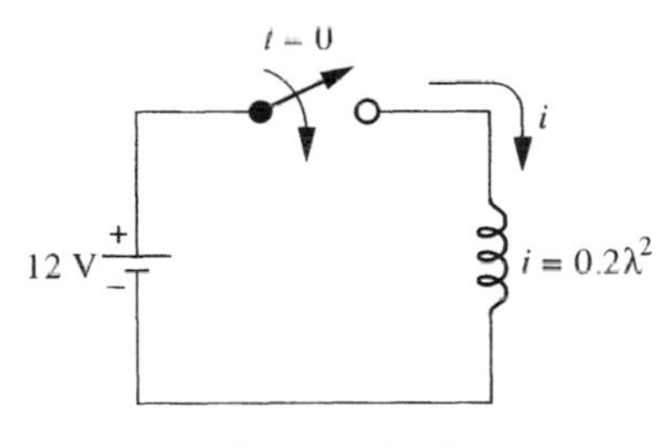

Figure P2.19

2.20. Figure P2.20 shows a nonlinear, lossless inductor with a flux linkage relationship $\lambda(i) = 3\sqrt{i}$ weber with i in amperes. The inductor is energized by a battery through a switch that is closed at $t = 0$.

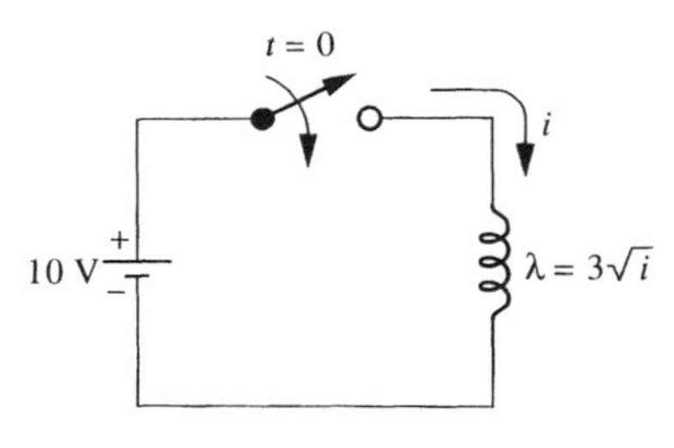

Figure P2.20

(a) Determine the time after switch closure when the power into the inductor is 12 W.

(b) At what time is the stored energy 100 J?

2.21. Two parallel infinite wires, each carrying a current i in the same direction, set up a magnetic field. The wires are a distance d apart, as shown in cross-section in Fig. P2.21. A conducting bar of length ℓ moves upward midway between the wires at a velocity $\vec{u}$. Derive a formula for the emf generated in the bar as a function of time, counting $t = 0$ when the bar passes the plane of the wires.

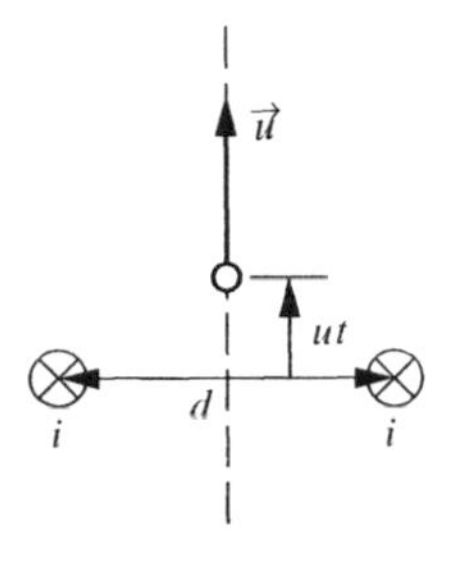

Figure P2.21

2.22. Figure P2.22 shows a permanent magnet that has a loop of wire around one of its poles.

(a) Draw the magnetic flux pattern, including the direction of the flux.

(b) As the loop of wire is lifted vertically, is voltage v_{ab} positive or negative?

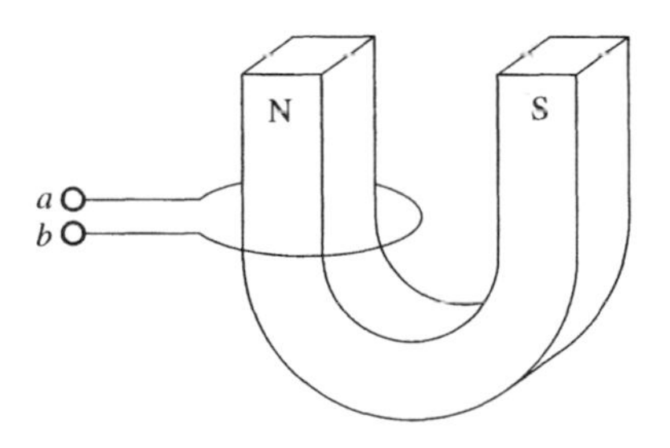

Figure P2.22

2.23. An infinite wire carrying a current i puts flux through a rectangular loop, as shown in Fig. P2.23. Note that the differential area of integration is $w\,dr$, where w is the width of the loop and r is the distance from the wire.

(a) Determine the flux in the loop, with direction, for $w = 10$ cm and $i = 1$ A. This flux should be independent of R_1.

(b) Determine the voltage v_{ab} if $di/dt = 100$ A/s.

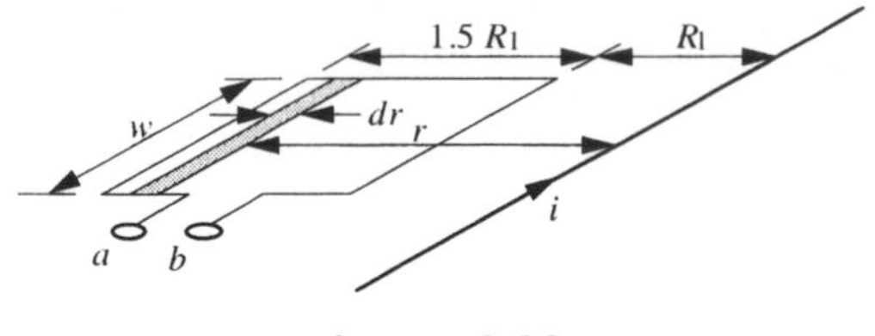

Figure P2.23

2.24. Figure P2.24 shows a homopolar generator, which consists of a conductor rotated in the presence of a magnetic flux. In this case, sliding contacts are made at the outer radius, r_o, and the radius of the axis, r_i. Determine the emf v_{ab} as a function of the magnetic flux density, B, the angular velocity, ω, r_i, and r_o.

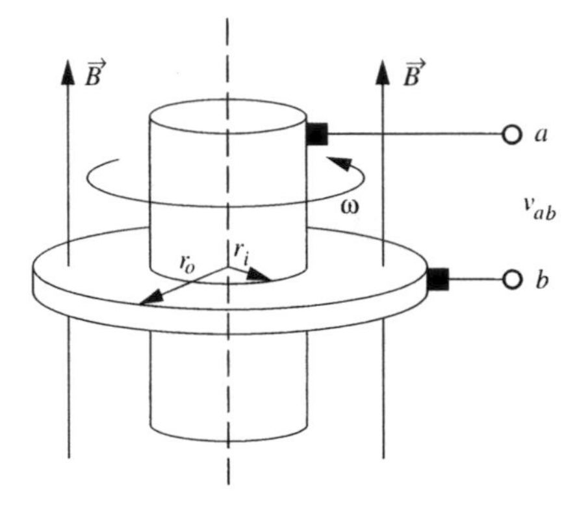

Figure P2.24

2.25. A wire is bent in a 90° angle and rotated in a uniform magnetic field with a flux density B, as shown in Fig. P2.25. Determine the emf produced, including the polarity.

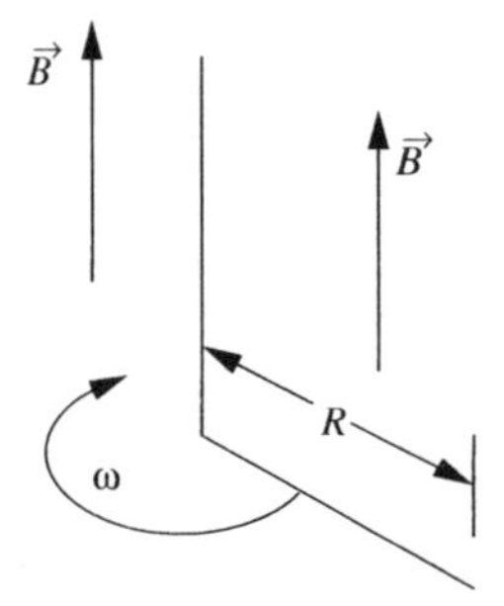

Figure P2.25

2.26. A circuit is shown is Fig. P2.26 that has a nonlinear inductor with a flux linkage-current relation $\lambda(i) = 5(1 - e^{-i})$ webers with i in amperes. The current source is OFF for a long time and then suddenly becomes a constant value of 2 amperes.

(a) Write a differential equation for the current in the inductor. You do not have to solve the equation.

(b) Determine the final stored energy in the inductor.

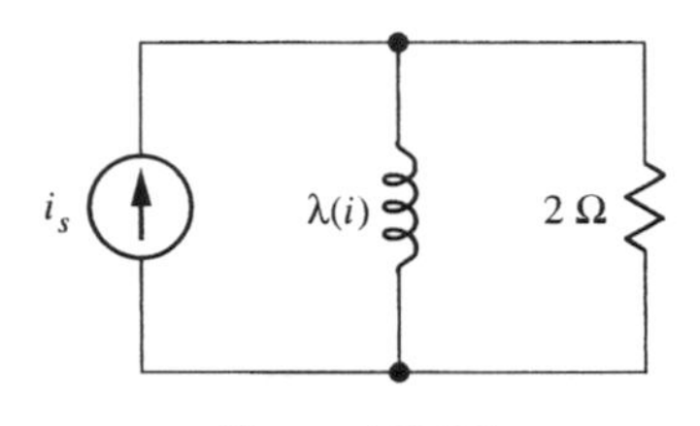

Figure P2.26

2.27. The relationship between the flux linkage and current for a magnetic system is plotted in Fig. P2.27 and is approximated by the equation $\lambda = 2 \ln(i + 1)$. Calculate the stored energy for $i = 0.4$ A.

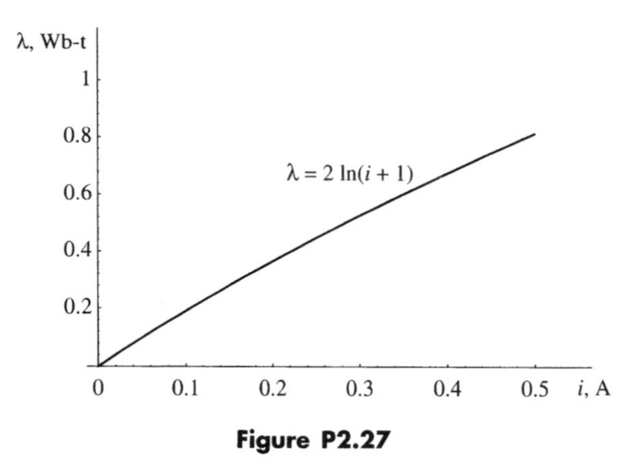

Figure P2.27

2.28. A nonlinear inductor is described by the equation $\lambda(i) = 1 - e^{-3i}$ webers, with i in amperes. This inductor has an initial current of 1 A, as shown in Fig. P2.28, and then is connected to a 3-Ω resistor.

(a) Derive a DE for the current, $i(t)$.

(b) During the course of the transient, find the total energy given to the resistor. *Hint*: This would be the energy stored initially in the magnetic system.

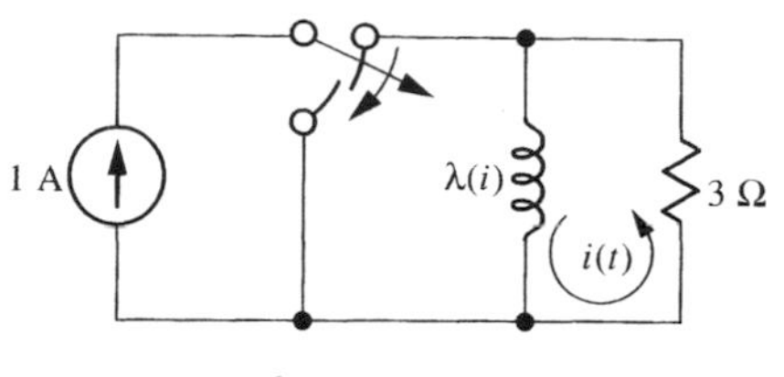

Figure P2.28

2.29. A lossless, nonlinear inductor has a characteristic $\lambda(i) = K\sqrt{i}$, where K is a constant and we consider only positive values of i. The inductor is energized by means of a battery and resistor, as shown in Fig. P2.29, with a final current of V/R_1. After a steady-state current is reached, at $t = 0$ the switch connects the inductor to a second resistor, giving its stored energy to the second resistor.

(a) What is the stored magnetic energy in the inductor at the instance the switch is thrown?

(b) Determine the flux linkage as a function of time, $\lambda(t)$, in the deenergization process.

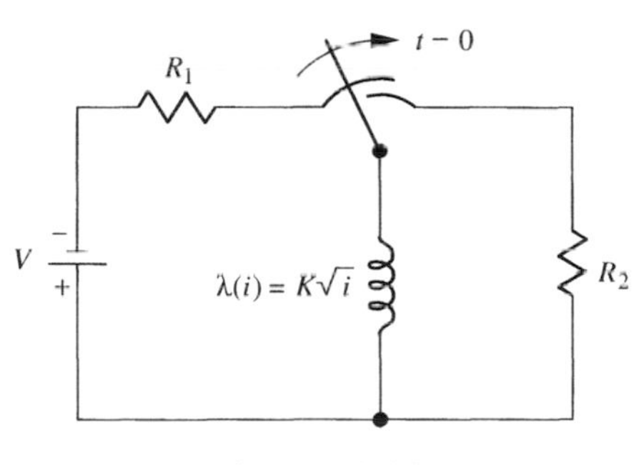

Figure P2.29

2.30. The circuit in Fig. P2.30 has a nonlinear inductor described by the equation $i = K\lambda^2$. The switch is closed at $t = 0$.

(a) Write the DE for the current in terms of V, R, K, and time.

(b) Find the current as time becomes very large.

(c) Find the stored energy as time becomes very large in terms of V, R, K, and time.

2.31. Figure P2.31 shows the magnetic flux linkage of a coil as a function of the current. The curve is nonlinear due to saturation of the iron in the system. The model uses straight-line approximations.

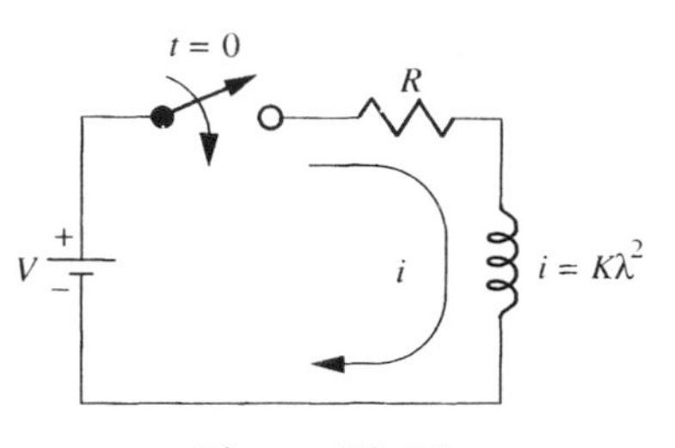

Figure P2.30

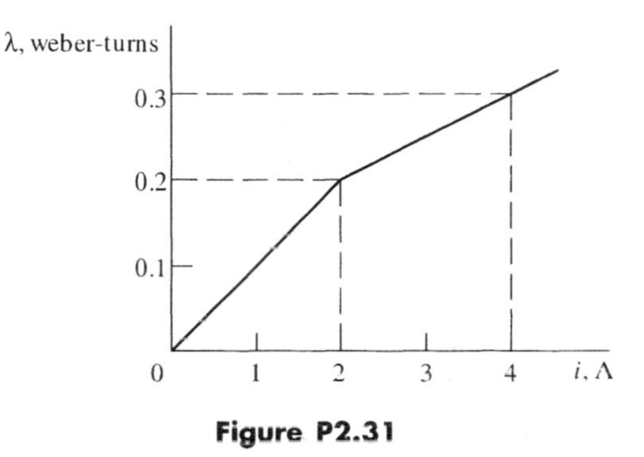

Figure P2.31

(a) What is the inductance of the coil for currents below 2 A?

(b) If the current starts at zero and increases at a rate of 100 A/s, determine the resulting voltage as a function of time during the first 40 ms. Plot the voltage during this time period.

(c) Determine the stored energy of the system for a current of 4 A. *Hint*: You don't need to do any integrals; just determine the areas geometrically.

2.32. Standard No. 12 wire has a diameter of 80.8 mils (thousandths of an inch), a weight of 19.77 pounds/1000 ft, and a capacity of 30 A under ideal conditions. In Fig. P2.32, we show two No. 12 wires of infinite length, each carrying the same unknown current I. The bottom wire is fixed, and the upper wire is free to float vertically.

(a) Mark the direction of current in the upper wire such that it will tend to float, and indicate the pattern of magnetic flux in the vicinity of the wires.

(b) Determine the minimum current I at which the

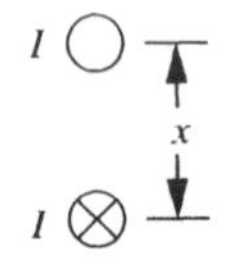

Figure P2.32

top wire just begins to float. (1 lb of force = 4.448 newton.)

2.33. A circuit consists of two parallel rails, shorted together at one end, and a movable bar, as shown in Fig. P2.33. A magnetic flux is perpendicular to the plane of the circuit and has a direction out of the paper, as shown. The magnetic flux density is uniform in space but increasing with time.

(a) Which way does induced current flow in the circuit?

(b) Which way does the bar tend to move?

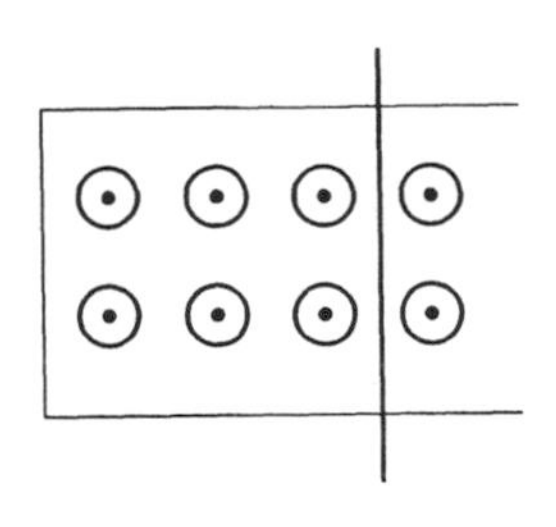

Figure P2.33

Answers to Odd-Numbered Problems

2.1. $\pm 0.707d$.

2.3. **(a)** Approximately 2.5×10^{-5}, depending on the size of the assumed path around the battery for Ampère's circuit law;
(b)

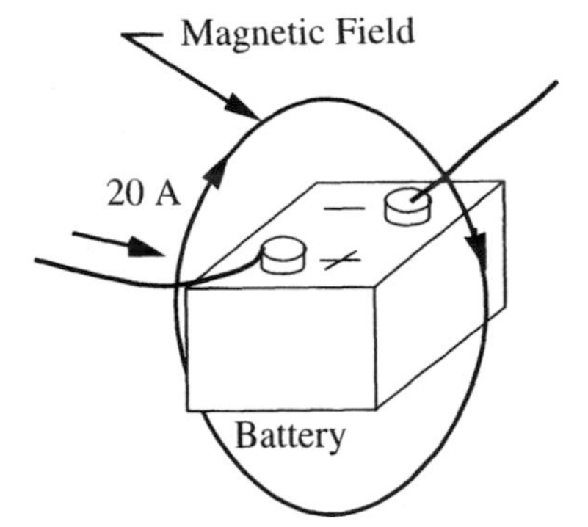

2.5. **(a)** 387 A/m neg z direction; **(b)** 3.60×10^{-4} N, repulsion; **(c)** 4.77×10^{-2} A/m.

2.7. **(a)** 1.73 N/m; **(b)** 1.15×10^{-3} T; **(c)** 47.7 A/m.

2.9. **(a)** 0.586; **(b)** 2 cm.

2.11. **(a)** along the line $y = -1.5x$;

(b)

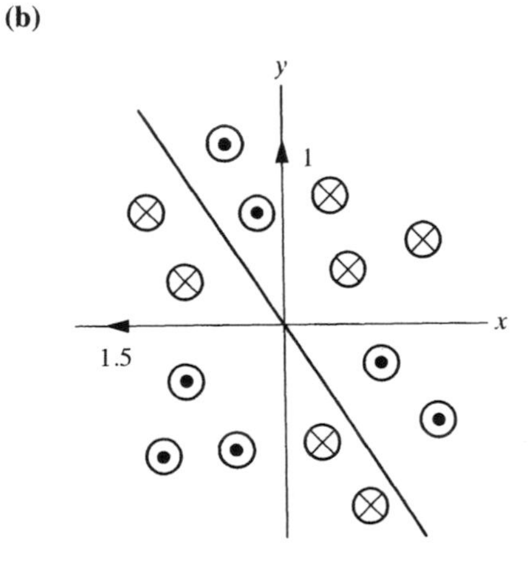

(c) a constant $1.5\mu_0/2\pi$ N-m/meter of torque.

2.13. **(a)** $-z$-direction; **(b)** $\frac{I}{2\pi w}\ln\left(\frac{a + w/2}{a - w/2}\right)$;

(c) $+x$-direction; **(d)** $\frac{I}{2\pi w}\tan^{-1}\left(\frac{b^2 + (w/2)^2}{b^2 - (w/2)^2}\right)$.

2.15. Down.

2.17. **(a)** Electrons to the bottom; **(b)** 3.68×10^{-2} V.

2.19. $115t^3$ J.

2.21. $2\mu_0 \ell i u^2 t/\pi\,[(ut)^2 + (d/2)^2]$.

2.23. **(a)** 1.83×10^{-8} Wb, upward; **(b)** 1.83×10^{-6} V, b positive.

2.25. $V = \omega BR^2/2$, plus at the axis of the wire.

2.27. 0.127 J.

2.29. **(a)** $\frac{K}{3}\left(\frac{V}{R_1}\right)^{1.5}$;

(b) $\lambda = \frac{1}{R_2 t/K^2 + (1/K)\sqrt{R_1/V}}$.

2.31. **(a)** 0.1 H;

(b)

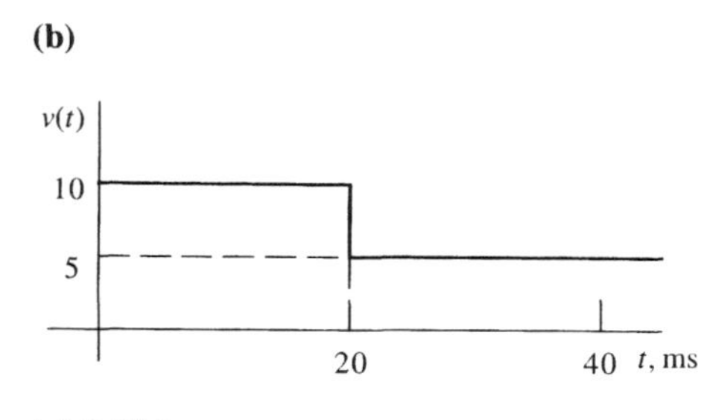

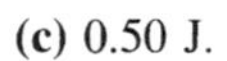

(c) 0.50 J.

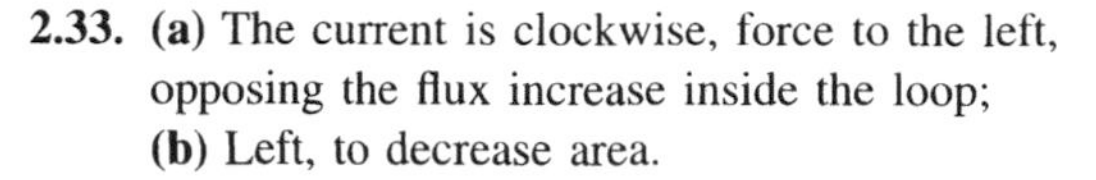

2.33. **(a)** The current is clockwise, force to the left, opposing the flux increase inside the loop; **(b)** Left, to decrease area.

CHAPTER

Magnetic Structures and Electrical Transformers

1. To understand the function and analysis of magnetic structures
2. To understand the physical basis for transformer models
3. To understand how to derive a transformer equivalent circuit from open-circuit/short-circuit measurements and use it to calculate the efficiency and percent regulation
4. To understand how to calculate the magnetic force in a magnetic structure

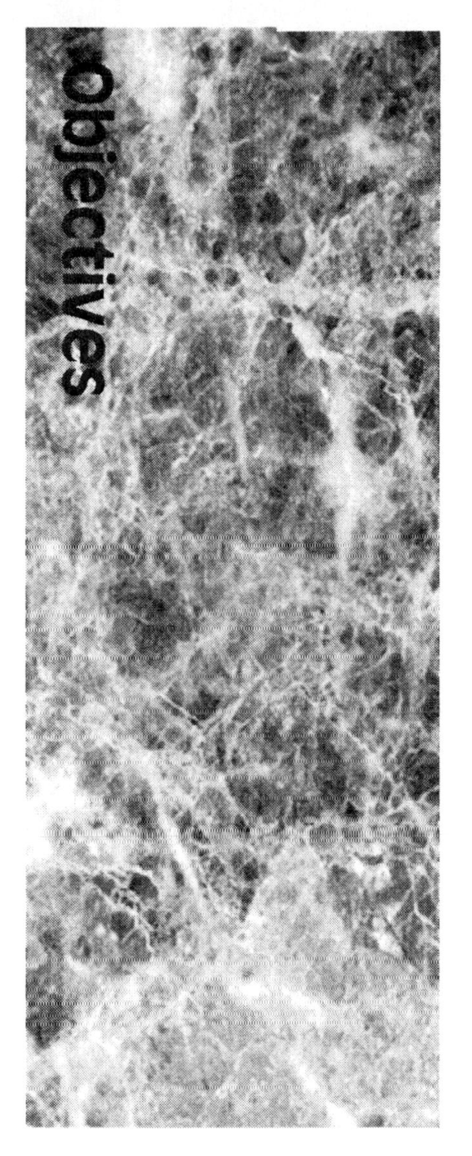

Electric motors and transformers are heavy because they contain iron structures to increase and control magnetic flux. This chapter uses the basic laws of magnetism to model inductors, transformers, and electromechanical transducers designed to produce force or torque.

3.1 ANALYSIS OF MAGNETIC STRUCTURES

Introduction to Magnetic Structures

magnetic structure, magnetic circuit

LEARNING OBJECTIVE 1.

To understand the function and analysis of magnetic structures

Magnetic structures A *magnetic structure*[1] is an iron structure that increases the amount of magnetic flux and controls its distribution in space. Figure 3.1 shows several types of magnetic structures. The transformer in Fig. 3.1(a) changes electrical voltage, current, and impedance levels. Relays, Fig. 3.1(b), are electrical actuators used in switches, locks, and the like. In electric motors and generators, Fig. 3.1(c), magnetic structures control the magnetic flux that produces torque. Transducers, such as the loudspeaker in Fig. 3.1(d), form yet another class of magnetic structures that includes phonograph pickups, tape heads, and tachometers. Finally, we should mention the ordinary inductors used as circuit elements, many of which require magnetic structures.

Although we stated before that magnetic structures are made out of iron, some use other magnetic materials. Materials research has produced ferrite ceramic materials with

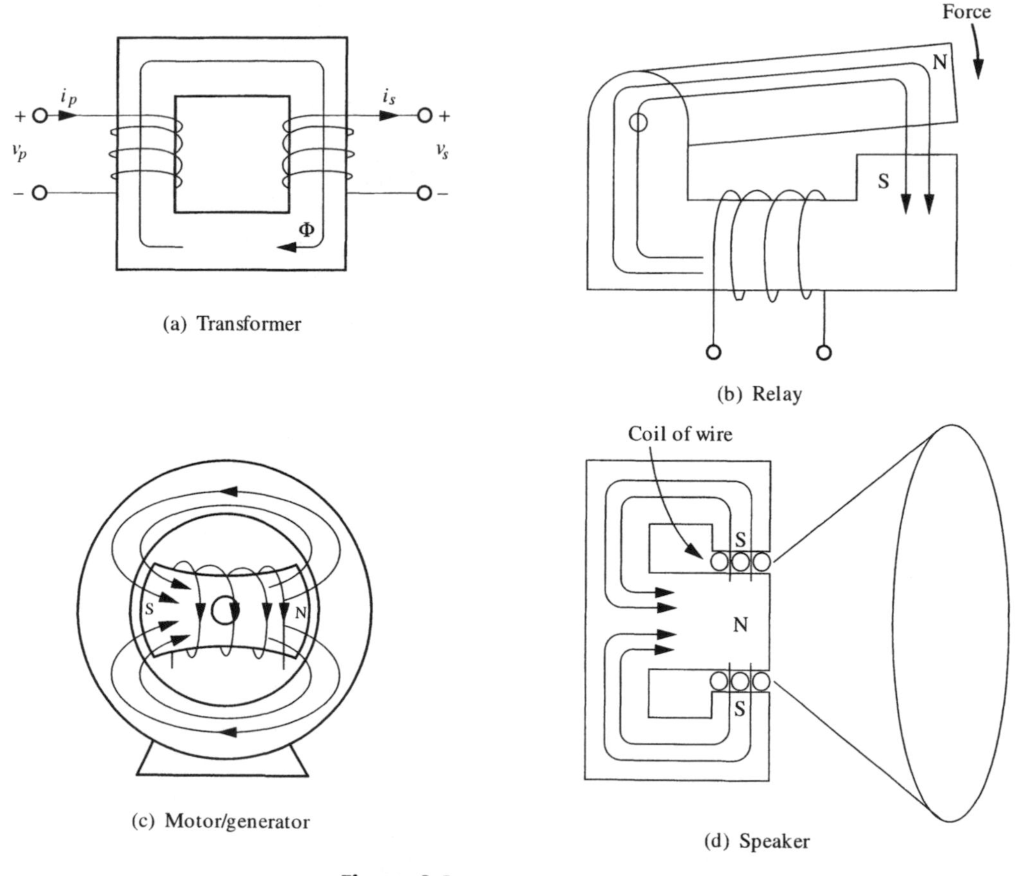

(a) Transformer

(b) Relay

(c) Motor/generator

(d) Speaker

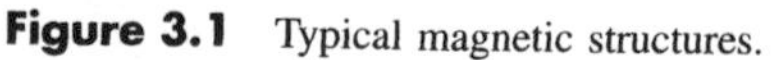

Figure 3.1 Typical magnetic structures.

[1] More commonly called a "magnetic circuit."

excellent magnetic properties, and a special class of electric motors, not to mention other devices, employ such materials in their magnetic structures. However, most magnetic structures are made of iron because of its low cost and excellent magnetic properties.

Chapter contents. First, we demonstrate methods of analysis for magnetic structures, continuing the themes of Chap. 2. We then analyze a transformer, beginning from the definition of an ideal transformer and then developing models to describe real transformers. Finally, we study forces generated by magnetic structures, using conservation of energy as a basis for calculating force and torque in electromechanical systems.

Toroidal Ring

Figure 3.2 shows a toroidal iron ring with a gap. We analyze this magnetic structure as an extended example. A coil of wire carrying current drives the magnetic system. We determine the magnetic field, flux density, and flux in the toroid and air gap.

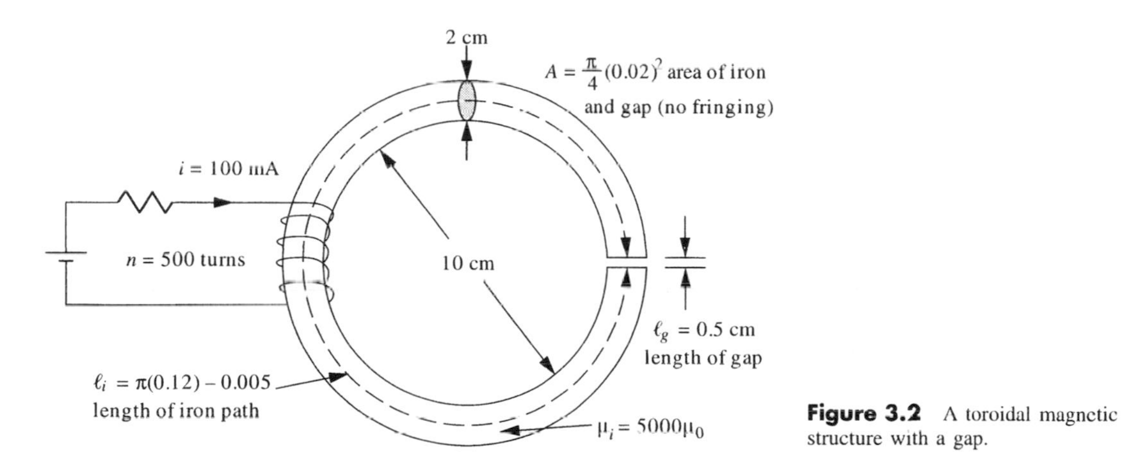

Figure 3.2 A toroidal magnetic structure with a gap.

Gaps in magnetic structures. You may wonder why we have included a gap in the iron ring. We include the gap because many electromechanical devices require such a gap. In Fig. 3.1, for example, the relay, motor, and loudspeaker must have gaps. The magnetic flux in such devices couples the electrical and mechanical systems, and the gap is required to permit motion and hence energy conversion.

Assumptions. Our analysis of the iron ring is based on several assumptions.

- We assume that the relative permeability of the iron is large and constant. Hence, we neglect magnetic saturation and hysteresis.
- We assume that the significant magnetic flux remains in the iron except at the gap, where the flux must pass through air. This reduces Ampère's circuit law to a scalar integral because the flux direction is assumed to be circumferential.
- We assume that the flux density is constant in magnitude, meaning that the flux density does not vary significantly across the cross-section of the toroid and does not vary around the ring.

- We assume that fringing in the gap is negligible. This means that the flux density of the gap does not vary over its cross-section and is circumferential, as in the iron.

These assumptions are reasonable in view of the discussion of the effect of the iron structure, as discussed on page 68.

magnetomotive force (mmf)

Magnetomotive force (mmf). The analysis begins with Ampère's circuital law Eq. (2.11), around the axis of the toroid

$$\oint H \, d\ell = \text{current enclosed} = ni \text{ ampere-turns} \tag{3.1}$$

The current enclosed by this path is the product of the current and the number of turns in the coil, which is called the *magnetomotive force*, or *mmf*. The name recognizes that the current-carrying coil energizes the magnetic system in the same way an electromotive force energizes an electric circuit.

Iron and gap. We separate the integral of the magnetic field into two terms

$$\oint H \, d\ell = \int_{\text{iron}} H_i \, d\ell + \int_{\text{gap}} H_g \, d\ell$$

$$= H_i \ell_i + H_g \ell_g = ni \quad \text{A-t} \tag{3.2}$$

where and H_i and H_g are the magnetic fields within the iron and gap, respectively, and ℓ_i and ℓ_g are the nominal distances within the iron and gap, respectively. The integral in Ampère's circuital law has contributions from both iron and gap. Because the field is constant along the path, H can be brought outside the integrals, leading to the last form of Eq. (3.2).

Magnetic flux. The magnetic fluxes in the iron and gap are also easily derived

$$\Phi_i = \int B_i \, da = B_i A_i \qquad \text{and} \qquad \Phi_g = \int B_g \, da = B_g A_g \tag{3.3}$$

where da is a differential area in the cross-section, B_i and B_g are the magnetic flux densities, A_i and A_g the cross-section areas, and Φ_i and Φ_g the fluxes in the iron and gap, respectively. Conservation of magnetic flux, Eq. (2.19), requires equal fluxes by integrating over the cross-section

$$\Phi_i = \Phi_g = \Phi \tag{3.4}$$

where Φ is the magnetic flux in the magnetic structure.

Iron and gap permeabilities. The magnetic fields and flux densities are related by the permeabilities of the iron, μ_i, and gap, μ_0

$$H_i = \frac{B_i}{\mu_i} \qquad \text{and} \qquad H_g = \frac{B_g}{\mu_0} \tag{3.5}$$

Because the direction of these vector fields is constrained by the magnetic structure, we can treat them as scalar quantities.

Analytical result. Combining Eqs. (3.2) through (3.5), we can eliminate all unknown quantities except the flux, with the result

$$\Phi = \frac{ni}{(\ell_i/\mu_i A_i) + (\ell_g/\mu_0 A_g)} \text{ Wb} \tag{3.6}$$

Numerical results. When we substitute the current of 0.1 A and the dimensions of the structure in Fig. 3.2 and assume that the relative permeability of the iron is 5000,[2] we obtain the numerical result

$$\Phi = \frac{500(0.1)}{1.88 \times 10^5 + 1.27 \times 10^7} = 3.89 \times 10^{-6} \text{ Wb} \tag{3.7}$$

We can now determine the values of the various field quantities

$$B_i = B_g = \frac{\Phi}{A} = \frac{3.89 \times 10^{-6}}{3.14 \times 10^{-4}} = 1.24 \times 10^{-2} \text{ T}$$

$$H_i = \frac{B_i}{\mu_i} = \frac{1.24 \times 10^{-2}}{5000(4\pi \times 10^{-7})} = 1.97 \text{ A/m} \tag{3.8}$$

$$H_g = \frac{B_g}{\mu_0} = 9850 \text{ A/m}$$

We also can determine the flux linkage and the inductance of the structure

$$\lambda = n\Phi = 1.95 \times 10^{-3} \quad \text{and} \quad L = \lambda/i = 19.5 \text{ mH} \tag{3.9}$$

In the next section, we discuss and generalize this analysis, using the calculated magnitudes to justify several approximations.

Principles of Analysis of Magnetic Structures

Magnetomotive force. Examination of the results of the previous section indicates that the magnetic flux depends on several factors. Notice first the role of the magnetomotive force, mmf $= ni$. The formal units of mmf are amperes, but ampere-turns, A-t, is commonly used to distinguish mmf from current. Magnetic systems are energized by current-carrying coils, and the mmf acts as the magnetic drive for the system. The "polarity" of the mmf is given by the right-hand rule. If there were more coils contributing to the flux, the mmfs would add or subtract according to this polarity.

Gain and loss of mmf. Equation (3.2) can be written in the form

$$ni - H_i\ell_i - H_g\ell_g = 0 \tag{3.10}$$

which suggests that mmf is gained in the coil and lost to the path around the toroid. For example, the mmf lost to, or required to magnetize, the gap is $H_g\ell_g$. In fact, the magnetic field can be considered as the loss in mmf per unit length along the field lines.

[2] That is, $\mu_i = 5000\mu_0$.

EXAMPLE 3.1 mmf in toroid

Find the fraction of the mmf required to magnetize the iron in the toroid.

SOLUTION:
The total mmf of the coil is 500 turns × 0.1 A = 50 A–t. The mmf lost to iron and the gap, respectively, is

$$H_i\ell_i = 1.97 \times (\pi \times 0.12 - 0.005) = 0.73 \text{ A-t}$$
$$H_g\ell_g = 9850 \times 0.005 = 49.27 \text{ A-t} \tag{3.11}$$

Thus, 1.47% of the coil mmf is used to magnetize the iron and 98.5% to magnetize the air in the gap.

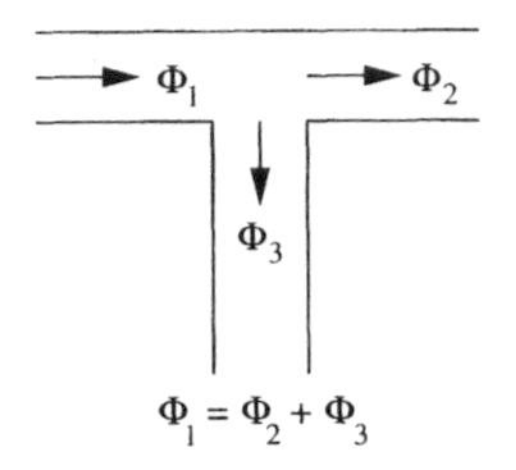

Figure 3.3
Conservation of magnetic flux at a junction.

Conservation of magnetic flux. Another important principle, although it applied trivially in our example, is conservation of magnetic flux. Had our magnetic structure contained branches and parallel paths, such as suggested in Figs. 3.3 and 3.15, we would have required conservation of magnetic flux at each such junction.

Reluctance. Equation (3.6) can be cast into the form

$$\Phi = \frac{ni}{\Re_i + \Re_g} = \frac{\text{mmf}}{\text{reluctance}} \tag{3.12}$$

where

$$\Re_i = \frac{\ell_i}{\mu_i A_i} = 1.88 \times 10^5 \ \frac{\text{A-t}}{\text{Wb}} \quad \text{and} \quad \Re_g = \frac{\ell_g}{\mu_0 A_g} = 1.27 \times 10^7 \ \frac{\text{A-t}}{\text{Wb}} \tag{3.13}$$

reluctance

are the reluctances of the iron and gap, respectively. *Reluctance* indicates the mmf required to magnetize a portion of a magnetic structure. Because the iron and gap are in series, have the same flux, their reluctances add to give the total reluctance of the path. The coil mmf is divided by this combined reluctance to give the flux in the system.

Role of iron. In Eq. (3.7), the reluctance of the iron path is small relative to the reluctance of the gap because of the large permeability of the iron. The total reluctance is approximately that of the air gap, and hence the magnetic flux of the system is in effect determined by the dimensions of the air gap. This is typical: The iron does such an effective job in guiding the magnetic flux that the magnetic system is limited by the air gap. As shown in the previous example, the mmf of the coil is used largely to magnetize the gap, and very little mmf is required to magnetize the iron.

Combining reluctances. Reluctances in series or parallel may be combined like resistances in series or parallel. Equation (3.12) illustrates a series addition. Parallel magnetic paths are sometimes used in transformers.[3]

[3] See Fig. 3.15.

Nonlinear effects. When saturation of the iron becomes a factor, several changes occur in the analysis. The equations become nonlinear and hence require numerical or graphical solution. Also, when the iron becomes saturated, the reluctance of the iron part of the path increases and becomes more of a factor in limiting the magnetic flux. This is frequently the case in practice, for practical devices often are operated partially saturated. However, our study of magnetic structures excludes the effects of saturation except for models of loss in inductors or transformers.

Inductance. The inductance is closely related to the reluctance of the system. The definition of inductance, Eq. (3.27), becomes

$$L = \frac{\lambda}{i} = \frac{n\Phi}{i} = \frac{n}{i} \times \frac{ni}{\Re} = \frac{n^2}{\Re} \text{ H} \tag{3.14}$$

Note that small reluctance corresponds to large inductance. The last two forms of Eq. (3.14) are valid only when the magnetic system is linear.

EXAMPLE 3.2 Winding an inductor

An inductor has 125 turns and an inductance of 150 mH. A 100-mH inductor is required. How many turns should be removed?

SOLUTION:

We may use scaling principles because $L \propto n^2$. Thus

$$\frac{100 \text{ mH}}{150 \text{ mH}} = \left(\frac{n'}{125}\right)^2 \Rightarrow n' = 102 \text{ turns} \tag{3.15}$$

Thus, 23 turns should be removed.

WHAT IF? What if, instead of removing turns, more turns are added but wound in the opposite direction? How many reverse turns should be added?[4]

Summary. In this section, we introduced the concept and function of a magnetic structure. We showed the role of magnetomotive force and reluctance in the analysis of magnetic structures. We now apply these concepts in the study of electrical transformers. In the next chapter, we introduce the cylindrical magnetic structures used in motors.

Check Your Understanding

1. A magnetic system is shown in Fig. 3.4.
   (a) For $i > 0$, mark the direction of the magnetic flux in the iron.
   (b) If i were decreasing at a rate of 20 A/s, find v_{ab}.

[4] Twenty-three turns again.

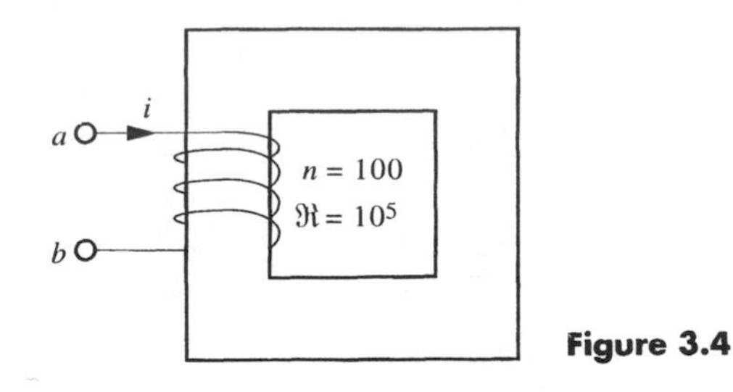

Figure 3.4

2. A magnetic structure has 1000 turns and an inductance of 12 H. Find the reluctance.
3. A magnetic structure has 1000 turns and an inductance of 12 H. It is scaled down by a factor of 2 in every physical dimension, but the number of turns is unchanged. Find the new inductance using scaling principles.
4. The flux is 0.1 Wb in a one-loop magnetic structure. It has an air gap with an area of 0.1 m^2 and a length of 1 mm.
   (a) What is the flux density in the air gap?
   (b) What is the reluctance of the air gap?
   (c) What mmf is required to force this flux across the air gap?

Answers. **(1) (a)** Up in the coil side, down in the other side; **(b)** –2 V; **(2)** 83,300 A-t/Wb; **(3)** 6 H; **(4) (a)** 1 T, **(b)** 7.96×10^3 A-t/Wb, **(c)** 796 A-t.

3.2 ELECTRICAL TRANSFORMERS

Introduction to Transformers

transformer

An electrical *transformer* consists of two or more coils, or windings, tightly coupled by magnetic flux that is guided by a magnetic structure. Transformers are used for voltage transformation, current transformation, and impedance transformation. Chapter 1 presented three-phase transformer connections and introduced applications in power distribution systems. In this chapter, we develop a physical understanding of transformers, leading to realistic models for transformers. Our treatment of transformers in this chapter is oriented toward power transformers.

Section contents. We begin with a brief review of ideal transformer relationships, and then analyze the transformer as a magnetic structure. From this analysis, supplemented by physical considerations, we develop equivalent circuit models for single- and three-phase transformers.

ideal transformer

Ideal transformers. The *ideal transformer* is represented on circuit diagrams with the symbol shown in Fig. 3.5(a) and Eqs. (3.16) and (3.17) relating the sinusoidal steady-state voltages and currents in the primary and secondary

$$\frac{\underline{\mathbf{V}}_p}{n_p} = \frac{\underline{\mathbf{V}}_s}{n_s} \tag{3.16}$$

$$n_p \underline{\mathbf{I}}_p = n_s \underline{\mathbf{I}}_s \tag{3.17}$$

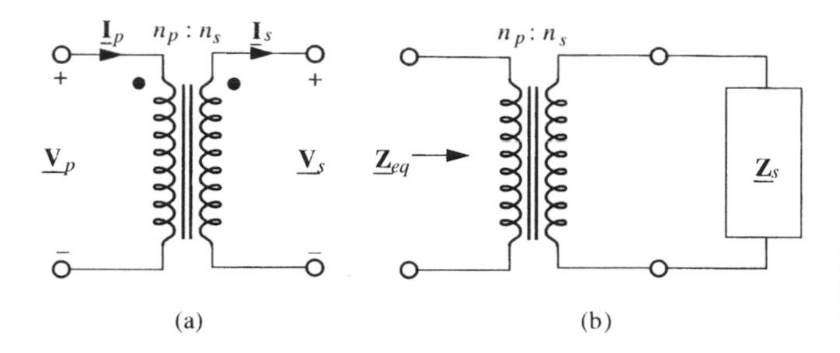

Figure 3.5 (a) Circuit symbol for an ideal transformer; (b) impedance transformation.

Impedance Level

Multiplication of Eqs. (3.16) and (3.17) gives an equation that shows the conservation of apparent power in the ideal transformers. It is easily shown that real and reactive power are also conserved. Division of Eq. (3.16) by Eq. (3.17) shows that the transformer changes impedances by the square of the turns ratio, Fig. 3.5(b)

$$\underline{Z}_{eq} = \frac{\underline{V}_p}{\underline{I}_p} = \left(\frac{n_p}{n_s}\right)^2 \frac{\underline{V}_s}{\underline{I}_s} = \left(\frac{n_p}{n_s}\right)^2 \underline{Z}_s \tag{3.18}$$

where $\underline{Z}_{eq}$ is the equivalent impedance into the primary. Our purpose in the following section is to show the physical basis for these relationships and to investigate the extent to which they describe real transformers. The analysis is based on the physical laws from Chap. 2 and the concepts of magnetic structures from the previous section.

Analysis of a Transformer as a Magnetic Structure

Learning Objective 2.

To understand the physical basis for transformer models

We now analyze the structure shown in Fig. 3.6, where turns, area, path length, permeability, and circuit variables are defined. We assume a sinusoidal steady state, and hence use phasors to represent voltage, current, field, and flux quantities. The voltage relationship in Eq. (3.16) follows from Faraday's law, Eq. (2.24), if we assume the same flux in the primary and secondary windings.

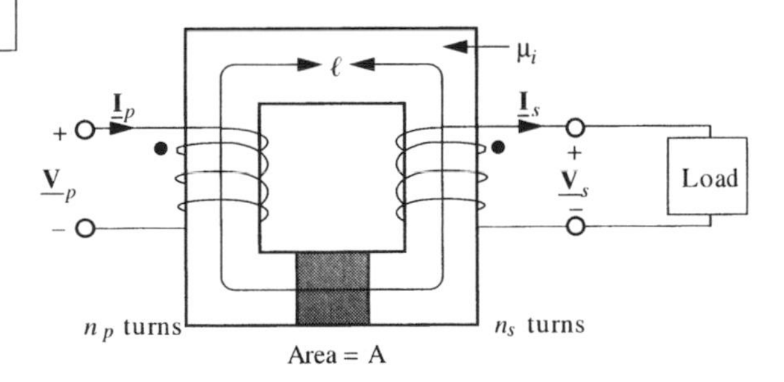

Figure 3.6 Transformer conventions and dimensions.

stray or leakage flux

Leakage flux. In a real transformer, not all flux follows the iron path, and hence the ideal voltage relationship of Eq. (3.16) does not apply strictly to a real transformer. However, in a well-designed magnetic structure, only a small amount of flux escapes from the iron to become *stray* or *leakage flux*; hence, the ideal voltage relationship is very nearly obeyed in a real transformer. In the following, we assume that all flux couples primary and secondary coils, and later we insert small series inductors into our equivalent circuit model to account for leakage flux.

Faraday's law. From Faraday's law, Eq. (2.24), we gain an important relationship between voltages and flux values. Applied to the primary, Faraday's law is

$$v_p(t) = n_p \frac{d\Phi(t)}{dt} \quad \Rightarrow \quad \underline{\mathbf{V}}_p = n_p(j\omega)\underline{\Phi} \tag{3.19}$$

where the first form is in the time domain and the second is a phasor relationship in the frequency domain. We discuss Eq. (3.19) after we discuss notation.

Voltage and flux. The quantity $\underline{\Phi}$ in Eq. (3.19) is the phasor representation of the flux, phase and rms amplitude. The j factor on the right side indicates that the flux lags the voltage by 90° in phase. Because the maximum value of the flux is frequently critical, the magnitude of Eq. (3.19) is often written

$$V_p = \frac{\omega n_p}{\sqrt{2}} \Phi_{max} \tag{3.20}$$

where Φ_{max} is the maximum flux and V_p is the rms magnitude of the primary voltage.[5] Clearly, Eqs. (3.19) and (3.20) also could have been written for secondary voltages. Equation (3.16) results when the common factors are eliminated between Eq. (3.19) and the corresponding equation involving the secondary voltage.

EXAMPLE 3.3 Doorbell ringer

A doorbell ringer has 200 turns and operates on 6-V, 60-Hz voltage. Find the maximum flux in the magnetic structure.

SOLUTION:
From Eq. (3.20)

$$6 = \frac{120\pi \times 200}{\sqrt{2}} \Phi_{max} \Rightarrow \Phi_{max} = 1.13 \times 10^{-4} \text{ Wb} \tag{3.21}$$

WHAT IF? What if the circular cross-section has a diameter of 1 cm? What is the maximum flux density?[6]

Dot convention

Polarities and the dot convention. The voltage polarity between transformer windings is indicated by the dots in Figs. 3.5(a) and 3.6. Once the voltage polarity is established with the + at the dots, the current reference directions are automatic. The reference direction of the *primary* current is *into* the dot, but the reference direction of the *secondary* current is *out of* the dot. Thus, the primary is represented as a load set, and the secondary is represented as a source set.

[5] We used V_p for peak voltage earlier, but here it means rms voltage in the *primary*.

[6] 1.43 tesla.

Lenz's law. The dot on the primary is assigned arbitrarily, and a load set of voltage and current reference directions is assigned with the + on the voltage corresponding to the dot. The + of the secondary voltage reference direction is then determined by the right-hand rule and Lenz's law in the following manner. Assume current flows into the dot on the primary and use the right-hand rule to establish the flux direction. Then take the secondary coil in your right hand with the thumb *opposite* to the flux from the primary. Your fingers then indicate the direction that secondary current would flow if allowed. The dot is placed on the end of the secondary coil where current would exit.[7] The dot corresponds to the + on the reference direction of the secondary voltage.

Ampère's circuital law. The currents in primary and secondary are related through Ampère's circuital law:

$$\oint \underline{\mathbf{H}}\, d\ell = n_p \underline{\mathbf{I}}_p - n_s \underline{\mathbf{I}}_s \tag{3.22}$$

where $\underline{\mathbf{H}}$ is the phasor magnetic field in the magnetic structure. The minus sign of the $n_s \underline{\mathbf{I}}_s$ term in Eq. (3.22) follows from Lenz's law. We placed the dots in Fig. 3.6 such that the flux from $\underline{\mathbf{I}}_s$ would *oppose* the flux created by $\underline{\mathbf{I}}_p$; hence, the mmfs of the two coils subtract. We can change the left-hand side of Eq. (3.22) by introducing the concepts of magnetic structures

$$\oint \underline{\mathbf{H}}\, d\ell = \underline{\mathbf{H}}\ell = \frac{\underline{\mathbf{B}}_i}{\mu_i} \times \ell = \frac{\underline{\Phi}}{A} \times \frac{\ell}{\mu_i} = \underline{\Phi} \times \Re_i \tag{3.23}$$

where $\Re_i$ is the reluctance of the iron, and $\underline{\Phi}$ is the phasor coupling flux. Thus, Eq. (3.22) can be rearranged to the form

$$\underline{\mathbf{I}}_p = \frac{n_s}{n_p} \underline{\mathbf{I}}_s + \frac{\underline{\Phi} \Re_i}{n_p} \tag{3.24}$$

We can introduce the primary voltage into Eq. (3.24) by introducing Eq. (3.19), with the result

$$\underline{\mathbf{I}}_p = \frac{n_s}{n_p} \underline{\mathbf{I}}_s + \frac{\underline{\mathbf{V}}_p}{j\omega L_i} \tag{3.25}$$

where $L_i = n_p^2 / \Re_i$ from Eq. (3.14).

IDEA 6 **Equivalent Circuits**

Equivalent circuit. Equation (3.25), interpreted as KCL, suggests the equivalent circuit shown in Fig. 3.7. Subject to the assumptions we have made, a real transformer is represented by an ideal transformer and an inductor. The inductor accounts for the magnetic energy of the flux; thus in this circuit model of the transformer, derived from Ampère's circuital law, real power is conserved, but reactive power is not because the transformer requires inductive current to magnetize the iron.

[7] A consequence of these conventions is that currents into the dots give flux in the same direction.

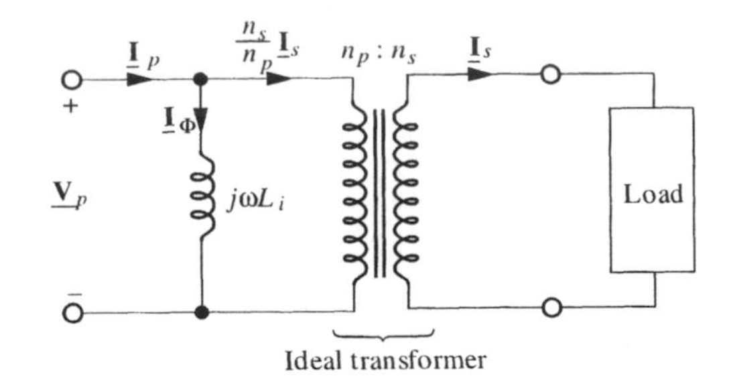

Figure 3.7 The inductor represents the energy stored in the magnetic flux. $\mathbf{I}_\Phi$ represents the current required to magnetize the transformer iron.

Causality in a transformer. Normally, the transformer primary is connected to an ac voltage source of constant voltage and the secondary to a load. The causal relationships for the transformer are shown in Fig. 3.8. By Faraday's law, the primary voltage must be matched by time-varying flux linkage in the primary winding, which in turn requires fields and the magnetizing current, $\mathbf{I}_\Phi$, represented by the inductor in Fig. 3.7. The changing flux in the secondary coil causes the secondary voltage, which causes load current in the secondary that is reflected back to the primary. If the secondary is open circuited, the lower path in Fig. 3.8 is eliminated and the only primary current is the magnetizing current.

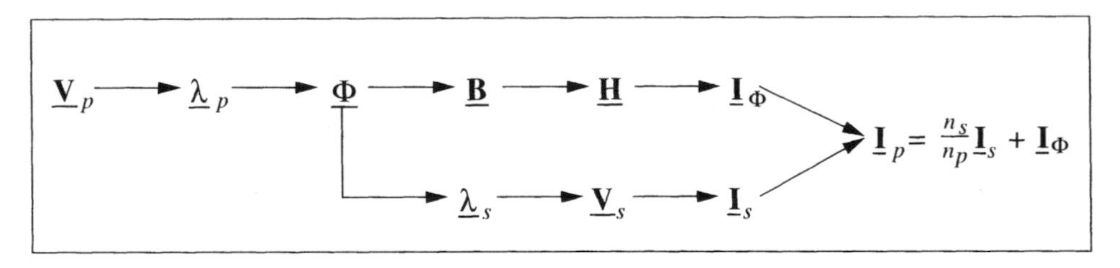

Figure 3.8 Causality diagram for a transformer.

LEARNING OBJECTIVE 3.

To understand how to derive a transformer equivalent circuit from open-circuit/short-circuit measurements and use it to calculate efficiency and percent regulation

Transformer Equivalent Circuits

To model a real transformer with an equivalent circuit, we need to account for the power loss and energy storage of the transformer, as well as the transformation properties represented by an ideal transformer. We showed that Ampère's circuital law introduces an inductor in the equivalent circuit of the transformer. Similarly, the other significant effects in the transformer are represented by circuit elements in the equivalent circuit.

iron losses, eddy currents, hysteresis loss

Iron losses. Iron loss comes from two effects. *Eddy currents* are induced by the changing flux in the iron and generate heat in the resistance of the iron. Eddy-current loss may be reduced but not eliminated by fabricating the magnetic structure out of thin, insulated laminations. As discussed in Chap. 2, page 67, *hysteresis loss* represents the work done in the iron to cyclically reorient the magnetic domains. Iron and eddy-current losses vary approximately as Φ_{max}^2 and hence are proportional to V_p^2, Eq. (3.20). These *iron losses*, therefore, may be modeled by a resistance in parallel with the primary voltage.

Copper losses. The losses in the resistance of primary and secondary windings are called *copper losses.*[8] These losses are modeled in the equivalent circuit by resistances, R_p and R_s, in series with the primary and secondary of the ideal transformer.

Leakage flux. Some flux escapes the iron path and thus fails to couple primary and secondary, as shown in Fig. 3.9. The energy associated with this flux is modeled by inductors, L_p and L_s, in series with the primary and secondary resistances. In the ac equivalent circuit, these inductances become primary and secondary leakage reactances, X_p and X_s.

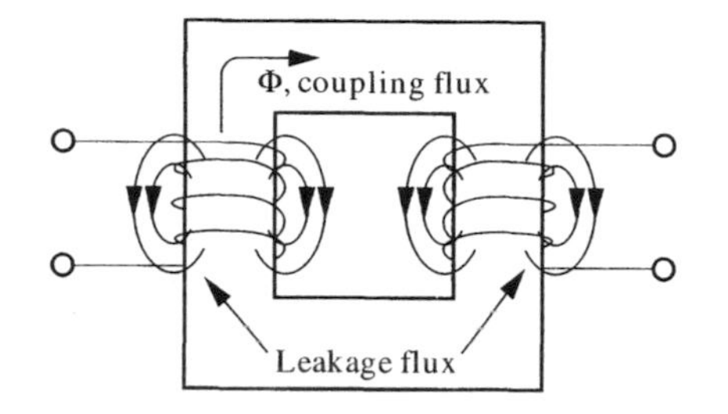

Figure 3.9 Some leakage flux escapes the iron path.

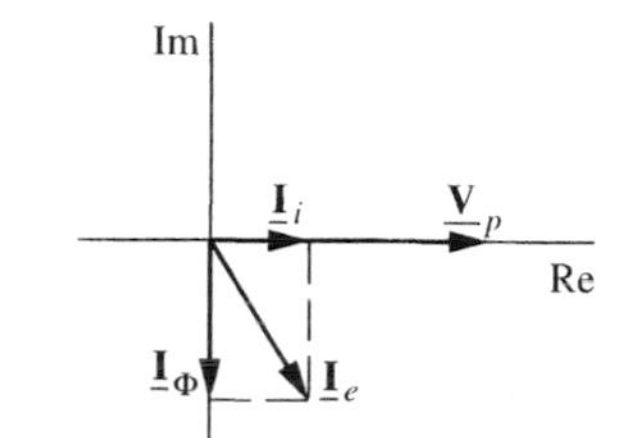

Figure 3.10 The exciting current consists of an in-phase component to supply iron losses and an out-of-phase component to supply energy stored in the magnetic structure.

exciting current

Exciting current. Figure 3.11 shows the *exciting current,* $\underline{\mathbf{I}}_e$, required to magnetize the iron, which is detailed in Fig. 3.10. The in-phase component, I_i, supplies the iron losses, and the out-of-phase magnetizing current, I_Φ, supplies the magnetic energy in the iron. If the secondary were an open circuit, the current in the primary would be the exciting current and the power into the primary would be the iron loss.

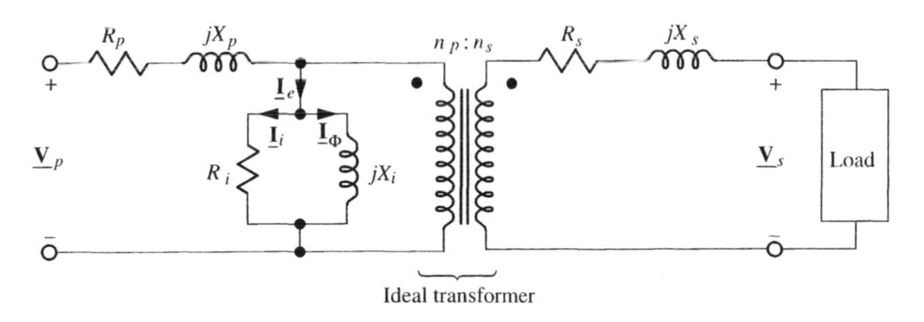

Figure 3.11 The resistors account for losses. The inductors account for stored magnetic energy. The series elements model losses and energy storage proportional to current squared, and the parallel components model losses and energy storage proportional to voltage squared.

[8] Although occasionally aluminum and other metals are now used for wires in transformers and other electrical apparatus, originally only copper was used. For this reason, resistive losses are called "copper losses," regardless of the metal.

Equivalent Circuits

series winding impedance

Equivalent circuits. We have now considered the significant loss and energy-storage effects in a real transformer. An equivalent circuit that accounts for these various effects in a real transformer is given in Fig. 3.11. Figure 3.12 transforms the secondary series impedance to the primary, where it is combined with the primary series impedance to give the *series winding impedance*, $R_w + jX_w$. In normal operation, the exciting current is small compared with the total input current, so we may move R_i and X_i to either position shown in Fig. 3.11 without significant change.

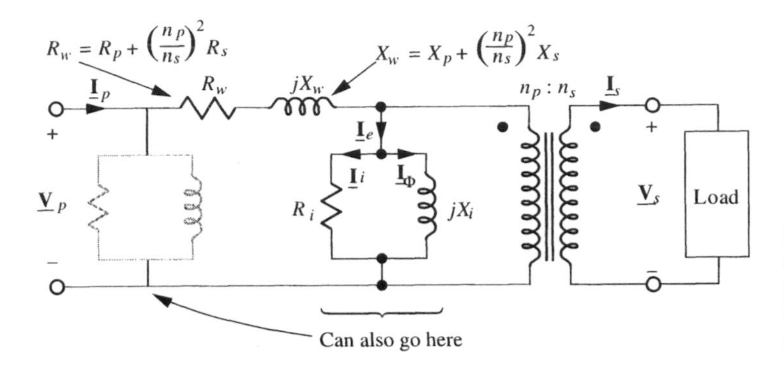

Figure 3.12 Combining the series elements causes insignificant error. The parallel elements can be put on either side of the series elements with negligible change.

Use of the equivalent circuit. Circuit models of transformers are used to predict power and voltage loss in the transformer under various operating conditions. Such models are used in analysis of power systems to determine load flows, fault conditions and system performance generally. We give examples after showing how to determine the series and parallel elements in the equivalent circuit.

Transformer Open-Circuit (OC)/Short-Circuit (SC) Test

Transformer nameplate. A transformer nameplate normally gives the nominal primary and secondary voltages, apparent power rating for output, and the series impedance, $|R_w + jX_w|$ in Fig. 3.12 in per-unit. The ratio of primary and secondary voltage gives the transformer turns ratio, but the other parameters in the equivalent circuit model must be measured.

EXAMPLE 3.4 Transformer ratings

A 10-kVA, 60-Hz, single-phase transformer is rated 2400/240 V. This transformer is used with the high-voltage side as the primary. Find the turns ratio and the nominal current ratings on the primary and secondary.

SOLUTION:

The turns ratio is derived from the nameplate voltages

$$\frac{2400}{n_p} = \frac{240}{n_s} \Rightarrow \frac{n_p}{n_s} = \frac{10}{1} \tag{3.26}$$

The nominal current ratings follow from the voltage and apparent power rating:

$$I_s = \frac{\text{apparent power}}{V_s} = \frac{10,000 \text{ VA}}{240 \text{ V}} = 41.7 \text{ A} \tag{3.27}$$

and similarly for the primary, $I_p = 4.17$ A.

WHAT IF?

What if the transformer is excited at 2400 V on the primary, but the actual load is 2000 VA with $PF = 0.87$, lagging? Find the nominal primary current.[9]

Open-circuit (OC) test. The exciting current and its associated parameters, R_i and X_i, are determined by an open-circuit measurement. For safety, the transformer is excited at rated voltage on the low-voltage side with the high-voltage side open.[10] The voltage, V_s, exciting-current magnitude, I'_e, and power, P'_i, into the low-voltage winding allows determination of R'_i and X'_i, where the primed quantities are referred to the low-voltage side of the transformer, or secondary. The power indicated in the open-circuit measurement is iron loss only because currents and thus copper losses are very small. The voltage must be at rated value so that the flux levels are normal and the iron loss is that associated with normal usage.

Determination of circuit parameters. The in-phase, I'_i, and out-of-phase I'_Φ, components of the exciting current can be determined as follows

$$I' = \frac{P'_i}{V_s} \quad \text{and} \quad I'_\Phi = \sqrt{(I'_e)^2 - (I'_i)^2} \tag{3.28}$$

These lead directly to the circuit parameters

$$R'_i = \frac{V_s}{I'_i} \quad \text{and} \quad X'_i = \frac{V_s}{I'_\Phi} \tag{3.29}$$

EXAMPLE 3.5 **OC test**

The 10-kVA, 2400/240-V, 60-Hz, single-phase transformer in the previous example has an OC test excited on the 240-V side. The measured exciting current is 1.2 A and the measured power is 170 W. Find R'_i and X'_i.

SOLUTION:

From Eqs. (3.28)

$$I'_i = \frac{170}{240} = 0.708 \text{ A} \quad \text{and} \quad I'_\Phi = \sqrt{(1.2)^2 - (0.708)^2} = 0.969 \text{ A} \tag{3.30}$$

[9] 0.833 A. The power factor does not matter since apparent power is given.

[10] The primary may be the low- or high-voltage side of the transformer, depending on the application. In this analysis, the primary is the high-voltage side.

Thus, from Eqs. (3.29)

$$R_i' = \frac{240}{0.708} = 339\ \Omega \qquad \text{and} \qquad X_i' = \frac{240}{0.969} = 248\ \Omega \tag{3.31}$$

WHAT IF? What if you transform the parameters to the primary?[11]

Short-circuit (SC) test. The combined winding resistance, R_w, and leakage reactance, X_w, in Fig. 3.12 are determined by shorting the low-voltage winding and exciting the high-voltage side at reduced voltage to produce the rated primary current, I_p. The primary voltage, V_{sc}, current, I_{sc}, and power, P_c, are measured; and R_w and X_w are determined as follows:

$$I_{sc}^2 R_w = P_c \Rightarrow R_w = \frac{P_c}{I_{sc}^2} \tag{3.32}$$

and

$$\frac{V_{sc}}{I_{sc}} = \sqrt{R_w^2 + X_w^2} \Rightarrow X_w = \sqrt{(V_{sc}/I_{sc})^2 - R_w^2} \tag{3.33}$$

Performing the SC test at rated current ensures that the measured power is the copper loss of the transformer under nameplate conditions because the flux levels and hence iron losses are extremely low at the reduced voltage.

EXAMPLE 3.6 **SC test**

The 10-kVA, 2400/240-V, 60-Hz, single-phase transformer in the previous example has the 240-V winding shorted. It is found that 72.0 V on the 2400-V winding produces the rated current of 4.17 A with an input power of 162 W. Find R_w and X_w.

SOLUTION:

Using Eq. (3.32)

$$R_w = \frac{162}{(4.17)^2} = 9.32\ \Omega \tag{3.34}$$

and using Eq. (3.33)

$$X_w = \sqrt{(72.0/4.17)^2 - (9.32)^2} = 14.5\ \Omega \tag{3.35}$$

Equivalent Circuits

Summary. The OC/SC test determines the series and parallel components in the transformer model. Voltage, current, and power measurements are made. Figure 3.13 shows the equivalent circuit derived in the previous three examples.

[11] The (turns ratio)^2 is $(10)^2$, so 33.9 kΩ and 24.8 kΩ.

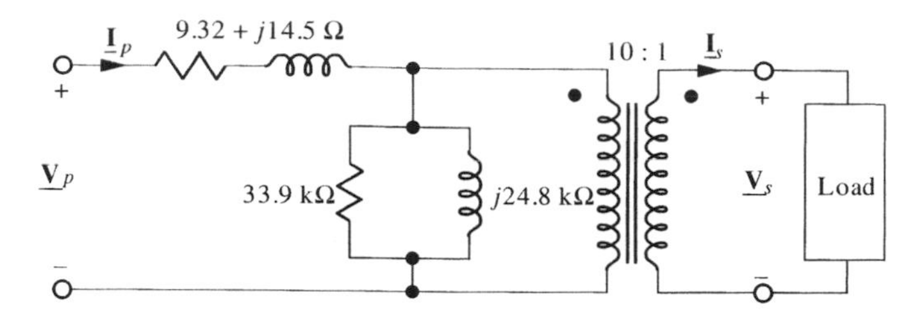

Figure 3.13 Model for 10-kVA, 2400/240-V transformer derived from the OC/SC tests.

Transformer calculations. The equivalent circuit model of a transformer permits the calculation of transformer efficiency and voltage regulation. Exact calculations may be made based upon load power and power factor, *PF*, requirements, but approximate calculations are usually adequate.

Efficiency. The efficiency of the transformer is based on the power passing through the transformer and its losses.

$$\eta = \frac{P_{out}}{P_{in}} = \frac{P_{out}}{P_{out} + P_c + P_i} \tag{3.36}$$

where η is the efficiency, usually given in percent. The output power is determined by the load requirements, and the transformer copper loss, P_c, and iron loss, P_i, can be derived from the equivalent circuit.

EXAMPLE 3.7 Efficiency calculation

The 10-kVA, 2400/240-V, 60-Hz, single-phase transformer in the previous examples, which is modeled by the equivalent circuit in Fig. 3.13, provides 8 kVA at 0.94 *PF* (lagging) and 230 V to a load. Find the transformer efficiency under these conditions.

SOLUTION:

The output power is $P_{out} = 8000 \times 0.94 = 7520$ W. The iron losses may be estimated from the nominal primary voltage and the parallel resistor in the equivalent circuit, $R_i = 33.9$ kΩ. The nominal primary voltage is derived from the secondary voltage and the turns ratio

$$V_p = \frac{10}{1} \times 230 = 2300 \text{ V} \tag{3.37}$$

Hence, the iron loss is

$$P_i = \frac{(2300)^2}{33,900} = 156 \text{ W} \tag{3.38}$$

To estimate the copper losses, we need the nominal primary current

$$I_s = \frac{8,000 \text{ VA}}{230 \text{ V}} = 34.8 \text{ A} \Rightarrow I_p = \frac{n_s}{n_p} \times I_s = \frac{1}{10} \times 34.8 = 3.48 \text{ A} \tag{3.39}$$

to find the power in the series resistance, $R_w = 9.32$ Ω:

$$P_c = (3.48)^2 \times 9.32 = 113 \text{ W} \tag{3.40}$$

We use nominal and not exact primary voltage and current in these approximate calculations. From Eq. (3.36), we determine the approximate transformer efficiency to be

$$\eta = \frac{P_{out}}{P_{out} + P_c + P_i} = \frac{7520}{7520 + 113 + 156} = 0.965(96.5\%) \tag{3.41}$$

WHAT IF?

What if you calculate the exact efficiency based on a full analysis of the equivalent circuit in Fig. 3.13?[12]

Voltage regulation. The percent regulation is defined as

$$\%\ \text{Reg} = \frac{V_s\ (\text{no load}) - V_s\ (\text{loaded})}{V_s\ (\text{loaded})} \times 100\% \tag{3.42}$$

where V_s (no load) is the magnitude of the secondary voltage with no load and V_s (loaded) is the magnitude of the secondary voltage under given load conditions. If the primary voltage is constant, the no-load secondary voltage will be approximately

$$V_s\ (\text{no load}) \approx \frac{n_s}{n_p} \times V_p \tag{3.43}$$

Thus, we need to use the equivalent circuit in Fig. 3.12 to calculate the primary voltage required to supply the specified secondary voltage and power.

Let $\underline{\mathbf{V}}_s$ and $\underline{\mathbf{I}}_s$ be the required phasor secondary voltage and current. Then KVL in the primary in Fig. 3.12 is

$$\underline{\mathbf{V}}_p = \frac{n_p}{n_s} \times \underline{\mathbf{V}}_s + \left(\frac{n_s}{n_p} \times \underline{\mathbf{I}}_s\right) \times (R_w + jX_w) \tag{3.44}$$

where $\underline{\mathbf{V}}_p$ is the actual (not nominal) primary voltage and we have again ignored the exciting current, $\underline{\mathbf{I}}_e$.

EXAMPLE 3.8 Percent regulation

The 10-kVA, 2400/240-V, 60-Hz, single-phase transformer in the previous example, which is modeled by the equivalent circuit in Fig. 3.13, provides 8 kVA at 0.94 *PF* (lagging) and 230 V to a load. Find the percent regulation under these conditions.

SOLUTION:

For 8000 VA at 230 V with 0.94 *PF* lagging, the current is

[12] Then you are in for a lot of unnecessary work. The answer is still 96.5% to three-place precision.

$$\underline{\mathbf{I}}_s = \frac{8,000}{230} \angle -\cos^{-1}(0.94) = 34.8 \angle -20.0^\circ \text{ A} \quad (3.45)$$

Using Eq. (3.44), we find the primary voltage to be

$$\underline{\mathbf{V}}_p = \frac{10}{1} \times 230 \angle 0^\circ + \left(\frac{34.8}{10} \angle -20.0^\circ\right) \times (9.32 + j14.5) = 2348 \angle 0.887^\circ \text{ V} \quad (3.46)$$

The no-load secondary voltage is thus 2348/10 = 234.8 V. Thus, Eq. (3.42) gives the percent regulation

$$\%\text{Reg} = \frac{234.8 - 230}{230} \times 100 = 2.09\% \quad (3.47)$$

Why Voltage and Apparent Power (kVA) Rating Are on the Nameplate

Voltage rating. Transformers are rated for a specific voltage and apparent power, kVA. The magnetic structure is designed for a specific maximum flux, which by Faraday's law implies a specific voltage rating because the frequency and number of turns are constant. To operate the transformer at a voltage considerably higher than the design value would give excessive losses and invite eventual failure. To operate at a lower voltage would underutilize the transformer.

Apparent power rating. Transformers are rated for a specific apparent power because operating voltage and current determine transformer losses, which heat the transformer winding. Excessive operating temperature of the transformer windings deteriorates wire insulation and hence reduces reliability and transformer lifetime. Thermal design, environment, and transformer loss combine to establish the apparent power limits of the transformer.

In addition to iron loss, we have copper losses in the primary and secondary windings; hence, the maximum current is limited likewise by the amount of heat the transformer can dissipate. Thus, the transformer is rated according to the product of rated voltage and rated current, or apparent power.

We may distinguish two apparent power limits:

1. The rated apparent power is the operating level that may be sustained under a worst-case thermal environment. For example, a small pole-hung transformer would be rated for a hot summer day, in full sun and with no wind. These would be the most severe conditions expected, and the transformer would be expected to perform satisfactorily under such conditions.
2. The actual limit for the transformer would depend on the operating conditions. For example, a transformer rated for conditions in Texas could be operated at higher levels of apparent power in Alaska. In a given application, the operating level of the transformer depends on the load and may be safely below the rated or actual limits of the transformer.

When transformers have multiple primary and/or secondary windings, the kVA rating is independent of the way in which the transformer is connected, as shown by the following example.

EXAMPLE 3.9

Transformer connections

Consider a 480:240/240:120 transformer with a 8.8-kVA rating. The voltage rating reveals that both primary and secondary have two identical windings, which may be connected in series or parallel. Find the individual winding voltages and currents for both connections at nameplate conditions.

SOLUTION:

Figure 3.14(a) shows the windings in series. The primary voltage is 480 V, the series connection of two 240-V windings, and the secondary voltage is 240 V. From the kVA rating, we calculate the primary current as 18.3 A and the secondary current as 36.7 A as shown.

Figure 3.14(b) shows the parallel connection of both windings. From the kVA calculation, we now have twice the current in primary and secondary, but with parallel connections, this current is divided between two windings to give the rated current in each winding, as shown.

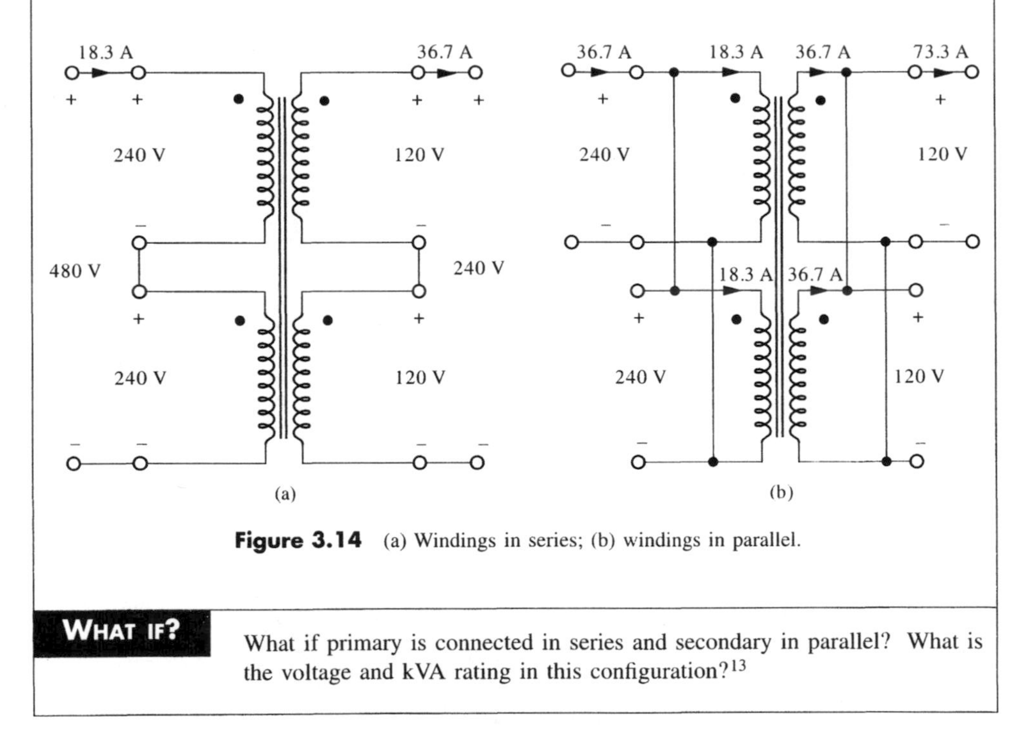

Figure 3.14 (a) Windings in series; (b) windings in parallel.

WHAT IF?

What if primary is connected in series and secondary in parallel? What is the voltage and kVA rating in this configuration?[13]

[13] The voltage ratio is 480/120 V and the kVA rating is unchanged.

Summary. The magnetic structure in the previous example operates at the same flux level in both connections, and the currents in the individual windings are unchanged; hence, the losses for the two cases are identical. Similar results are obtained for three-phase transformer connections. Consequently, the transformer is rated according to its apparent power; the power factor and hence the real power flowing through the transformer are not directly a limiting factor.

Models of Three-phase Transformers

Three-phase transformers. Three-phase systems constitute the overwhelming majority of all power generation and distribution systems. Chapter 1 discussed the use of transformers in such systems. Transformation of three-phase power from one voltage level to another may be accomplished by three single-phase transformers or by one three-phase transformer. Figure 3.15 shows one possible configuration for a three-phase transformer.

A three-phase transformer is cheaper, smaller, and more efficient than three single-phase transformers, but the latter configuration is more versatile. For example, should a single transformer fail, only one transformer would have to be replaced, and in certain cases, the system could continue operation at reduced load until the replacement arrived.

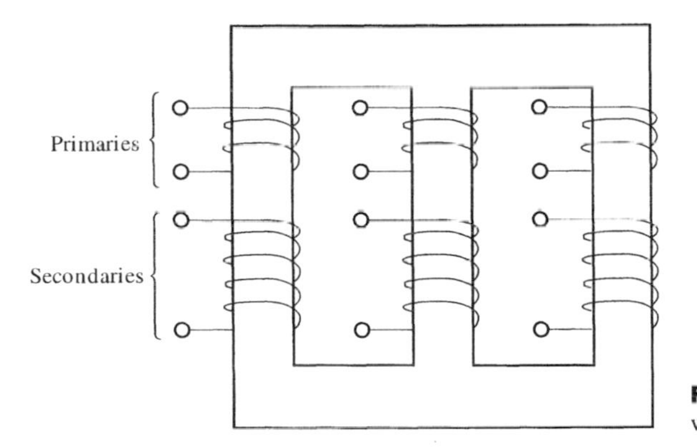

Figure 3.15 A three-phase transformer wound on a common magnetic structure.

Per-phase equivalent circuit model for three-phase transformers. Three-phase transformers, whether realized in a single unit or with three single-phase transformers, may be modeled by a per-phase equivalent circuit[14] identical to that shown in Fig. 3.12. In this case, all power quantities represent one-third the actual quantities, and voltages are reduced by $\sqrt{3}$ from three-phase values. The per-phase circuit represents transformer loss and energy storage and does not depend on whether windings are connected in wye or delta.

EXAMPLE 3.10 **Per-phase calculations**

A 2400/480-V, 60-kVA, 60-Hz, three-phase transformer is modeled by the equivalent circuit shown in Fig. 3.16. Find the efficiency of the transformer at nameplate kVA with 0.9 *PF*, lagging.

[14] Per-phase equivalent circuits are explained on page 19.

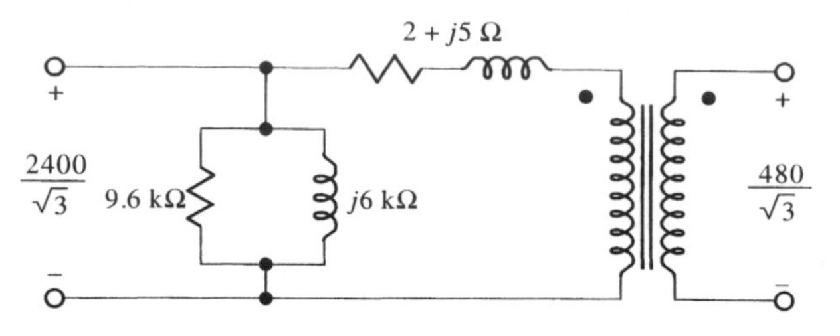

Figure 3.16 Per-phase model for a three-phase transformer.

SOLUTION:
Efficiency is given by Eq. (3.36), which may be applied to the entire transformer or to its per-phase representation. Choosing the former, we find the output power to be

$$P_{out} = S \times PF = 60{,}000 \times 0.9 = 54{,}000 \text{ W} \tag{3.48}$$

The iron loss per phase are represented by the parallel resistance in Fig. 3.16. Thus, the iron loss in the transformer is

$$P_i = 3 \times \frac{(2400/\sqrt{3})^2}{9600} = 600 \text{ W} \tag{3.49}$$

To find the copper loss, we must find the nominal current in the transformer primary. Equation (1.20) gives

$$I_p = \frac{S}{\sqrt{3}V} = \frac{60{,}000 \text{ VA}}{\sqrt{3} \times 2400 \text{ V}} = 14.4 \text{ A} \tag{3.50}$$

Because per-phase current levels are the same as the actual transformer, the nominal current in the 2-Ω series resistor in Fig. 3.16 is 14.4 A; hence, the copper loss in the three-phase transformer is

$$P_c = 3 \times (14.4)^2 \times 2\ \Omega = 1250 \text{ V} \tag{3.51}$$

As before, we multiply by 3 because the 2 Ω resistor is a per-phase resistance. Using Eq. (3.36), we find the efficiency to be

$$\eta = \frac{P_{out}}{P_{out} + P_c + P_i} = \frac{54{,}000}{54{,}000 + 1250 + 600} = 96.7\% \tag{3.52}$$

Check Your Understanding

1. Some basic electrical laws are Coulomb's law, Ampère's circuital law, Faraday's law, and/or Ohm's law. Which two describe transformer action?
2. Many turns on a winding go with high or low voltage on a transformer. Which ones?
3. A transformer has 500 turns on the side connected to the power source and 25 turns on the side connected to the load. If the load requires 120 V, what should be the voltage of the power source?

4. A single-phase transformer has 300 turns on the side connected to the source of power and 27 turns on the side connected to the load. If the load requires 100 A, what would be the current capacity on the other side?
5. In the transformer circuit shown in Fig. 3.17, what do the following measure?
   (a) An ammeter in the secondary.
   (b) A voltmeter in the primary.
   (c) A voltmeter in the secondary.
   (d) A wattmeter in the primary.
   (e) A wattmeter in the secondary.

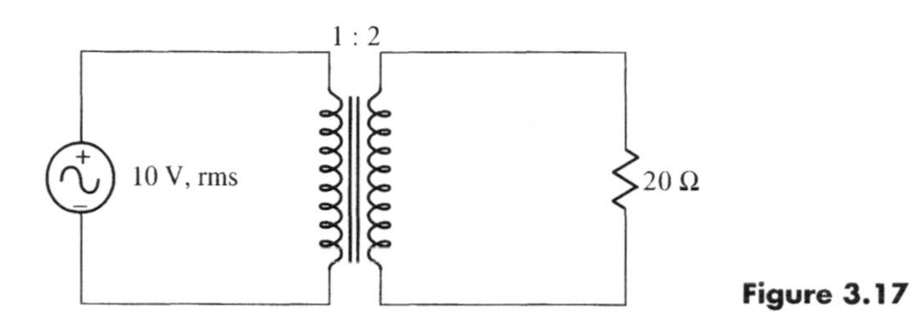

Figure 3.17

6. A 560-V rms, 20-kVA, three-phase transformer operates fully loaded with a power factor of 0.85, lagging.
   (a) What is the per-phase voltage?
   (b) What is the per-phase current?

Answers. **(1)** Ampère's circuital law and Faraday's law; **(2)** high voltage; **(3)** 2400 V; **(4)** 9.0 A; **(5) (a)** 1 A, **(b)** 10 V, **(c)** 20 V, **(d)** and **(e)** 20 W; **(6) (a)** 323 V, **(b)** 20.6 A.

3.3 FORCES IN MAGNETIC SYSTEMS

Figure 3.18 shows a magnetically driven mechanical actuator, which might operate a lock or ring a bell. The magnetic structure is iron, with a hinged member. The electrical input is produced by a coil with n turns. Our goal in this section is to develop means for determining the magnitude and direction of the force produced by this electromechanical system.

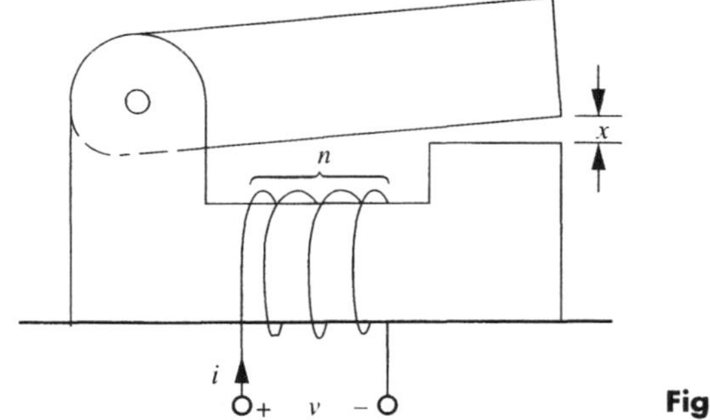

Figure 3.18 A magnetic actuator.

Magnetic-Pole Approach

Everyone has handled permanent magnets and knows about the magnetic compass. The properties of such can be described in terms of magnetic poles. We can use the concept of magnetic poles to build an intuitive understanding of the force in our actuator and to determine the direction of the force.

Relationship between magnetic poles and flux. We already discussed magnetic poles and terrestrial magnetism on page 65 in defining the direction of a magnetic field. In speaking of a magnetic pole, we assume that there is a physical body, a piece of iron or perhaps the Earth, as shown in Fig. 3.19(a) with magnetic flux leaving or entering its surface. If magnetic flux is leaving the surface, as at the south *geographic* pole, we attribute this flux to the presence of a north *magnetic* pole within the surface, which is acting as a source for the flux. Alternatively, if flux lines are entering the surface, as at the north *geographic* pole, we attribute this flux to the presence of a south *magnetic* pole within the surface.

Because opposite poles attract and like poles repel, we can use the pole concept to determine the direction of forces and torques. Consider again the device in Fig. 3.18. If i is positive, then by the right-hand rule the direction of the magnetic flux is as shown in Fig. 3.19(b). The top member thus contains a north magnetic pole because *from the viewpoint of the gap* it produces the magnetic flux. Likewise, the bottom of the gap contains a south magnetic pole because the flux enters the bottom surface. We conclude that a force of attraction exists between the north and south magnetic poles, tending to close the gap. It is easily shown that the force is independent of the direction of the current.

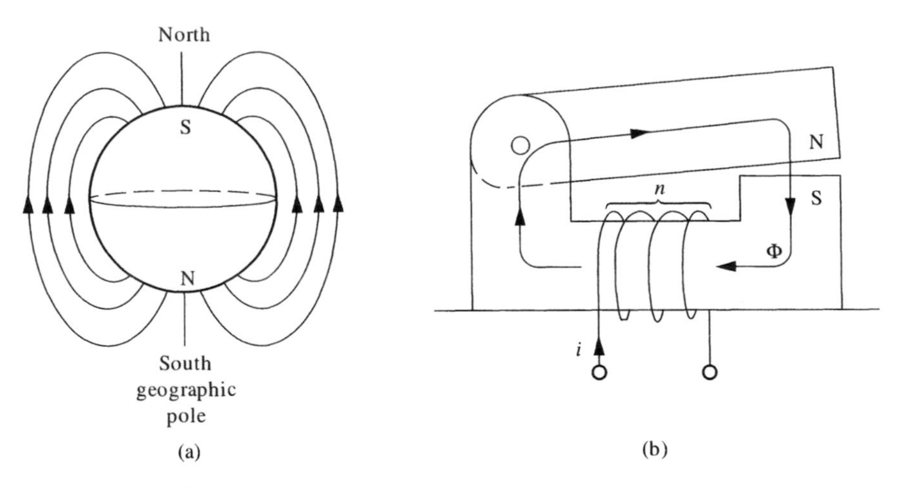

Figure 3.19 (a) The Earth's magnetic poles and flux; (b) the actuator with magnetic poles shown.

Conclusion. This viewpoint gives us a qualitative understanding of the magnetic forces in the system but no quantitative information. Soon we will develop a method, based on conservation of energy, that gives both qualitative and quantitative information, albeit at the cost of some abstraction.

Analysis from Current–Flux Interaction

In Eq. (2.18), we presented Ampère's force law for the force per meter on a current-carrying conductor in a magnetic flux

$$\vec{f} = \vec{i} \times \vec{B} \text{ N/m} \tag{3.53}$$

This expression may be used to determine the force on a system of current-carrying conductors, as in a motor, if the magnetic flux density and currents are known. However, magnetic structures such as shown in Fig. 3.19(b) do not permit force calculation by Eq. (3.53) because the force is produced by magnetic dipoles in the magnetic material.

Conservation of Energy

Analysis from Energy Considerations

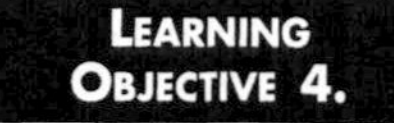

To understand how to calculate the magnetic force in a magnetic structure

Modeling an electromechanical transducer. The basic approach of this section is suggested in Fig. 3.20. Here we model the actuator with an electrical input, and we include resistive losses as part of the electrical part of the system. The electrical circuit terminates into a box labeled "ideal transducer." The ideal transducer is assumed to be lossless. Thus, it stores magnetic energy, which can be returned to the electrical circuit, and converts the remainder of the input electrical energy to mechanical energy. The output of the transducer is a developed force and a mechanical displacement. We show the developed mechanical energy going into another box representing mechanical losses, and output energy; but this box does not concern us here. We consider only the conversion of electrical into mechanical energy in the middle box and, from the conservation of energy, determine the developed mechanical force.

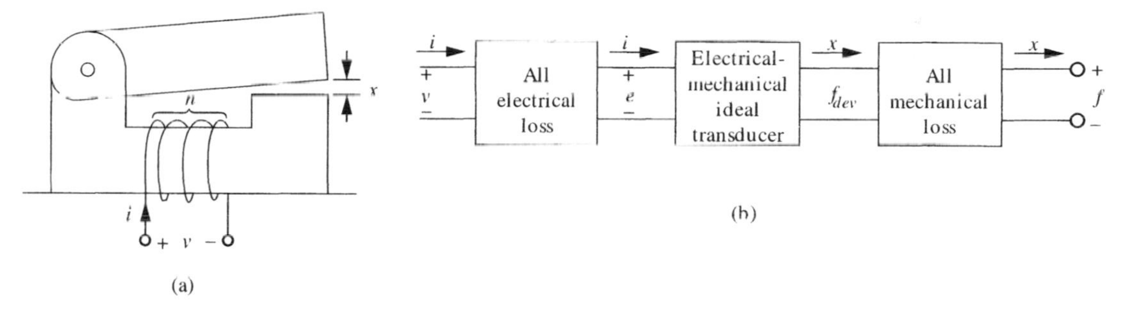

Figure 3.20 (a) A magnetic transducer; (b) a model for energy analysis.

Equivalent Circuits

Electrical model. Figure 3.21 models the actuator electrically. We separate the copper loss in the resistance from the magnetic energy storage in the inductance. The emf, e, produced by the changing inductance flux linkages, λ, represents electrical energy entering the magnetic system.

Figure 3.21 Electrical model for the actuator.

Analysis. Our assumptions and the conservation of energy require that the input electrical energy into the ideal transducer is either stored or converted to mechanical energy. Equation (3.54) expresses conservation of energy on an incremental scale

$$\underbrace{ei\,dt}_{\substack{\text{electrical}\\ \text{energy}\\ \text{in}}} = \underbrace{dW_m(\lambda,x)}_{\substack{\text{increase}\\ \text{in stored}\\ \text{energy}}} + \underbrace{f_{dev}(\lambda,x)\,dx}_{\substack{\text{mechanical}\\ \text{energy out}}} \tag{3.54}$$

where e is the emf, $dW_m(\lambda,x)$ is the stored magnetic energy as a function of flux linkage and mechanical displacement, and f_{dev} is the magnetic force developed in the mechanical system. Faraday's law requires

Conservation of Energy

$$e\,dt = \frac{d\lambda}{dt}dt = d\lambda \tag{3.55}$$

Hence Eqs. (3.54) and (3.55) combine to

$$i\,d\lambda = dW_m(\lambda, x) + f_{dev}(\lambda, x)dx \tag{3.56}$$

Subject to certain mathematical assumptions to be discussed in what follows, the incremental stored energy change must satisfy the chain rule of calculus

$$dW_m(\lambda, x) = \underbrace{\frac{\partial W_m(\lambda, x)}{\partial \lambda}}_{x = \text{constant}} d\lambda + \underbrace{\frac{\partial W_m(\lambda, x)}{\partial x}}_{\lambda = \text{constant}} dx \tag{3.57}$$

But Eq. (3.56) can be rearranged to the form

$$dW_m(\lambda, x) = i(\lambda, x)d\lambda - f_{dev}(\lambda, x)dx \tag{3.58}$$

Because Eqs. (3.57) and (3.58) are valid for arbitrary $d\lambda$ and dx, the coefficients of $d\lambda$ and dx must be equal

$$i(\lambda, x) = \underbrace{\frac{\partial W_m(\lambda, x)}{\partial \lambda}}_{x = \text{constant}} \qquad \text{and} \qquad f_{dev}(\lambda, x) = -\underbrace{\frac{\partial W_m(\lambda, x)}{\partial x}}_{\lambda = \text{constant}} \tag{3.59}$$

The second equation in Eq. (3.59) gives the developed magnetic force as a function of the stored energy. The first equation in Fig. 3.59 give a means for calculating the stored magnetic energy.

Stored-energy-state function. We can determine the stored magnetic-energy-state function, $W_m(\lambda, x)$, by integrating Eq. (3.56), provided we keep x constant ($dx = 0$). Thus, our strategy is to start with a system with no energy storage, move x to the position at which we want to know the force (say, $x = 1$ mm), and then integrate the electrical energy input as we energize the system electrically. This determines the magnetic stored energy as

$$W_m(\lambda, x) = \int_0^{\lambda} \underbrace{i(\lambda', x)}_{x = \text{fixed}} d\lambda' \tag{3.60}$$

where λ' is a dummy variable for integration.

EXAMPLE 3.11 Stored energy

Determine the state function for the stored magnetic energy of the system shown in Fig. 3.20(a).

SOLUTION:

Consider x as a constant throughout the following development. The total flux is given by Eq. (3.12)

$$\Phi = \frac{ni}{\mathfrak{R}_i + \mathfrak{R}_g(x)} \text{ Wb} \tag{3.61}$$

where the reluctances of iron and gap are

$$\Re_i = \frac{\ell_i}{\mu_i A_i} \qquad \text{and} \qquad \Re_g(x) = \frac{x}{\mu_0 A_g} \tag{3.62}$$

Note that the mechanical displacement enters the analysis through its effect on the reluctance of the gap. The flux linkage is derived from Eq. (3.9)

$$\lambda = n\Phi = \frac{n^2 i}{\Re_i + \Re_g(x)} \text{ Wb-t} \tag{3.63}$$

Because Eq. (3.60) requires the current as a function of flux linkage, we rearrange Eq. (3.63) to the form

$$i(\lambda, x) = \frac{1}{n^2}[\Re_i + \Re_g(x)]\lambda \tag{3.64}$$

and integrate:

$$\begin{aligned} W_m(\lambda, x) &= \int_0^\lambda i(\lambda', x)d\lambda' = \frac{1}{n^2}[\Re_i + \Re_g(x)]\int_0^\lambda \lambda' d\lambda' \\ &= \frac{1}{2n^2}[\Re_i + \Re_g(x)]\lambda^2 \end{aligned} \tag{3.65}$$

Force determination. Equation (3.60) gives the energy-state function required for calculating the force. To compute the developed force, Eq. (3.59) requires that we take the partial derivative of $W_m(\lambda, x)$ with respect to x, with λ held constant.

$$f_{dev}(\lambda, x) = -\left.\frac{\partial W_m(\lambda, x)}{\partial x}\right|_{\lambda = \text{constant}} \tag{3.66}$$

Equation (3.66) gives the force on the mechanical system as a function of the flux linkage.

EXAMPLE 3.12 Force calculation

Compute the force on the actuator in Fig. 3.20(a).

SOLUTION:

Equation (3.66) requires that we take the partial derivative of $W_m(\lambda, x)$ with respect to x, with λ held constant. Using the results from the previous example, Eq. (3.65), we find

$$f_{dev} = -\frac{\partial}{\partial x}\frac{1}{2n^2}\left[\Re_i + \frac{x}{\mu_0 A g}\right]\lambda^2 = -\frac{\lambda^2}{2n^2\mu_0 A_g} \tag{3.67}$$

Equation (3.67) gives the force on the mechanical system as a function of the flux linkage.

WHAT IF? What if we wish to have the force as a function of the current?[15]

[15] We may reintroduce Eq. (3.63) to yield $f_{dev} = -(ni)^2/[\Re_i + \Re_g(x)]^2 \times (1/2\mu_0 A_g)$.

Torque in rotational systems. Clearly, this theory can be adapted to a rotational system, such as shown in Fig. 3.22. Specifically, $W_m(\lambda, x)$ becomes $W_m(\lambda, \theta)$, and the torque is

$$T_{dev}(\lambda, \theta) = -\left.\frac{\partial W_m(\lambda, \theta)}{\partial \theta}\right|_{\lambda = \text{constant}} \tag{3.68}$$

where $T_{dev}(\lambda, \theta)$ represents the developed torque. In this case, the dependence of the reluctance on θ becomes the important factor.

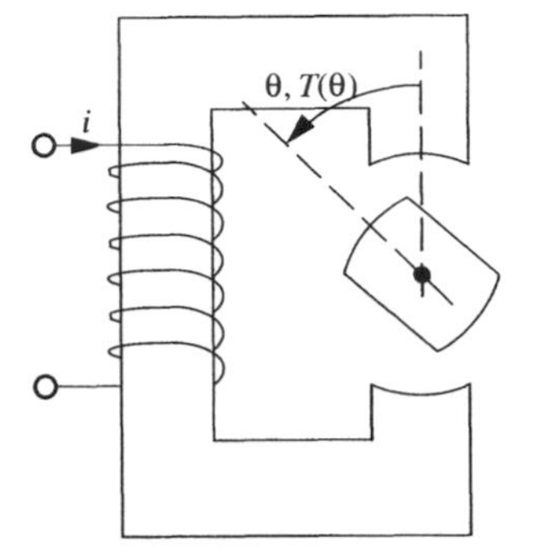

Figure 3.22 A magnetic structure involving rotation.

Discussion of examples. The minus sign means that the force is directed opposite to the direction of increasing x and tends therefore to decrease x to close the gap. This we anticipated from the argument based on magnetic poles. If the flux linkage is held constant,[16] the force is independent of the gap width.

Constant-current excitation. The form of the force with constant-current excitation is given in footnote 15. The force depends on the square of the mmf. To see how the force depends on the width of the gap, we assume that the reluctance of the iron is negligible compared with the reluctance of the gap. Under this assumption, the developed force, f_{dev}, reduces to

$$f_{dev} = -\frac{1}{2}\frac{(ni)^2}{(x/\mu_0 A_g)^2}\frac{1}{\mu_0 A_g} = -\frac{(ni)^2 \mu_0 A_g}{2x^2} \tag{3.69}$$

and hence the force varies as the inverse square of the gap width.

Zero gap? Although Eq. (3.69) leads to infinite forces as x approaches zero, this is unrealistic for two reasons. For one, the gap cannot have zero width because of surface roughness; hence, a minimum effective value of x exists. Also the finite permeability of the iron limits the force. If we were to determine the maximum force, we would have to consider carefully the properties of the iron, both the finite value of permeability and also the saturation effects.

Mathematical and physical assumptions. The assumptions underlying the derivation of Eqs. (3.60) and (3.66) are as follows:

[16] As in Example 3.3.

- The magnetic system must be lossless. Because most magnetic systems do have some loss, this assumption weakens the analysis slightly, meaning that the accuracy of the results is compromised by magnetic losses.
- The stored energy must be zero if the current or flux linkage is zero. This assumption rules out forces due to permanent magnets.[17]
- Only magnetic energy is stored in the system. This means no capacitors or springs can be hidden in the system, and velocities have to be small. This is no problem, because we are dealing with known systems and virtual displacements.
- The stored-energy function must be a single-valued state function. This requires that the stored energy depends only on the final value of λ and x, and not on how the system is energized. Because most magnetic systems exhibit some hysteresis, this assumption also weakens the analysis.
- The requirements that flux linkage be constant corresponds, by Eq. (2.24), to $e = 0$, but this is a requirement of the mathematical form, not the physical excitation. The force depends only upon the state of the excitation; the dx's and $d\lambda$'s are mental, not physical, changes.

Summary. We derived and applied an expression for the force developed in a magnetic system. The formula is based upon the conservation of energy and the mathematical properties of the energy-state function.

Coenergy and Magnetic Force

Changing $\lambda \rightarrow i$. The requirement that energy and force be expressed in terms of flux linkage is inconvenient because flux linkage is difficult to measure and control in a circuit. For this reason, we now develop a modification of the analysis based upon a change of electrical variables and the concept of magnetic coenergy. We change variables beginning with a basic rule of calculus:

$$d(i\lambda) = i d\lambda + \lambda di \quad \Rightarrow \quad i d\lambda = d(i\lambda) - \lambda di \tag{3.70}$$

In the transformation, $W_m(\lambda, x)$ is merely changed to $W_m(i, x)$ symbolically. After some rearrangement, Eq. (3.56), which expresses conservation of energy, becomes

$$d[i\lambda - W_m(i,x)] = \lambda(i,x)\, di + f_{dev}(i,x) dx \tag{3.71}$$

coenergy

We now define a new type of energy function called the *coenergy*, $W'_m(i,x)$ as

$$W'_m(i,x) = i\lambda(i,x) - W_m(i,x) \tag{3.72}$$

and thus Eq. (3.71) becomes

$$dW'_m(i,x) = \lambda\,(i,x) di + f_{dev}(i,x) dx \tag{3.73}$$

The chain rule of calculus requires

$$dW'_m(i,x) = \underbrace{\frac{\partial W'_m(i,x)}{\partial i}}_{x\,=\,\text{constant}} di + \underbrace{\frac{\partial W'_m(i,x)}{\partial x}}_{i\,=\,\text{constant}} dx \tag{3.74}$$

[17] For an analysis that includes permanent magnets, see *Electrical Machines*, 5th ed., A. E. Fitzgerald, Charles Kingsley, Jr., and Stephen D. Umans. New York: McGraw-Hill, pp. 32f.

Equations (3.73) and (3.74), which must be valid for arbitrary di and dx, require

$$\lambda(i,x) = \underbrace{\frac{\partial W'_m(i,x)}{\partial i}}_{x\,=\,\text{constant}} \qquad \text{and} \qquad f_{dev}(i,x) = +\underbrace{\frac{\partial W'_m(i,x)}{\partial x}}_{i\,=\,\text{constant}} \tag{3.75}$$

Force determination. We determine the force by the same procedure as before. First, hold x constant and calculate the coenergy by integrating Eq. (3.73):

$$W'_m(i,x) = \int_0^i \lambda(i',x)di' \tag{3.76}$$

and then calculate the developed force from Eq. (3.75):

$$f_{dev}(i,x) = +\frac{\partial W'_m(i,x)}{\partial x}\bigg|_{i\,=\,\text{constant}} \tag{3.77}$$

EXAMPLE 3.13 Force calculation using coenergy

Find the force in the magnetic structure in Fig. 3.20(a) using coenergy.

SOLUTION:

The flux linkage is given by Eq. (3.63). Thus, the coenergy function is

$$W'_m(i,x) = \int_0^i \lambda(i',x)di' = \frac{n^2i^2}{2[\mathfrak{R}_i + \mathfrak{R}_g(x)]} \tag{3.78}$$

Equation (3.77) gives the force as

$$\begin{aligned} f_{dev} &= +\frac{\partial W'_m(i,x)}{\partial x} = -\frac{(ni)^2}{2[\mathfrak{R}_i + \mathfrak{R}_g(x)]^2}\frac{\partial \mathfrak{R}_g(x)}{\partial x} \\ &= -\frac{(ni)^2}{2\mu_0 A_g[\mathfrak{R}_i + \mathfrak{R}_g(x)]^2} \end{aligned} \tag{3.79}$$

This is the same answer as in footnote 15.

Conservation of Energy

What is coenergy? Mathematically, coenergy is the function calculated by Eq. (3.76). Graphically, coenergy is the area under the λ–i curve, as shown in Fig. 3.23(a). But these answers do not explain coenergy. Let us begin with the question, "What is energy?" The answer is not obvious; certainly, many brilliant minds pondered the physical creation before coming on the concept of energy and postulating its conservation. Our frequent use of the concept of energy has perhaps dulled us to its abstract quality.

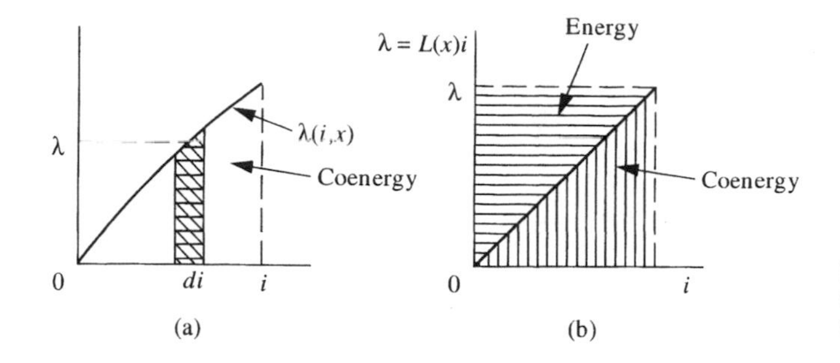

Figure 3.23 (a) Coenergy is the area under the curve; (b) for a linear magnetic system, energy and coenergy are equal.

energy

Whatever we say about energy from a physical point of view, *energy* is a descriptor of a system that has useful conceptual and mathematical properties. Conservation of energy leads to important relationships between variables, sometimes allows shortcuts in an analysis, and usually gives insight into the operation of a physical system. In the present context, for example, conservation of energy allows us to calculate the magnetically generated force on a mechanical system.

Similarly, coenergy is a mathematical function with useful properties. Here coenergy gives us an alternative means to compute the magnetically generated force. But as a basic principle of science, coenergy has minor importance compared with energy.

Energy is the area above the λ–i curve, Fig. 2.17, and coenergy is the area below the curve in Fig. 3.23(a). Energy and coenergy are equal when we have a linear relationship between λ and i, as in Fig. 3.23(b). This is the case that follows, where we apply these ideas to circuit theory.

Circuit Approach

From an electrical point of view, the mechanical actuator pictured in Fig. 3.20(a) is simply an inductor, as shown in Fig. 3.21. The inductance depends on the reluctance of the magnetic structure,[18] which depends on x, the width of the gap. Thus, we may write the flux linkage, $\lambda = L(x)i$, where $L(x)$ is the inductance as a function of the mechanical variable, and the coenergy is thus

$$W'_m(i,x) = \int_0^i L(x)i'\,di' = \tfrac{1}{2}L(x)i^2 \tag{3.80}$$

The force is

$$f_{dev} = \partial W'_m(i,x) = \frac{1}{2}i^2\frac{dL(x)}{dx} \tag{3.81}$$

For an ac current, the time-average force is determined from the time average of the square of the ac current, which is by definition the square of the rms current:

$$\langle f_{dev}\rangle = \frac{1}{2}\langle i(t)^2\rangle\frac{dL(x)}{dt} = \frac{1}{2}I_e^2\frac{dL(x)}{dx} \tag{3.82}$$

where < > denotes the time average and I_e is the effective value of the current.

[18] See Eq. (3.14).

EXAMPLE 3.14 Rotational transducer

The rotational magnetic structure in Fig. 3.22 has n turns and a reluctance that depends on angle as

$$\mathfrak{R}(\theta_m) = \frac{\mathfrak{R}_o}{|\theta_m| + 0.01} \tag{3.83}$$

where $0 < |\theta_m| < 0.2$ radians. The structure is excited by an ac current $i(t) = I_p \cos(\omega t)$. Find the time-average torque at $\theta_m = 0.1$ radian.

SOLUTION:
The inductance for $\theta_m > 0$ is

$$L(\theta_m) = \frac{n^2}{\mathfrak{R}(\theta_m)} = \frac{n^2}{\mathfrak{R}_o}(\theta_m + 0.01) \tag{3.84}$$

Adapting Eq. (3.82) for rotational motion,

$$\langle T_{dev} \rangle = \frac{1}{2}\left(\frac{I_p}{\sqrt{2}}\right)^2 \frac{d}{d\theta_m}\left[\frac{n^2}{\mathfrak{R}_o}(\theta_m + 0.01)\right] = \frac{n^2 I_p^2}{4\mathfrak{R}_o} \tag{3.85}$$

Thus, the torque is independent of angle for $0 < |\theta_m| < 0.2$ radian.

WHAT IF? What if $\theta_m = -0.1$ radian?[19]

Where is the magnetic energy? As Eq. (3.65) implies, the total magnetic energy divides between iron and the gap in proportion to the reluctances. Because the relative permeability for iron is so large, most of the magnetic energy is stored in the air gap, even though the volume of the gap is relatively small.

Check Your Understanding

1. In a magnetic system in which the reluctance of the magnetic structure decreases with increasing x, does the magnetic force tend to increase or decrease x?
2. The sum of the energy and coenergy of a magnetic system is constant if the current into the system is constant, even if a mechanical part of the magnetic structure is moved. True or false?

Answers. **(1)** Increase; **(2)** false.

CHAPTER SUMMARY

In this chapter, we apply the physical laws from Chap. 2 to magnetic structures. We analyze the electrical transformer and present equivalent circuit models based on these laws, supplemented with reasoning based on energy considerations. We derive expres-

[19] $\langle T_{dev} \rangle = -n^2 I_p^2 / 4\mathfrak{R}_o$.

sions for magnetically generated forces and torques in magnetic structures that have movable members.

The energy in a magnetic system is stored in the space comprising that system, primarily in the gaps. We visualize the magnetic fields as exerting a stress on the material and the space, and hence being the vehicle for force and energy storage. This analysis is applied in the next chapter, where we determine the torque developed in a cylindrical magnetic structure.

Objective 1: To understand the function and analysis of magnetic structures. The function of a magnetic structure is to increase the amount of magnetic flux and to direct its path through space. The analysis of a magnetic structure is based on Ampère's circuital law and leads to the definition of reluctance as the ratio between magnetomotive force and the flux in the system. The inductance is determined by the reluctance and the number of turns in the coil driving the system.

Objective 2: To understand the physical basis for transformer models. A transformer consists of two or more coils tightly coupled by a magnetic structure. The voltage ratio of an ideal transformer is derived from Faraday's law, and the current ratio is derived from Ampère's circuital law. Real transformers have stray and magnetizing flux, iron losses, and winding resistance.

Objective 3: To understand how to derive a transformer equivalent circuit from open-circuit/short-circuit measurements and use it to calculate the efficiency and percent regulation. A transformer equivalent circuit model includes, in addition to an ideal transformer, resistors to represent iron and copper loss and inductors to represent stray and magnetizing flux. The circuit elements in the equivalent circuit can be determined by voltage, current, and power measurements on the transformer with the output open and, with reduced voltage, with the output shorted.

Objective 4: To understand how to calculate the magnetic force in a magnetic structure. Among the various methods for calculating forces in magnetic structures, a method based on virtual displacement is the most general. In this method, the energy of the system is calculated as a function of the flux linkage and the mechanical variable, and then the force or torque is determined as the partial derivative with respect to the mechanical variable with flux linkage constant. An alternative method uses the coenergy function in a similar procedure.

In Chap. 4, we apply the principles of Chap. 3 to cylindrical structures to determine flux and torque related to motor operation.

GLOSSARY

Coenergy, p. 122, a state function of a magnetic system, similar to energy, used to calculate magnetic forces.

Copper losses, p. 105, the losses in the resistance of the windings of a transformer.

Dot convention, p. 102, a means for representing relative polarities of magnetically coupled coils on a circuit diagram.

Eddy currents, p. 104, currents induced by changing flux that generate heat in a transformer core.

Exciting current, p. 105, the current required to magnetize and supply the losses in a magnetic structure.

Hysteresis loss, p. 104, loss due to the cyclic magnetization of a magnetic structure.

Ideal transformer, p. 100, a circuit model for a transformer that models transformation of voltage, current, and impedance but ignores nonlinearities, losses, and imperfect coupling between primary and secondary.

Iron losses, p. 104, losses caused in iron by changing magnetic flux.

Leakage flux, p. 105, magnetic flux that does not couple primary and secondary in a transformer, also called stray flux.

Magnetic structure, p. 94, an iron structure that increases the amount of magnetic flux and controls its distribution in space, also called a magnetic circuit.

Magnetomotive force (mmf), p. 97, the product of the current and the number of turns in the coil, which energizes a magnetic system in the same way that an electromotive force energizes an electric circuit.

Reluctance, p. 98, a measure of the mmf required to magnetize a portion of a magnetic structure to a specified degree.

Series winding impedance, p. 106, an impedance in either the primary or secondary of a transformer equivalent circuit to model the copper losses and leakage flux of the transformer.

Transformer, p. 100, a device consisting of two or more coils, or windings, tightly coupled by magnetic flux in a magnetic structure, used for voltage transformation, current transformation, and impedance transformation.

PROBLEMS

Section 3.1: Analysis of Magnetic Structures

3.1. A toroidal inductor has a circular cross-section and the dimensions shown in Fig. P3.1, and no gap. Assume $\mu_r = 6000$ and H is constant in the iron. Use the average radius for computing the path length. Find the inductance.

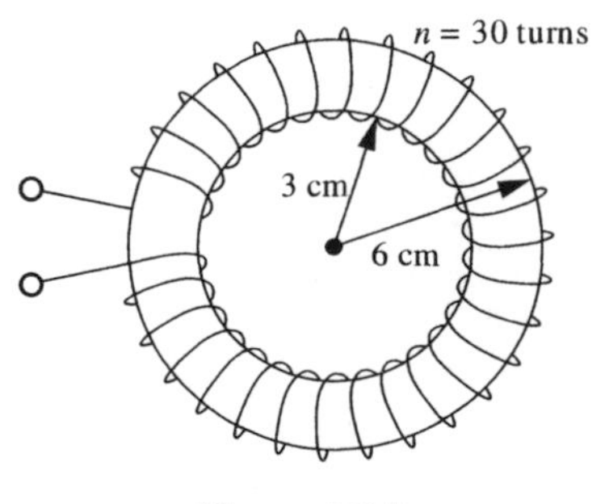

Figure P3.1

3.2. The iron structure shown in Fig. P3.2 has a relative permeability of 2500. The structure has a winding (not shown) of 1500 turns.

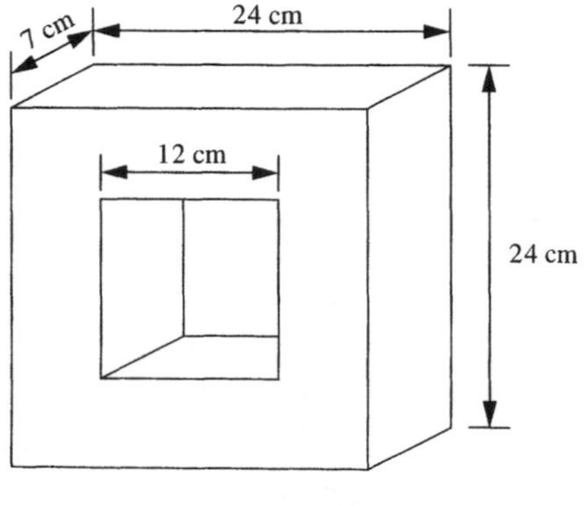

Figure P3.2

(a) Find the reluctance of the iron structure.
(b) Find the stored energy in the system if excited at a steady value of $B = 1.1$ T.

3.3. A lossless toroidal inductor with a gap is represented in Fig. P3.3. The reluctances of the iron and air gap are 10,000 and 100,000 A-t/Wb, respectively.
(a) If $i = 1$ A dc, what is the flux in the system?
(b) Under these conditions, an electron passes through the gap traveling outward from the center. What would be the effect on its motion?
(c) If $i = 1$ A (rms) ac at 60 Hz, what would be the rms voltage per turn in the coil?
(d) What is the impedance of the inductor at 60 Hz?

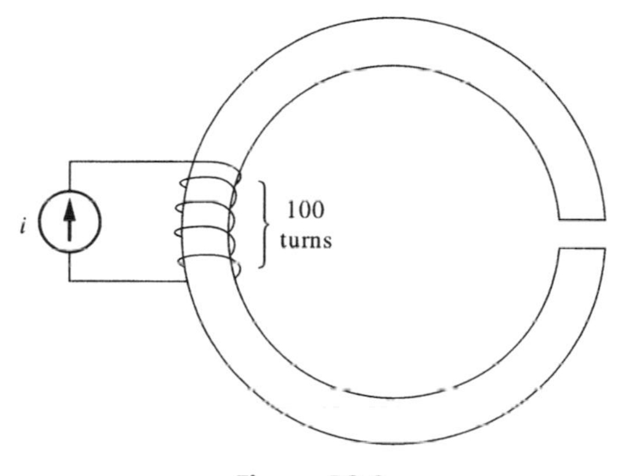

Figure P3.3

3.4. A 12-V (rms), 60-Hz electrical relay has 1000 turns and draws 12 mA (rms) of current. Assume that the relay magnetic system acts as a linear inductor, and neglect resistance.
(a) What is the mmf operating in the magnetic structure?
(b) What is the peak flux in the relay?
(c) What is the reluctance in the magnetic structure?

3.5. An iron ring weighing 3.65 kg and having a relative permeability of 5000 has some wire wrapped around it and is excited by an ideal ac voltage source. The current is measured. A small gap is then cut in the ring, and the current is observed to increase by a factor of 10. How much do the filings weigh? The density of iron is 7.65 g/cm³.

3.6. An inductor is shown in Fig. P3.6.
(a) What is the reluctance of the iron path?
(b) A 60-Hz ac voltage of 120 V rms is applied to the coil input. What is the maximum of the magnetic flux in the iron?
(c) What is the rms current required to supply this flux?
(d) The iron hysteresis and eddy-current losses are found to be 15 W. Give a parallel equivalent circuit for the inductor that accounts for the loss and energy storage.

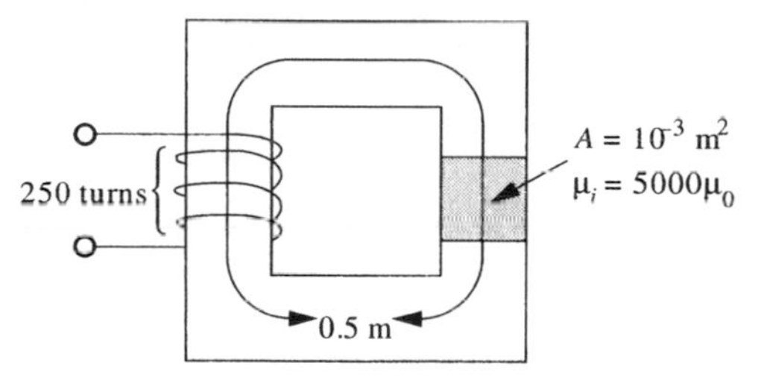

Figure P3.6

3.7. The ferrite toroid shown in Fig. P3.7 has a circular cross-section, an ID of 6 mm, an OD of 8 mm, and $\mu_i = 1600\mu_0$. The inductance is maximized by winding No. 32 wire (diameter = 8 mils = 8×10^{-3}

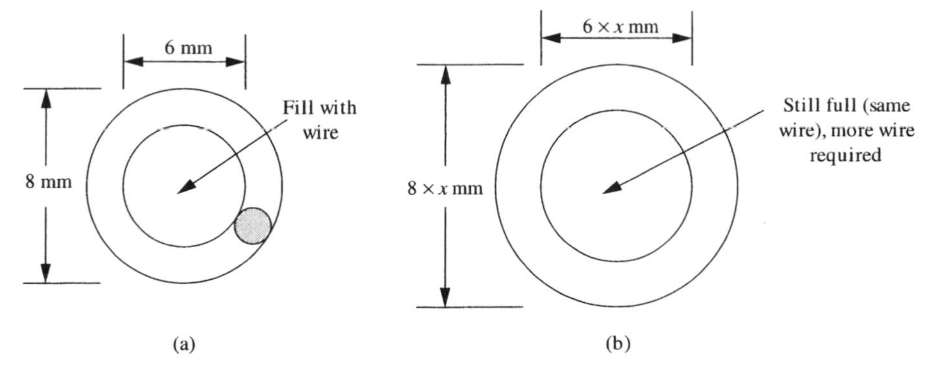

Figure P3.7

inches) in the hole until the inner hole of the toroid is completely filled.

(a) Estimate the inductance.

(b) The inductor is constructed according to the procedure given and the inductance is measured. The measured value is found to be only 60% of the required value. The inductor is redesigned by physically scaling up the toroid by a factor x ($x =$ scaling factor > 1) but keeping the wire diameter the same, as in Fig. P3.7(b). Estimate x to give the correct inductance.

3.8. A toroidal inductor is wound for maximum inductance by filling the "hole of the donut" totally with wire. The inductance is found to be 200% of the required value, so it is decided to make a smaller inductor. If the same wire is used and again the hole is filled, determine the scaling factor to make the new scaled-down version have the required inductance.

3.9. An iron ring with a relative permeability of 2000 has a coil with n turns wound on it. The inductance of the structure is measured at 60 mH. Give the new inductance as the following changes are made one at a time (and then restored before the next change).

(a) The number of turns is changed to $n/2$.

(b) The relative permeability increases to 4000.

(c) The entire structure is scaled up in every dimension by 25%, but the number of turns is constant.

(d) The windings are replaced by wires with half the original diameter, but the number of turns remains at n.

(e) We cut a gap in the iron such that 0.5% of the path length is now an air gap.

3.10. An iron-core inductor has a length of 0.5 m, a cross-sectional area of 0.01 m^2, and is wound with 100 turns of No. 23 wire, which has 20.4 Ω/1000 ft. The density of the iron is 7.65 g/cm^3. The loss in the iron is 1 watt/kilogram at a maximum flux density of 1.4 tesla and is proportional to flux squared. The inductor is excited at 250 V rms, 60 Hz. Find the iron and copper losses and, from these, a circuit model that is valid at 60 Hz. Ignore flux that escapes the iron, and assume the ac resistance is 10% higher than the dc. Let $\mu_i = 6000\mu_0$.

Section 3.2: Electrical Transformers

3.11. Figure P3.11 shows an ideal magnetic structure ($\mu_i = \infty$), which thus becomes an ideal transformer, except that it has two secondaries, each with a resistor attached.

(a) Mark polarity dots on the two secondaries.

(b) Determine the equivalent resistance into the primary. *Hint:* Use fundamental laws.

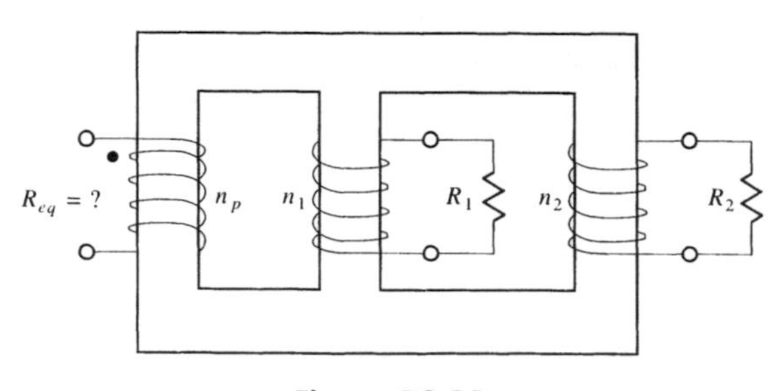

Figure P3.11

3.12. Figure P3.12 shows an ideal magnetic structure ($\mu_i = \infty$), which thus becomes an ideal transformer, except that it has two secondaries, each with a resistor attached.

(a) Mark polarity dots on the two secondaries.

(b) Determine the equivalent resistance into the primary. *Hint:* Use fundamental laws.

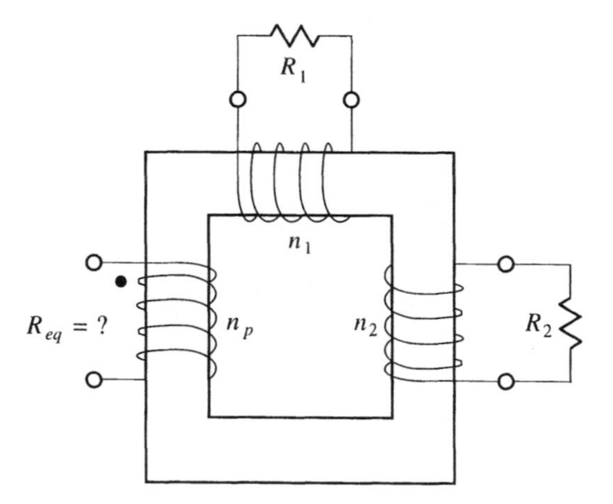

Figure P3.12

3.13. A transformer has 1000 primary turns and 100 secondary turns. The reluctance of the magnetic structure is $\Re = 10^5$ A-t/Wb. Assume no leakage flux, no iron losses, and ignore wire resistance.

(a) If the primary voltage is 120 V, 60 Hz, and the secondary is open-circuited, how much current will flow in the primary?

(b) If the secondary has a resistance of 30 Ω

connected to it, what will be the current in the resistor?

(c) For the circuit in part (b), what would be the magnitude of the current in the primary?

3.14. Figure P3.14 shows the approximate model for a nominal 880/220-V, 60-Hz, single-phase transformer with all internal impedances referred to the primary.

(a) What term(s) in the model account for the following:

  i. Secondary copper losses?

  ii. Magnetic energy stored in the iron?

  iii. Faraday's law?

  iv. Magnetic energy stored in the primary leakage fields?

  v. Core losses due to eddy currents and hysteresis losses?

(b) If the transformer heat exchanger is capable of dissipating 300 W without undue temperature rise, estimate the rated kVA of the transformer.

(c) What would be the approximate efficiency at rated kVA if the output power factor were 0.92?

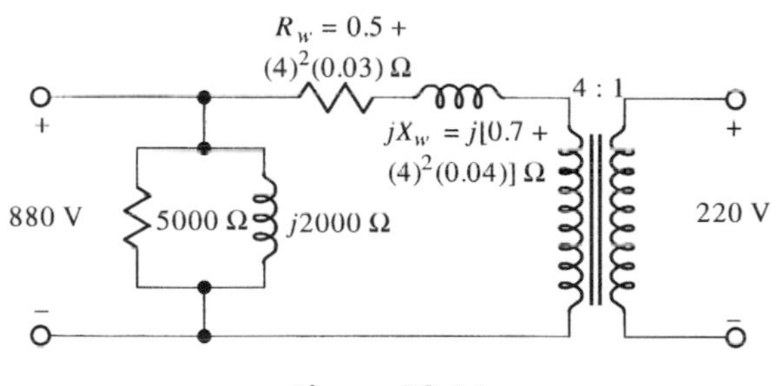

Figure P3.14

3.15. A 2400/120-V, 30-kVA, 60-Hz, single-phase transformer draws 2 kVA at a 0.2 power factor (lagging) at no load. The series winding resistance and reactance due to stray inductance are $3.5 + j5.0\ \Omega$, referred to the primary.

(a) Determine the equivalent circuit for the transformer.

(b) Estimate the efficiency of the transformer if it serves a 30-kVA load at a 0.9 power factor (lagging) with rated secondary voltage.

(c) Estimate the voltage regulation under these load conditions.

3.16. Consider a 60-Hz, single-phase transformer with the following specifications—core: area = 0.1 m^2, length 2 m, $\mu_r = 5000$; windings: 150 primary and 15 secondary turns; coil impedances: primary = $0.5 + j0.8\ \Omega$; secondary = $0.005 + j0.008\ \Omega$; allowable losses: 4000 W, divided equally between iron and windings. Also, we know that for 2000-W loss in the iron, the maximum magnetic flux density is $B_m = 1.4$ T.

(a) Find the allowable kVA for the transformer.

(b) Determine the equivalent circuit for the transformer.

3.17. Two coils are wound on a piece of iron, as shown in Fig. P3.17. A 120-V, 60-Hz source is connected to one of the coils and a 10-Ω resistor to the other.

(a) Put polarity marks on the resistor such that the voltages on both coils are in phase.

(b) Draw an equivalent circuit without using an ideal transformer that permits calculation of in-phase and out-of-phase components of the input current. Ignore iron and copper losses and leakage inductance.

(c) In this situation, are the coils attracted or repelled by the currents in the wires? Explain.

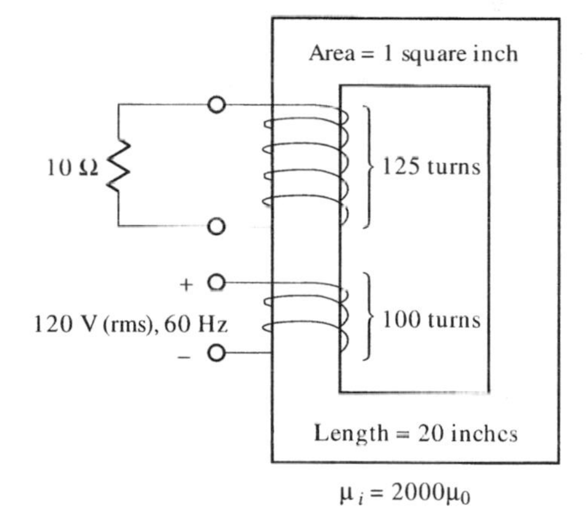

Figure P3.17

3.18. A 72-VA, 60-Hz, single-phase transformer has the equivalent circuit shown in Fig. P3.18.

(a) What is the no-load current if excited on the 120-V side?

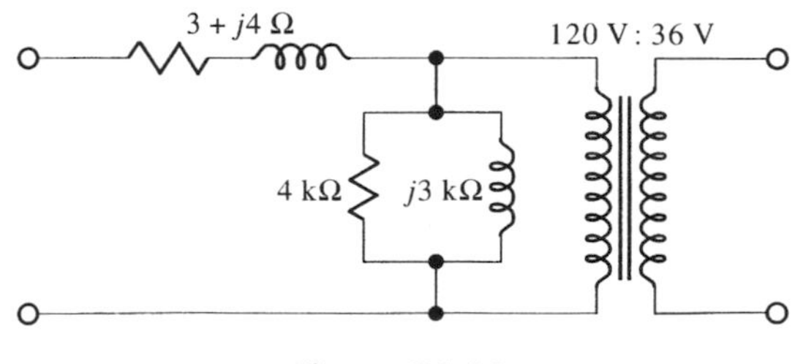

Figure P3.18

(b) Estimate the losses in the transformer at full load, resistive load.

(c) Estimate the input voltage required for full load if the output voltage is 36 V and the output current lags the voltage by 25°.

3.19. A 60-Hz, single-phase, iron-core transformer with $\mu_i = 4200\mu_0$, area = 0.05 m^2, and nominal length of 2 m has 1000 turns on the high-voltage (HV) and 200 turns on the low-voltage (LV) side. At the rated voltage, the peak flux density in the iron is 1.2 tesla.

(a) Find the rated voltage (rms value) at both the HV and LV sides.

(b) What rms magnetizing current is required on the HV side?

(c) The rated current in the transformer is 50 A on the HV side. The losses at rated current are 3000 watts. Give an equivalent circuit for the transformer, neglecting iron losses and leakage inductance.

3.20. Figure P3.20 shows a 20-kVA, single-phase, 8600/240-V (center-tapped), 60-Hz pole transformer.

(a) What is in the cylinder physically, and what is the electrical model for it? No values are required.

(b) At rated load and unity power factor, the iron losses are 4% and the copper losses 6% of the output power. What does that tell you about the equivalent circuit? Refer to the high-voltage side.

(c) Find the input real power if we have 5 kVA at $PF = 1$, lagging, between A and N, 3 kVA at $PF = 0.75$, lagging, between B and N, and 6 kVA at PF = 0.95, lagging, between A and B.

(d) Discuss the kVA limits of the transformer for New Year's Day in Detroit.

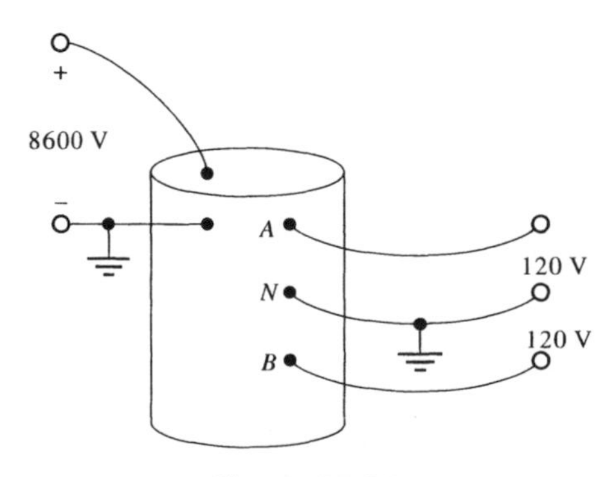

Figure P3.20

3.21. A 50-kVA, 2300/230-V, 60-Hz, single-phase transformer has an iron structure 0.03 m^2 in area, 1 m in length, and $\mu_i = 3000\mu_0$. The peak flux density in the iron is 0.6 T. With no load and rated voltage, the input current on the high-voltage side is 0.3 A rms.

(a) Estimate the turns on the high-voltage and low-voltage windings.

(b) Find the dc current required on the high-voltage side to magnetize the iron to 1.2 T, assuming no saturation occurs.

(c) Estimate the iron loss at rated voltage.

(d) Estimate the current magnitude, rms, on the high-voltage side with full output kVA and rated voltage at $PF = 1$.

3.22. An open-circuit (OC)/short-circuit (SC) test is performed on a single-phase transformer in the standard manner, with the results shown in Table P3.22 Quantities not measured are noted "NM."

TABLE P3.22

	OC	*OC*	*SC*	*SC*
Quantity	Side 1	Side 2	Side 1	Side 2
Voltage, V	7500	240	372	0, NM
Current, A	0, NM	3.62	2	NM
Power, W	NM	290	245	NM

(a) What is the transformer apparent power rating?

(b) Give an equivalent circuit for the transformer, referred to the HV side.

(c) Find the magnetizing current, I_ϕ, if excited on the high-voltage side

(d) Estimate the transformer efficiency if supplying 12 kVA at $PF = 1$ and rated voltage.

3.23. (Requires material in Appendix A.) A transformer is modeled by the simplified equivalent circuit shown in Fig. P3.23. The ac source is connected at $t = 0$.

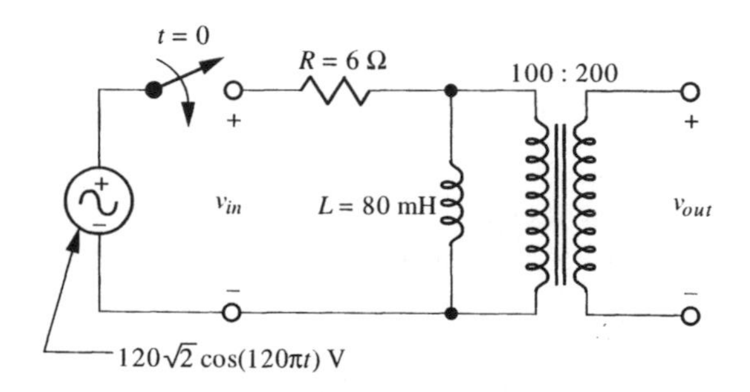

Figure P3.23

(a) Derive the transfer function of the transformer, $\underline{\mathbf{T}}(\underline{s}) = \underline{\mathbf{V}}_{out}(\underline{s})/\underline{\mathbf{V}}_{in}(\underline{s})$.
(b) Find the steady-state response. You may use the transfer function or ordinary ac circuit analysis.
(c) Find the natural frequencies and the form of the natural response.
(d) Apply the initial conditions and determine the total response.

3.24. (Requires material in Appendix A.) The equivalent circuit for a single-phase transformer is shown in Fig. P3.24. Although normally such equivalent circuits are valid in a narrow range of frequencies, we consider this model valid at all frequencies, including dc and complex frequencies.
(a) Find the transfer function,
$\underline{\mathbf{T}}(\underline{s}) = \underline{\mathbf{V}}_{out}(\underline{s})/\underline{\mathbf{V}}_{in}(\underline{s})$.
(b) If an ideal 10-V dc battery were connected to the input at $t = 0$, find the output voltage as a function of time for positive t. For this part, consider $n_1 = 100$, $n_2 = 200$, $R_w = 2\ \Omega$, $L_w = 0.2$ H, $R_i = 1000\ \Omega$, and $L_i = 10$ H. The initial condition on the derivative of the output voltage is 10^5 V/s.

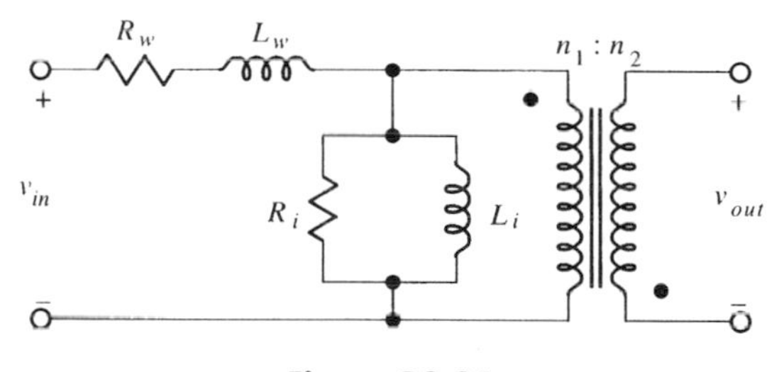

Figure P3.24

3.25. An iron-core inductor with 1000 turns draws 12 mA rms, if excited by 120-V rms, 60-Hz source.
(a) Find the peak flux in the inductor.
(b) Find the reluctance of the iron magnetic structure.
(c) Find the equivalent circuit if the copper loss due to the resistance of the wire is 0.1 W under these conditions. Use a series model.
(d) Find the current if 1000 more turns are added to the inductor.

3.26. A 60-Hz, 240/120-V single-phase, transformer with $\mu_i = 4500\mu_0$ is shown in Fig. P3.26. The primary resistance is 0.3 Ω and the primary leakage reactance is 0.8 Ω. The secondary resistance is 0.075 Ω and the secondary leakage reactance is 0.15 Ω.

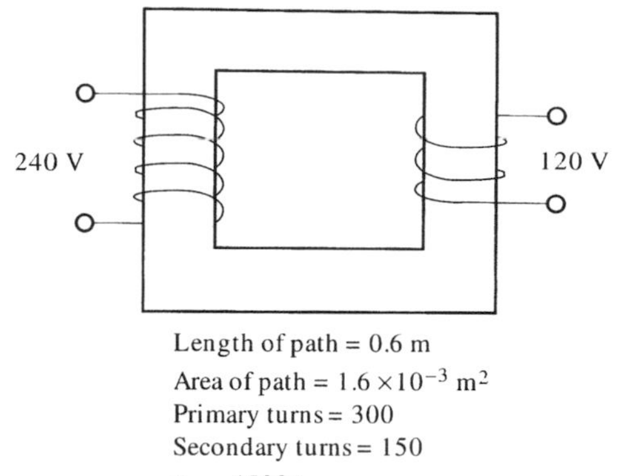

Length of path = 0.6 m
Area of path = 1.6×10^{-3} m^2
Primary turns = 300
Secondary turns = 150

$\mu_i = 4500\mu_0$

Figure P3.26

(a) Find the reluctance of the magnetic structure.
(b) Determine the magnetizing current, I_Φ, required on the high-voltage side to magnetize the iron for the voltages shown.
(c) If the hysteresis and eddy-current losses are 60 W, what is the in-phase current in the primary required to supply these losses? There is no load on the secondary.
(d) What is the no-load current?
(e) Give the equivalent circuit for the transformer with all circuit elements referred to the primary.
(f) If the total losses of the transformer must be kept below 135 W, what should be the kVA rating of the device?

3.27. A single-phase, 60-Hz transformer is 20 kVA, 1320/440, 85% efficient at full load, unity power factor. At no-load, the transformer current on the HV side is 1.5 A and the power is 1000 W.
(a) There are 120 turns on the 440 V side. Find the turns on the 1320 V winding.
(b) What would be the no-load current if the transformer were excited from the LV side?
(c) Estimate the copper losses of the transformer at the operating level of 15 kVA.
(d) Find the reluctance, $\Re$, of the iron core of the transformer.

3.28. A 60-Hz, single-phase transformer has 100 turns and 240 V rms on the primary. The graph of primary current as a function of secondary current for a resistive load is shown in Fig. P3.28.
(a) What is the primary/secondary turns ratio?
(b) What is the exciting current?
(c) Estimate the reluctance of the magnetic structure.

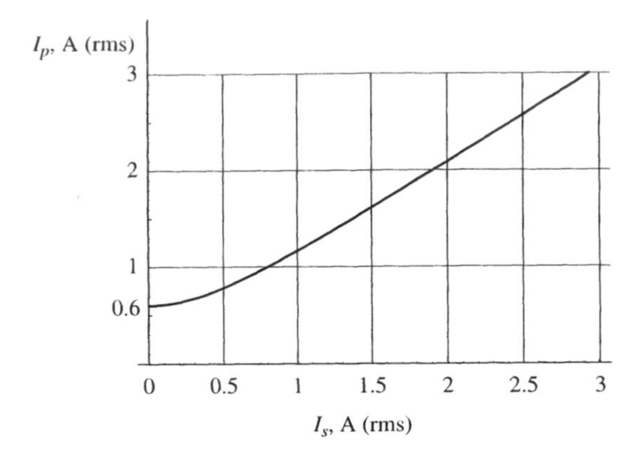

Figure P3.28

3.29. Figure P3.29 shows the equivalent circuit for a 2400/240-V, 60-Hz, 75-kVA, single-phase transformer. Assume that the transformer is operating at rated secondary voltage and apparent power, with a 0.9 power factor (lagging).

(a) What is the turns ratio?
(b) Estimate the iron losses.
(c) Estimate the copper losses.
(d) What is the efficiency?
(e) What would be the exciting current, I_e?
(f) Estimate the input power factor. *Hint:* Determine the reactive powers since you know the real powers.

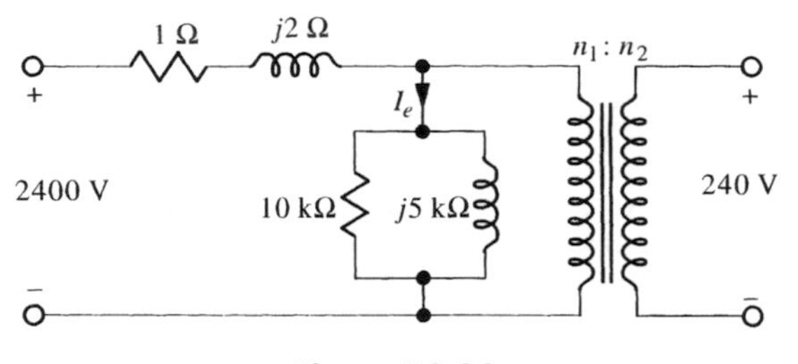

Figure P3.29

3.30. Figure P3.30 shows the equivalent circuit for a 1-MVA, 60-Hz, single-phase transformer. The voltages given are nameplate values.

(a) What current would flow on the high-voltage side if the low-voltage side were open-circuited?
(b) Estimate the copper losses at full load.
(c) Estimate the iron losses at full load.
(d) Estimate the efficiency at full load and unity power factor.

3.31. A single-phase, 60-Hz transformer has an iron structure length of 1 m and a cross-sectional area of

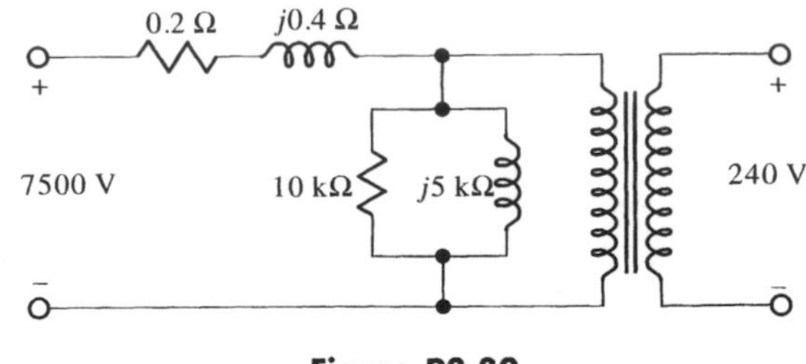

Figure P3.30

4×10^{-3} m^2. The iron has a permeability of $5000\mu_0$ and is at rated voltage when operated to a maximum flux density of 1.4 tesla. Side 1 has 500 turns and side 2 has 1000 turns. We ignore iron and copper losses.

(a) Find the rms magnetizing current if excited on the low-voltage side at rated voltage.
(b) Find the rms magnetizing current if excited on the high voltage side at rated voltage.
(c) Find the rated rms voltage on the high-voltage side.
(d) Find the rated rms voltage on the low-voltage side.
(e) Find the volt-amperes required to excite the transformer.

3.32. The equivalent circuit in Fig. P3.32 represents a 240/120-V, 60-Hz, 10-kVA, single-phase transformer.

(a) What is the exciting current for the transformer referred to the high-voltage side?
(b) Estimate the transformer losses if the transformer is operating at 80% capacity.
(c) If operating at 80% rated capacity and a power factor of 0.9 lagging, estimate the input voltage and power to the transformer.

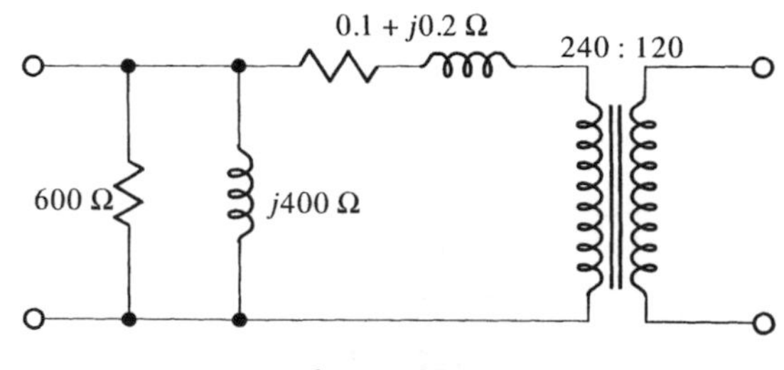

Figure P3.32

3.33. The equivalent circuit for a 460/230-V, 9.2-kVA, 60-Hz, single-phase, transformer is shown in Fig. P3.33(a).

(a) Figure P3.33(b) shows a table to record the

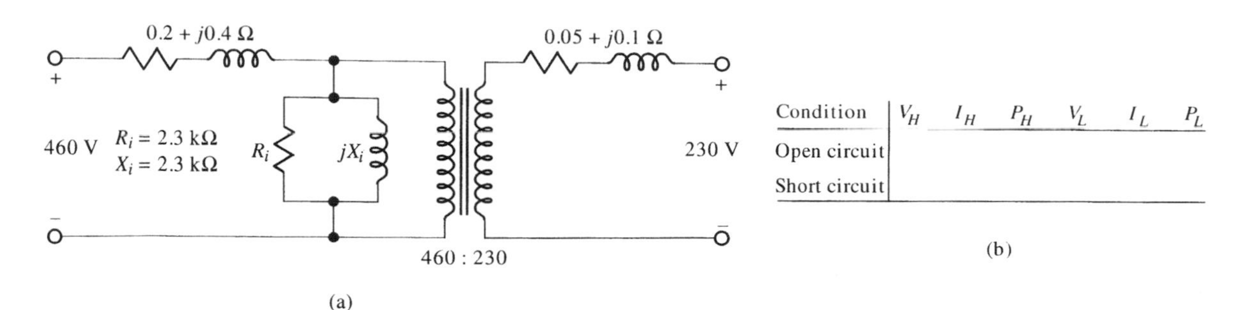

Figure P3.33

results of a standard open-circuit/short-circuit test. Fill in every slot, even if the quantity normally would not be measured. Assume that the wattmeters are connected to measure the input power on the side supplying the power.

(b) If the secondary voltage is 230 V and the current 10 A with unity power factor, find the primary voltage and the efficiency.

3.34. A 60-Hz, single-phase, transformer is tested with a standard open-circuit/short-circuit test at nameplate voltage and current values, with the results shown in Table P3.34.

TABLE P3.34

Test	*Primary*			*Secondary*		
	V	*I*	*P*	*V*	*I*	*P*
SC	28	20	350	0	N/A	N/A
OC	480	0	N/A	240	1.93	290

(a) Find the exciting current, as seen from the primary.

(b) Find the apparent power rating of the transformer.

(c) Estimate the efficiency if operated at 80% full-load capacity and 0.95 *PF*.

(d) Estimate the voltage regulation at this load.

3.35. A single-phase transformer operated at unity power factor has a maximum efficiency of 94%, which occurs at a load of 30 kVA. What is its efficiency at a load of 20 kVA, unity power factor? *Hint*: It can be shown that the maximum transformer efficiency, if power factor is constant, occurs when iron and copper losses are equal.

3.36. A 480/120-V, 20-kVA, 60-Hz, single-phase transformer has 400-W iron losses and 500-W copper losses when operating at rated voltage and current. Because the transformer is used in a tropical country, the total losses must be kept less than 90% of the rated losses. What is the derated kVA of the transformer for operating in that location, assuming nameplate voltage?

3.37. A 240/120-V, 720-VA, 60-Hz, single-phase transformer has a standard OC/SC test. The power into the transformer is 25 W in the open-circuit test and 30 W on the short-circuit test.

(a) Find the current in the low-voltage winding during the short-circuit test, assuming that the voltage is applied to the high-voltage side.

(b) Find the voltage across the high-voltage winding during the open-circuit test, assuming that the voltage is applied to the low-voltage side.

(c) Estimate the efficiency if operated with a 0.95 power factor and full load.

3.38. A 50-kVA, 2400/240-V, 60-Hz, single-phase, transformer is tested in an open-circuit/short-circuit test, with the results shown in Table P3.38.

TABLE P3.38

Test	*Primary*			*Secondary*		
	V	*I*	*P*	*V*	*I*	*P*
OC	—	—	—	240	4.22	650
SC	72	20.8	800	—	—	—

(a) Derive an equivalent circuit for the transformer with all components referred to the primary.

(b) The load on the transformer is 50 kVA with a power factor of 0.9, lagging. The output voltage is 240 V. Find the efficiency.

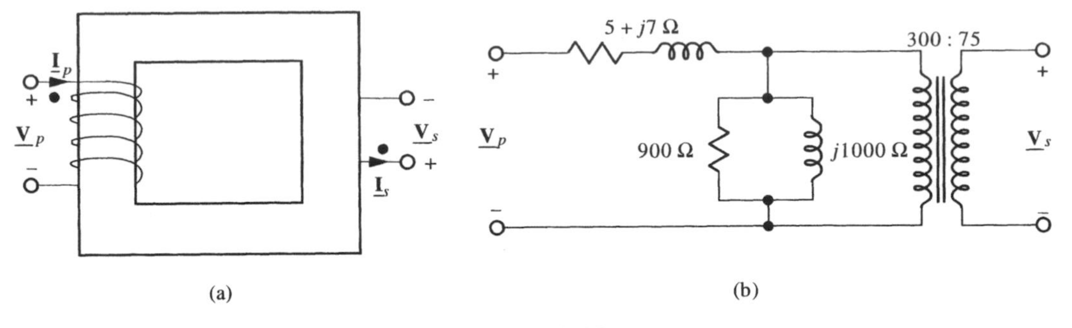

Figure P3.39

(c) Estimate the input voltage required for the condition described in part (b).

3.39. A 60-Hz transformer has an area of 0.01 m^2 and is made of iron with $\mu_i = 2400\mu_0$ and mass density 7.65 g/cm^3. The transformer is shown in Fig. P3.39(a) and the equivalent circuit is shown in Fig. P3.39(b). The primary has 300 turns and the secondary 75 turns. At the rated kVA, the maximum flux density is 1.1 T and the total losses are 2000 W.

(a) Show the winding direction for the secondary side in accordance with the dots shown.

(b) Find the nominal voltages at primary and secondary sides, effective values.

(c) Estimate the transformer efficiency if operated at rated kVA and $PF = 1$.

(d) Estimate the mass of the transformer if 70% of the weight is in the iron core.

3.40. Figure P3.40 shows (a) the circuit symbol, (b) the wiring connection, and (c) the equivalent circuit for the same transformer connected as a normal transformer and (d) as an *autotransformer.* This transformer is a 720-VA, 120/120-V, 60-Hz transformer when connected in the normal way. Assume the same flux levels in both connections.

(a) What are the primary and secondary voltages of the autotransformer?

(b) What are the parameters in the equivalent circuit of the autotransformer shown in Fig. P3.40(d)?

(c) What is the apparent power rating of the autotransformer, assuming the same losses in both cases?

3.41. A single-phase, 60-Hz, 20-kVA transformer has two 460-V primary windings and two 230-V secondary windings, as shown in Fig. P3.41.

(a) Find the nominal primary and secondary voltages and rated current for each of the four possible connections shown in Table P3.41.

TABLE P3.41

Primary/Secondary Connection	V_p	I_p	V_s	I_s
Series/series				
Series/parallel				
Parallel/series				
Parallel/parallel				

(b) Show the connections of the windings for the parallel/series connection and show the rated current in each winding.

3.42. A transformer, Fig. P3.42(a), has the following nameplate information: 10 kVA, 60 Hz, and 1380/440 V. Tests show the transformer efficiency at 92.0% at nameplate operation with unity power factor, and 89.3% at half-load (5 kVA) and unity power factor.

(a) What resistive load on the 440-V output gives the rated load?

(b) Find the no-load losses of the transformer.

(c) The transformer is connected as an autotransformer as shown in Fig P3.42(b). Under these conditions, what is the output voltage?

(d) Still connected as an autotransformer, what is the kVA rating of the transformer in this connection. (The answer is *not* 10 kVA.)

3.43. A three-phase 60-Hz transformer is used to reduce 4160 V to 240 V. The rating of the transformer is 30 kVA. When operated at rated kVA, the

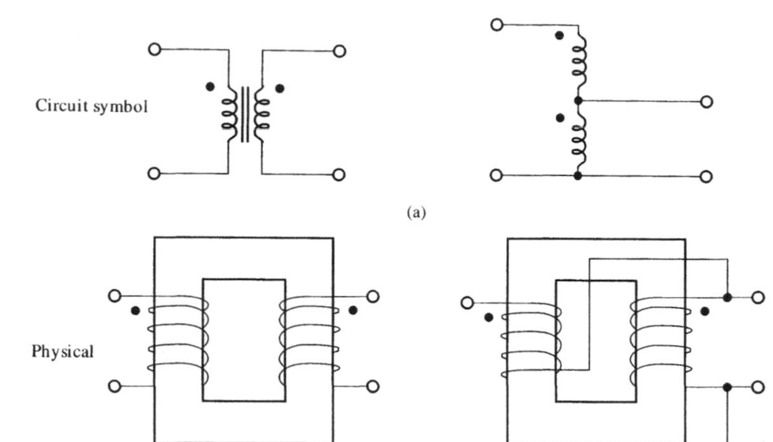

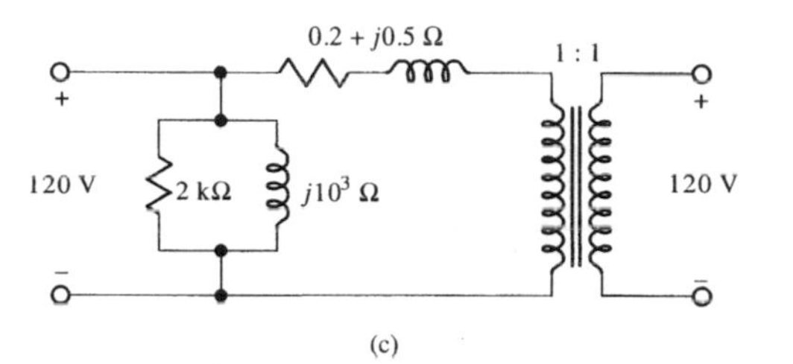

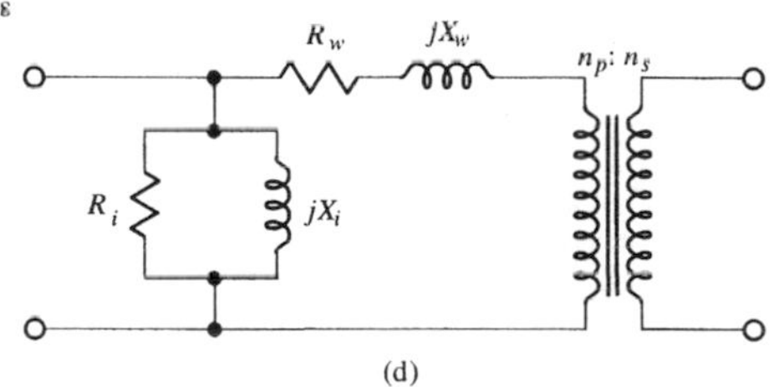

Figure P3.40

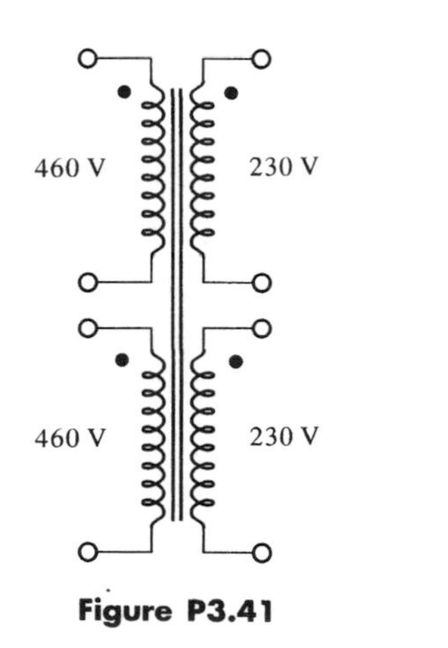

Figure P3.41

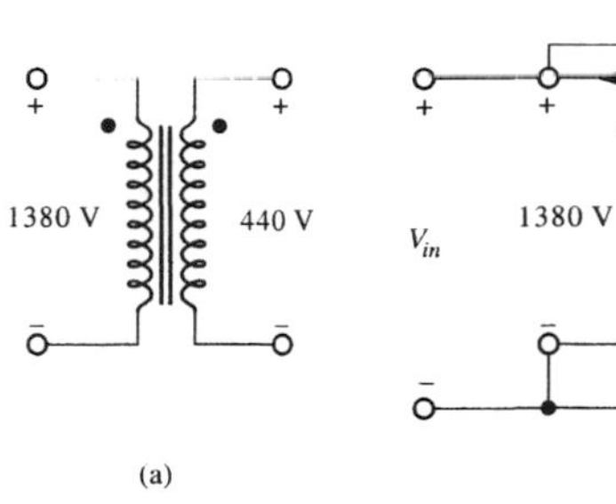

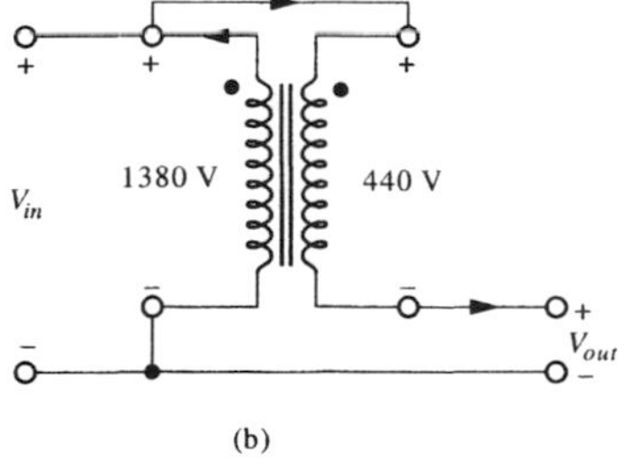

Figure P3.42

transformer losses are 1150 W, which are 30% iron losses and 35% each primary and secondary copper losses. At a load of 20 kW, 0.85 power factor, lagging, estimate the efficiency of the transformer.

3.44. A 30-kVA, 60-Hz, three-phase transformer has an efficiency of 94% at full load and 93% at half load, unity *PF*, and the same primary voltage in both cases. Find the iron loss of the transformer.

3.45. Three 25-kVA, 8600/1200-V, 60-Hz, single-phase transformers are connected in a delta-wye connection for three-phase operation, with the high-voltage side in wye and the low-voltage in delta. With the transformers operating at rated kVA, the iron loss/transformer is 550 W and the copper loss/transformer is 600 W.

(a) Draw a circuit diagram showing the single-phase transformer connections. No voltages are required.
(b) Find the low-voltage and high-voltage three-phase line voltages if the single-phase transformers are operated at their rated voltages.
(c) If 50 kVA is the load on the three-phase system, find the nominal line current on the high-voltage side. Ignore the exciting current for the transformers.
(d) Estimate the efficiency under the condition in part (c) and a *PF* of 0.7.
(e) In a per-phase equivalent circuit, what would be the value of the resistance representing the copper losses if placed on the low-voltage side?

3.46. The model shown in Fig. P3.46 is the per-phase model for a three-phase, 60-Hz transformer. The transformer is operating at an output load of 30 kW at 0.8 *PF*, lagging, with 240-V output voltage.

(a) Estimate the copper losses of the transformer.
(b) Estimate the iron losses of the transformer.
(c) Find the exciting current on the low-voltage side if power is applied to that side.

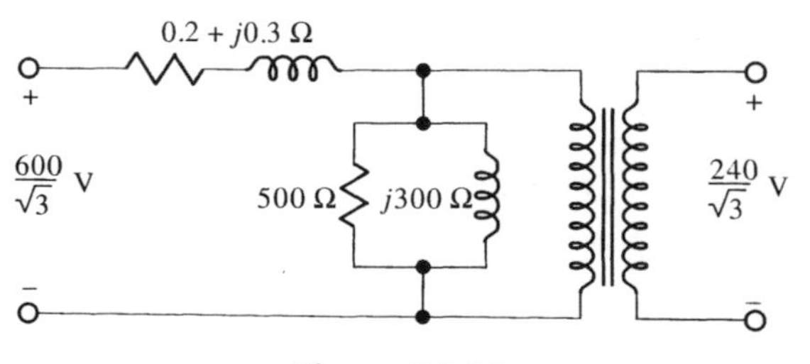

Figure P3.46

(d) Estimate the transformer efficiency.
(e) If the transformer has been designed for maximum efficiency at the nameplate apparent power and unity power factor, what is the nameplate kVA rating. *Hint*: For fixed *PF*, transformer efficiency is maximum when iron and copper losses are equal.

3.47. A three-phase, 60-Hz transformer is connected Y–Δ and rated 13 kV/480 V, 25 kVA. The efficiency is 95% at full-load, unity power factor, and losses are divided 60% iron, 40% copper. Determine the per-phase equivalent circuit for the transformer with numerical values for the turns ratio, the resistors in the circuit, and the voltage levels. No reactance values are required. Refer impedance values to the HV side of the transformer.

3.48. Three single-phase transformers are connected as a three-phase transformer. The high-voltage side is in wye and the low-voltage side is in delta. The per-phase equivalent circuit is shown in Fig. P3.48; also it is known that at full load (rated kVA), the iron and copper losses are equal.

(a) Find the turns ratio of the single-phase transformers.
(b) Find the ratio (full load current)/(exciting current), magnitudes only.
(c) Estimate the transformer efficiency at 80% full load and 0.75 power factor, lagging.
(d) What voltage in Fig. P3.48 corresponds to a physical voltage in the real system?

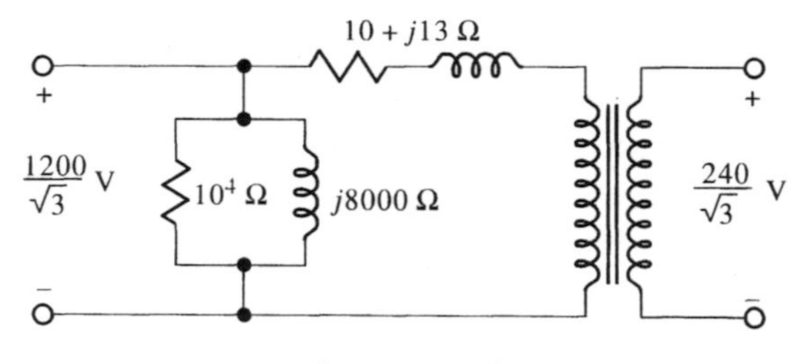

Figure P3.48

3.49. The per-phase equivalent circuit for a three-phase, 60-Hz transformer is shown in Fig. P3.49, along with rated voltage and current.

(a) Find the nameplate kVA.
(b) Find the nameplate primary voltage.
(c) Find the exciting current on the high-voltage primary.
(d) Estimate the transformer losses at 80% nameplate kVA.

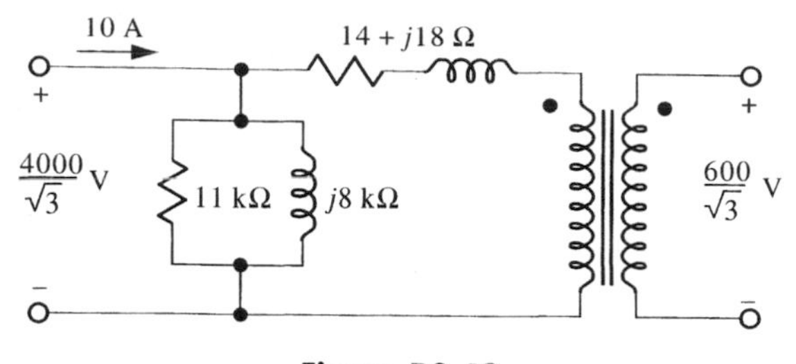

Figure P3.49

3.50. A three-phase transformer has the following information: 60 Hz, 50 kVA, 4160/600 V, 95% efficiency at full load, and $PF = 1$, no-load current is 4% of full-load current, and at full load, losses divide equally between iron and copper losses.

(a) Determine a per-phase equivalent circuit for the transformer, evaluating numerically all the components you can, plus rated voltages and currents.

(b) Estimate the efficiency of the transformer at a load of 30 kVA, 0.95 *PF*, lagging.

3.51. Figure P3.51 shows the per-phase equivalent circuit for a 20-kVA, three-phase transformer. The rated voltages for the transformer are 4000/2300 V.

(a) Find the magnitude of the exciting current for the three phase transformer if excited on the high-voltage side.

(b) Estimate the efficiency of the transformer at rated kVA and a *PF* of 0.82 lagging.

(c) Estimate the voltage regulation of the transformer under the conditions in part (b).

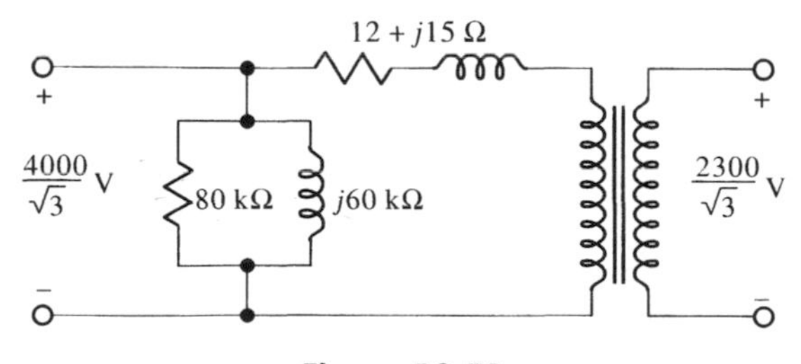

Figure P3.51

3.52. A three-phase transformer has the following parameters: 2400/480 V, 60 Hz, 60 kVA, primary resistance of 1 Ω per phase, primary leakage reactance of 1.3 Ω per phase, secondary resistance of 0.04 Ω per phase, secondary leakage reactance of 0.06 Ω per phase, iron losses of 600 W, and no-load current of 0.240 A on high-voltage side, which is the primary.

(a) Give the per-phase equivalent circuit for the transformer with all circuit quantities referred to the primary.

(b) For rated output voltage and kVA with 0.9 power factor, lagging, determine the input line voltage.

(c) Estimate the efficiency for the conditions in part (b).

(d) Estimate the voltage regulation for the conditions in part (b).

3.53. A three-phase, 60-Hz transformer is required to connect a 1320-V distribution line to a 460-V motor. We require a Y on the low-voltage side to provide a grounded neutral. The input is to be Δ-connected. The motor requires 10 kW at a power factor of 0.82, lagging.

(a) What are the phase voltages and currents on both sides of the transformer? Ignore losses.

(b) Give the per-phase circuit of the transformer and load, assuming an ideal transformer.

3.54. In Chap. 1, we showed how to transform three-phase power with the use of three single-phase transformers. There are two ways to transform three-phase power with *two* single-phase transformers. This problem investigates these methods. In this problem, we transform 460 V three-phase to 230 V three-phase; hence, the transformers have a turns ratio of 2:1. *Hint*: In both figures, the geometric orientation hints of the phasor relationships.

(a) The configuration shown in Fig. P3.54(a) is called the "open-delta" or V connection, for obvious reasons. Identical transformers are used.

(1) Show that if *ABC* is three-phase, *abc* is also three-phase. Consider the *ABC* voltages to be a three-phase set and prove the *abc* set is three-phase.

(2) If the load is 30 kVA, find the required rating of the transformers to avoid overload.

(b) The configuration shown in Fig. P3.54(b) is called the T connection. For this connection, the transformers are not identical but have different voltage and kVA ratings. The bottom transformer is center-tapped so as to have equal, in-phase voltages for each half.

(1) Find the voltage, V_{AX}, to make this transform three-phase.

(2) If the load is 30 kVA, find the required kVA rating of each transformer to avoid overload.

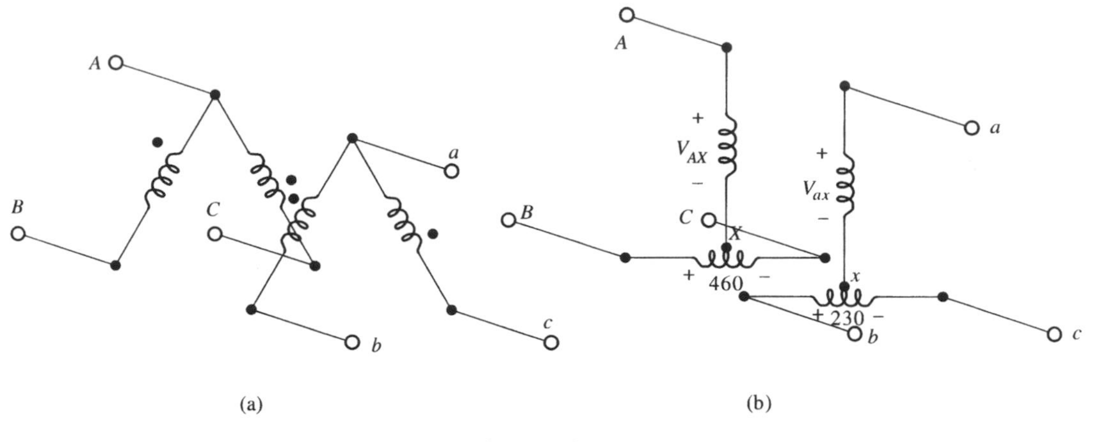

Figure P3.54

Section 3.3: Forces in Magnetic Systems

3.55. The reluctance of a magnetic system is given by the formula

$$\Re(x) = \frac{10^6}{x + 0.001}$$

with x in meters. Find the force generated by the system at 1000 A-t of mmf as a function of x.

3.56. The current and flux linkage of a magnetic system are related by the formula

$$i = \lambda[(1 + \sin(2\,|\theta|)], \qquad |\theta| < \pi/2$$

with θ in radians. If $i = 1.2$ A, find the angle at which the torque is -0.3 N-m.

3.57. A magnetic structure with ideal iron properties is used to magnetize an air gap with an area of 2 cm^2 and a gap width of 2 mm. The mmf of the coil wound on the iron is 1000 ampere-turns.

(a) Find H in the air gap.

(b) What is the force between pole faces?

3.58. A magnetic structure has a reluctance given by the expression. $\Re(x) = 10^5 + 10^4 x$ henries^{-1}, where x is a mechanical displacement in cm. The structure has 100 turns of wire.

(a) If the coil has a current $i(t) = 0.01 \cos(120\pi t)$ A and $x = 0$, find the rms voltage across the coil.

(b) Find the time-average force at $x = 0$ for the current in part (a).

(c) If $i = 10$ A dc and $x = 1 + 0.1 \sin(120\pi t)$ cm, estimate the resulting voltage.

3.59. An ac relay has a cross-sectional area of 1 cm^2 and a path length of 10 cm in iron with $\mu_i = 5000\mu_0$. With the relay open, the gap is 2 mm, and with the relay energized, the minimum average gap is 0.1 mm due to roughness and misalignment. The relay coil has 3000 turns and operates from 24 V (rms), 60 Hz ac.

(a) Find the current into the relay with the gap closed.

(b) Find the time-average force on the movable member at maximum gap width.

(c) Find the time-average force on the movable member at minimum gap width.

3.60. The device shown in Fig. P3.60 is a doorbell ringer, where a magnetic force operates against a spring (not shown). The ringer is excited by 12 V rms, 60-Hz ac. The inductance of the structure is $L(x) = 2e^{-x/2}$ H, with x in inches.

(a) At $x = 0$, what is the rms current in the coil?

(b) At $x = 0$, what is the time-average force on the plunger? Consider the force positive if it is in the direction to increase x. *Hint:* Distance must be expressed in meters.

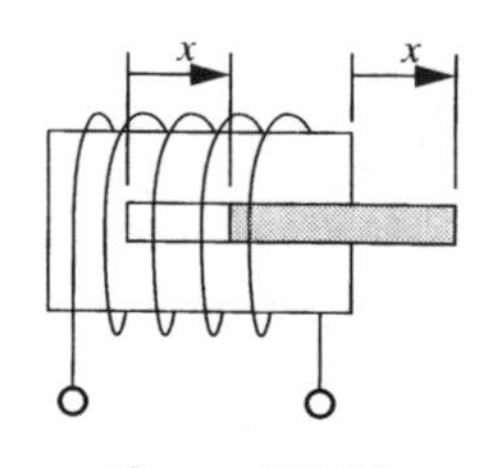

Figure P3.60

3.61. The magnetic structure shown in Fig. P3.61(a) has an inductance given by the graph in Fig. P3.61(b). Find the force at $d = 0.5$ cm and $i = 1.0$ A.

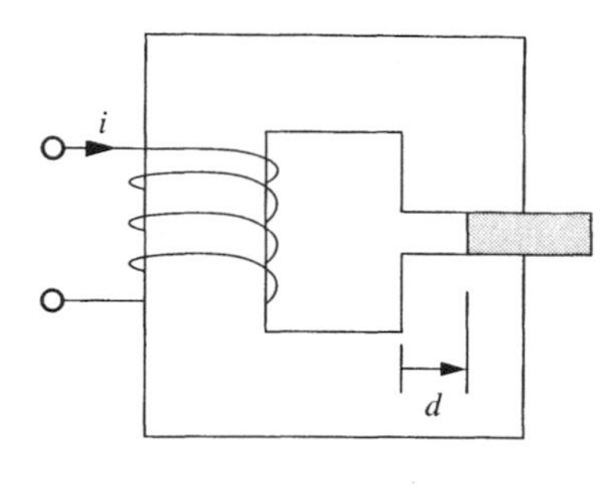

(a)

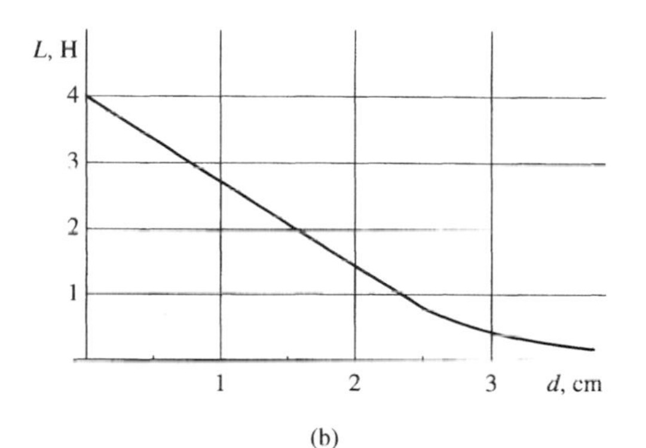

(b)

Figure P3.61

3.62. An electromechanical transducer is shown in Fig. P3.62(a). The coil has 1000 turns, and you may neglect stray flux and losses. An experiment is performed in which the coil is excited with 120-V (rms) ac and the current is measured as the gap length is changed. The results are given in Fig. P3.62(b). From these data, you are to determine the time-average force developed across the gap at $x = 2$ mm. The frequency is 60 Hz.

3.63. The flux linkage created by a current i is given by the equation $\lambda = (1 + x/10)i$, where x is a mechanical displacement in the system in centimeters.

(a) Find the energy required to increase the current from 0 to 2 A for $x = 0$.

(b) What magnetically generated force operates on the mechanical system in the direction of increasing x for $i = 2$ A?

3.64. A magnetic structure, with its flux-linkage-current curve, is shown in Fig. P3.64.

(a) Find the inductance for $i < 1$ A.

(b) Determine the energy required to increase the current from 0 to 1 A.

(c) Determine the energy required to increase the current from 1 to 2 A.

(d) The current is kept at 2 A and a 1-mm gap is sawed in the iron path. The filings are found to weigh 10 g and the density of the iron is 7.65 g/cm^3.

(1) What is the final flux linkage after the gap is opened?

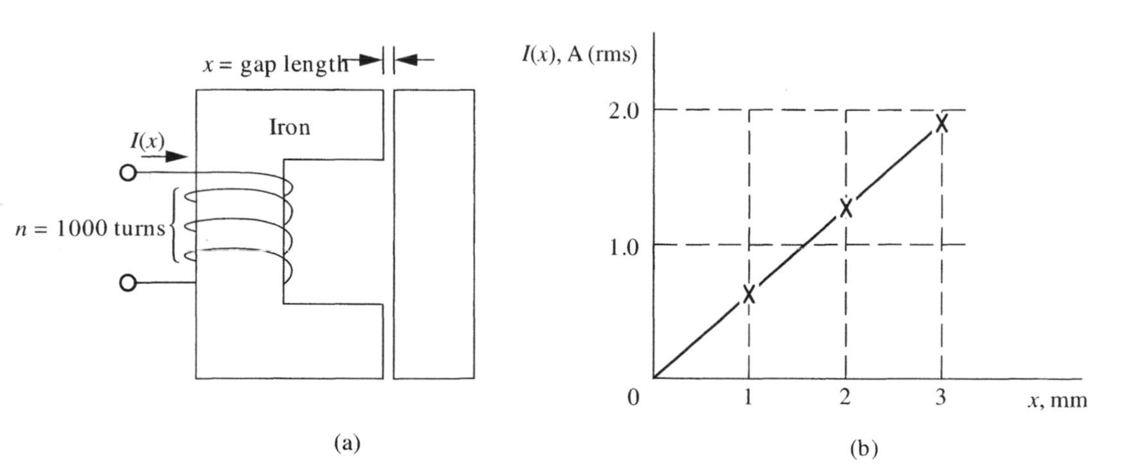

Figure P3.62

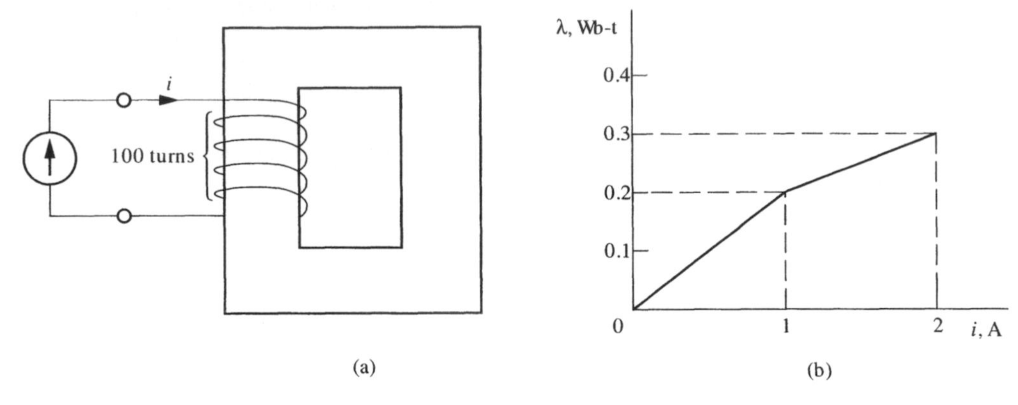

Figure P3.64

(2) As the sawing is going on, does the source maintaining the current receive or give energy to the magnetic system?

3.65. Consider a magnetic structure with a moving mechanical member. Figure P3.65 shows the flux linkage–current characteristic for the structure for two values of mechanical displacement. The areas of several parts of the diagram are given.

(a) Estimate the mechanical force generated by the structure at $\lambda = 1.8$ Wb–t, $i = 1.0$ A, and $x = 1.0$ cm.

(b) Estimate the stored energy at the same point.

(c) Estimate the inductance at $x = 1.0$ cm for small currents.

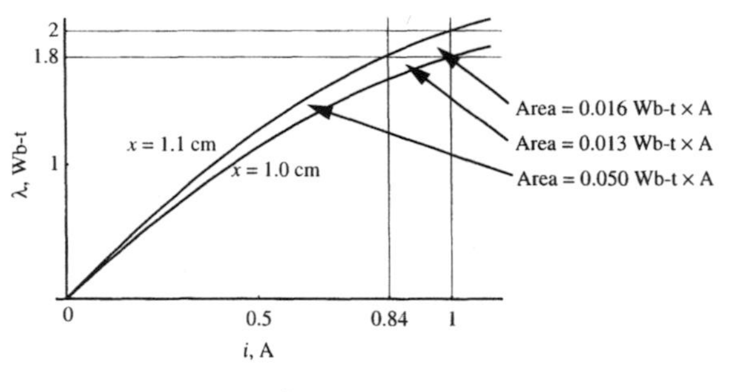

Figure P3.65

3.66. A 100-turn linear inductor has an inductance $L = 5\sqrt{x} + 0.001$ H, where x is a mechanical displacement in meters. The inductor is excited by a 24-V, 60-Hz voltage source.

(a) Find the peak flux in the inductor.

(b) Find the time-average force tending to increase x at 1 mm.

3.67. For the rotational device in Fig. P3.67, the reluctance of the air gap is $\Re(\theta) = 10^5(1 + |\theta|)$, where θ is expressed in radians and $\theta < \pi/2$. Neglect mmf losses in the iron.

(a) Find the torque at $\theta = +\pi/4$ for a current of 10 A and 100 turns.

(b) Find the inductance of the system at the same angle.

(c) How much coenergy is stored at this angle and current?

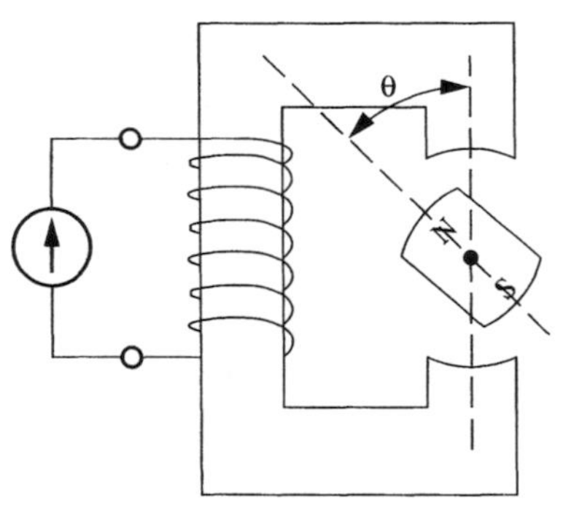

Figure P3.67

3.68. A doorbell circuit uses 12 V rms for safety, a simple switch on the front porch, and a magnet to vibrate a clapper for a small bell, as shown in Fig. P3.68. The magnet is wound with 2500 turns of fine enamel-insulated wire. The iron in the magnet and clapper has $6000\mu_0$, but the system reluctance is limited mainly by the two 0.1-inch gaps, each of which with an area of 1 cm^2.

(a) Find the maximum flux density in the iron.

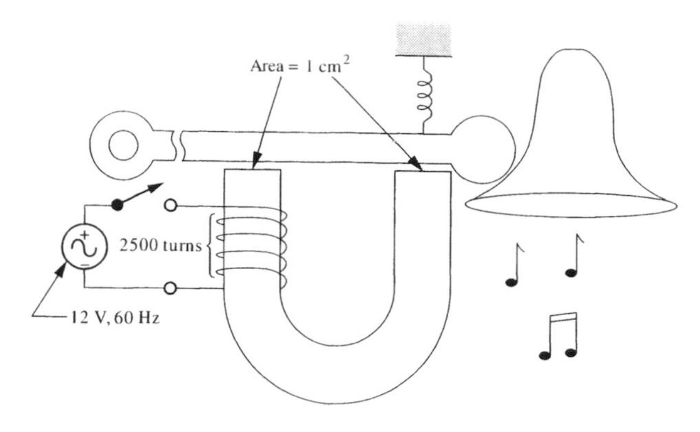

Figure P3.68

(b) Find the rms current through the switch when closed.

3.69. A 120-V/6-V, 60-Hz transformer is to be wound on a square iron core with a 5-cm outer size and a 3-cm inner hole. The area is 1 cm^2, and the nominal length of the path is 16 cm. To make it easy to wind, the top or the core is bonded after winding, and the bonding substance has a thickness of 0.05 mm. Assume these gaps have a permeability of μ_0. The permeability of the iron is $6000\mu_0$, and the maximum flux density is to be 1.0 tesla at rated voltage.

(a) Find the number of turns on the primary and secondary windings.

(b) Calculate the magnetizing current on the 120-V side, rms value.

(c) Estimate the time-average force developed on the bonding substance at rated voltage.

3.70. A magnetic transducer has a 2 cm × 2 cm member that slides out, as shown in Fig. P3.70. The effective gap on the member is 0.1 mm on each side. The mmf is 1000 A-t. Find the force exerted on the moving member at $x = 1$ cm if the reluctance of the iron is ignored.

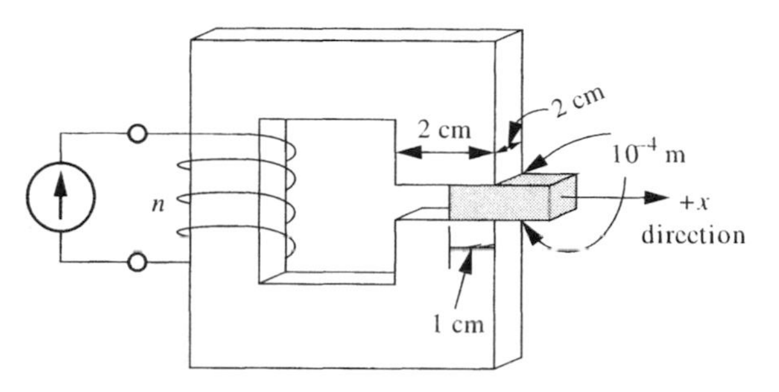

Figure P3.70

3.71. A 120-V, 60-Hz hair clipper, Fig. P3.71(a), has the vibration motor dimensioned in Fig. P3.71(b). The spring is adjusted to resonate with the mass of the vibrator for maximum motion. The coil has 5000 turns, the average gap size is about 0.08 in., and the areas are as shown. Ignore mmf losses in the iron. Assume lateral motion.

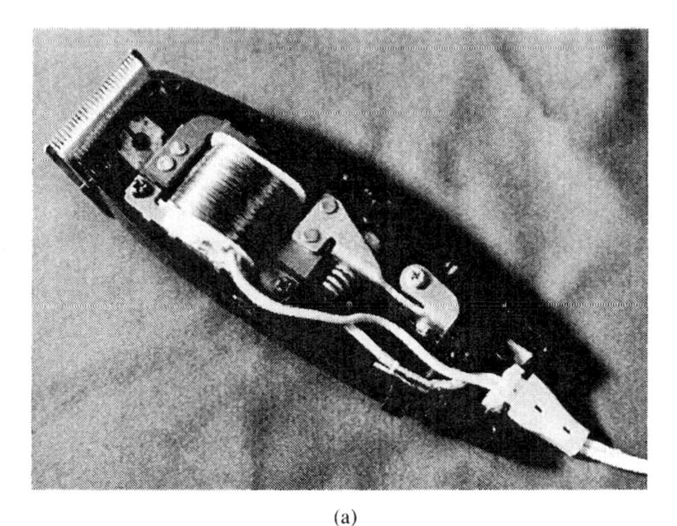

(a)

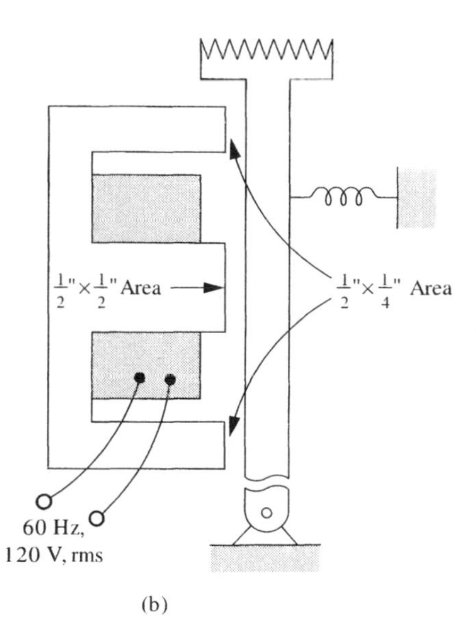

(b)

Figure P3.71

(a) At what frequency should the spring–mass system resonate?

(b) Estimate the input current required, assuming the real power into the motor is 25 W. Assume that the in-phase and magentizing current add in parallel.

(c) Give a parallel equivalent circuit for the vibrator motor.

3.72. Figure P3.72 shows a magnetic structure with a rotational permanent magnet in the gap. A steady current is applied to the coil. Experimentally, it is determined that the torque that must be applied to keep the magnet in equilibrium at θ_m is $T(\theta_m) = 5 \sin\theta_m + 1.5 \sin(2\theta_m)$ N-m.

(a) Identify which term in the torque expression comes from the permanent magnetism and which from the induced magnetism.

(b) Determine the polarity of the electromagnet and show the corresponding current direction on the coil.

(c) If the permanent magnet is moved from $\theta_m = 0°$ to $\theta_m = 180°$ in the positive θ_m direction, how much work is done on the electrical system? Assume lossless and linear magnetic properties.

(d) Mark the polarity of the induced voltage on the coil for the rotation in part (c).

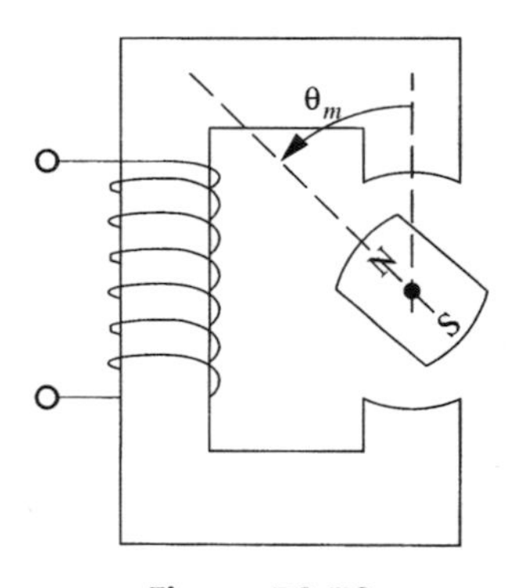

Figure P3.72

3.73. A magnetic structure has a rotating member that is permanently magnetized. With 1-A dc current in the direction shown, the torque in the positive θ_m direction is found to be

$T(\theta_m) = -\ 5 \sin\theta_m - 0.3 \sin(2\theta_m)$N-m.

(a) What would be the torque equation if the dc current were doubled?

(b) What would be the torque equation if the dc current were reversed?

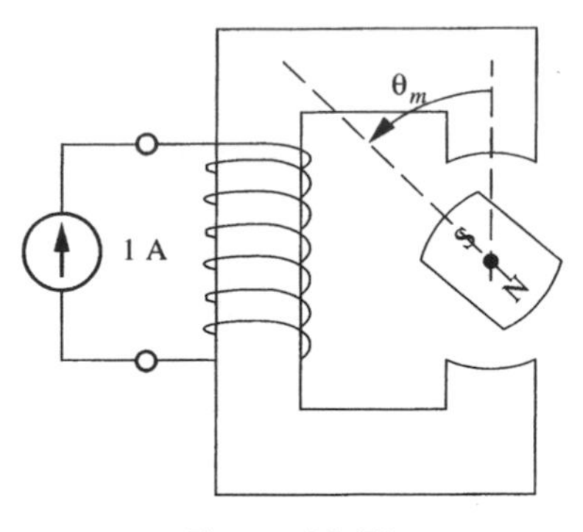

Figure P3.73

(c) What would be the equation for the time-average torque if the current were $i(t) = 1 \cos(120\pi t)$ A?

3.74. The λ–i curve shown in Fig. P3.74 describes the magnetic characteristics of a magnetic structure that has a movable member. The curves for two positions of that member are shown. The movable member is set at $x = 0$ and the current is increased from zero to 10 A.

(a) How much energy was put into the system by increasing the current from zero to 10 A?

(b) With the current maintained at 10 A, the movable member moves from $x = 0$ to $x = 1$ mm. How much work was done by the electrical system in this change?

(c) What was the average force exerted by the electrical system on the movable member as the value of x was changed?

(d) Is the force at $x = 0.5$ mm greater than, less than, equal to, or cannot tell relative to the average force? Choose one answer and explain.

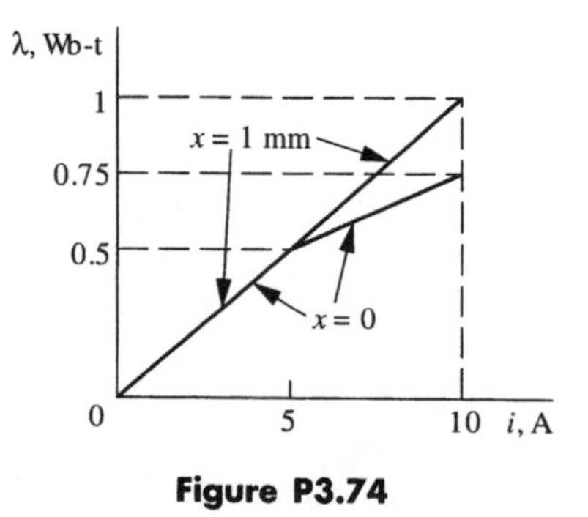

Figure P3.74

3.75. An electrical weighing device is constructed by using a magnetic structure with a gap, as shown in Fig. P3.75. The force puts a strain on the iron,

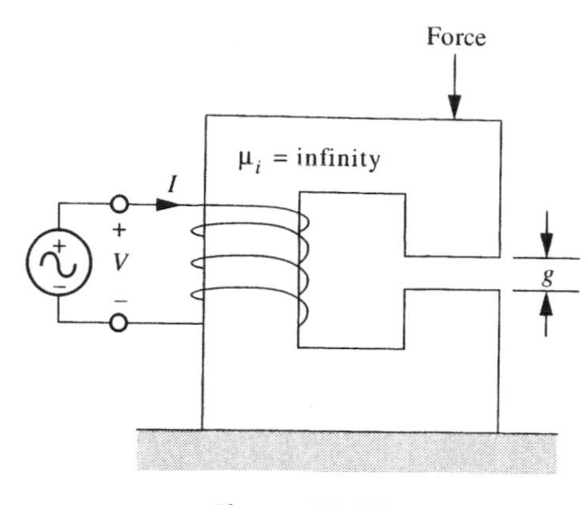

Figure P3.75

which changes the gap width, which affects inductance. The circuit uses an ac constant voltage source and uses the current as an indication of the force acting on the system. Consider the iron ideal ($\mu_i = \infty$) and neglect any losses in the wire.

(a) Define the variables you need and derive a formula for the current in terms of the gap width and whatever variables you introduce.

(b) Discuss the effect of the magnetic forces on the accuracy of the system.

3.76. Figure P3.76 shows an electromechanical transducer. When the coil is excited, the movable member

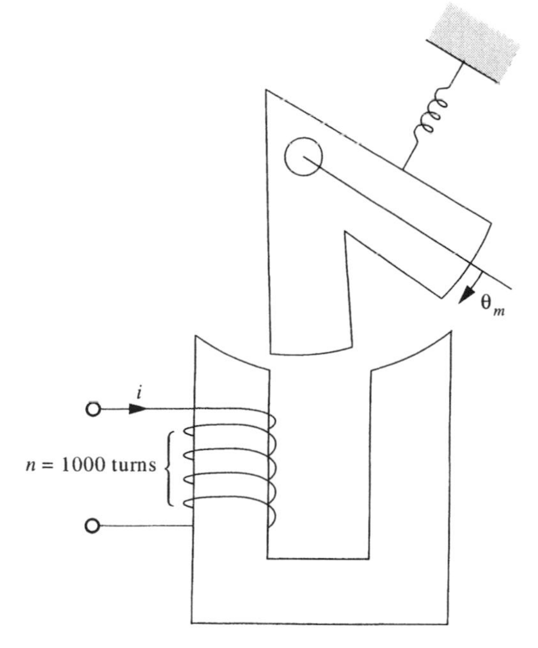

Figure P3.76

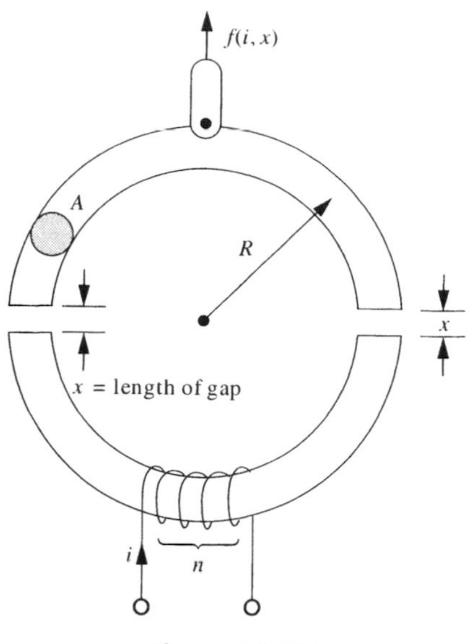

Figure P3.77

rotates, as measured by an angle θ_m. The reluctance of the iron structure is given by $\Re(\theta_m^\circ) = 9000/(\theta_m^\circ + 3^\circ)$, where θ_m° is the angle in degrees and must be in the range from 5° to 20°. Find the torque generated at $\theta_m^\circ = 12^\circ$ and a current of 1.5 A dc.

3.77. The toroidal ring in Fig. P3.77 has two gaps, which remain equal. The nominal radius is R, there are n turns on the coil, and the cross-sectional area is A. The permeability of the iron is μ_i. Determine the force required to hold the top in equilibrium as a function of current and gap distance x.

3.78. A magnetic structure is shown in Fig. P3.78. The iron cross-section is square, $a \times a$. The iron is ideal. The coil has n turns and a dc current of I. Due to fringing, the effective area of the gap depends on g, such that the gap area is approximately $(a + g)^2$. Find the force on the top pole face.

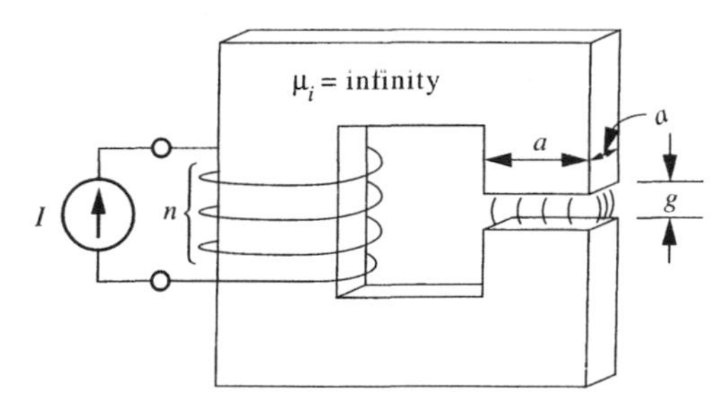

Figure P3.78

Answers to Odd-Numbered Problems

3.1. 17.0 mH.

3.3. **(a)** 9.09×10^{-4} Wb; **(b)** deflected out of the paper; **(c)** 0.343 V, rms; **(d)** 34.3 Ω.

3.5. 6.57 g.

3.7. **(a)** 54.6 mH; **(b)** 1.19.

3.9. **(a)** 15 mH; **(b)** 120 mH; **(c)** 75 mH; **(d)** no change; **(e)** 5.46 mH.

3.11. **(a)** Dots are at the bottom
(b) $R_{eq} = (n_p/n_1)^2 R_1 + (n_p/n_2)^2 R_2$.

3.13. **(a)** 31.8 mA; **(b)** 400 mA; **(c)** 51.1 mA.

3.15. **(a)**

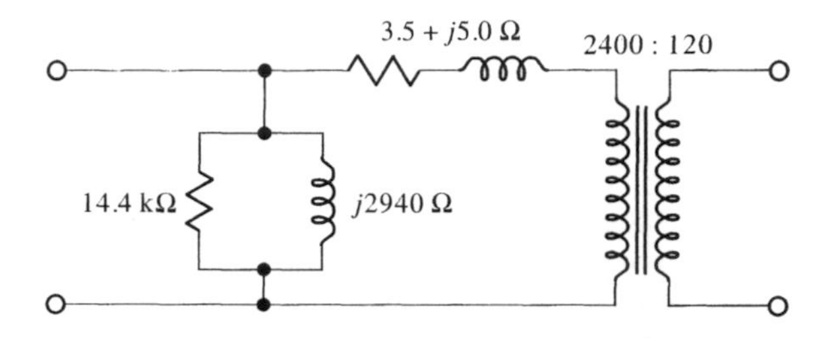

(b) 96.6%; **(c)** 2.55%.

3.17. **(a)** Both dots at the top or both at the bottom;
(b)

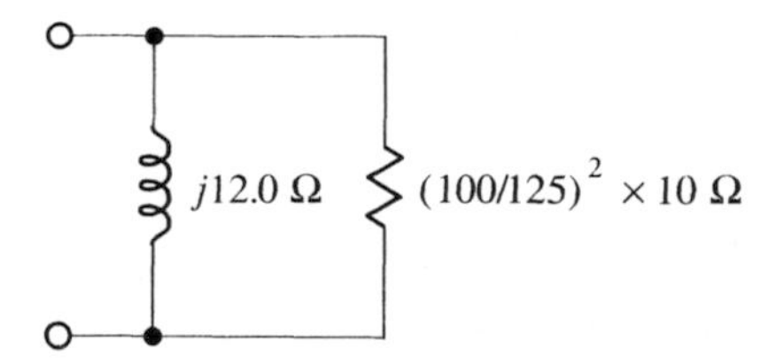

(c) repelled.

3.19. **(a)** 16,000 V and 3200 V; **(b)** 0.322 A;
(c)

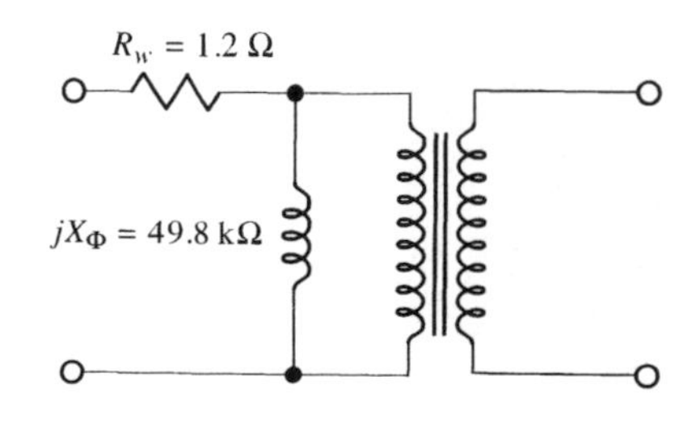

3.21. **(a)** 479 and 48; **(b)** 0.655 A; **(c)** 430 W; **(d)** 21.9 A.

3.23. **(a)** $\mathbf{T}(\underline{s}), = 2\underline{s}/(\underline{s} + 75.0)$; **(b)** $v(t) = 235\sqrt{2}\cos(120\pi t + 11.3°)$ V; **(c)** $-75.0\ \text{s}^{-1}$; **(d)** $12.9e^{-75t} + 235\sqrt{2}\cos(120\pi t + 11.3°)$ V.

3.25. **(a)** 4.50×10^{-4} Wb; **(b)** $3.77 \times 10^{4}\ \text{H}^{-1}$; **(c)** 694 Ω in series with 26.5H; **(d)** 3.00 mA.

3.27. **(a)** 360 turns; **(b)** 4.5 A; **(c)** 1420 W; **(d)** $47.9 \times 10^{4}\ \text{H}^{-1}$.

3.29. **(a)** 10:1; **(b)** 576 W; **(c)** 977 W; **(d)** 97.8%; **(e)** 0.537 A; **(f)** 0.888 lagging.

3.31. **(a)** 0.315 A; **(b)** 0.158 A; **(c)** 1490 V; **(d)** 746 V; **(e)** 235 VA.

3.33. **(a)**

Condition	V_H	I_H	P_H	V_L	I_L	P_L
OC	460 V	0	0	230 V	0.566 A	92 W
SC	17.9 V	20 A	160W	0	40 A	0

(b) 461 V and 95.8%.

3.35. 93.5%.

3.37. **(a)** 6 A; **(b)** 240 V; **(c)** 92.6%.

3.39. **(a)**

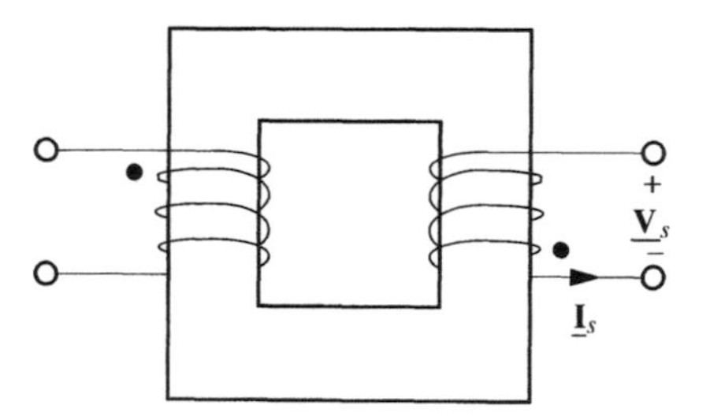

(b) 880 V and 220 V; **(c)** 86.9% at 13.3 kVA; **(d)** 111.8 kg.

3.41. **(a)**

Primary/secondary connection	V_p	I_p	V_s	I_s
series/series	920 V	21.7 A	460 V	43.5 A
series/parallel	920 V	21.7 A	230 V	87.0 A
parallel/series	460 V	43.5 A	460 V	43.5 A
parallel/parallel	460 V	43.5 A	230 V	87.0 A

(b)

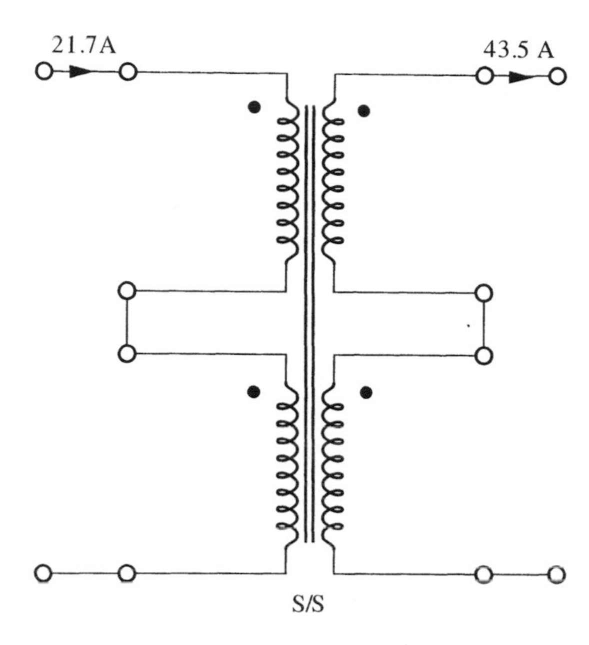
21.7A
43.5 A
S/S

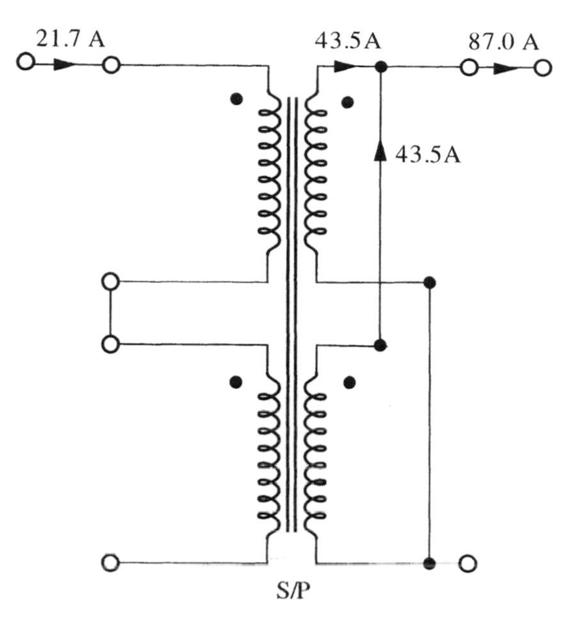
21.7 A
43.5A
87.0 A
43.5A
S/P

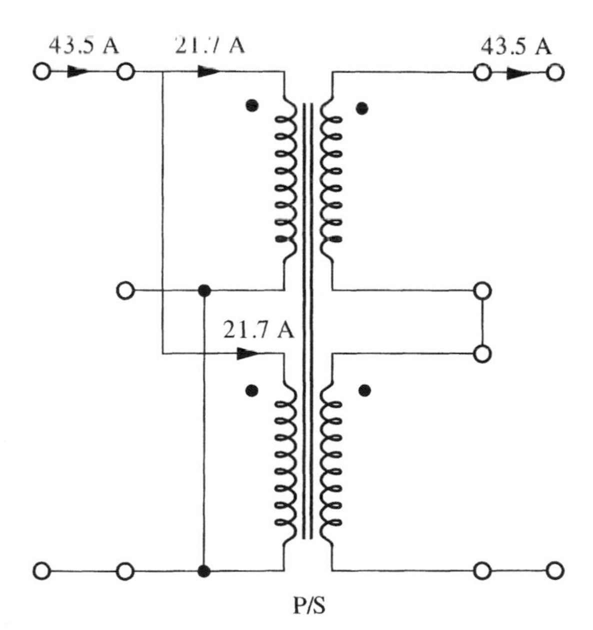
43.5 A
21.7 A
43.5 A
21.7 A
P/S

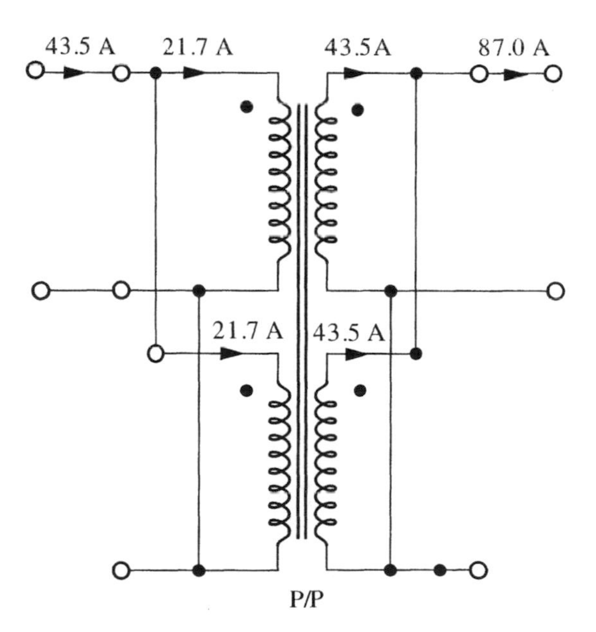
43.5 A
21.7 A
43.5A
87.0 A
21.7 A
43.5 A
P/P

3.43. 96.0%.

3.45. **(a)**

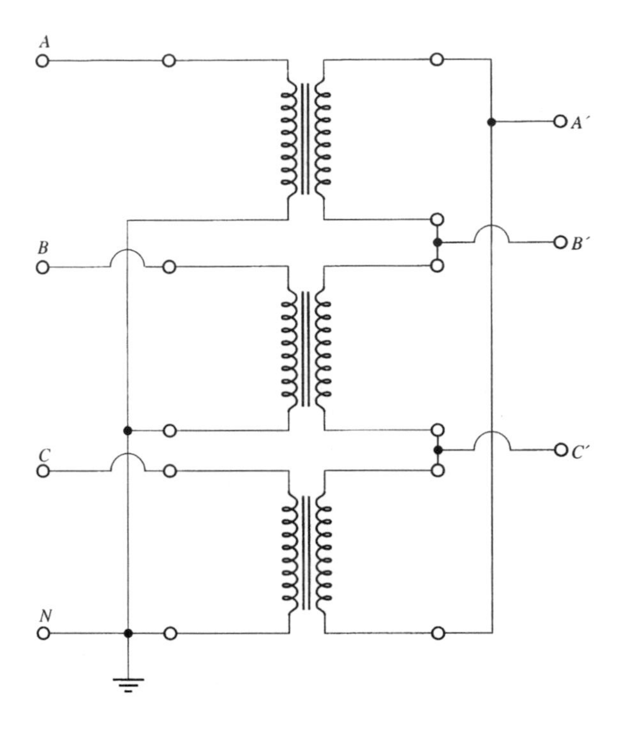

(b) 14,900 V, 1200 V; **(c)** 1.94 A; **(d)** 93.5%; **(e)** 0.461 Ω/phase.

3.47.

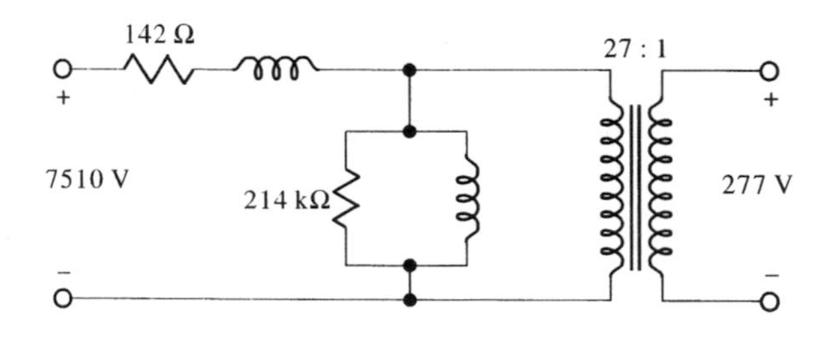

3.49. **(a)** 69.3 kVA; **(b)** 4000 V; **(c)** 0.357 A, rms; **(d)** 4140 W.

3.51. **(a)** 0.0481 A, rms; **(b)** 97.0%; **(c)** 2.31%.

3.53. **(a)** 1320 V on the primary, 266 V on the secondary, 5.33 A on primary, 15.3 A on secondary; **(b)**

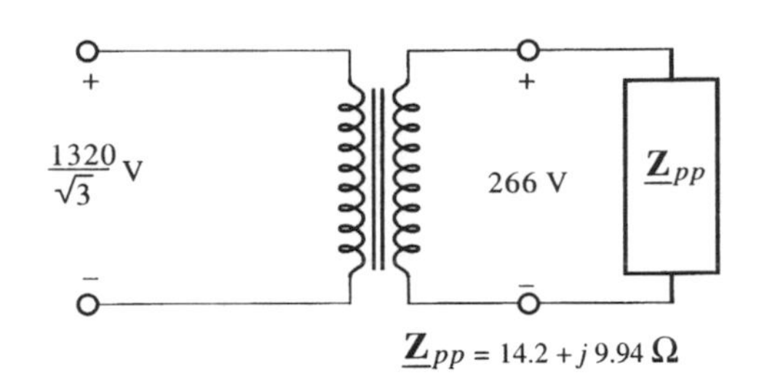

3.55. 0.5 N independent of x.

3.57. **(a)** 5×10^5 A-t/m; **(b)** – 31.4 N.

3.59. **(a)** 6.75 mA; **(b)** 1.79 N; **(c)** 1.79 N.

3.61. – 66.7 N.

3.63. **(a)** 2 J; **(b)** +20 N.

3.65. **(a)** + 63.0 N on the basis of energy, 79.0 N on the basis of coenergy, average = 71.0 N; **(b)** 0.8 J approximate; **(c)** 34 approximate.

3.67. **(a)** −1.57 N-m; **(b)** 56.0 mH; **(c)** 2.80 J.

3.69. **(a)** 4500 and 225 turns; **(b)** 15.8 mA; **(c)** 3.98×10^{-2} N.

3.71. **(a)** 120 Hz; **(b)** 0.330 A; **(c)** 576 Ω in parallel with $j471$ Ω.

3.73. **(a)** $T(\theta_m) = -1.0 \sin \theta_m - 1.2 \sin(2\theta_m)$ N-m; **(b)** $T(\theta_m) = +0.5 \sin \theta_m - 0.3 \sin(2\theta_m)$ N-m; **(c)** $T(\theta_m) = -0.15 \sin(\theta_m)$ N-m.

3.75. **(a)** A = area, g = gap, n = turns, V = rms voltage, and ω = electrical radian frequency $I = \dfrac{V}{\omega n^2 \mu_0 A} g$;

(b) $f = -V^2/2\omega^2 n^2 \mu_0 A$, independent of g, so magnetic force does not affect the scheme.

3.77. $f_{dev} = -\dfrac{1}{2}(ni)^2\left(\dfrac{1}{\Re_i + 2x/\mu_0 A}\right)^2 \times \dfrac{2}{\mu_0 A}$.

The external force is "+".

CHAPTER

The Synchronous Machine

Flux and Torque in Cylindrical Magnetic Structures

Rotating Magnetic Flux for AC Motors

Synchronous Generator Principles and Characteristics

Characteristics of the Synchronous Motor

Chapter Summary

Glossary

Problems

1. To understand how magnetic flux is created and distributed in the air gap of a cylindrical magnetic structure
2. To understand how magnetic flux is made to rotate in cylindrical magnetic structures
3. To understand the role of the rotor and stator flux in the operation of synchronous generators
4. To understand the characteristics of the synchronous generator when operated into a large power system
5. To understand the characteristics of the synchronous motor, particularly the effect of field current on the power factor

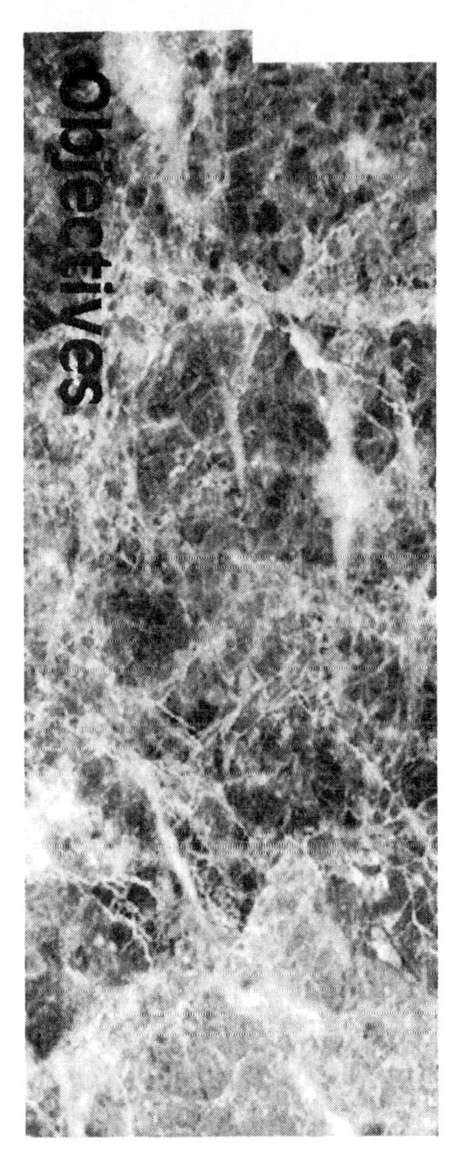

The synchronous machine produces the ac power that we use in home, office, and factory. To understand and develop circuit models for this important machine, we must analyze the means by which suitable magnetic flux is established in cylindrical structures, and by which this flux is made to rotate. Rotating flux is required for steady torque in both synchronous and induction machines.

Introduction to Motor/Generators

Importance of the synchronous machine. The three-phase synchronous machine can be used as a generator or motor. As a motor, it has highly specialized properties and serves a narrow range of applications. The synchronous generator is the workhorse of the electric power industry for generating ac electric power. On a smaller scale, the automotive alternator is a small three-phase synchronous generator with six or eight diodes to rectify the ac to dc for charging the battery and supplying the electrical system.

General characteristics of the synchronous machine. Figure 4.1 shows the external connections of a synchronous machine. Energy can be converted from mechanical to electrical form, generator action, or from electrical to mechanical form, motor action. The machine has two electrical connections: a dc field connection and a three-phase ac armature connection. The mechanical connection is a rotating shaft. As we will establish later, the machine has the following characteristics:

- The dc field circuit, placed on the rotor, is essentially a rotating electromagnet whose flux is controlled by the dc field current.
- The armature circuit is placed on the stator and carries three-phase currents.

Conservation of Energy

- The flow of real power through the system is determined by the mechanical input because the mechanical system exchanges real power only.[1] The power supplied to the dc field circuit supplies ohmic losses in the field winding but does not enter directly into the energy-conversion process. For generator action, the mechanical input power, minus the losses, becomes three-phase electrical output power. For motor action, the real power from the electrical input, minus the losses, becomes mechanical output power.
- When the load on a generator is a large power system, the reactive power flow is controlled by the dc field current. When the synchronous machine is operated as a motor, the reactive power required, and thus the power factor is con-

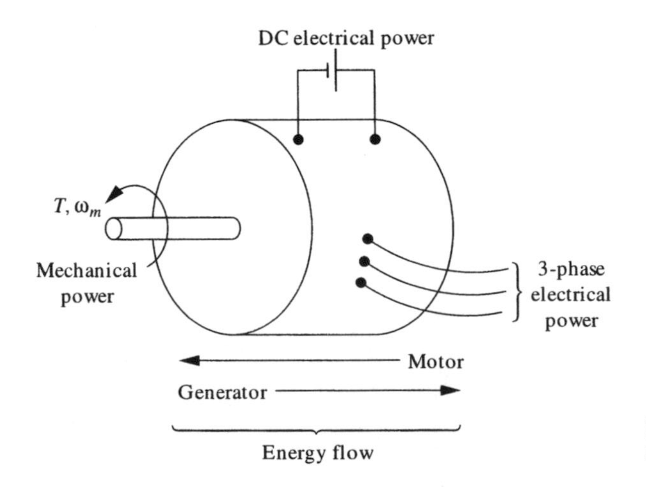

Figure 4.1 A synchronous machine can act as generator or motor.

[1] Mechanical energy flows smoothly because the inertia and general stiffness of the system eliminate fluctuations in mechanical stored energy.

trolled by the field current. This direct control of reactive power flow is a surprising, and extremely useful, property of the synchronous machine.

Conditions for Motor/Generator Action

Torque generation. The steady transformation of electrical into rotational mechanical power or vice versa requires both torque and rotation.

$$P_{dev} = T_{dev} \times \omega_m \tag{4.1}$$

stable equilibrium, unstable equilibrium

where P_{dev} is the developed power, T_{dev} is the developed torque, and ω_m is the mechanical angular speed of the rotor. Let us consider first the means for producing torque. Figure 4.2(a) shows a compass at an angle θ_m relative to a magnetic flux. At $\theta_m = 0°$, the compass is in *stable equilibrium*: no torque is produced. For θ_m in the first quadrant, as shown in Fig. 4.2(a), a torque in the negative θ_m direction tends to restore equilibrium. This torque increases up to $\theta_m = 90°$ and then decreases to zero at $\theta_m = 180°$, where the compass has an *unstable equilibrium*. The equilibrium is unstable because a small displacement, either way, produces torque tending to increase the displacement. The $T_{dev}(\theta_m)$ curve shown in Fig. 4.2(b) is sinusoidal because the lever arm for the torque varies as $\sin\theta_m$.

We consider now the torque generated by the stator and rotor magnetic poles in Fig. 4.3. Because opposite poles attract, we also have a stable equilibrium at $\theta_m = 0$, an unstable equilibrium at $\theta_m = 180°$, and a maximum restoring torque around $\theta_m = \pm 90°$. Although it is not obvious, the torque characteristic for this structure is that shown in Fig. 4.2(b), as is demonstrated later in this chapter. Thus, torque can be generated magnetically through a displacement between stator and rotor poles.

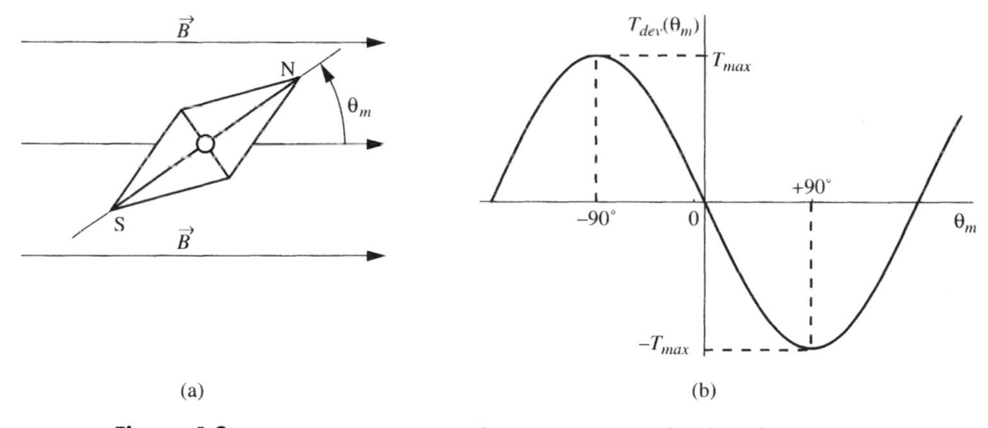

Figure 4.2 (a) Compass in magnetic flux; (b) torque as a function of displacement angle.

Ways to achieve motor action. As indicated by Eq. (4.1), sustained torque and rotation are required to convert electrical energy to mechanical energy and vice versa. Electrical motors operate from this principle but differ in (1) how the rotor and stator poles are produced and (2) whether the rotor or stator poles cause rotation. For a synchronous machine, the stator flux (poles) rotates due to the three-phase currents and torque is developed when the electromagnet on the rotor is rotating at the same speed.

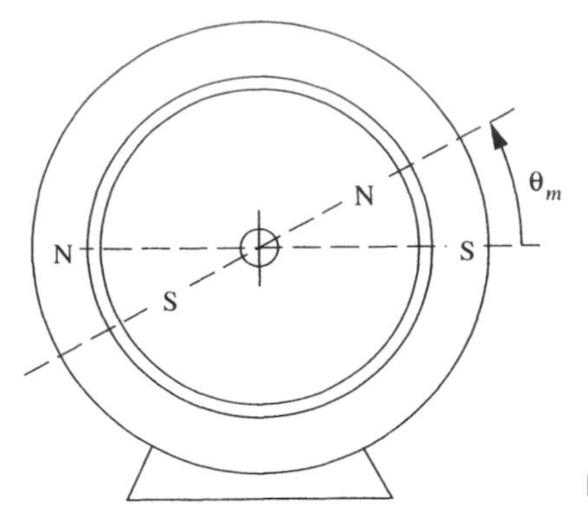

Figure 4.3 Stator and rotor poles.

Chapter contents. We begin by showing how stator magnetic flux is established by coils in the stator of cylindrical magnetic structures. We then introduce rotor flux and investigate the torque between stator and rotor. Then we show how the stator flux is made to rotate in two- and three-phase stators. A model for a synchronous machine is then presented. Performance characteristics for synchronous generators and motors are then explored.

4.1 FLUX AND TORQUE IN CYLINDRICAL MAGNETIC STRUCTURES

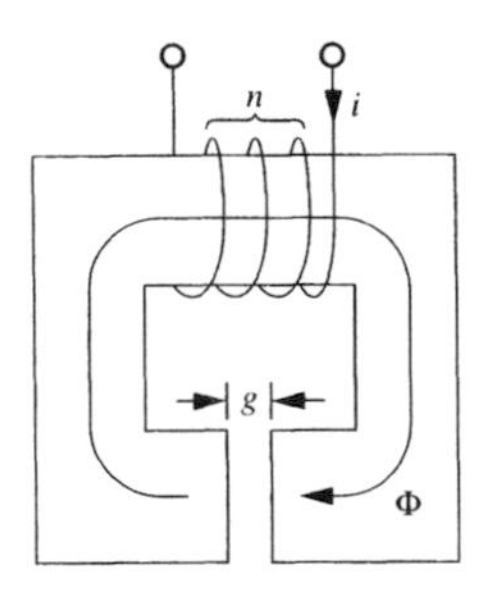

Figure 4.4 Magnetic structure with a gap.

Analysis of Cylindrical Magnetic Structures

Role of iron and the air gap. In this section, we show how currents on the inner surface of the stator produce magnetic flux in the air gap. We begin by reviewing magnetic-structure concepts from the previous chapter. Figure 4.4 shows a simple magnetic structure with a gap. The mmf of the coil produces a magnetic flux that is limited by the reluctances of the iron and the air gap. As derived in Sec. 3.1, the total flux is

$$\Phi = BA = \frac{\text{mmf}}{\text{reluctance}} = \frac{ni}{\mathfrak{R}_i + \mathfrak{R}_g} \tag{4.2}$$

where Φ is the flux, B the flux density, A the area, ni the mmf, and the $\mathfrak{R}$'s the reluctances of the iron and the gap. Because the relative permeability of iron is large, the flux is limited mainly by the air gap.

LEARNING OBJECTIVE 1.

To understand how magnetic flux is created and distributed in the air gap of a cylindrical magnetic structure

Description of structure. Most alternating-current motors have an iron structure with cylindrical symmetry. Figure 4.5(a) shows an axial view of such a cylindrical magnetic structure with the current-carrying wires on the stator side of the air gap. The currents in the coil go from front to back on the bottom, cross over to the top at the back, and return on top, as shown in Fig. 4.5(b). The wires lie in slots that have been milled in the inner circumference of the stator. The stator structure for a small motor is shown in Fig. 4.6. The rotor is assumed to be a coaxial iron cylinder in our analysis. We will determine the air-gap flux density established by the stator currents.

Right-hand rule. We may apply the right-hand rule to the currents in Fig. 4.5(a) by putting the fingers of the right hand around the rotor in the direction of the stator currents and noting that the thumb is horizontal and pointing to the right. Hence, the magnetic flux

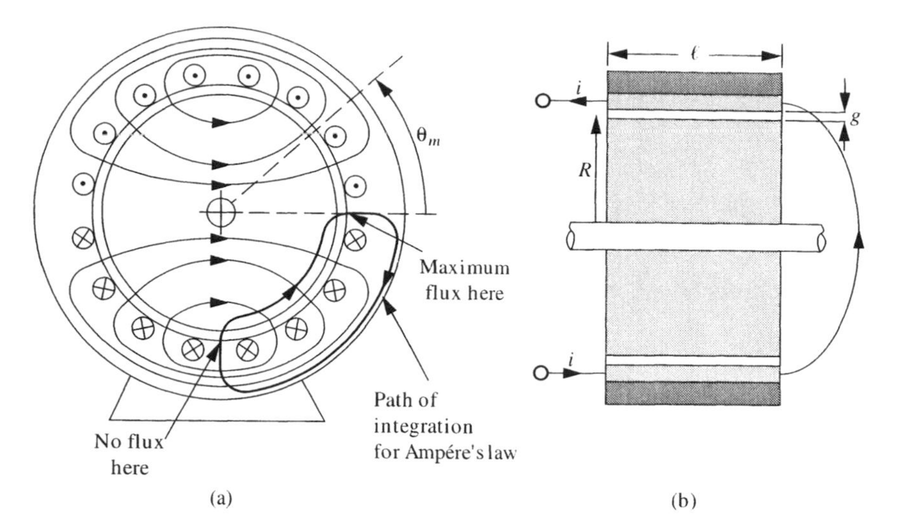

Figure 4.5 Cylindrical magnetic structure: (a) axial view; (b) side view. The mechanical angle is θ_m, and the width, radius, and length of the air gap are g, R, and ℓ, respectively.

Figure 4.6 Four-pole stator structure for a single-phase induction motor. The main windings for the motor are horizontal and vertical. The smaller starter windings are on the diagonals. The windings may be connected in parallel for 120-V operation or in series for 240-V operation.

in the rotor and air gap is directed to the right, as shown. By symmetry the magnetic flux density is maximum at $\theta_m = 0°$ and 180° and zero at $\theta_m = -90°$ and +90°, as indicated.

Flux distribution. Although Fig. 4.5 suggests one conductor in each slot, in practice there are many, as shown in Fig. 4.6. The mmf of the coil, which is distributed around the circumference, is tapered sinusoidally. This sinusoidal distribution is accomplished by having heavy concentrations in the center of the coil, tapered to few wires[2] at the edge of the windings. The resulting flux distribution, shown in Fig. 4.5(a), has no flux crossing the air gap at the top and bottom, maximum flux density in the air gap in the positive radial direction at $\theta_m = 0°$, and maximum flux density in the air gap in the negative radial direction at $\theta_m = 180°$. This pattern may be interpreted as a distributed south magnetic pole in the stator centered at $\theta_m = 0°$ and a distributed north magnetic pole in the stator

[2] Do not worry about those partially filled slots; we fill them later with more coils.

centered at $\theta_m = 180°$. Thus, we are describing a two-pole structure. Because of the sinusoidal taper of the mmf, the flux density is sinusoidal in form:

$$B(\theta_m) = B_c \cos \theta_m \tag{4.3}$$

where B_c is the maximum flux density at $\theta_m = 0°$. The flux crosses the air gap radially.

Maximum flux density. We may determine the maximum flux density, B_c, by applying Ampère's circuital law around the indicated path of integration in Fig. 4.5(a). The path encloses half the wires in the coil, so the result is

$$\oint \vec{H} \cdot \vec{dl} = \text{enclosed current} = \frac{n}{2} i \tag{4.4}$$

where n is the number of turns in the coil and i is the current in each turn. The line integral has four contributions, two from crossing the air gap twice and two from the paths within the rotor and stator iron. We assume that the magnetic field in the iron makes a negligible contribution to the line integral because the iron has a large μ_i. Because of symmetry, the flux density at the bottom is zero and hence no contribution is made by crossing the air gap at $\theta_m = -90°$. This leaves the contribution from crossing the air gap at $\theta_m = 0°$. Because we are crossing in the same direction as the flux, Eq. (4.4) reduces to

$$H_{gap} g = \frac{B_c}{\mu_0} g = \frac{n}{2} i \tag{4.5}$$

and thus the maximum flux density due to a single coil is

$$B_c = \frac{\mu_0 n i}{2g} \tag{4.6}$$

Thus, the maximum flux density depends on the mmf of the coil and the gap width if we neglect the mmf required to magnetize the iron.

EXAMPLE 4.1 Maximum flux density

A small motor has an air-gap radius of 5 cm and a gap width of 1.0 mm. We wish to establish a maximum flux density of 1.4 T in the machine. Find the required rms mmf.

SOLUTION:

We can determine the peak mmf directly from Eq. (4.6):

$$1.4 = \frac{4\pi \times 10^{-7} n I_p}{2 \times 1.0 \times 10^{-3}} \Rightarrow n I_p = 2230 \text{ A-t, peak} \tag{4.7}$$

Thus, the rms mmf would be $2230/\sqrt{2} = 1580$ A-t.

WHAT IF?

What if there are 36 slots and the rms current is 5 A? How many wires/slot on the average are required?[3]

[3] 18 wires/slot.

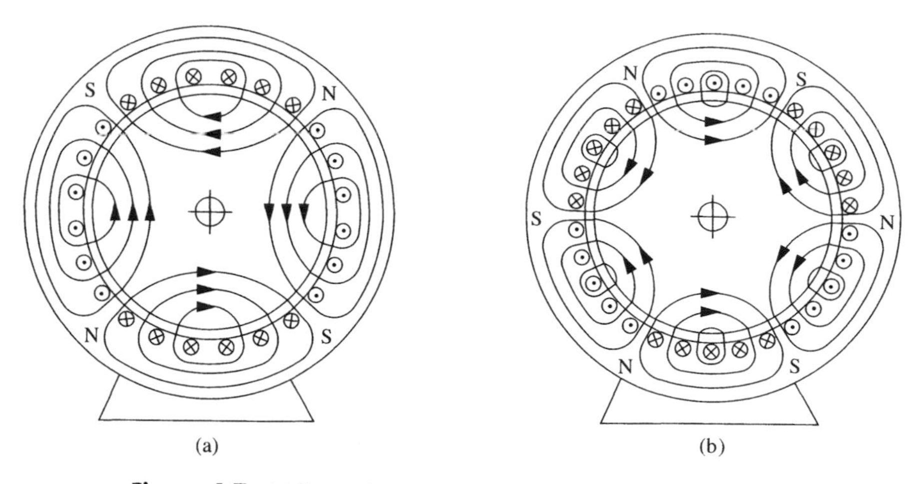

Figure 4.7 (a) Four-pole, two-coil stator; (b) six-pole, three-coil stator.

winding

For *P* poles. We now consider magnetic structures with more than two poles. Figures 4.7(a) and 4.7(b) show stators wound for four and six poles, respectively. As more coils are added to the winding, two additional poles are added for each additional coil. The general form of Eq. (4.3) is

$$B(\theta_m) = B_c \cos\left(\frac{P}{2}\,\theta_m\right) \tag{4.8}$$

where P is the number of poles and $P/2$ is the number of coils in the *winding*.[4] Note that B_c is still given by Eq. (4.6), with n still being the number of turns in one coil.

Torque Development between Rotor and Stator Fluxes

Introduction. We showed before that we need rotor and stator poles to have torque between the rotor and stator. In this section, we assume that we have both rotor and stator fluxes, that each varies sinusoidally in space, and that they combine to produce the total flux in the air gap. In the next section, we show the means by which the rotor flux is established. Here we investigate the developed torque and introduce the various power angles that play a role in synchronous motor characteristics.

Stop the rotation. During motor or generator operation, the rotor and stator fluxes rotate in synchronism around the cylindrical structure. However, the developed torque depends on the flux magnitudes, the angles between the fluxes, and the geometry of the machine. In the following, we assume for simplicity that the fluxes are stationary.

rotor–stator power angle

Combining fluxes. We assume the stator flux, $B_S(\theta_m)$, is horizontal and has two poles:

$$B_S(\theta_m) = B_S \cos\theta_m \tag{4.9}$$

[4] A winding is a series connection of coils.

where B_S is the maximum stator–flux density. As shown in Fig. 4.8, the rotor flux density, $B_R(\theta_m)$, is displaced from it by a physical angle δ_{RS}, which is called the *rotor–stator power angle*, measured positive from the stator magnetic axis to the rotor magnetic axis.

$$B_R(\theta_m) = B_R \cos(\theta_m - \delta_{RS}) \tag{4.10}$$

where B_R is the maximum rotor-flux density. The total air–gap flux density is the sum of the two, which may be combined like phasors[5] to produce

$$\mathbf{\underline{B}}_{RS} = \mathbf{\underline{B}}_R \angle -\delta_{RS} + \mathbf{\underline{B}}_S \angle 0° = \underbrace{\sqrt{B_R^2 + B_S^2 + 2B_R B_S \cos \delta_{RS}}}_{B_{RS}} \angle -\delta_S \tag{4.11}$$

rotor power angle

where $\theta_m = \delta_S$ is the angle of the maximum of the combined flux density, B_{RS}. The phasorlike addition of the rotor and stator fluxes is shown in Fig. 4.9,[6] where we also define the *rotor power angle*, δ_R, as the angle between the rotor pole and the maximum of the total flux density. Because the sum of two spatial sinusoids is still sinusoidal, the total flux density is

$$B_{RS}(\theta_m) = B_{RS} \cos(\theta_m - \delta_S) \tag{4.12}$$

This flux density is used to calculate the coenergy of the system, from which the developed torque is derived.

Coenergy determination. As shown in Sec. 3.3, we may determine the developed torque on the rotor by taking the partial derivative with respect to angle of either the en-

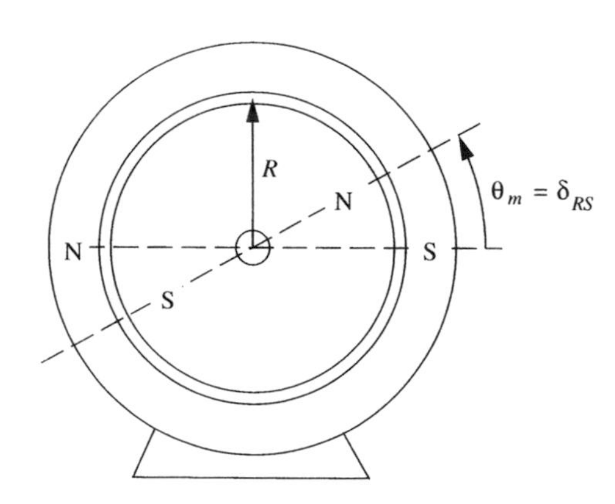

Figure 4.8 Rotor and stator poles are displaced by the rotor–stator power angle, δ_{RS}.

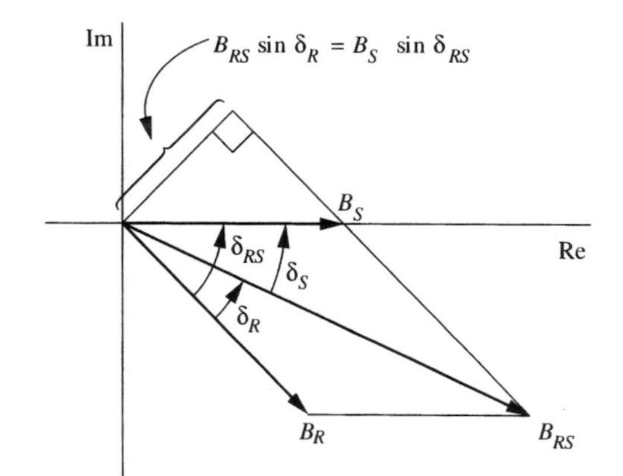

Figure 4.9 The rotor and stator fluxes combine by the law of cosines to produce the total (rotor–stator) flux density. The rotor–stator power angle, δ_{RS}, is the angle between the rotor and stator poles. The rotor power angle, δ_R, is the angle between the rotor poles and the total flux density maximum.

[5] These are stationary fluxes distributed sinusoidally in space. We are merely using the phasor technique to add two sinusoids.

[6] The angles appear reversed because of the way we define phasors. We marked them to correspond to spatial angles.

ergy or the coenergy functions. In the present case, we are assuming that the rotor and stator poles (fluxes) are constant in magnitude and vary only in relative position (angle). This requires constant current excitation, which suggests using the coenergy function. However, in this linear magnetic system, the coenergy and the energy are equal and are both proportional to the square of the total flux density. As shown in Sec. 3.3, almost all the energy is stored in the air gap, so we must substitute Eq. (4.11) into Eqs. (2.41) and (2.42) and integrate over the air-gap volume. The result is

$$W'_m(\delta_{RS}) = \frac{\pi R \ell g}{2\mu_0}(B_R^2 + B_S^2 + 2B_R B_S \cos \delta_{RS}) \tag{4.13}$$

where W'_m is the coenergy of the system and R, ℓ, and g are the air-gap radius, length, and width, respectively, as defined in Fig. 4.5(b).

Torque. We consider the stator poles stationary and determine the torque on the rotor poles. Equation (3.77) gives the developed torque as the partial derivative of the coenergy function with respect to the rotor–stator power angle, with the rotor and stator flux-density magnitudes kept constant.

$$T_{dev}(\delta_{RS}) = +\frac{\partial W'_m}{\partial \delta_{RS}} = -\frac{\pi R \ell g}{\mu_0} B_R B_S \sin \delta_{RS} \tag{4.14}$$

Figure 4.9 shows

$$B_S \sin \delta_{RS} - B_{RS} \sin \delta_R \tag{4.15}$$

which allows an alternative form of Eq. (4.14):

$$T_{dev}(\delta_{RS}) = -\frac{\pi R \ell g}{\mu_0} B_R B_{RS} \sin \delta_R \tag{4.16}$$

Equation (4.14) confirms that the torque is zero for $\delta_{RS} = 0$ and negative if δ_{RS} is positive, meaning the magnetic interaction tends to align the rotor poles with the stator poles.

Summary. Equation (4.14) gives the magnetically generated torque between two sets of poles with an angle δ_{RS} between them. The maximum torque depends on the product of the individually contributing fluxes and the geometric factors. The torque characteristic given in Eq. (4.14) is shown in Fig. 4.2(b). In the next section, we show how the stator poles are made to rotate in space so that torque between the stator and the rotating rotor can be sustained.

For *P* poles. For more than two poles, Eq. (4.8) requires that we place a $P/2$ in front of all mechanical angles relating to magnetic fluxes. In particular, the torque derived in Eq. (4.14) increases by $P/2$ because

$$\frac{\partial}{\partial \delta_{RS}} \cos\left(\frac{P}{2}\delta_{RS}\right) = -\frac{P}{2} \sin\left(\frac{P}{2}\delta_{RS}\right) \tag{4.17}$$

Thus, the general torque is

$$T_{dev}(\delta_{RS}) = -\frac{\pi R\ell gP}{2\mu_0}B_R B_S \sin\left(\frac{P}{2}\delta_{RS}\right) = -\frac{\pi R\ell gP}{2\mu_0}B_R B_{RS} \sin\left(\frac{P}{2}\delta_R\right) \quad (4.18)$$

where δ_{RS} is the mechanical angle between the rotor and stator poles, measured positive from stator poles to rotor poles, and δ_R is the mechanical angle between the total flux and the rotor poles.

electrical units, electrical degrees

Mechanical angles expressed in electrical units. Often mechanical angles are expressed in *electrical units,*[7] which is mechanical angle × *P*/2. In electrical angle units, the (*P*/2)'s go away in the various flux equations, and after the *P*/2 is placed in the torque equation, all machines are equivalent to a two-pole machine. Converting mechanical angle to electrical angle has the added advantage that physical angles in electrical units correspond directly to electrical phase angles.

Check Your Understanding

1. A cylindrical magnetic structure cannot have an odd number of magnetic poles. True or false?
2. In Fig. 4.6, the wires that can be seen are not the wires that produce the motor flux, but are the crossover wires between poles. True or false?
3. For large flux, the air gap in a machine should be as large or as small as possible. Which?
4. If the maximum torque in Fig. 4.2(b) is 5 N-m, determine the work required to move the rotor from its position of stable equilibrium to the position of unstable equilibrium.
5. For maximum torque, the rotor and stator fluxes should be aligned. True or false?
6. The rotor–stator power angle is 10° in a six-pole machine. How many electrical degrees is that?

Answers. **(1)** True; **(2)** true; **(3)** small; **(4)** 10 J; **(5)** false; **(6)** 30°.

4.2 ROTATING MAGNETIC FLUX FOR AC MOTORS

LEARNING OBJECTIVE 2.

To understand how magnetic flux is made to rotate in cylindrical magnetic structures

Two-Phase Rotating Flux

The previous section showed how magnetic flux is produced in a cylindrical magnetic structure and how torque is developed between rotor and stator flux. This section shows how the stator flux, or the stator magnetic poles, is (are) made to rotate. Although the three-phase motor is more common than the two-phase, the latter offers a convenient place to start. We therefore consider a two-pole stator, like that in Fig. 4.5, with two coils: an *h* coil that creates a magnetic flux pattern with its maximum in the horizontal plane (shown), and a *v* coil that creates a magnetic flux pattern with its maximum in the vertical plane (not shown).

Sinusoidal mmfs. The wires/slot distributions for both coils are tapered to produce flux patterns that are sinusoidal in space. Hitherto, we have been considering only the *h*

[7] As *electrical degrees*.

coil, for which the wires are most dense on the top and bottom of the stator; the wires for the v coil are most dense on the sides of the stator and share intermediate slots with the h coil. The two coils are identical except for their orientation in space. If $i_h(t)$ is the current in the h coil and $i_v(t)$ the current in the v coil, then the magnetic flux densities in the air gap by Eqs. (4.6) and (4.3) would be

$$B_h(t, \theta_m) = \frac{\mu_0 n}{2g} i_h(t) \cos \theta_m \quad \text{and} \quad B_v(t, \theta_m) = \frac{\mu_0 n}{2g} i_v(t) \sin \theta_m \tag{4.19}$$

Both fluxes are radial across the air gap and have their maximum densities at $\theta_m = 0°$ and $\theta_m = +90°$, respectively.

The two-phase currents are

$$\begin{aligned} i_h(t) &= I_p \cos(\omega t), \quad \text{excites } B_c \text{ in the } h\text{-direction} \\ i_v(t) &= I_p \sin(\omega t), \quad \text{excites } B_c \text{ in the } v\text{-direction} \end{aligned} \tag{4.20}$$

where I_p is the peak current in each coil, ω is the electrical angular frequency, and a 90° phase shift between the two phases is indicated by the sine and cosine functions shown in Fig. 4.10. The two flux-density components are

$$\begin{aligned} B_h(t, \theta_m) &= B_c \cos(\omega t) \cos \theta_m \\ B_v(t, \theta_m) &= B_c \sin(\omega t) \sin \theta_m \end{aligned} \tag{4.21}$$

where B_c is the peak flux density created by each coil separately, Eq. (4.6).

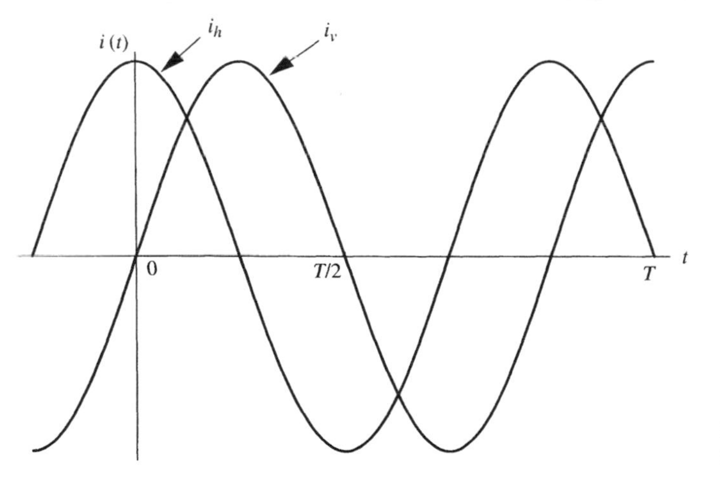

Figure 4.10 The two-phase system has a 90° phase shift between the two phases.

Rotating flux. When both coils are excited simultaneously, the fluxes add as scalars because both are radial, and the flux pattern moves in the counterclockwise direction, as demonstrated mathematically in Eq. (4.22):[8]

[8] Using a common trigonometric identity: $\cos(A - B) = \cos A \cos B + \sin A \sin B$.

$$
\begin{aligned}
B(t, \theta_m) &= B_h(t, \theta_m) + B_v(t, \theta_m) \\
&= B_c[\cos(\omega t) \cos \theta_m + \sin(\omega t) \sin \theta_m] \\
&= B_r \cos(\omega t - \theta_m)
\end{aligned} \tag{4.22}
$$

where $B_r = B_c$ is the magnitude of the rotating flux density. The flux density retains a stable pattern that rotates in space with the following properties:

- The magnitude is equal to the peak flux density from each contributing coil.
- If time is fixed, the flux is sinusoidal in space with the maximum flux density at $\theta_{\max} = \omega t$ if $t = 0$ when the maximum flux is at $\theta_m = 0°$.
- At a fixed θ_m, the flux-density magnitude is sinusoidal in time. The time of peak flux density is $t_p = \theta_m / \omega$.
- The flux, therefore, exhibits angular wave motion, a rotating wave of flux.

synchronous speed

For P poles. For more than two poles, the pattern of the magnetic flux advances one pole pair for each electrical cycle; hence, the mechanical speed of the flux pattern is slower than for two poles. If there are P poles, the flux density pattern will be

$$
B(t, \theta_m) = B_r \cos\left(\omega t - \frac{P}{2}\theta_m\right) \tag{4.23}
$$

and the *synchronous speed*, ω_s, therefore, will be

$$
\omega_s = \frac{\omega}{P/2} \text{ spatial rad/s} \tag{4.24}
$$

where ω is the electrical angular frequency in radians/second, and P the number of stator poles. Table 4.1 gives common synchronous speeds for 60 Hz.

TABLE 4.1 Synchronous Speeds for 60 Hz

P	ω_s	*rpm*
2	$2\pi \times 60$	3600
4	$2\pi \times 30$	1800
6	$2\pi \times 20$	1200

EXAMPLE 4.2 50-Hz motor

A two-phase, four-pole motor is operated at 50 Hz. The motor windings create counterclockwise rotation with a maximum flux density of 1.3 tesla. Find the synchronous speed and give the equation of the flux density, assuming the maximum flux density is at $\theta_m = 0°$ at $t = 0$.

SOLUTION:

For 50 Hz, $\omega = 2\pi \times 50$ and Eq (4.23) becomes

$$B(t, \theta_m) = 1.3 \cos\left(100\pi t - \frac{4}{2}\theta_m\right) \tag{4.25}$$

The synchronous speed comes from Eq. (4.24) as 50π rad/s.

WHAT IF? What if you want the speed in rpm?[9]

Phase reversal. If either the horizontal or vertical coil is reversed in polarity, the direction of flux rotation will change to clockwise; thus, the general expression for the flux density is

$$B(t, \theta_m) = B_r \cos\left(\omega t \mp \frac{P}{2}\theta_m\right) \tag{4.26}$$

where the minus sign goes with rotation in the positive θ_m direction and the plus with rotation in the negative θ_m direction.

Three-Phase Rotating Flux

Three-phase windings. Figure 4.11 shows a two-pole magnetic structure wound with three coils separated, by 120° in space. Note that coil a corresponds to our h coil from before, with the currents entering on the bottom and returning on the top, but coil b currents enter in the first quadrant and those of coil c enter in the second quadrant.[10] As before, the coil mmfs are tapered sinusoidally and share slots between their most

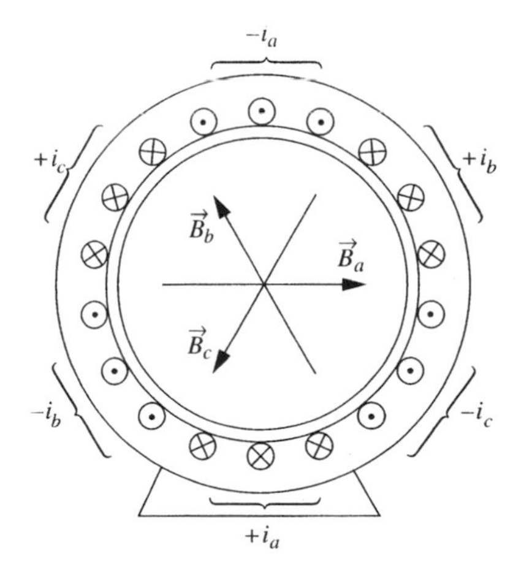

$$B_a = Ki_a(t)\cos\theta_m$$
$$B_b = Ki_b(t)\cos(\theta_m - 120°)$$
$$B_c = Ki_c(t)\cos(\theta_m - 240°)$$

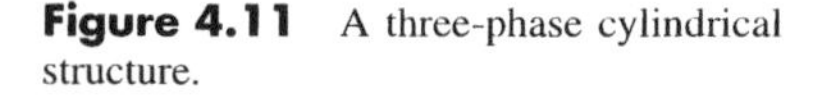

Figure 4.11 A three-phase cylindrical structure.

[9] $50\,\pi \times 60/2\pi = 1500$ rpm.

[10] In practice, the three coils are internally connected in wye or delta, and only three wires are connected externally to the three-phase system.

dense regions. The equations of the flux patterns from the coils, given in Fig. 4.11, show that the maximum from coil a lies in the horizontal plane, and the maxima from coils b and c are displaced 120° and 240° in space, respectively.

Analysis of three-phase system. The three-phase currents are shown in Fig. 1.1:

$$\begin{aligned} i_a(t) &= I_p \cos(\omega t) \\ i_b(t) &= I_p \cos(\omega t - 120°) \\ i_c(t) &= I_p \cos(\omega t - 240°) \end{aligned} \tag{4.27}$$

so, for example, the flux density from coil a is

$$B_a(t, \theta_m) = B_c \cos(\omega t) \cos \theta_m \tag{4.28}$$

where B_c is the peak flux density from one coil alone, Eq. (4.6). The combined flux density is

$$\begin{aligned} B(t, \theta_m) = {} & \underbrace{B_c[\cos(\omega t)\cos\theta_m}_{\text{phase } a} + \underbrace{\cos(\omega t - 120°)\cos(\theta_m - 120°)}_{\text{phase } b} \\ & + \underbrace{\cos(\omega t - 240°)\cos(\theta_m - 240°)]}_{\text{phase } c} \end{aligned} \tag{4.29}$$

We use the trigonometric identity in Footnote 8 to obtain the results

$$\begin{aligned} B(t, \theta_m) = B_c\{\tfrac{3}{2}\cos(\omega t - \theta_m) + \tfrac{1}{2}[\cos(\omega t + \theta_m) + \cos(\omega t + \theta_m - 240°) \\ + \cos(\omega t + \theta_m - 480°)]\} \end{aligned} \tag{4.30}$$

The three bracketed terms add to zero at all times because they form a balanced three-phase set.[11] Hence, Eq. (4.30) reduces to

$$B(t, \theta_m) = \tfrac{3}{2} B_c \cos(\omega t - \theta_m) = B_r \cos(\omega t - \theta_m) \tag{4.31}$$

where $B_r = \frac{3}{2} B_c$ is the magnitude of the rotating flux density with all coils operating together, a 50% increase in the flux density above what each gives individually. Equation (4.31) is identical to Eq. (4.22) and represents a rotating flux pattern. It is easily shown that changing the three-phase sequence from *abc* to *acb* reverses the direction of rotation to give Eq. (4.31) with $-\theta_m$ for *abc* rotation and $+\theta_m$ for *acb*.

Single Phase Expressed as Rotating Flux

Figure 4.12 shows a cylindrical magnetic structure with one winding excited by a single-phase current. The magnetic flux density in the air gap is given by

$$B(t, \theta_m) = B_c \cos \theta_m \cos(\omega t) \tag{4.32}$$

[11] Considering that −480° is the same as −120°.

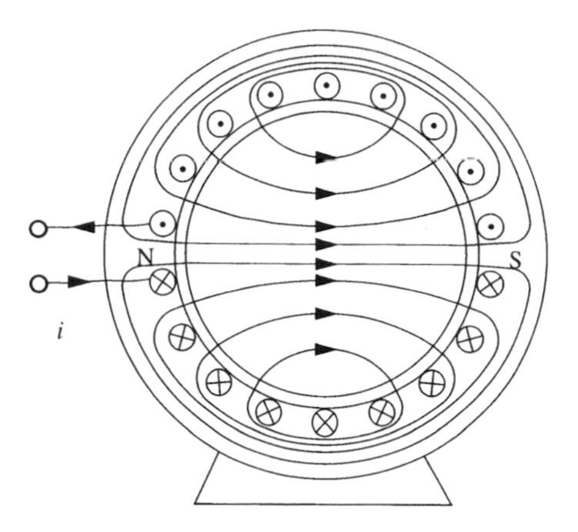

Figure 4.12 Magnetic flux due to a single coil.

Using a trigonometric identity[12] with $a = \omega t$ and $b = \theta_m$, we may expand this into two terms:

$$B(t, \theta_m) = \underbrace{B_r \cos(\omega t - \theta_m)}_{\text{positive rotation}} + \underbrace{B_r \cos(\omega t + \theta_m)}_{\text{reverse rotation}} \tag{4.33}$$

where $B_r = B_c/2$ is the rotating flux density in each direction. Therefore, we may consider a single-phase oscillating flux as composed of two counter-rotating fluxes. This viewpoint is useful in understanding the characteristics of single-phase induction motors in the next chapter.

Coils, windings, phases, and poles. We have now discussed cylindrical magnetic structures for one-, two-, and three-phase systems. Two- and three-phase synchronous machines exist,[13] although two-phase is rare. Equation (4.6) gives the maximum flux density produced in the air gap by a single coil, where n is the number of turns in the coil. Table 4.2 shows the magnitude of the rotating flux density and the number of windings and coils required in such machines. The number of poles is P.

TABLE 4.2 Rotating Flux Magnitude, Windings, and Coils for Single-, Two-, and Three-Phase Machines

Phases	*Br*	*Windings*	*Coils*
1	$B_c/2$	1	$P/2$
2	B_c	2	P
3	$1.5B_c$	3	$3P/2$

[12] $\cos a \cos b = \frac{1}{2}[\cos(a - b) + \cos(a + b)]$.

[13] Of course, some ac electric clocks and timers are single-phase synchronous machines. But these operate on different principles than those discussed in this section.

General results. The results of the previous sections are summarized by the general equation for a rotating flux wave:

$$B(t, \theta_m) = B_r \cos\left[\omega t \mp \frac{P}{2}(\theta_m - \theta_{m0})\right] \tag{4.34}$$

where all the quantities have been defined previously except θ_{m0}, which is the position of the flux maximum at $t = 0$.

EXAMPLE 4.3 Rotating flux

A six-pole magnetic structure produces a clockwise rotating flux that has a north pole at $\theta_m = -90°$ at $t = 0$. The synchronous speed is 800 rpm. Give the equation of the flux density, counting flux positive in the outward radial direction. The maximum flux density is 0.92 T.

SOLUTION:
The synchronous speed is $\omega_s = 800 \times (2\pi/60) = 83.8$ rad/s, so for $P = 6$, Eq. (4.24) gives the electrical frequency as $\omega = (6/2) \times 83.8 = 251$ rad/s. For clockwise rotation, we use the bottom sign in Eq. (4.34). Because the north pole on the stator represents flux in the negative direction, across the air gap, we have

$$\begin{aligned} B(t,\theta_m) &= -0.92 \cos\{251t + (6/2)[\theta_m - (-90°)]\} \\ &= -0.92 \cos(251t + \theta_m + 270°)\ \text{T} \end{aligned} \tag{4.35}$$

WHAT IF? What if you want to eliminate the minus sign?[14]

Check Your Understanding

1. The vector magnetic flux density in the gap of a machine changes only in magnitude, not in direction. True or false?
2. If we reverse the connections to both windings of a two-phase cylindrical structure, the direction of rotation of the magnetic flux density will reverse. True or false?
3. If we change the phase sequence of a three-phase cylindrical structure, the direction of rotation of the magnetic flux will reverse. True or false?
4. Three-phase rotating fluxes go faster than two-phase. True or false?

Answers. **(1)** True; it is always radial across the air gap; **(2)** false; **(3)** true; **(4)** false.

[14] $B(t, \theta_m) = +0.92 \cos(251t + \theta_m + 90°)$ T.

4.3 SYNCHRONOUS GENERATOR PRINCIPLES AND CHARACTERISTICS

Synchronous Generator Construction and Equivalent Circuit

Stator construction. The stator of the synchronous machine was described and analyzed in the previous section. Three-phase stator currents produce magnetic flux that rotates at synchronous speed, which depends on the electrical frequency and the number of poles.

cylindrical rotor, salient-pole rotor

Rotor construction. The rotor of a synchronous machine is a dc electromagnet. The dc current may be supplied to the rotor through a brush–slip ring assembly[15] or may be supplied by rectification of an ac voltage that is induced in a separate winding on the rotor. Figure 4.13 shows *cylindrical* and *salient-pole*[16] rotor construction. The magnetic structure of the cylindrical rotor is symmetric, and a sinusoidally tapered rotor flux is produced by tapered distribution of the rotor mmf. The magnetic structure of the salient–pole rotor is unsymmetrical: The dc coil mmf is concentrated by coils wound around the pole pieces, and a sinusoidal flux distribution is accomplished by tapering the width of the air gap. To develop torque, the same number of poles must be used on stator and rotor. Electrically, the rotor carries the field circuit, in the sense that it holds many turns of small wire carrying relatively small current.

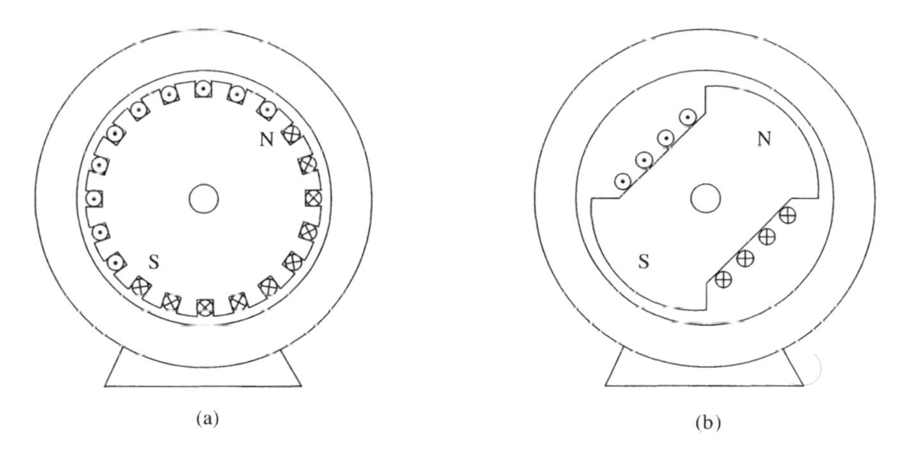

Figure 4.13 (a) Cylindrical rotor; (b) salient-pole rotor.

LEARNING OBJECTIVE 3.

To understand the role of the rotor and stator flux in the operation of synchronous generators

The field-generated emf. Assume the rotor is rotating in the positive θ_m direction at synchronous speed. Because the rotor flux density is sinusoidal in space, the flux density at $\theta_m = 0°$ is a sinusoidal time function representable by a phasor. We count time from the moment when the rotor N magnetic axis passes $\theta_m = 0°$, so the rotor flux density phasor, $\underline{\mathbf{B}}_R$, is

$$\underline{\mathbf{B}}_R = B_R \angle 0° \tag{4.36}$$

[15] The brush on the stator makes sliding contact with a cylindrical surface that rotates on the rotor.

[16] Salient means, among other things, outstanding or jutting out.

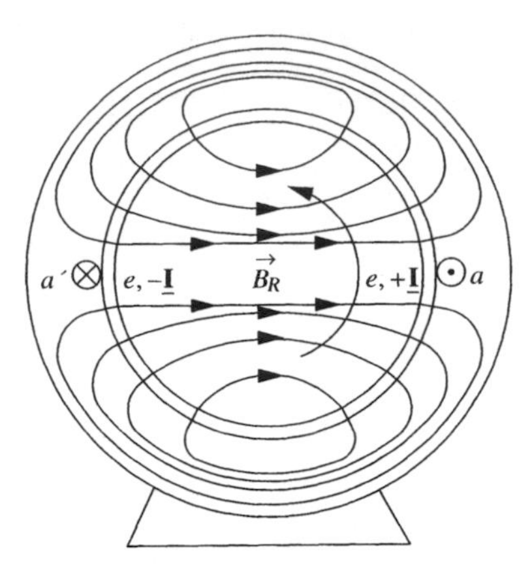

Figure 4.14 The loop a–a' has voltage induced by the rotating flux.

where B_R is the maximum rotor flux density and the electrical frequency is $(P/2)\omega_s$. The flux waves generate voltage in the stator coils, which in turn produce current in the stator and load. To fix ideas, we concentrate on the loop of wire a–a' in Fig. 4.14. From Eq. (2.43), we determine the emf in this loop to be

$$e = 2\vec{\ell} \cdot \vec{u} \times \vec{B} = 2\ell \omega R B_R \text{ V} \tag{4.37}$$

where ℓ is the axial length of the machine, and R is the nominal radius of the air gap, as shown in Fig. 4.8. The polarity of e is $+$ at a and $-$ at a'.[17] The phasor emf, $\underline{\mathbf{E}}_f$, of the entire coil is approximately

$$\underline{\mathbf{E}}_f = 2n\ell \omega R B_R \angle 0^\circ \tag{4.38}$$

where n is the number of turns in the coil centered on a–a'. If there were no stator currents, $\underline{\mathbf{E}}_f$ would be the voltage in the coil[18] due to the rotation of the rotor field.

Effect of stator currents. The phase of the current in a–a', $\underline{\mathbf{I}}$, depends on the load impedance, but the flux caused by the stator current, $\underline{\mathbf{B}}_s$, lags by 90° ($= -j$) due to the right-hand rule.[19] Equations (4.6) and (4.31) give this flux density as

$$\underline{\mathbf{B}}_s = -j\frac{3}{2}\frac{\mu_0 n \underline{\mathbf{I}}}{2g} \tag{4.39}$$

where g is the width of the air gap and the $\frac{3}{2}$ comes from the increase of rotating flux due to the other two phases. Thus, if stator current exists, the flux density in the air gap, $\underline{\mathbf{B}}_{RS}$, will be the sum of the rotor and stator fluxes:

$$\underline{\mathbf{B}}_{RS} = \underline{\mathbf{B}}_R + \underline{\mathbf{B}}_S \tag{4.40}$$

[17] Recall that $\vec{u}$ is the velocity of the wire *relative to the flux*, which is downward at the right and upward at the left.

[18] Later we use $\underline{\mathbf{E}}_f$ as the per-phase excitation voltage, which might differ from the definition in Eq. (4.38) by $\sqrt{3}$.

[19] Put your right-hand fingers around the rotor in the direction of the current in a–a', and your thumb points downward. This 90° spatial angle is transformed into a phase shift by the rotation.

We may convert Eq. (4.40) to a voltage equation by multiplying by $2\ell\omega Rn$:

$$\underbrace{2\ell\omega Rn\underline{\mathbf{B}}_{RS}}_{\underline{\mathbf{V}}} = \underbrace{2\ell\omega Rn\underline{\mathbf{B}}_{R}}_{\underline{\mathbf{E}}_f} - j\omega \underbrace{\frac{n^2}{2g/3\mu_0\ell R}}_{L_s}\underline{\mathbf{I}} \tag{4.41}$$

Circuit interpretation of Eq. (4.41). If $\underline{\mathbf{I}}$ were zero, the voltage produced by the rotation of the rotor flux would be $\underline{\mathbf{E}}_f$; think of it as the Thèvenin voltage of the generator.[20] The $\underline{\mathbf{V}}$ term is the output voltage of the generator with stator current flowing because $\underline{\mathbf{B}}_{RS}$ is the flux density[21] in the air gap. The last term in Eq. (4.41) suggests an inductive reactance:

$$X_s = \omega\frac{n^2}{\Re_g} = \omega L_s, \text{ where } \Re_g = \frac{2g}{3\mu_0\ell R} \tag{4.42}$$

where $\Re_g$ is the effective reluctance of the cylindrical air gap, L_s is the effective inductance of the stator winding, and X_s is its effective reactance at the ac frequency. Thus Eq. (4.41) can be interpreted as KVL in the ac circuit shown in Fig. 4.15(a).

excitation voltage, synchronous reactance

Per-phase equivalent circuit. The derivation of Eq. (4.41) has been approximate in part, but the form and interpretation are correct and lead to a valid per-phase equivalent circuit, as shown in Fig. 4.15(a). The effect of the rotor flux is represented by the *excitation voltage*, $\underline{\mathbf{E}}_f$, which is controlled by the dc field current, I_f, as shown in Fig. 4.15(b). However, we should not consider the excitation voltage as physically present in the machine because the rotor flux occupies the same space as the flux due to the stator currents represented by the per-phase *synchronous reactance*, X_s. The flux

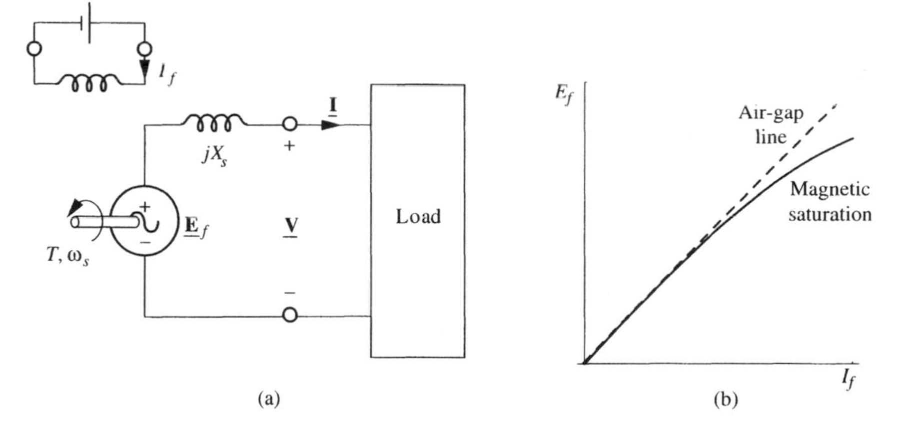

Figure 4.15 (a) Per-phase equivalent circuit for a synchronous motor; (b) the field determines the magnitude of the excitation voltage, $\underline{\mathbf{E}}_f$.

[20] In practice, large synchronous generators are operated with some magnetic saturation. The attendant nonlinear behavior weakens the strict applicability of Thèvenin's theorem. Our brief theory assumes linearity.

[21] There is only one flux pattern in the machine. We conceptually divide it into rotor and stator components for analysis.

that exists physically in the machine is the rotor–stator flux. By Faraday's law, the per-phase voltage induced in the stator coils by the motion of the total flux around the structure must be the per-phase terminal voltage, $\underline{\mathbf{V}}$.

Summary. Figure 4.16 shows the causal factors at work in the synchronous generator. The circuit aspects are shown in the per-phase equivalent circuit in Fig. 4.15(a), which represents the rotating rotor flux by the excitation voltage and represents the rotating stator flux by the synchronous reactance times the stator current. The sum of the rotor and stator fluxes, which is the actual physical flux in the machine, is represented in the equivalent circuit by the sum of the excitation voltage and the voltage drop across the synchronous reactance. Faraday's law becomes

$$\underline{\mathbf{V}} = \underline{\mathbf{E}}_f - jX_s\underline{\mathbf{I}} \tag{4.43}$$

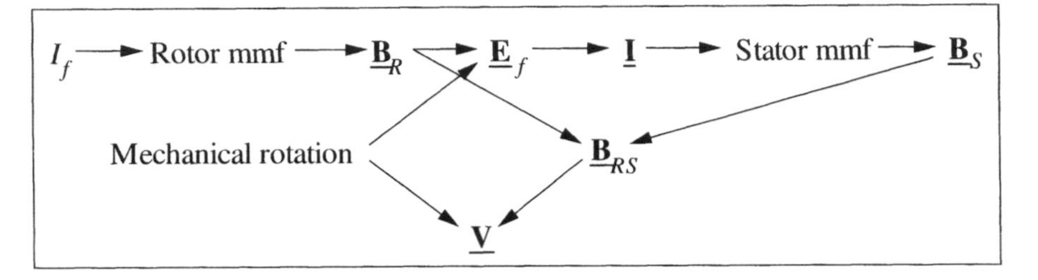

Figure 4.16 Causal relationships in the synchronous generator. The real voltage in the generator, $\underline{\mathbf{V}}$, is generated by the real flux density, $\underline{\mathbf{B}}_{RS}$.

Losses and leakage flux in the stator circuit have been ignored. The magnitude of the excitation voltage is controlled by a dc field current. The equivalent circuit in Fig. 4.15(a) applies only to cylindrical-rotor machines because it ignores the asymmetry in the magnetic structure of a salient-pole rotor.

Generator Operation in Stand-alone Power Systems

System characteristics. Most synchronous generators are used in large power systems and hence are used in tandem with many other generators. However, a synchronous generator may serve as the sole generator in an isolated power system. In this section, we examine the characteristics of the synchronous generator in such stand-alone applications.

Figure 4.17 suggests a stand-alone system. We note two generator inputs: the throttle on the mechanical drive and the dc field current. This power system has the following characteristics:

- The frequency is determined by the speed of the mechanical drive. The electrical frequency is $(P/2)\omega_m$, where ω_m is the mechanical speed of the rotor. Thus, tight control of the speed is required if the system frequency needs to be constant, say, for time keeping. Otherwise, the system frequency could vary somewhat without degrading the performance of most loads.
- The voltage of the system is controlled in part by the field current. Increasing the field current produces an increase in the generated voltage. However, the

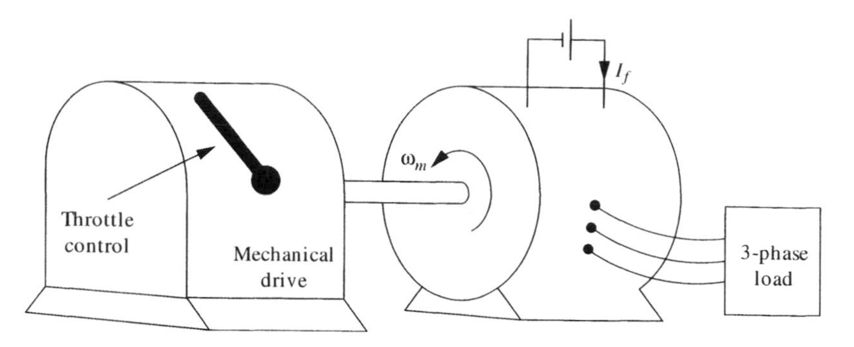

Figure 4.17 A stand-alone power system.

following example shows that the voltage of the system also depends on the load impedance.

EXAMPLE 4.4 **Stand-alone generator**

A 230-V, four-pole, 60-Hz, three-phase generator is required to supply 10 kW to a load at a *PF* of 0.9, lagging. The synchronous reactance of the generator is 3.0 Ω and we neglect all losses and leakage reactance.[22] What would be the output voltage of the generator if the load were disconnected?

SOLUTION:

The output voltage with no load would be the magnitude of $\underline{\mathbf{E}}_f$, which we can determine from the equivalent circuit in Fig. 4.15(a). The per-phase voltage and current are

$$\underline{\mathbf{V}} = \frac{230}{\sqrt{3}} \angle 0^\circ$$

and

$$\underline{\mathbf{I}} = \frac{10,000}{\sqrt{3} \times 230 \times 0.9} \angle -\cos^{-1}(0.9) = 27.9 \angle -25.8^\circ \text{ A} \tag{4.44}$$

The excitation voltage is determined by solving Eq. (4.43):

$$\underline{\mathbf{E}}_f = \underline{\mathbf{V}} + jX_s\underline{\mathbf{I}} = \frac{230}{\sqrt{3}} + j3.0 \times 27.9 \angle -25.8^\circ = 185.3 \angle 24.0^\circ \tag{4.45}$$

This per-phase voltage is equivalent to a three-phase voltage of 321 V, which would be the open-circuit voltage of the system.

WHAT IF? What if you want to know the torque requirement on the engine driving the generator?[23]

[22] Here, as in transformers, not all flux couples rotor and stator.

[23] 53.1 N-m.

Summary. A stand-alone synchronous generator requires control systems to regulate both generator speed, which determines electrical frequency, and output voltage, which is strongly dependent on the load.

Power Angles

Definitions of power angles. In Sec. 4.1, we investigated the combination of rotor and stator fluxes to produce a rotor–stator flux. In the course of that development, we introduced the angles δ_R and δ_S as the physical angles of the rotor and stator fluxes relative to the total flux. Here we investigate more fully these power angles, which correspond to physical angles between the various contributors to the total flux and correspond also to the phase angles between these sinusoidal quantities considered as phasors.

Power angles and phasor diagrams. Figure 4.18(a) shows the various power angles as physical angles and Fig. 4.18(b) shows the phasor fluxes at $\theta_m = 0°$. The rotor flux-density phasor is aligned with the N rotor pole because this pole puts flux across the air gap in the positive (outward) direction. The stator flux-density phasor, however, is aligned with the S stator pole, because this draws flux across the air gap in the positive (outward) direction. The rotor and stator fluxes combine to produce the rotor–stator flux. The rotor power angle, δ_R, is positive when the rotor poles lead the air-gap flux maximum.

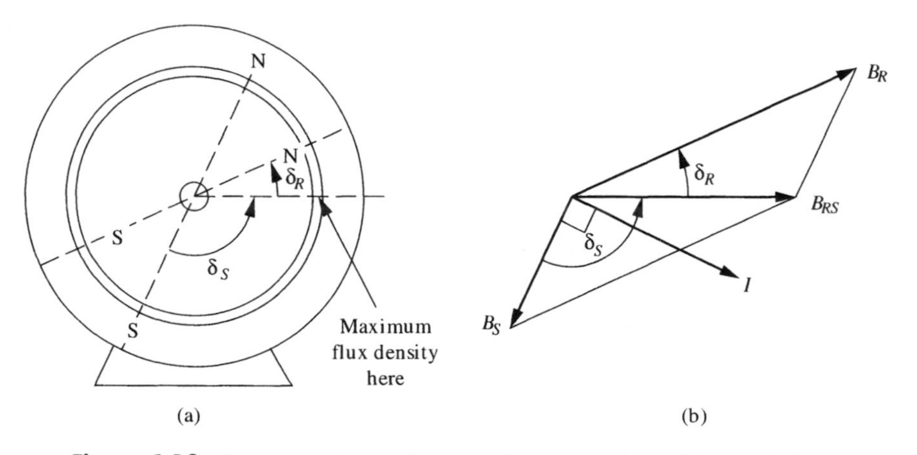

Figure 4.18 The rotor and stator fluxes combine to give the total flux. This is the pole arrangement in the previous example. (a) Pole arrangement; (b) phasor representation of current and flux density at $\theta_m = 0°$.

Torque and power angles. Equation (4.18), repeated here,

$$T_{dev}(\delta_{RS}) = -\frac{\pi R \ell g P}{2\mu_0} B_R B_S \sin\left(\frac{P}{2}\delta_{RS}\right) = -\frac{\pi R \ell g P}{2\mu_0} B_R B_{RS} \sin\left(\frac{P}{2}\delta_R\right) \tag{4.46}$$

gives the output torque of the machine in terms of the rotor power angle. The second form is useful because its two flux quantities are tied to controlled quantities: The rotor flux is controlled by the dc field current, and the rotor–stator flux is linked to the output voltage. For fixed field excitation and output voltage, the torque depends on the rotor power angle, δ_R.

We show the correspondence between the fluxes and the phasor diagram of the stand-alone operation in the previous example. The circuit and its phasor diagram are shown in Fig. 4.19. The voltage generated in the stator windings, $\underline{\mathbf{V}}$, is the phase reference. The current and excitation voltage are shown. The excitation voltage is in phase with the rotor flux density, and the generated voltage in the stator is in phase with the rotor–stator flux density; thus, the phase angle between $\underline{\mathbf{E}}_f$ and $\underline{\mathbf{V}}$ is the rotor power angle, δ_R. The rotor power angle is particularly important in understanding the operation of a synchronous generator in a large power system. The negative of $jX_s\underline{\mathbf{I}}$ represents the stator flux density.

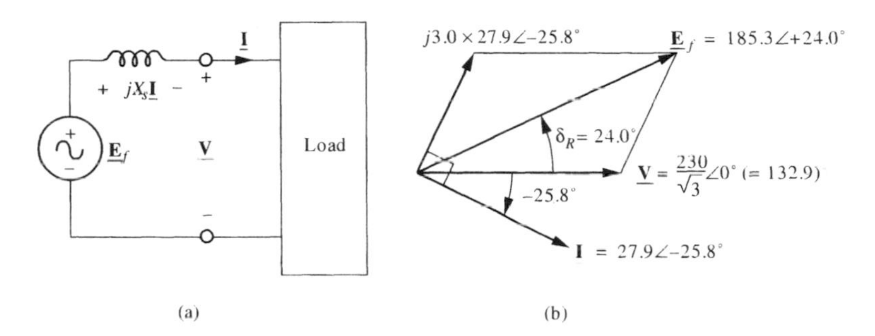

Figure 4.19 (a) Circuit and (b) phase relationships, for example, of a stand-alone generator. The pole arrangement and flux density phasors are shown in Fig. 4.18.

Synchronous Generator Operation in a Large Power System

LEARNING OBJECTIVE 4.

To understand the characteristics of the synchronous generator when operated into a large power system

In this section, we examine the characteristics of synchronous generators when operated in large power systems where many loads and many generators are interconnected over a geographic region. Such interconnections are used to enhance reliability, to permit maintenance on individual generators, and to permit exchange of electric power between participating power companies, as discussed in Sec. 1.2.

Power grids. In such power grids, many generators are operated in tandem. All generators are synchronized and interconnected by long-distance transmission lines. From the viewpoint of an individual generator, the power system is the load into which the generator delivers real and reactive power. The characteristics of such a load, however, are quite different from those of a passive load.

infinite bus

The infinite bus. The power system is modeled as *infinite bus* in that it maintains constant frequency and constant voltage, both amplitude and phase, independent of the operation of the generator under consideration. The frequency of the system may be considered constant, being established by the controlled rotation of many generators. Thus, if we increase the mechanical drive to an individual generator, we do not increase frequency as in stand-alone operation; rather, we contribute more real power to the grid. Likewise, if we increase the dc field current on an individual generator, we do not increase the output voltage as in stand-alone operation; rather, we change the reactive power contributed to the system. Such independent control over real and reactive output power is demonstrated in the following analysis.

Operation into an infinite bus. Figure 4.20 shows the per-phase circuit of a synchronous generator operating into an infinite bus. The real power out of the generator is

$$P_{out} = 3VI \cos \theta \tag{4.47}$$

where θ is the angle between the per-phase output voltage and current, and V and I are rms per-phase values. The phasor diagram for such a system is shown in Fig. 4.21.

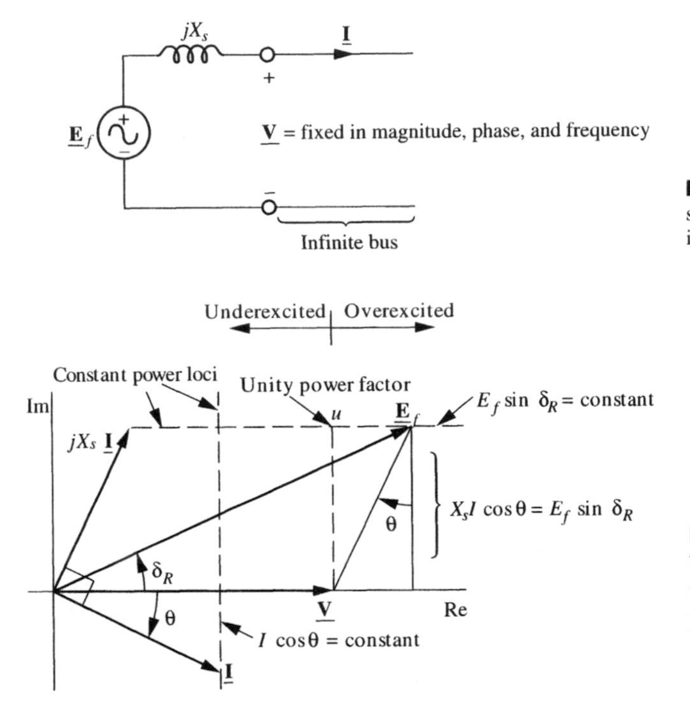

Figure 4.20 Per-phase model of synchronous generator operating into an infinite bus.

Figure 4.21 Per-phase phasor diagram for a synchronous generator operating into an infinite bus. The dashed lines show the loci of the excitation voltage and current phasors for constant output power as the magnitude of the excitation voltage is varied.

We now consider the changes in the phasor diagram as the dc field current is varied, thus changing the *magnitude* of the excitation voltage. We assume that the mechanical drive torque and speed are kept constant, and hence the input and output real powers of the generator remain constant. The line voltage of the infinite bus is fixed; and, hence, from Eq. (4.47), we see that $I \cos \theta$ must remain constant. Thus, the tip of the current phasor follows the dashed vertical line passing through the real axis at the in-phase current required to deliver the real power. As the dc field current is changed, the magnitude and phase of the output current vary, but the component of the current *in phase* with the bus voltage remains constant.

Figure 4.21 shows the rotor power angle, δ_R, as the phase angle by which the excitation voltage, $\underline{\mathbf{E}}_f$, leads the bus voltage, $\underline{\mathbf{V}}$. By KVL, the excitation voltage is the phasor sum of $\underline{\mathbf{V}}$ and $jX_s\underline{\mathbf{I}}$; hence, from the geometry shown in Fig. 4.21, it follows that

$$X_s I \cos \theta = E_f \sin \delta_R \tag{4.48}$$

where E_f is the magnitude of the per-phase excitation voltage. We may therefore modify Eq. (4.47) to the form

$$P_{out} = \frac{3VE_f}{X_s} \sin \delta_R \tag{4.49}$$

Thus, for constant power (drive throttle) and constant line voltage, $E_f \sin \delta_R$ remains constant. The tip of the excitation voltage phasor, therefore, follows the dashed horizontal line shown in Fig. 4.21.

The effect of dc field current on power factor. With these two loci in mind, we can estimate the effects of changing the dc field current for a synchronous generator operating into an infinite bus. Consider, for example, the effect of decreasing the dc field current on the phasor diagram in Fig. 4.21. Decreasing the field current reduces the magnitude of E_f, which moves the tip of $\underline{\mathbf{E}}_f$ back along the horizontal dashed line toward point u. At the same time, the output voltage, $\underline{\mathbf{V}}$, is fixed; and, hence, the phasor $jX_s\underline{\mathbf{I}}$, which is parallel to the line connecting $\underline{\mathbf{V}}$ and $\underline{\mathbf{E}}_f$, must swing toward a vertical position. Hence, current $\underline{\mathbf{I}}$ must swing toward a horizontal position, with its tip remaining on the vertical dashed line. Consequently, as the dc field current is decreased, the current becomes more in phase with the output voltage, the power factor moves toward unity, and the magnitude of the current decreases. At point u, $\underline{\mathbf{E}}_f$ lies directly above $\underline{\mathbf{V}}$, and the current is in phase with the per-phase output voltage. At this point, we have a right triangle, and the excitation voltage is

$$E_{fu}^2 = V^2 + (X_s I)^2 \tag{4.50}$$

where E_{fu} is the magnitude of the excitation voltage required for unity power factor. As the magnitude of $\underline{\mathbf{E}}_f$ is further decreased, the current swings ahead of the output voltage and begins to lead. The angle of the current relative to the output voltage is controlled, therefore, by the magnitude of the excitation voltage, which is controlled in turn by the dc field current. The in-phase component of the current is constant but the out-of-phase component varies in magnitude and sign. Hence, the reactive power exchanged between the generator and power system is controlled by the dc field current.

underexcited, overexcited generator

Figure 4.22 shows the effect of the dc field current on the current magnitude and power factor of a synchronous generator with constant output power. The output current leads the output voltage for small field current, an *underexcited* generator, and decreases in magnitude as field current increases. A minimum is reached for unity power factor at u, and then the current increases and begins to lag the output voltage, an *overexcited* generator.

Summary. The real power output is controlled by the throttle on the mechanical drive to the generator. The sign and magnitude of the reactive power are controlled by the dc field current. With an overexcited generator, with $\underline{\mathbf{E}}_f$ beyond point u in Fig. 4.21, the generator supplies positive reactive power. With an underexcited generator, with $\underline{\mathbf{E}}_f$ to the left of point u in Fig. 4.21, the generator supplies negative reactive power. Equation (4.50) defines the boundary between overexcited and underexcited behavior and corresponds to unity power factor.

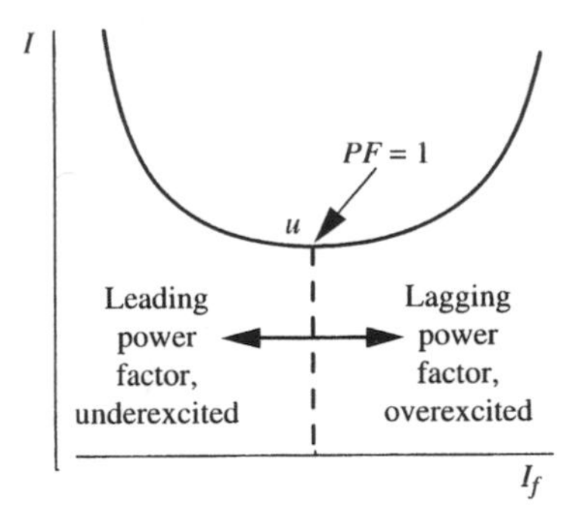

Figure 4.22 Effect of field current on output current of a synchronous generator with output power kept constant.

EXAMPLE 4.5 Supplying reactive power

A 600-V (line voltage), three-phase synchronous generator supplies 25 kW. The synchronous reactance is 10.0 Ω. Calculate the per-phase excitation voltage to give unity power factor and determine what excitation voltage causes the generator to supply +10 kVAR of reactive power in addition to the real power.

SOLUTION:

The required line current for unity power factor is

$$I = \frac{P}{3V} = \frac{25 \times 10^3}{3(600/\sqrt{3})} = 24.1 \text{ A} \tag{4.51}$$

and, hence, from Eq. (4.50), the required excitation voltage

$$E_{fu} = \sqrt{V^2 + (X_s I)^2} = \sqrt{(600/\sqrt{3})^2 + (10.0 \times 24.1)^2} = 422 \text{ V} \tag{4.52}$$

Note that we use the per-phase voltage in the calculations. For the second part of the problem, we find the apparent power:

$$S = \sqrt{P^2 + Q^2} = \sqrt{(25)^2 + (10)^2} = 26.9 \text{ kVA} \tag{4.53}$$

The power factor is $PF = 25/26.9 = 0.928$ and the phase angle of the current is $\theta = \tan^{-1}(10/25) = 21.8°$, lagging, and the magnitude of the new current, I', is

$$I' = \frac{I}{PF} = \frac{24.1}{0.928} = 25.9 \text{ A} \tag{4.54}$$

From Eq. (4.43), the new phasor excitation voltage $\underline{\mathbf{E}}'_f$ is

$$\underline{\mathbf{E}}'_f = \underline{\mathbf{V}} + jX_s\underline{\mathbf{I}}' = \frac{600}{\sqrt{3}} + j10.0 \times 25.9 \angle -21.8° = 504 \angle 28.5° \text{ V} \tag{4.55}$$

Thus, a 19% increase in the excitation voltage over the value required for unity power factor adds +10 kVAR of reactive power.

Output power and power angle. Figure 4.23 shows Eq. (4.49) and the effect of rotor power angle on generated power with output voltage and field excitation constant. The rotor power angle, δ_R, is the angle in space, measured in electrical units, by which the physical axis of the rotor leads the total flux in the air gap. As the drive torque is increased, the rotor pulls more on the flux, and more energy is converted from mechanical to electrical form. The output power can be increased up to a limit, as seen in Fig. 4.23, beyond which the motor-generator loses synchronism and runs away. In practice, generators are operated with power angles around 15–25° and thus stay well away from runaway.

Figure 4.23 indicates that negative power angles correspond to negative output power, to motor action. In this case, the air-gap flux leads the axis of the rotor, the electrical system pulls on the mechanical system, and electrical energy is converted to mechanical energy. The characteristics of the synchronous motor are investigated in the next section.

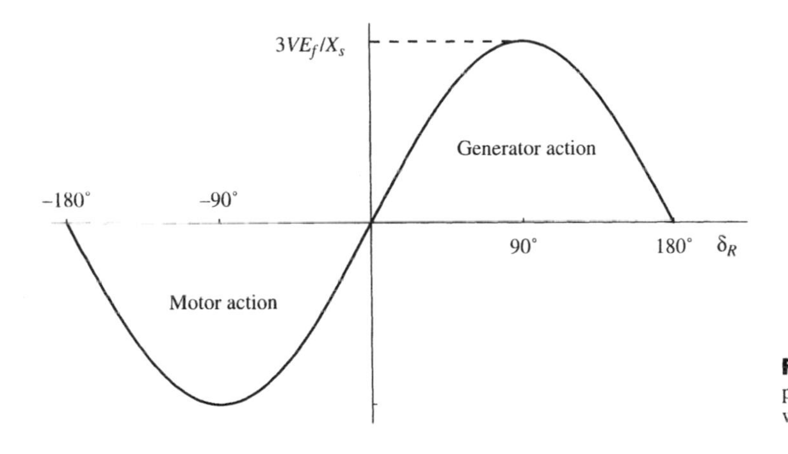

Figure 4.23 Output power versus rotor power angle for a synchronous generator with constant excitation and output voltage.

Check Your Understanding

1. The stator current in a loaded synchronous generator will be minimum with the excitation voltage smaller than, equal to, or greater than the per-phase output voltage. Which?
2. A permanent magnet could be used as the rotor of a synchronous generator. True or false?
3. In a cylindrical-rotor synchronous generator, increasing the dc field current increases the magnitude of the output current. Is the current leading or lagging?
4. It is impossible for a cylindrical-rotor synchronous generator to operate with a rotor power angle beyond +90°. True or false?

Answers. **(1)** Greater than; **(2)** true; **(3)** lagging; **(4)** true.

4.4 CHARACTERISTICS OF THE SYNCHRONOUS MOTOR

Generator and Motor Comparison

IDEA 6 **Equivalent Circuits**

Changed conventions. Figure 4.24 shows reference directions appropriate for (a) generator and (b) motor operation. We note that a load set of voltage–current reference directions are used for the motor, with appropriate changes in the circuit equation and power formula.

In the formula for power in Fig. 4.24(b), we changed the *meaning* of the current variable (from output to input current) and we accordingly changed the *meaning* of the power from output electrical power to output mechanical power. However, the meaning of the power angle is unchanged, and hence the power angle must be negative for positive output power as a motor. Put another way, we fixed the power equation in Fig. 4.24(b) by negating power and current variables in the power equation in Fig. 4.24(a). In the parts involving the power factor, we absorb the minus sign by changing the *meaning* of the power and current variables. But no meanings were changed in the form involving the power angle, and the minus sign therefore remains. Thus, the power angle must be negative for motor action.

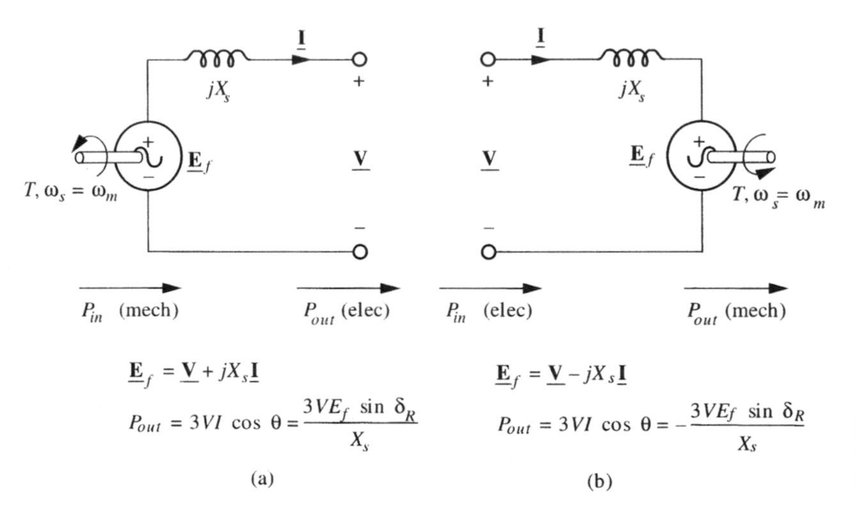

Figure 4.24 Reference conventions and resulting equations for (a) generator and (b) motor operation.

Phasor diagrams for generator and motor. Figure 4.25 compares phasor diagrams for generator and motor operation. The $jX_s\underline{I}$ phasor leads the stator current for a generator but lags for a motor, a consequence of the change of the current reference direction for the motor in Fig. 4.24(b).

LEARNING OBJECTIVE 5.

To understand the characteristics of the synchronous motor, particularly the effect of field current on the power factor

Analysis of the equivalent circuit. Kirchhoff's voltage law for the stator circuit of the motor, Fig. 4.24(b), is

$$\underline{V} = \underline{E}_f + jX_s\underline{I} \Rightarrow \underline{E}_f = \underline{V} - jX_s\underline{I} \tag{4.56}$$

The phasor diagram reflecting Eq. (4.56) is shown in Fig. 4.26, where θ is the phase angle between per-phase voltage and current and is shown negative in Fig. 4.26 (lagging current). The angle δ_R is the rotor power angle defined in Fig. 4.18 and is the physical

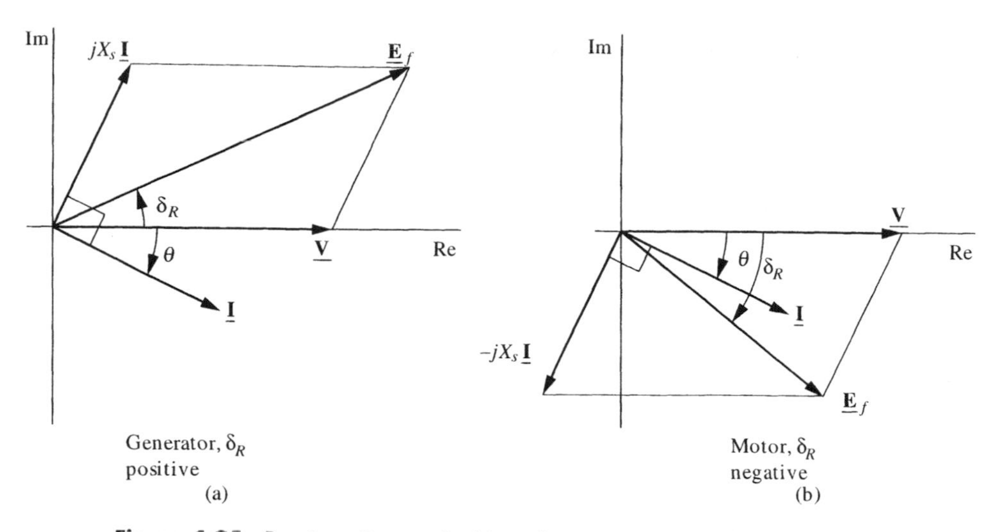

Figure 4.25 Per-phase diagrams for (a) synchronous generator and (b) synchronous motor.

angle, multiplied by $P/2$ for machines with more than two poles, between the rotor–stator flux and the rotor flux. This power angle appears between the per-phase voltage and the excitation voltage because the former represents the total flux and the latter the rotor flux.

The input electrical power to the machine is

$$P = 3VI\cos\theta = -\frac{3VE_f}{X_s}\sin\delta_R \tag{4.57}$$

Using geometric reasoning on Fig. 4.26, we may determine the reactive power into the machine to be

$$Q = 3VI\sin(-\theta) = \frac{3V^2}{X_s} - \frac{3VE_f}{X_s}\cos\delta_R \tag{4.58}$$

The excitation voltage may be eliminated between Eqs. (4.57) and (4.58) to give the rotor power angle in terms of the real and reactive power into the machine.

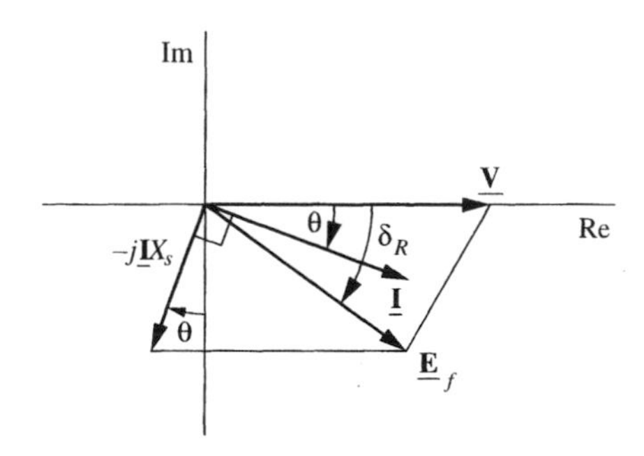

Figure 4.26 Phasor diagram for the synchronous motor. The applied per-phase voltage is the phase reference, and the current is shown lagging by a phase angle θ. The voltage across the synchronous reactance is shown lagging the current by 90°.

$$\tan \delta_R = \frac{P}{Q - (3V^2/X_s)} = \frac{P}{Q - (V_L^2/X_s)} \tag{4.59}$$

where V_L is the three-phase voltage ($= \sqrt{3}\,V$).

EXAMPLE 4.6

Synchronous motor

A 480-V, three-phase, six-pole synchronous motor has 50-hp output power at unity power factor with a field current of 5 A dc. The field current is increased to 6 A dc. Find the new power factor and the new input current. Ignore losses, assume that the synchronous reactance is 3.2 Ω, and assume that the excitation voltage is proportional to field current.

SOLUTION:

First, we find the excitation voltage for unity power factor. For 50 hp out, no losses, and unity power factor, the current is

$$I = \frac{P}{3V} = \frac{50 \times 746}{3(480/\sqrt{3})} = 44.9 \text{ A} \tag{4.60}$$

and is in phase with the input voltage. By KVL in Eq. (4.56), the phasor excitation voltage is

$$\underline{\mathbf{E}}_{fu} = \frac{480}{\sqrt{3}} - j3.2 \times 44.9 \angle 0^\circ = 312.1 \angle -27.4^\circ \tag{4.61}$$

where $\underline{\mathbf{E}}_{fu}$ is the excitation voltage for unity power factor. Hence, for unity power factor, the magnitude of the per-phase excitation voltage is 312.1 V and the rotor power angle is -27.4°. When we increase the dc field current from 5 to 6 A dc, the magnitude of the excitation voltage increases proportionally:

$$E_f' = \frac{6}{5} \times 312.1 = 374.5 \text{ V} \tag{4.62}$$

where the prime refers to conditions after the field current is changed. Changing the field current does not change the real power output of the motor because that is established by the speed, which is constant, and the torque requirement of the load at that speed. Thus, we may determine the new rotor power angle from Eq. (4.57):

$$312.1 \sin(-27.4^\circ) = 374.5 \sin(\delta_R') \Rightarrow \delta_R' = -22.5^\circ \tag{4.63}$$

and the new reactive power can be determined from Eq. (4.58):

$$\begin{aligned} Q' &= \frac{3(480/\sqrt{3})^2}{3.2} - \frac{3(480/\sqrt{3})(374.5)\cos(-22.5^\circ)}{3.2} \\ &= -17{,}900 \text{ VAR} \end{aligned} \tag{4.64}$$

The reactive power is negative, meaning that the new current leads the per-phase voltage. The new power factor may be determined from the real and reactive powers:

$$PF = \frac{P}{\sqrt{P^2 + Q^2}} = \frac{50 \times 746}{\sqrt{(50 \times 746)^2 + (-17,900)^2}} = 0.902 \quad (4.65)$$

Hence, the new current magnitude is 44.9/0.902 = 49.7 A, and the new phase angle is $\cos^{-1} 0.902 = 25.6°$ with current leading voltage.

WHAT IF? What if you use Eq. (4.56) instead of Eq. (4.64)?[24]

overexcited motor

Effect of varying the field current with constant output power. Figure 4.27 shows a phasor diagram for a synchronous motor that is *overexcited* because the excitation voltage is greater than that required to give unity power factor. Let us consider the effect of reducing the excitation voltage, reducing field current, while the output power and per-phase voltage are kept constant. By Eq. (4.57), $E_f \sin \delta_R$ must remain constant, so the tip of $\underline{\mathbf{E}}_f$ must move along a horizontal line. Likewise, for constant power, the in-phase component of the current, $I \cos \theta$ must remain constant; hence, the tip of the current phasor moves along a vertical line. At point u, the tip of $\underline{\mathbf{E}}_f$ (E_{fu}) lies below $\underline{\mathbf{V}}$ and the current is in phase with the per-phase voltage. If the excitation voltage is larger than E_{fu}, the current phase must swing ahead of the per-phase voltage and a leading power factor is produced. If the excitation voltage is less than E_{fu}, the current phase must lag the per-phase voltage and a lagging power factor is produced. The out-of-phase component of the current of the synchronous motor thus can be controlled by the field current, as with a synchronous generator.

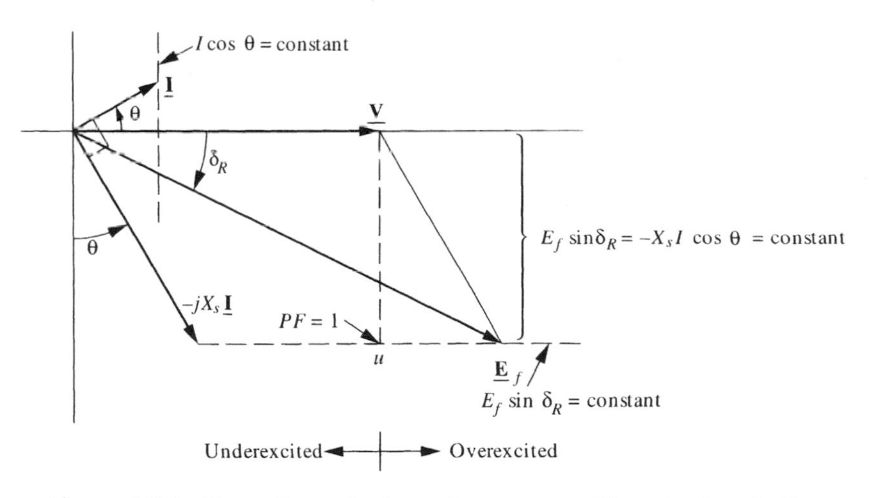

Figure 4.27 Phasor diagram for the synchronous motor. The motor draws leading or lagging current, depending on the magnitude of the excitation voltage.

[24] $I' = (374.5 \angle -22.5° - 480/\sqrt{3})/(-j3.2) = 49.7 \angle +25.7°$. This current leads to the same P, Q, and PF.

Power-factor control. The ability to draw leading current allows the synchronous motor to improve the overall power factor of a collection of loads that tend to draw lagging current. For example, in a factory with many induction motors, employment of overexcited synchronous motors is desirable to improve the power factor of the overall load. Overexcited synchronous machines with no mechanical output, indeed, no output shaft, are employed as synchronous condensers[25] to control the power factor.

EXAMPLE 4.7

Increasing the power

For the previous example, the field current remains at 6 A dc and the load is now increased to 65 hp. Find the new input current.

SOLUTION:
The increase in real power increases the power angle. From Eq. (4.57), we calculate the new power angle to be

$$65 \times 746 = -\frac{3(480/\sqrt{3})(374.5)}{3.2} \sin \delta_R'' \quad \Rightarrow \quad \delta_R'' = -29.9^\circ \tag{4.66}$$

where double-primed quantities refer to conditions under the second change. Thus, the new excitation voltage is $\underline{\mathbf{E}}_f'' = 374.5\angle{-29.9^\circ}$ V. Kirchhoff's voltage law, Eq. (4.56), now requires the current to be

$$\underline{\mathbf{I}} = \frac{\underline{\mathbf{V}} - \underline{\mathbf{E}}_f''}{jX_s} = \frac{(480/\sqrt{3}) \angle 0^\circ - 374.5 \angle -29.9^\circ}{j3.2} \tag{4.67}$$
$$= 58.3 + j14.9 = 60.2 \angle 14.3^\circ \text{ A}$$

The principal effect is an increase in the in-phase component of the current.

Starting a synchronous motor. A synchronous motor has no starting torque. To develop a steady torque, the rotor must be rotating at the synchronous speed. This appears to be a major defect of the synchronous motor, but the difficulty can be remedied by placing some shorted turns on the rotor. These shorted turns produce torque by induction-motor action[26] and accelerate the rotor to a speed just below the synchronous speed, at which the rotor field is energized and initiates synchronism. After the rotor field is energized, the shorted turns have no effect on the steady-state operation of the motor, except to improve the transient performance of the motor by damping out oscillations in the power angle caused by variations in the mechanical load.

Salient-Pole Motors

Salient-pole rotors, Fig. 4.13(b), are used in slow-turning machines, such as hydroelectric generators and high-torque motors. In such cases where many poles are required,

[25] Condenser is an antique word for capacitor.

[26] The next chapter begins with an explanation of how such torque is developed.

engineering and manufacturing considerations favor the salient-pole geometry. The asymmetry of the magnetic structure in this machine introduces two reactances in place of the synchronous reactance of the cylindrical rotor machine.

$$X_s \rightarrow X_d \text{ and } X_q \tag{4.68}$$

direct axis, quadrature axis

where X_d is the direct-axis reactance and X_q is the quadrature-axis reactance. The *direct-axis* reactance applies to the axis aligned with the rotor poles; the *quadrature-axis* reactance applies to the axis between the rotor poles and is smaller because the equivalent air gap is larger between the poles.

Analysis of the salient-pole motor. Space limitations prohibit an analysis of the salient-pole machine, but equations describing the performance of the motor[27] are given. No equivalent circuit exists, but the equations of the cylindrical-motor machine may be modified to describe the characteristics of the salient-pole machine. Equation (4.57) becomes

$$P = -\frac{3VE_f}{X_d}\sin\delta_R - \frac{3}{2}V^2\left(\frac{1}{X_q} - \frac{1}{X_d}\right)\sin(2\delta_R) \tag{4.69}$$

and Eq. (4.58) becomes

$$Q = -\frac{3VE_f\cos\delta_R}{X_d} + 3V^2\left(\frac{\cos^2\delta_R}{X_d} + \frac{\sin^2\delta_R}{X_q}\right) \tag{4.70}$$

and Eq. (4.59) is still valid using X_q for X_s.

EXAMPLE 4.8 Motor with and without load

A 2300-V, 60-Hz, 250-hp salient-pole synchronous motor has direct- and quadrature-axis reactances of 15 and 8 Ω, respectively. Ignore losses. The excitation voltage is adjusted for operation at unity power factor with full load. Find the rotor power angle under this condition.

SOLUTION:

For unity power factor, the reactive power is zero, and we may determine the rotor power angle directly from Eq. (4.59):

$$\tan\delta_R = \frac{250 \times 746}{0 - (2300)^2/8} = -0.282 \quad \Rightarrow \quad \delta_R = -15.8^\circ \tag{4.71}$$

The load is then removed, but the excitation voltage is unchanged. Find the reactive power into the motor under this new condition.

SOLUTION:

Under the unloaded condition and with no losses, the power angle and real power are zero, but the machine still interacts with the electrical system through reactive power flow, as indicated by Eq. (4.70) with $\delta_R = 0^\circ$. The excitation voltage is unchanged from the loaded condition, so

[27] The salient-pole generator analysis is similar; just change a few signs.

we may determine E_f from Eq. (4.69) with $\delta_R = -15.8°$.

$$250 \times 746 = -\frac{3(2300/\sqrt{3})E_f \sin(-15.8°)}{15} - \frac{3}{2}\left(\frac{2300}{\sqrt{3}}\right)^2 \left(\frac{1}{8} - \frac{1}{15}\right) \sin(-2 \times 15.8°) \tag{4.72}$$

Thus, $E_f = 1470$ V. We may determine the reactive power from Eq. (4.70) with $\delta_R = 0°$:

$$Q = -\frac{3(2300/\sqrt{3})(1470)}{15} + 3\left(\frac{2300}{\sqrt{3}}\right)^2 \left(\frac{1}{15} + 0\right) = -37{,}400 \text{ VAR} \tag{4.73}$$

The machine is overexcited and draws negative reactive power.

Reluctance torque. We may determine the developed torque from Eq. (4.69):

$$T_{dev} = \frac{P_{dev}}{\omega_s} = -\frac{3VE_f}{\omega_s X_d} \sin \delta_R - \frac{3}{2}\frac{V^2}{\omega_s}\left(\frac{1}{X_q} - \frac{1}{X_d}\right) \sin(2\delta_R) = -T_f \sin \delta_R - T_{\Re} \sin(2\delta_R) \tag{4.74}$$

reluctance torque

The sin $(2\delta_R)$ term in Eq. (4.74) reveals the presence of reluctance torque due to the salient poles. In Eq. (4.74), T_f is the maximum torque[28] due to the rotor field and $T_{\Re}$ is the maximum reluctance torque due to the asymmetry of the rotor magnetic structure. This *reluctance torque* arises because the rotor magnetic structure tries to come into alignment with the stator poles. Reluctance torque is due to induced magnetic poles on the rotor protrusions and would be present if the rotor had no field coils.

EXAMPLE 4.9 Reluctance torque

The 60-Hz motor in the previous example has 12 poles. Find the equation of the torque and identify the terms due to rotor field and reluctance.

SOLUTION:

For 12 poles and 60 Hz, the synchronous speed is 20π rad/s. Substitution of the excitation voltage into the first term in Eq. (4.74) fields a maximum torque of 6210 N-m due to the rotor dc current. The second term yields 2460 N-m for the maximum reluctance torque. For the rotor power angle calculated, the output torques due to the two components are 1690 N-m and 1280 N-m, respectively, and hence about 43% of the output torque is due to reluctance torque. The reluctance torque improves the characteristics of the motor because it increases the maximum possible torque and causes the motor to run with a smaller rotor power angle. Figure 4.28 shows the torque characteristic of this machine.

[28] Recall that δ_R is negative.

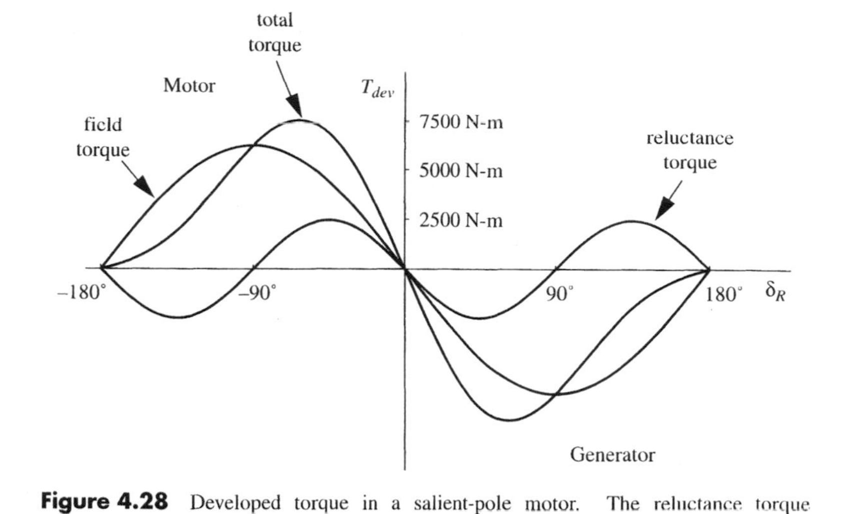

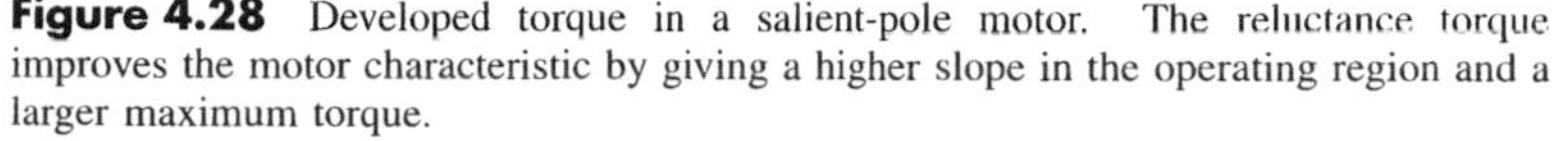

Figure 4.28 Developed torque in a salient-pole motor. The reluctance torque improves the motor characteristic by giving a higher slope in the operating region and a larger maximum torque.

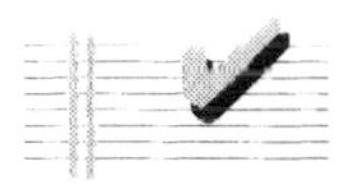

Check Your Understanding

1. For a cylindrical-rotor synchronous motor, the torque is 500 N-m with a rotor power angle of -30°. What is the maximum torque available from this machine without changing the excitation voltage?
2. The effect of the reluctance torque is to degrade the performance of a salient-pole synchronous motor. True or false?
3. An important virtue of a synchronous motor is (a) easy speed control, (b) high efficiency, or (c) adjustable power factor. Which?
4. A 25-hp synchronous motor with eight poles has a rated output torque twice that of a 25-hp synchronous motor with four poles. True or false?
5. Explain why a factory uses a combination of synchronous and induction motors.

Answers. **(1)** 1000 N-m; **(2)** false; **(3)** adjustable power factor; **(4)** true; **(5)** induction motors have lagging current and synchronous motors can draw leading current.

CHAPTER SUMMARY

Chapter 4 begins by showing that steady energy transformation requires both torque and rotation. Torque is produced between misaligned rotor and stator magnetic poles. Rotating flux is produced in cylindrical structures by ac excitation. The synchronous machine is studied as a generator, with emphasis on operating into a large power grid. We show that real and reactive output powers can be controlled independently. Similarly, the real power to the synchronous motor is determined by load demand but the reactive power is controlled by field current.

Objective 1: To understand how magnetic flux is created and distributed in the air gap of a cylindrical magnetic structure. Magnetic flux is created in a cylindrical structure by distributed coils placed in slots on the inner surface of the stator. The magnitude of the maximum flux density can be determined through Ampère's circuital law. The number of poles is twice the coils/phase. The wire density is tapered to produce a sinusoidally tapered flux pattern.

Objective 2: To understand how magnetic flux is made to rotate in cylindrical magnetic structures. The flux pattern is rotated by exciting spatially separated coils with two- or three-phase currents. The direction of rotation depends on the phase relationships between the exciting currents. The speed of flux rotation equals the ac frequency divided by the number of pole pairs.

Objective 3: To understand the role of the rotor and stator flux in the operation of synchronous generators. The rotor of the synchronous machine is an electromagnet whose poles match the stator poles. When the rotor is rotated, a voltage is induced in the stator coils. The output voltage depends upon this voltage and the effect of stator currents. The machine has a large inductive output impedance.

Objective 4: To understand the characteristics of the synchronous generator when operated into a large power system. When operated into an infinite bus, the real power output of the synchronous generator is determined by the input mechanical drive to the rotor. The reactive power is controlled in sign and magnitude by the strength of the rotor flux, which is established by the dc field current to the rotor. Normally, the field is overexcited to produce the lagging current required by magnetic loads.

Objective 5: To understand the characteristics of the synchronous motor, particularly the effect of field current on the power factor. The real power into the synchronous motor is determined by the torque requirement of the load. The reactive power, sign and magnitude, is controlled by the dc field current. Normally, the field is overexcited to draw leading current to balance the lagging current required by induction motors.

Synchronous motors are specialized devices with limited applications. Chapter 5 takes up three- and single-phase induction motors, which constitute the vast majority of ac motors.

GLOSSARY

Cylindrical rotor, p. 165, a rotor that has cylindrical symmetry in which magnetic flux is produced by wires in slots milled in an axial direction on the surface of the rotor.

Developed power, p. 151, power converted from electrical to mechanical form in a motor, or vice versa in a generator.

Developed torque, p. 151, the torque produced by magnetic forces on the rotor of a motor before any losses are subtracted.

Excitation voltage, p. 167, a fictitious voltage that represents the effect of rotor flux in the stator windings, which is controlled by the rotor dc field current; corresponds roughly to the open-circuit voltage.

Field current, p. 150, the current that controls the rotor field in a synchronous machine; controls the phase of the current in the machine.

Infinite bus, p. 171, an electrical bus (system of conductors) that maintains constant frequency and constant voltage, both amplitude and phase, independent of the operation of any generator.

Overexcited generator, p. 173, a generator that produces lagging current, the usual case.

Overexcited motor, p. 179, draws leading current, often used to improve the power factor of a group of loads.

Reluctance torque, p. 182, torque produced in a salient-pole synchronous motor due to inducted magnetic poles on the rotor.

Rotor, p. 165, the part of an electric motor that rotates physically.

Rotor power angle, p. 156, the angle between the rotor pole and the maximum of the total flux density.

Rotor–stator power angle, p. 155, for a two-pole machine, the physical angle measured positive from the stator magnetic axis to the rotor magnetic axis. For this and all power angles, the physical angle is multplied by $P/2$ for machines with P poles to convert to angle in electrical measure, such as electrical degrees, which correspond to electrical phase angles.

Salient-pole rotor, p. 165, a rotor that has bar-type electromagnets to produce patterns of magnetic flux.

Stable equilibrium, p. 151, an equilibrium state in which forces tend to correct small displacements from the equilibrium position.

Synchronous reactance, p. 167, a reactance that models the effect of the stator flux on the stator windings, acts roughly as the output impedance of the synchronous generator.

Synchronous speed, p. 160, the angular speed of the magnetic flux rotation in rpm or radians/s.

Two-phase system, p. 158, a three-wire electrical power system in which one wire is neutral and the other two are hot, with voltages separated 90° in phase angle.

Underexcited generator, p. 173, a generator that produces leading current, rarely used.

Underexcited motor, p. 158, a motor that draws lagging current, rarely used.

Unstable equilibrium, p. 151, an equilibrium state in which forces tend to increase small displacements from the equilibrium position.

Winding, p. 155, a series or parallel connection of coils to increase voltage or current in an electrical machine or transformer.

PROBLEMS

Section 4.1: Flux and Torque in Cylindrical Magnetic Structures

4.1. If you wanted the magnetic flux density to be 0.4 T at $\theta_m = 42°$ in Fig. 4.5(a), what would the mmf of the coil have to be? The gap is 0.5 mm wide.

4.2. The cylindrical magnetic structure shown in Fig. P4.2(a) has 24 slots with tapered windings, as shown in the table in Fig. 4.2(b). The gap width is 1 mm and the mmf loss in the iron may be neglected. The current is 20 A into the paper on the bottom half and out on the top half. Determine the maximum magnetic flux density in the air gap and determine the location of the maximum.

4.3. Two magnetic structures with cylindrical geometry are shown in Fig. P4.3. A dc current i flows in at the cross and out at the dot, with one wire in each slot. The rotor is a permanent magnet with poles and orientation shown.

(a) How many stator poles are there in each case?

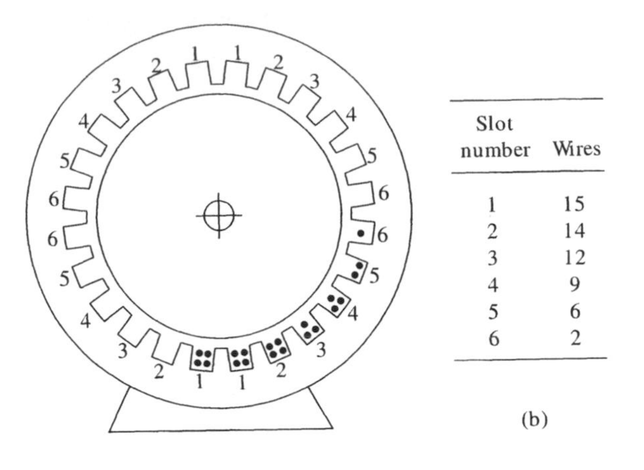

Slot number	Wires
1	15
2	14
3	12
4	9
5	6
6	2

(b)

Figure P4.2

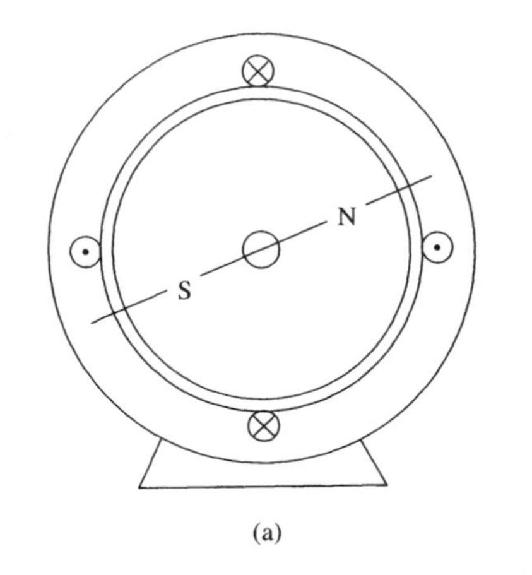

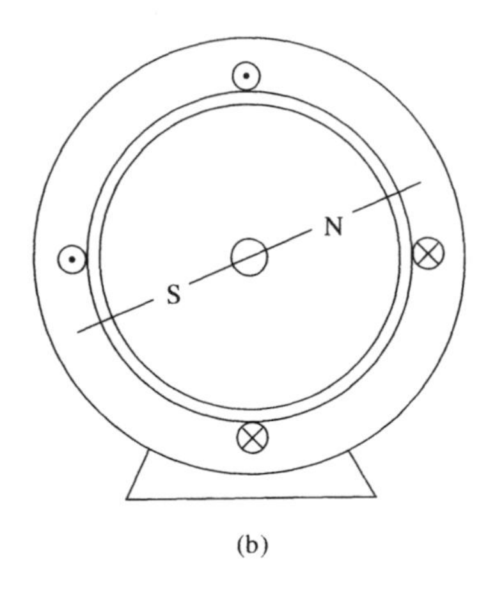

Figure P4.3

(b) In each case, determine the direction of the torque on the rotor, if there is any torque.

4.4. We wish to design a cylindrical magnetic structure to produce a four-pole flux pattern. We require N poles at $\theta_m = -45^\circ$ and $+135^\circ$. The peak value of the flux density is 1.0 T, the gap width is 0.9 mm.

(a) Draw the magnetic structure required and show the locations of the coils and currents on the stator. Mark $\times + i$ for where i enters the paper, etc.

(b) How many turns/coil are required if the peak current is 10 A?

(c) If there are 24 slots in the stator, give the number of wires in each slot to approximate a sinusoidal mmf.

Section 4.2: Rotating Magnetic Flux for AC Motors

4.5. A cylindrical rotor and a three-phase, two-pole stator are excited by dc and 60-Hz ac currents, respectively. The rotor is locked and stationary in the position shown in Fig. P4.5. The rotor-flux density is constant with 1.0 T maximum, sinusoidally distributed in space. The stator produces a flux density with 0.5 T maximum, sinusoidally distributed in space and rotating counterclockwise, as shown in Fig. P4.5. The maximum of the flux density due to the stator currents is at $\theta_m = 0^\circ$ in space at $t = 0$.

(a) At $t = 1/360$ s, what are the magnitude and

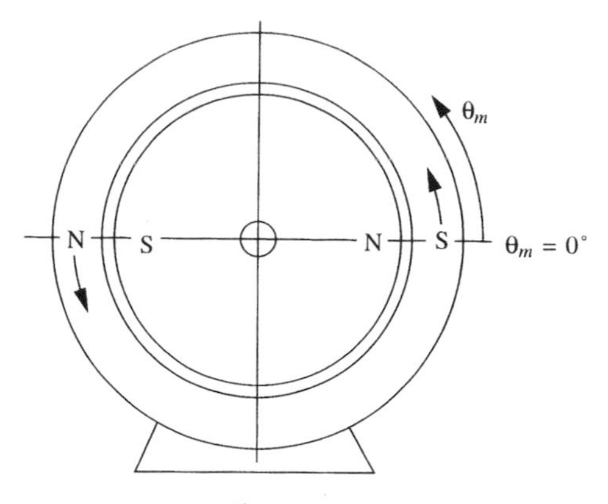

Figure P4.5

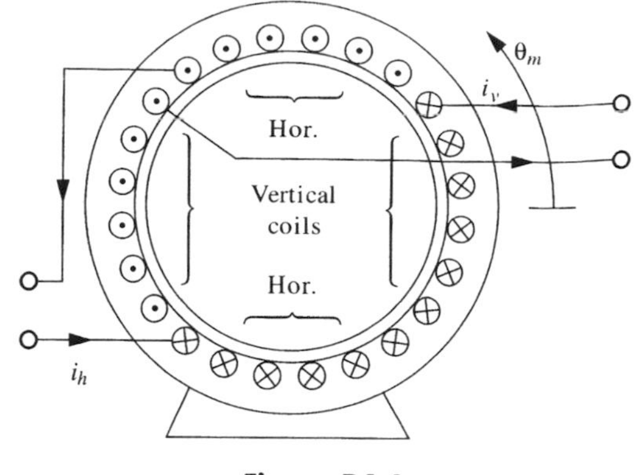

Figure P4.6

direction (in space) of the air-gap vector flux density at $\theta_m = 0°$?

(b) At $t = 1/360$ s, at what angle in space (θ_m) is the magnetic-flux density a maximum?

(c) At $t = 1/360$ s, what direction would the rotor tend to move if freed?

4.6. A cylindrical magnetic structure is wound with two coils, as shown in Fig. P4.6. One, excited by i_h, produces a horizontal flux, and the other, excited by i_v, produces a vertical flux. The horizontal flux density is

$$B_h(t, \theta_m) = \frac{\mu_0 n i_h(t)}{2g} \cos \theta_m$$

and the vertical flux density is

$$B_v(t, \theta_m) = \frac{\mu_0 n i_v(t)}{2g} \sin \theta_m$$

where n is the number of turns in each coil and g is the gap width. Find the magnetic flux density at a mechanical angle of 30°, that is, find $B(t, \theta_m = 30°)$ for the following conditions. In each case, explain the nature of the flux (rotating? which way? oscillating? direction? etc.).

(a) $i_h = 2I_p$ dc and $i_v = -I_p$ dc.

(b) $i_v = 0$ and $i_h = I_p \cos(\omega t)$.

(c) $i_v = I_p \sin(\omega t)$ and $i_h = 0$.

(d) The currents in parts (b) and (c) at the same time.

4.7. Table P4.7 has columns for time-domain expressions for magnetic flux densities in an air gap, as a function of angle in space, and a description of the field or the source of the field. Fill in the missing information.

4.8. Write mathematical expressions for the following air-gap flux densities. Assume 60 Hz.

(a) A two-pole, single-phase oscillating flux with peaks in the vertical direction and nulls in the horizontal direction. Assume that the flux density has a maximum of 0.5 T at $t = 0$.

(b) A two-pole rotating flux that moves counterclockwise and has its positive peak of 0.5 T at $\theta_m = 180°$ at $t = 0$.

(c) A four-pole rotating flux that rotates clockwise and has its negative peak of 0.6 T at $\theta_m = -90°$ at $t = 0$.

4.9. A cylindrical magnetic structure has a flux pattern in its air gap given by the expression

$$B(t, \theta_m) = 1.5 \cos(800\pi t + \theta_m - 60°)$$

(a) How many magnetic poles are indicated?

(b) Where is the maximum flux density at $t = 0$?

TABLE P4.7

Flux density	*Poles*	*Synchronous speed*	*Description*
$1.5 \cos(200\pi t + \theta_m)$			
$1.2 \cos(30t - 3\theta_m)$			
$1.0 \cos(100\pi t) \sin(2\theta_m)$			

(c) What are the rotation direction and speed in rpm?
(d) Sketch a pattern of currents on the stator that produces this flux at $t = 0$.

4.10. Give the mathematical expressions in the time domain for the following fluxes. The frequency is 60 Hz. The radially outward direction is considered positive in the air gap.
(a) A four-pole oscillating flux. One maximum flux density of 1.2 T occurs at $\theta_m = -45^\circ$ at $t = 0$.
(b) A two-pole rotating flux, clockwise rotation, north magnetic pole in the stator crosses $\theta_m = 0^\circ$ at $t = 0$. The maximum flux density is 0.6 T.

4.11. A rotating flux is created with 16 coils, spaced evenly around a cylindrical stator. The coils are connected to two-phase current so as to create a rotating flux in the CCW direction. The flux rotates at 600 rpm and has a maximum of 1.0 T in the vertical direction, $\theta_m = 90^\circ$ at $t = 0$. Find the magnitude and sign of the outward-directed flux density at $\theta_m = 234^\circ$ and $t = 10$ ms.

4.12. A three-phase stator produces a rotating flux that is described by the equation

$$B(t, \theta_m) = 1.5 \cos(\omega t + 3\theta_m + 60^\circ)\ \text{T}$$

(a) How many poles does the stator have?
(b) Where ($\theta_m = ?$) is the flux density maximum at $t = 0$?
(c) If two wires of the three-phase system were reversed, what would change in the flux-density expression?
(d) If the electrical frequency of the stator currents were 30 Hz, what would be the speed in rpm of the stator flux rotation?
(e) If the stator windings were reconnected into Y instead of Δ, what would be the resulting rotating flux? Assume constant voltage. Consider the phase rotation as original.

4.13. Figure P4.13(b) shows a two-phase, two-pole stator with current directions. Figure P4.13(a) shows a two-phase current with the period given.
(a) Where ($\theta_m = ?$), is the flux density maximum at $t = 0$? Count the positive as maximum where the flux is outward across the air gap.
(b) Which way does the flux rotate?
(c) Give the equation of the flux density as a function of time and mechanical angle, assuming that the maximum flux density is 0.72 T.

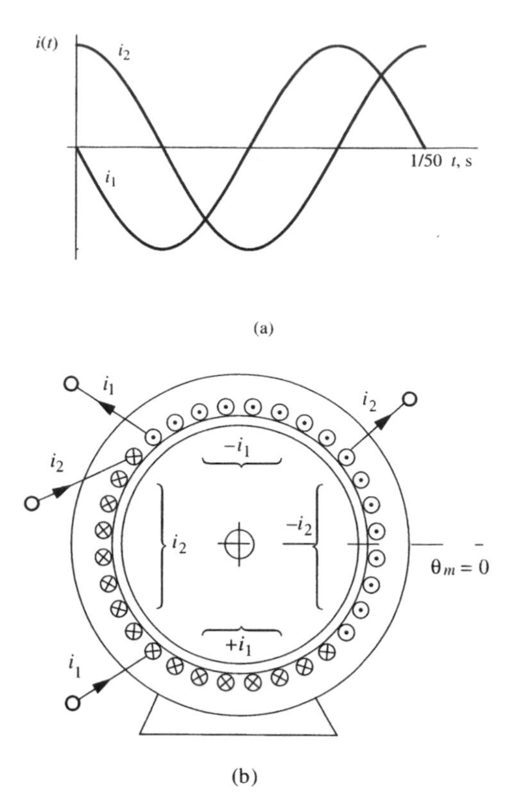

Figure P4.13

(d) Find the synchronous speed of the flux pattern in rpm.

4.14. Figure P4.14(b) shows a three-phase, two-pole stator with current directions. Figure P4.14(a) shows a three-phase current with the period given.
(a) Where ($\theta_m = ?$), is the flux density maximum at $t = 0$? Count the positive as maximum where the flux is outward across the air gap.
(b) Which way does the flux rotate?
(c) Give the equation of the flux density as a function of time and mechanical angle, assuming that the maximum flux density is 0.72 T.
(d) Find the synchronous speed of the flux pattern in radians/second.

4.15. A cylindrical magnetic structure has four coils on it, excited by 60-Hz, two-phase currents such that four counterclockwise rotating magnetic poles are produced. At $t = 0$, the N pole is at $\theta_m = 0^\circ$. The maximum value of the magnetic flux density in the air gap is 0.85 T.
(a) Draw a cylindrical magnetic structure and

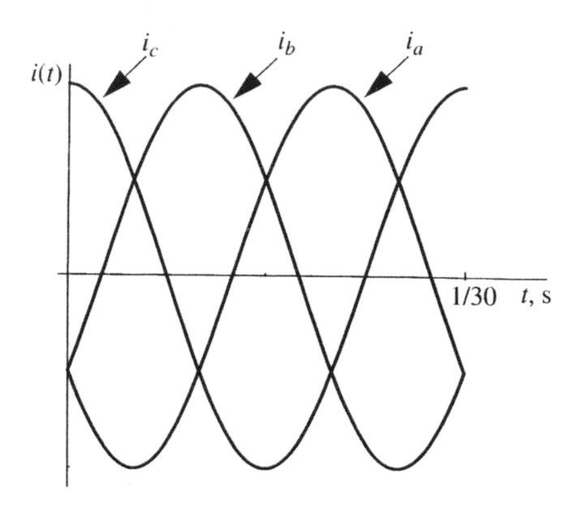

(a)

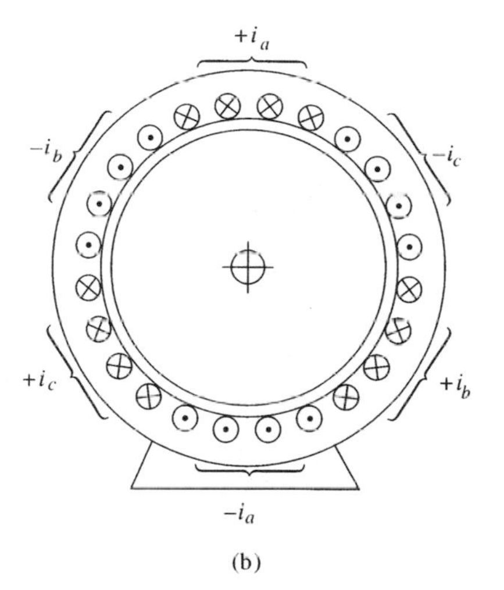

(b)

Figure P4.14

indicate the magnetic poles and sketch the magnetic flux patterns at $t = 0$.

(b) Write an expression for $B(t, \theta_m)$.

(c) Find a time when the flux density at $\theta_m = 45^\circ$ is $+\ 0.6$ T.

4.16. A cylindrical magnetic structure is wound with two coils, which we have symbolized with a single wire and labeled "1" and "2" in Fig. P4.16(a). The coils have 90 turns each and the air gap is 1 mm. Currents i_1 and i_2 are shown in Fig. P4.16(b). The coils overlap and are sinusoidally tapered.

(a) Is this a single-phase, two-phase, or three-phase system?

(b) Determine the location of the magnetic flux density maximum at $t = 0$.

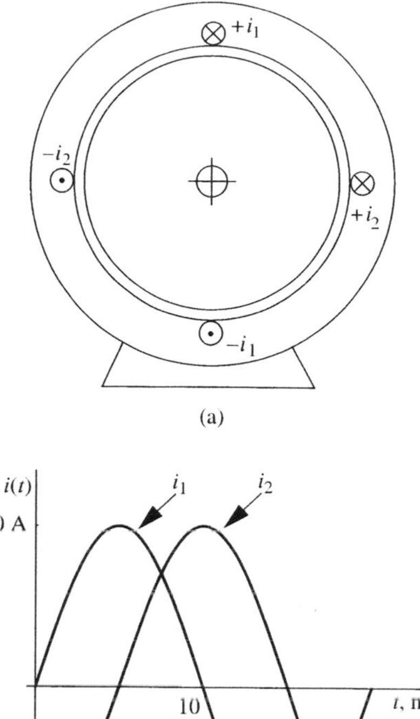

(a)

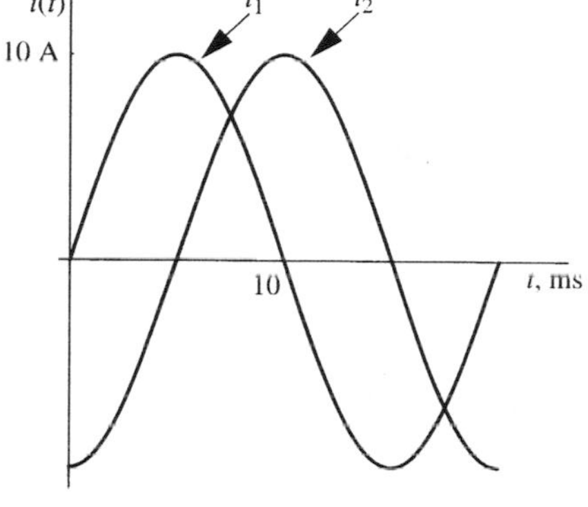

(b)

Figure P4.16

(c) Determine the direction of rotation of the flux.

(d) Determine the synchronous speed in rpm.

(e) Determine the maximum value of the magnetic flux density.

(f) Write an expression for $B(t,\theta_m)$.

4.17. Write the equation for the magnetic flux density in the air gap of a cylindrical magnetic structure having the following properties:

- The machine has 18 coils, excited by three-phase ac currents to produce a rotating flux.
- The synchronous speed of the flux pattern is 1500 rpm in the CCW direction.
- The maximum flux-density is 0.8 T.
- The stator coils are tapered to produce a sinusoidal spatial flux pattern.
- The flux-density maximum is horizontal at $t = 0$.

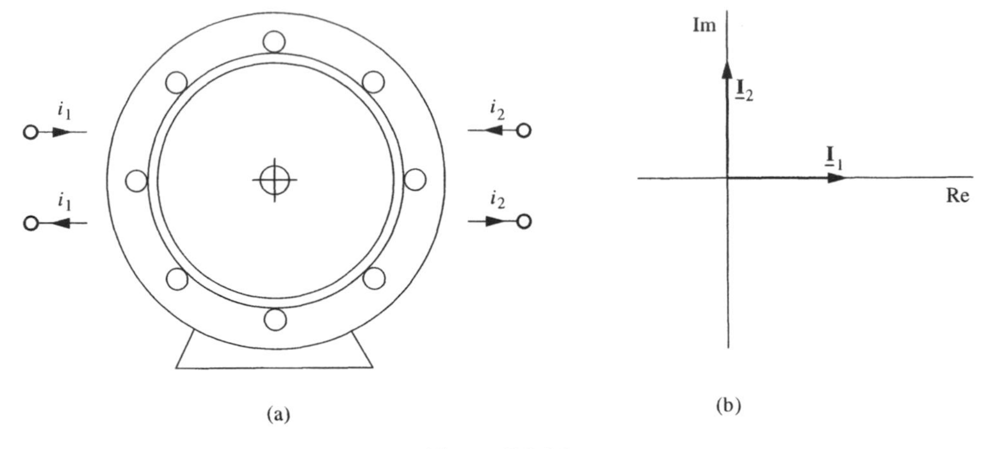

Figure P4.18

4.18. A cylindrical magnetic structure is shown in Fig. P4.18(a) with the windings simplified, showing a single wire for each distributed winding. You are to show the external connections so that two-phase currents produce a four-pole flux rotating in the clockwise direction. The phasors for i_1 and i_2 are shown in Fig. P4.18(b).

(a) On the diagram place by the wires $+\ i_1$ where the current enters the paper and $-\ i_1$ where the current comes out of the paper, and similarly for i_2. Connect the wires at the side.

(b) Draw the flux pattern at $t = 0$.

4.19. A three-phase cylindrical magnetic structure has $P = 2$ poles.

(a) The windings are excited with balanced three-phase 50-Hz currents. Describe in words the character and behavior of the flux in the machine.

(b) Give a mathematical expression for the flux density in the time domain as a function of mechanical angle, using the standard notation. Assume the maximum air-gap flux density at $\theta_m = 0°$ at $t = 0$. Let B_r be the maximum flux density in the air gap, and assume clockwise rotation.

(c) Give a phasor representing the flux density at $\theta_m = 30°$.

(d) If each coil has 100 turns and carries a current of 2 A and the gap width is 0.5 mm, what is the maximum flux density for the machine?

4.20. A counterclockwise rotating wave of flux density is represented at $\theta_m = 0°$ by the expression $B(t, 0°) = 0.5\cos(120\pi t - 45°)$ T. What is $B(t, \theta_m)$ generally, assuming a four-pole machine?

4.21. A rotating flux density is
$B(t, \theta_m) = 1.2\cos[100\pi t + 4\,(\theta_m\ - 30°)]$ T
The machine is three-phase, excited by 50-Hz currents.

(a) How many magnetic poles does the machine have?

(b) In which direction does the field rotate?

(c) Find the first time after $t = 0$ when a positive flux maximum passes the angle $\theta_m = 0°$.

4.22. A cylindrical magnetic structure is shown in Fig. P4.22(a). The slots are numbered, and Table P4.22 gives the number of wires and current

TABLE P4.22

Slot	*Coil for i_a Wires, direction*	*Coil for i_b Wires, direction*
1	19, +	5, +
2	14, +	14, +
3	5, +	19, +
4	5, −	19, +
5	14, −	14, +
6	19, −	5, +
7	19, −	5, −
8	14, −	14, −
9	5, −	19, −
10	5, +	19, −
11	14, +	14, −
12	19, +	5, −

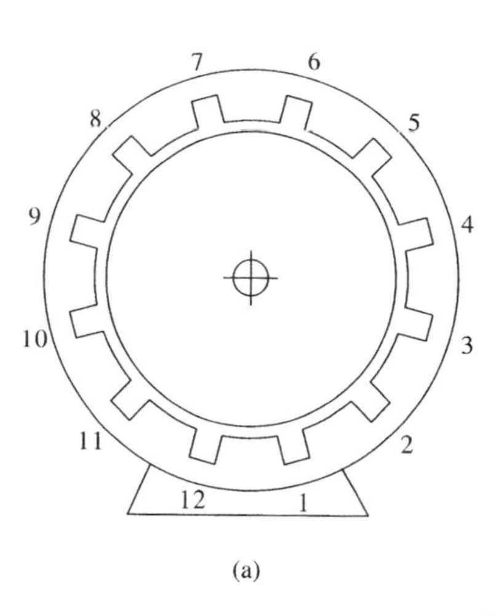

(a)

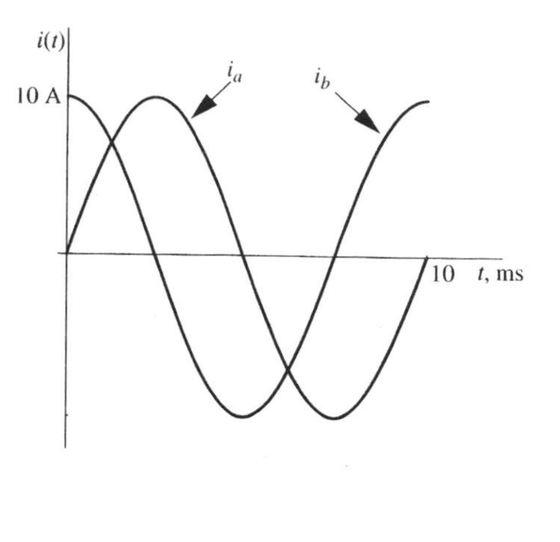

(b)

Figure P4.22

direction (+ into the paper) for the two coils that are wound on the stator. The currents are shown in Fig. P4.22(b).

(a) What is the maximum magnetic flux density if the gap is 1 mm wide?
(b) How many poles in this structure?
(c) What is the direction of rotation?
(d) What is the synchronous speed of rotation?
(e) Write the equation for $B(t, \theta_m)$.

4.23. A magnetic structure has a grand total of 600 turns. These are organized into two windings for two-phase operation, for four rotating poles. The OD of the rotor is 3.110 ± 0.001 inches and the ID of the stator is 3.136 ± 0.001 inches. The two-phase currents are 60 Hz with the same peak current in both phases. The windings are sinusoidally tapered. There are 40 slots on the stator.

(a) What is the rotational speed of the flux in radians/second?
(b) To reverse the direction of flux rotation, would one or both of the two-phase currents have to be reversed?
(c) If the least acceptable value of the peak magnetic flux density required were 1.1 T, what would be the rms current required in each winding?
(d) Find the maximum number of wires in any slot.

Section 4.3: Synchronous Generator Principles and Characteristics

4.24. A three-phase, four-pole, 60-Hz synchronous generator is operated in a stand-alone system. The load impedance is 2 Ω real per phase and requires 208 V (line voltage). The synchronous reactance of the generator is 2 Ω per phase. Stator resistance may be ignored.

(a) What value of open-circuit line voltage is produced to yield the required value with the load in place? Assume a linear magnetic structure.
(b) What is the load current and power?
(c) What is the required input torque, ignoring losses?

4.25. A six-pole, three-phase, 60-Hz, Y-connected synchronous generator produces a line voltage of 8600 V. The stator windings are capable of carrying a current of 12 A. The synchronous reactance is 60 Ω. At a given moment, the field current is 10 A dc, and the generator is producing nameplate output at 0.8 *PF*, lagging, into a passive load. Neglect losses.

(a) Find the input torque and speed in rpm.
(b) Draw a phasor diagram showing $\underline{\mathbf{I}}$, $jX_s\underline{\mathbf{I}}$, $\underline{\mathbf{V}}$, δ_R, and $\underline{\mathbf{E}}_f$.
(c) If the load were suddenly removed, to what would the line voltage jump? Assume the magnetic structure is linear.

4.26. A three-phase 12-pole, 60-Hz, synchronous generator operates into a 600-V infinite bus and delivers 20 kW at unity power factor. The synchronous reactance is 10 Ω. Ignore resistive losses.

(a) What is the per-phase current?

(b) What is the excitation voltage, E_f, magnitude only?

(c) What is the rotor power angle?

(d) Determine the input torque.

4.27. An open-circuit/short-circuit test is performed on a three-phase synchronous generator, with the results shown in Fig. P4.27. The nameplate voltage is 240 V, and the nameplate current is 12 A, as shown.

(a) Explain why the voltage curve is curved but the current curve is straight.

(b) Estimate the per-phase synchronous reactance of the machine.

(c) Is the field current for nameplate operation, $PF = 1$, less than, equal, or greater than 4 A?

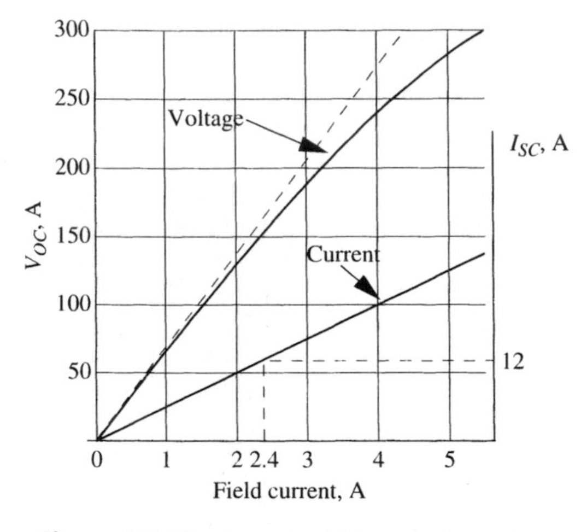

Figure P4.27 Open-circuit/short-circuit test data.

4.28. A three-phase synchronous generator has a no-load output voltage of 2400 V at a field current of 12 A dc. Connected to a 2300-V infinite bus, the mechanical drive is adjusted such that the generator puts out 48 kW and + 12 kVAR with this same field current. Assume a linear magnetic structure.

(a) Calculate the synchronous reactance per phase.

(b) To increase the reactive power output to 24 kVAR, what should be the field current? The real power output stays the same.

4.29. A three-phase synchronous generator produces a real power of 100 kW and a reactive power of + 50 kVAR into an infinite bus at 1200 V. The synchronous reactance of the generator is 10 Ω, and the losses may be ignored.

(a) Draw a per-phase phasor diagram showing the bus voltage, $\underline{\mathbf{V}}$, the line current, $\underline{\mathbf{I}}$, the power factor angle, θ, the rotor power angle, δ_R, and the excitation voltage, $\underline{\mathbf{E}}_f$,

(b) The drive torque is increased 10%. Indicate which of the quantities in part(a) increase in magnitude, which remain the same, and which decrease.

4.30. The phasor diagram for a 60-Hz, three-phase synchronous generator is shown in Fig. P4.30. Note that all sides and two angles of the triangle are shown. The current is 21 A, and the generator has four poles. The diagram shows per-phase values.

(a) Is the generator overexcited or underexcited?

(b) What is the rotor power angle?

(c) What is the power factor and is it leading or lagging?

(d) Determine the synchronous reactance.

(e) Determine the input power and torque, ignoring losses.

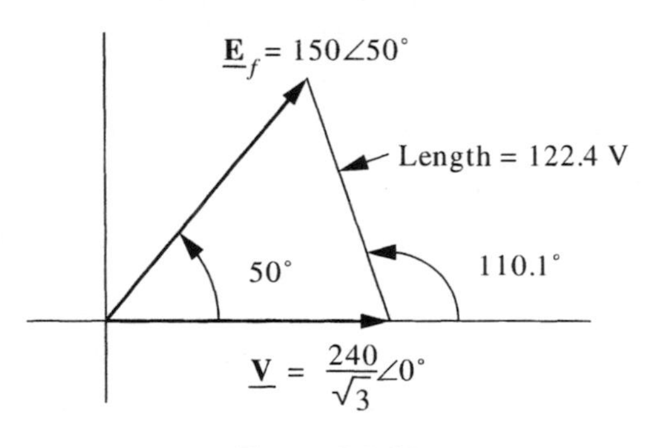

Figure P4.30

4.31. A 60-Hz, six-pole, three-phase synchronous generator is operated at nameplate kVA with the per-phase phasor diagram shown in Fig. P4.31. Ignore losses.

(a) Find the kVA of the machine.

(b) Find the input torque to the generator.

(c) Find the synchronous reactance per phase of the machine.

4.32. A three-phase, two-pole, 60-Hz, 10-MVA, 13-kV synchronous generator has a synchronous reactance of 10.0 Ω. Neglect losses.

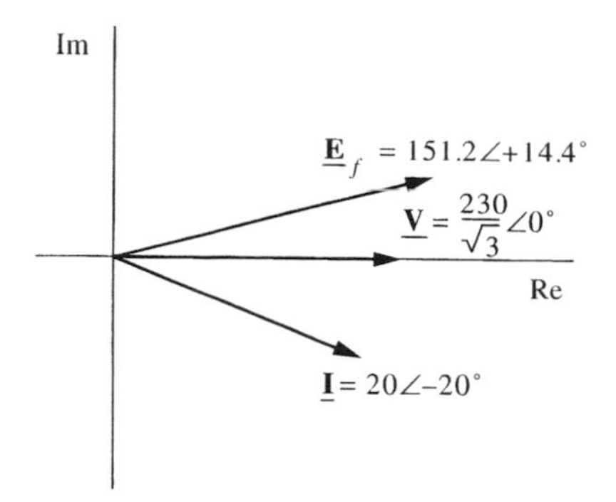

Figure P4.31

(a) At nameplate kVA and 0.9 *PF*, lagging, what is the torque required to drive the machine?
(b) If the condition described in part (a) requires a field current of 50 A, what field current would correspond to unity power factor and the same real power out?
(c) Estimate how low the field current could be reduced without the generator losing synchronism under the same real power load.

Section 4.4: Characteristics of the Synchronous Motor

4.33. Upper Slobovia runs on 60 Hz and Lower Slobovia runs on 50 Hz. Your assignment as an engineer in the Slobovia Light and Power Company is to design an interconnection between the two three-phase systems.
(a) Show how you would accomplish this.
(b) Specify the type of machine or machines, how many poles, rotational speeds, how to connect electrically and mechanically, and whatever else you wish to add.

4.34. A three-phase synchronous motor is rated at 6 kVA. The per-phase voltage is $240/\sqrt{3}$ V. Tests were performed with no load, with the following results: Minimum (essentially zero) input current was observed at a field current of 8 A, and nameplate kVA was drawn at 10 A. Neglect magnetic saturation and losses throughout this problem.
(a) What is the nameplate per-phase current and what was the phase of this current during the second test? Draw a phasor diagram of this situation. Assume $\underline{V}$ at an angle of 0°.
(b) A motor is now loaded. If the field current is then adjusted for nameplate kVA at unity power factor, what current is required in the field? Draw a phasor diagram for this situation.

4.35. A three-phase, 20-hp, four-pole, 60-Hz, 600-V synchronous motor has 60 N-m of output torque with a leading power factor of 0.92 and an efficiency of 89.5%. Find the input current.

4.36. A 460-V, three-phase, six-pole, 60-Hz synchronous motor has a nameplate output power of 10 hp. The per-phase synchronous reactance is 15 Ω. At nameplate output power, the motor requires a dc current of 3 A in the rotor field for unity power factor. Ignore losses.
(a) Find the input current per phase for unity power-factor operation at $P_{out} = 10$ hp.
(b) It is desirable to have the motor draw leading current from the bus. Find the field current for a leading angle of 30° at $P_{out} = 10$ hp. Assume that the magnetic structure is linear, that is, assume that the excitation voltage is proportional to the field current.
(c) For the condition in part (a), the load is suddenly dropped (say, the output shaft breaks!). Find the resulting per-phase current and the power factor.

4.37. Two identical synchronous machines are operated as a motor–generator set, as shown in Fig. P4.37. The rotor fields are connected in series so that they have the same field currents. The synchronous reactance of both machines is 2 Ω, and ignore all losses. The machine that is operated as a motor is connected to a 240-V infinite bus. The field current is adjusted to give minimum motor current (ideally zero) when the generator has no load.
(a) What is the output line voltage with no electrical load?
(b) Without a change in the field currents, the generator is now loaded with real 5 Ω/phase. Find the output line voltage and current.
(c) Assuming that the output voltage with no load is in phase with the input voltage to the motor, what is the phase of the output voltage in part (b) relative to the voltage on the infinite bus?

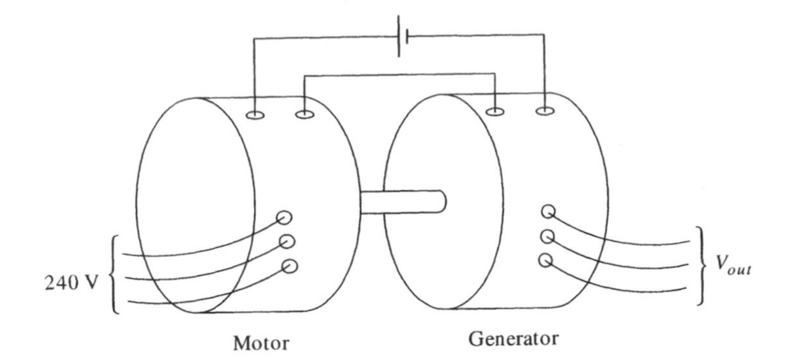

Figure P4.37 A motor–generator set. The dc field currents are equal.

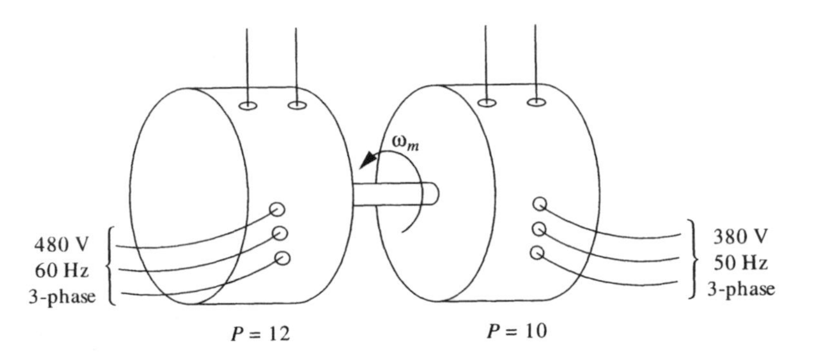

Figure P4.38 A 60-to-50-Hz converter.

4.38. Two 50-kVA[29] three-phase synchronous machines (with 12 and 10 poles, respectively) are used to tie together a 60-Hz system and a 50-Hz system, as shown in Fig. P4.38. The 60-Hz system is considered an infinite bus, and the 50-Hz machine is the sole generator in its system. The synchronous reactances are 1.7 and 1.2 Ω, respectively, at the appropriate frequencies. The field currents are set to give the nameplate voltages with unity power factor and nameplate kVA on both machines. Neglect losses.

(a) What is the mechanical speed of the machines in rpm?

(b) For nameplate operation at unity power factor, what is the per-phase resistance of the load on the 50-Hz system?

(c) For the situation described in part (b), draw a per-phase phasor diagram showing $\underline{\mathbf{E}}_f, \underline{\mathbf{I}}, \underline{\mathbf{V}}$, and the rotor power angle, δ_R, for the motor.

(d) What would be the no-load line voltage on the 50-Hz system, assuming linear magnetic structures?

4.39. Figure P4.39 shows a per-phase phasor diagram for a 12-pole, three-phase synchronous machine.

(a) Is the machine operating as a motor or a generator?

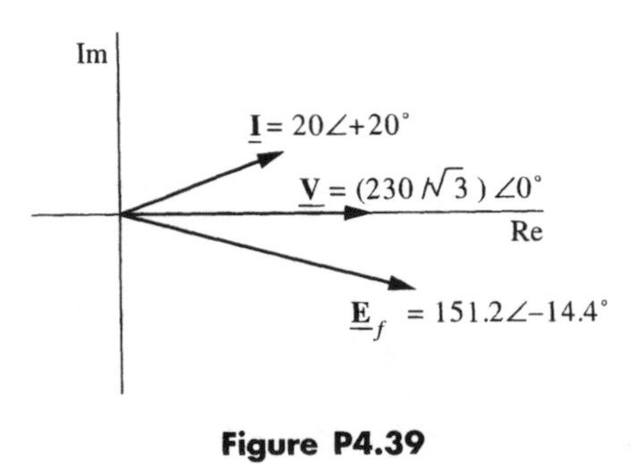

Figure P4.39

[29]The apparent power rating of a machine is frequency-dependent. These machines are rated at 60 and 50 Hz, respectively.

(b) What is the voltage and apparent power into/out of the machine?
(c) Determine the synchronous reactance of the machine.
(d) For the same real power, what magnitude of excitation voltage yields unity power factor?
(e) If the motor were stopped and used as a brake, find the maximum torque it could hold. Assume both rotor and stator currents must be derated by 50% from those implied in the phasor diagram due to loss of cooling by convection.

4.40. A 50-hp, 480-V, 8-pole, 60-Hz, three-phase synchronous motor has a power factor of 0.8, leading, at full-load output. When the mechanical load is removed, the current into the motor is 40 A. Neglect losses. Find the synchronous reactance.

4.41. A cylindrical-rotor, 60-Hz, three phase, 12-pole synchronous motor operates from 2300 V and produces 500 hp. The motor operates with unity power factor with an excitation voltage of $E_f = 1620$ V per phase. Neglect losses. Determine the following:
(a) The current.
(b) The synchronous reactance.
(c) The torque.
(d) The rotor power angle.

4.42. A 100-hp, 480-V, 8-pole, 60-Hz, three-phase synchronous motor has nameplate power at unity power factor with a field current of 5 A dc. What field current corresponds to 80 hp out and a 0.9 *PF*, leading? The synchronous reactance is 1 Ω. Neglect losses. Assume that the excitation voltage is proportional to the field current.

4.43. The per-phase phasor diagram for a three-phase, 60-Hz, 8-pole synchronous motor is shown in Fig. P4.43. Note that all sides and two angles of the triangle are shown. The current/phase is 21 A (rms).
(a) Is the motor overexcited or underexcited?
(b) What is the rotor power angle?
(c) What is the power factor and is it leading or lagging?
(d) Determine the synchronous reactance per phase.
(e) Determine the output power and torque, neglecting mechanical losses.

4.44. A 10-kVA, 600-V, three-phase synchronous motor operates at nameplate kVA with 10-hp output power.

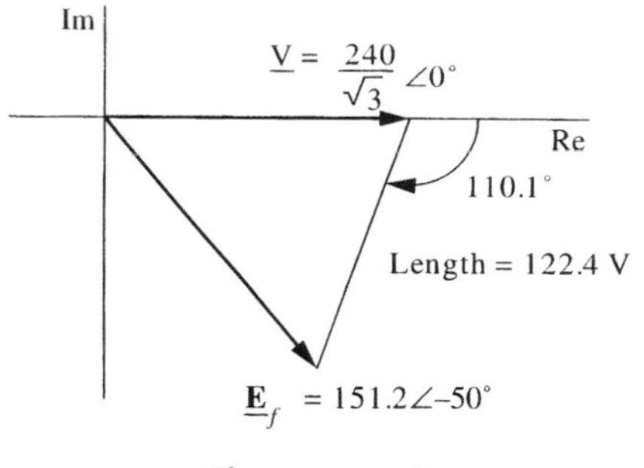

Figure P4.43

The field current is 5 A and the synchronous reactance is 30 Ω/phase. Find the field current for unity power factor, assuming that 5 A overexcites the machine. Ignore losses and assume excitation voltage proportional to field current.

4.45. A three-phase, 30-hp, 12-pole, cylindrical rotor, 600-V synchronous motor has a synchronous reactance of 8 Ω. At full load, the field current is adjusted for minimum motor current and unity power factor. Ignore losses.
(a) Find the rotor power angle under these conditions.
(b) With the same field current, what is the greatest torque surge the motor can handle without losing synchronism?
(c) If the field current is increased 10% (assume a linear increase in excitation voltage) and the load torque is held constant, what is the reactive power drawn by the motor?

4.46. The diagram in Fig. P4.46 shows the per-phase phasor diagram for an 8-pole, 460-V, 60-Hz, three-phase synchronous motor. With the per-phase voltage as the phase reference, the excitation voltage

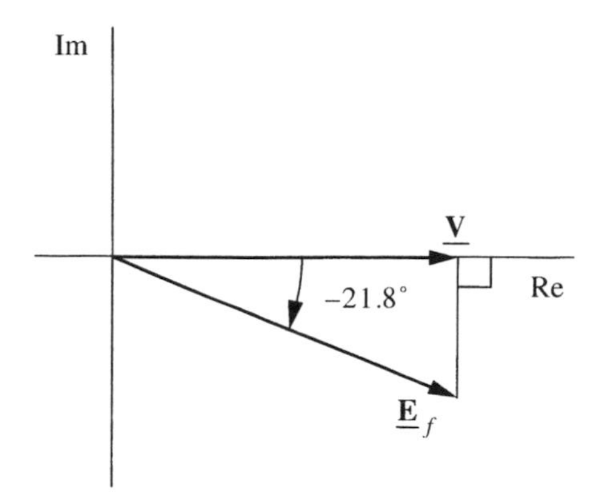

Figure P4.46

is 286 ∠ −21.8° V. The synchronous reactance is 6 Ω. Ignore losses.

(a) Find the per-phase current.
(b) Find the power factor.
(c) Determine the developed torque.

4.47. A three-phase synchronous motor operates under the following conditions: 208 V, 20 A, 0.866 *PF*, leading, and 3.8-A dc field current. The synchronous reactance is 8.5 Ω. For the same mechanical load, find the field current for a power factor of unity and the corresponding output power in hp.

4.48. A three-phase, 60-Hz, 460-V, 6-pole, 10-hp synchronous motor has a synchronous reactance of 30 Ω. Ignore losses.

(a) The motor output power is 8 hp and the field current is 2 A for a power factor of unity. Find the input current under these conditions.
(b) If the output power increases to 10 hp and the field current is not changed:
  (1) Does the current increase, decrease, or stay the same?
  (2) Does the current lag, lead, or remain in phase?
(c) Find the new field current to give unity power again if the magnetic flux is proportional to the field current.

4.49. A test is performed on a three-phase, 460-V synchronous motor. The load is kept constant at rated load, and input current is measured while the field current is varied. The results are shown in Fig. P4.49. The nameplate current is marked on the graph at 24 A. Ignore losses and assume that the excitation voltage is proportional to field current.

(a) What is the apparent power rating of the motor?
(b) What is the nameplate output power in hp?
(c) What range of field currents corresponds to overexcited operation?
(d) Determine the synchronous reactance of the motor.

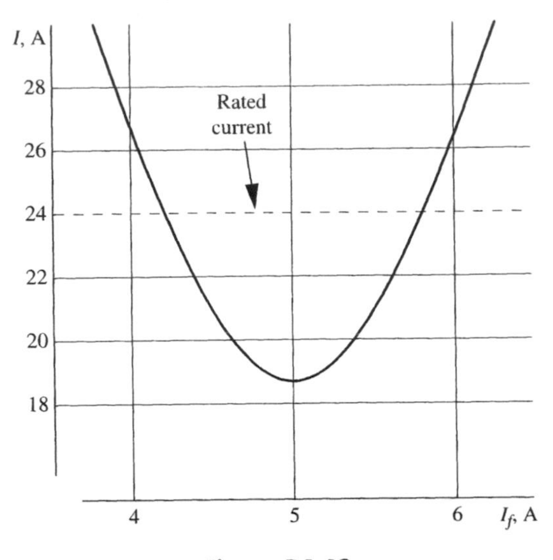

Figure P4.49

4.50. A 240-V, three-phase, 60-Hz synchronous machine has an open-circuit voltage described by the equation $V_{OC} = 100I_f$, where I_f is the field dc current. The short-circuit current is $I_{SC} = 34I_f$, where I_{SC} is the short-circuit line current.

(a) What is the per-phase synchronous reactance, X_s? Ignore any resistance in the stator windings.
(b) What field current is required to generate 20 kW at unity power factor with nameplate voltage?
(c) What is the power angle for part(b)?
(d) The machine is run as a motor. The field current is adjusted to give a leading power factor of 0.707, with an output power of 25 hp at 240 V. Sketch the per-phase phasor diagram, showing $\underline{\mathbf{V}}, \underline{\mathbf{I}}$, and $\underline{\mathbf{E}}_f$. What field current is required?

4.51. A 60-Hz, three-phase synchronous motor has four poles and a 5-hp rating. Ignore losses. The motor operates at 440 V with the power factor adjustable, according to the field current.

(a) If the field current is adjusted for minimum input current with the nameplate output power, what would be that current?
(b) If the field current is increased to change the current 10% above the minimum value, what then would be the reactive power into the motor? The output real power is unchanged.
(c) Give the output torque at full load in metric units.
(d) If it took a 21% increase in field current to produce the 10% increase in input current, what would be the synchronous reactance? Assume that the excitation voltage is proportional to the field current.

4.52. An eight-pole, 480-V, 68-kVA, 60-Hz, 75-hp synchronous motor is connected to a load. The field current is varied, and the line current is measured to follow the curve shown in Fig. P4.52. The

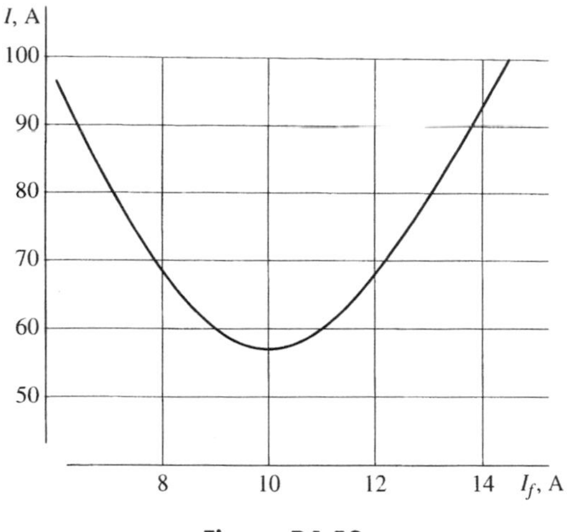

Figure P4.52

minimum current is 57 A, and at 14 A of field current, the current is 93 A. (*Note:* The kVA rating is higher than the real power rating because the motor is required to handle a reactive power load in addition to the real power load.) Assume linear magnetic structure and neglect losses.

(a) What is the load on the motor?
(b) Which part of the graph corresponds to leading current and which to lagging?
(c) Find the output torque of the motor.
(d) At full load, what is the limiting power factor of the motor?
(e) At the load represented by the graph, what is the reactive power the motor can handle?
(f) Find the synchronous reactance of the machine.
(g) What field current corresponds to a reactive power of –30 kVAR with the load on the graph?

4.53. A three-phase, 12-pole, 460-V, 60-Hz synchronous motor has a synchronous reactance of 4 Ω, an output power of 25 hp, and a leading power factor of 0.866. Ignore losses.
(a) Draw a per-phase phasor diagram showing per-phase voltage, current, and excitation voltage.
(b) Find the torque the motor can develop before losing synchronism.

4.54. A three-phase, 60-Hz, 460-V, 12-pole, 15-hp synchronous motor operates at nameplate power and the field current is adjusted for unity power factor. Ignore electrical and mechanical losses in this problem.
(a) Find the motor speed in rpm.
(b) Find the motor output torque.
(c) Find the input current/phase.
(d) If the mechanical load is removed, the current changes to 8 A. Find the phase of the current relative to the per-phase voltage and the synchronous reactance of the motor.

4.55. Adapt Eqs. (4.69) and (4.70) for cylindrical rotor *generators* by substituting X_s for X_d and X_q. Determine the power factor in Example 4.5 using only these adapted formulas based on complex power and Eq. (4.59), which is derived from them.

4.56. Calculate the maximum torque available from the motor in Example 4.9 on reluctance torque.

Answers to Odd-Numbered Problems

4.1. 428 A-t.

4.3. Left: **(a)** 4 poles on the stator; **(b)** no torque; right: **(a)** 2 poles on the stator; **(b)** CCW (+ direction) torque.

4.5. **(a)** 1.25 T; **(b)** 19.1°; **(c)** CCW.

4.7. **(a)**

Flux density	*Poles*	*Synchronous speed*	*Description*
$1.5 \cos(200\pi t + \theta_m)$	2	6000 rpm	Rotating CW wave
$1.2 \cos(30t - 3\theta_m)$	6	$300/\pi$ rpm	Rotating in CCW direction
$1.0 \cos(100\pi t) \sin(2\theta_m)$	4	—	No rotation, oscillation at 50 Hz

4.9. **(a)** 2; **(b)** 60°; **(c)** 24,000 rpm in CW direction; **(d)**

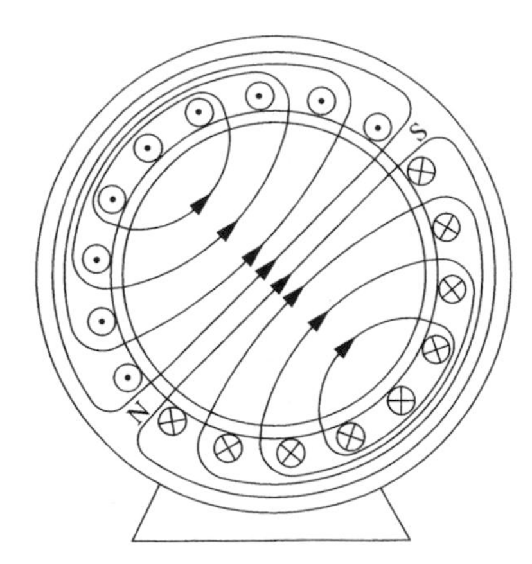

4.11. -0.809 T.

4.13. **(a)** $-90°$; **(b)** CW ($-$ direction); **(c)** $0.72 \cos[100\pi t + (\theta_m + 90°)]$; **(d)** 3000 rpm.

4.15. **(a)**

(b) $0.85 \cos[120\pi t - 2(\theta_m - 90°)]$ T; **(c)** -0.006255 s, and can add multiples of 1/120 s.

4.17. $B(t,\theta_m) = 0.8 \cos(942t - 6\theta_m)$ T.

4.19. **(a)** Rotating wave of flux; **(b)** $B_r \cos(\omega t + \theta_m)$; **(c)** $B_r \angle 30°$; **(d)** 0.377 T.

4.21. **(a)** 8; **(b)** CW ($-$direction); **(c)** 1/600 s.

4.23. **(a)** 60π rad/s; **(b)** just one; **(c)** 7.12 A; **(d)** 23 wires.

4.25. **(a)** 1140 N-m at 1200 rpm; **(b)**

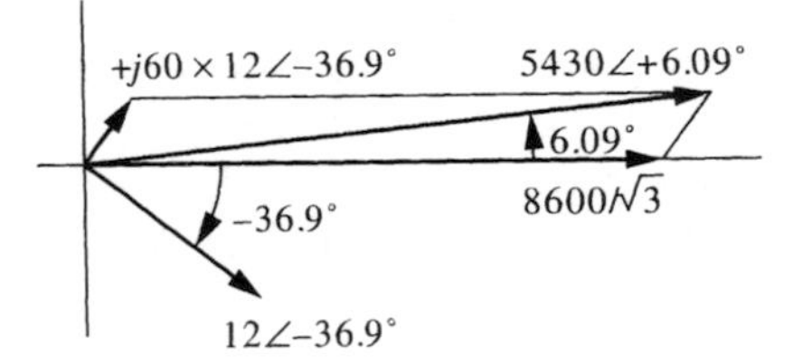

(c) 9400 V.

4.27. **(a)** Flux levels are high for the voltage measurement but low for the current measurement; **(b)** 7.46 Ω; **(c)** greater.

4.29. **(a)**

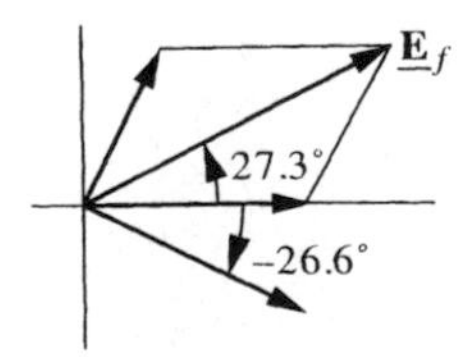

(b) $\underline{\mathbf{V}}$ is the same, $\underline{\mathbf{I}}$ increases, θ decreases, δ_R increases, $\underline{\mathbf{E}}_f$ same.

4.31. **(a)** 7.97 kVA; **(b)** 59.6 N-m; **(c)** 2.00 Ω.

4.33. **(a)**

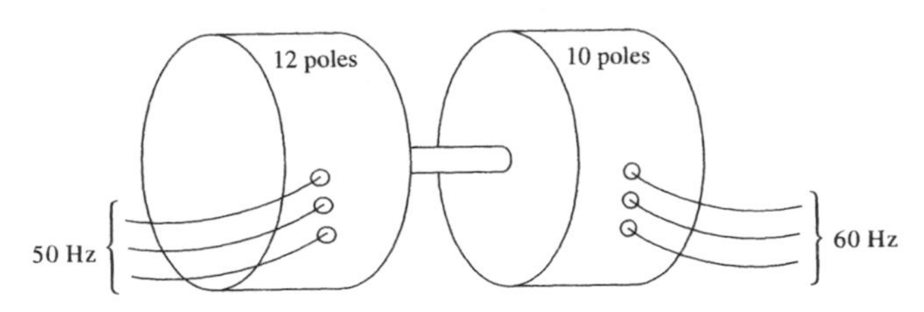

(b) rotational speed = 720 rpm; field currents are adjusted to make reactive powers balance.

4.35. 13.2 A.

4.37. **(a)** 240 V; **(b)** 25.7 A, 223 V; **(c)** $-42.0°$.

4.39. **(a)** Motor; **(b)** 230 V, 7970 VA; **(c)** 2.00 Ω; **(d)** 138 V; **(e)** 120 N-m.

4.41. **(a)** 93.6 A; **(b)** 9.91 Ω; **(c)** 5940 N-m; **(d)** $-34.9°$.

4.43. **(a)** Underexcited; **(b)** $-50.0°$; **(c)** 0.937, lagging; **(d)** 5.83 Ω; **(e)** 87.0 N-m.

4.45. **(a)** $-26.4°$; **(b)** 800 N-m; **(c)** 426 V.

4.47. 2.86 V, 8.35 hp.

4.49. **(a)** 19.1 kVA; **(b)** 14.8 kW; **(c)** > 5 A; **(d)** 3.72 Ω.

4.51. **(a)** 4.89 A; **(b)** -1710 VAR; **(c)** 19.8 N-m; **(d)** 31.6 Ω.

4.53. **(a)**

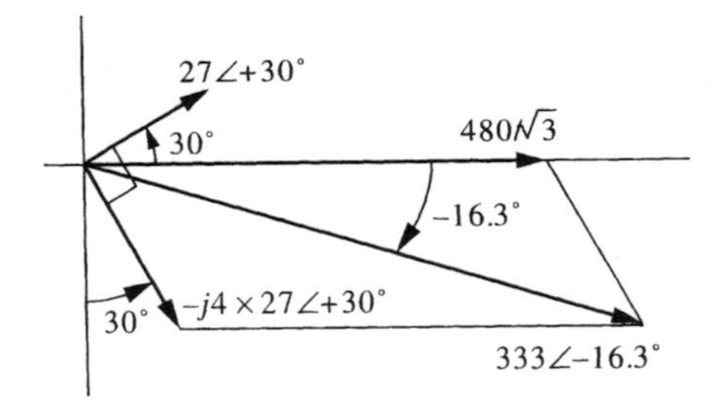

(b) 1060 N-m.

4.55. The same answers.

Induction Motors

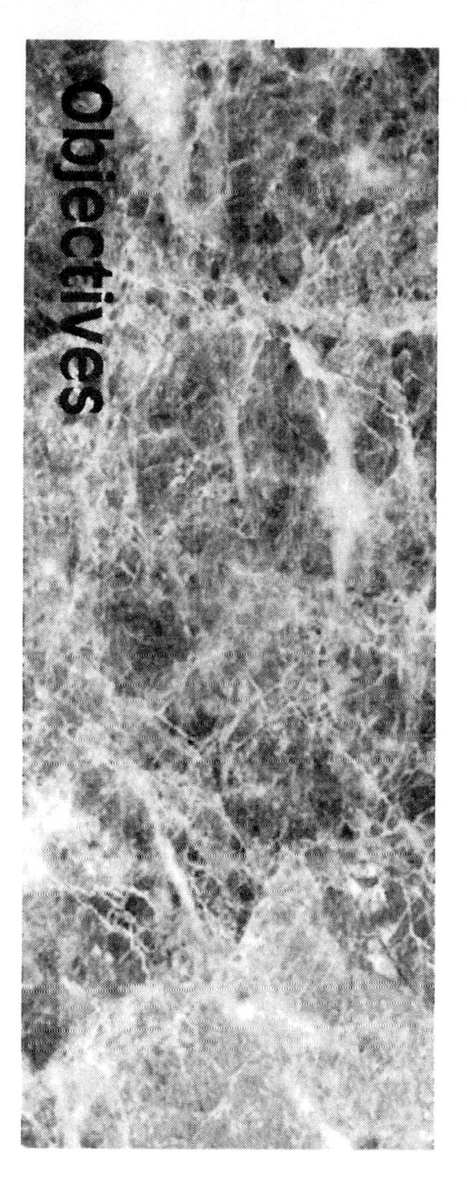

1. To understand the induction principle that produces torque in induction motors
2. To understand the physical origin of all elements in the equivalent circuit of an induction motor, and be able to use the circuit to predict motor performance
3. To understand the characteristics and applications of the various types of single-phase induction motors

The induction motor is the workhorse of electric motors. The induction effect drives motors from a small fan circulating air in a refrigerator to a 1000-hp motor driving a compressor in an airport air-conditioning system. This chapter explains this induction effect and its utilization in both three-phase and single-phase motors.

Introduction to Induction Motors

The three-phase induction motor was patented by Nikola Tesla in 1888 and currently accounts for over 90% of the motors used in industry. In an induction motor, the stator poles rotate at synchronous speed and the rotor poles are induced by transformer action and also rotate at synchronous speed. The rotor rotates physically at a speed slightly slower than the synchronous speed and slows down slightly as the load torque and power requirements increase. In Chap. 1, we discussed the interpretation of nameplate information for three- and single-phase induction motors. This led to a simple model for output torque and power that is adequate for many applications. This chapter explains the physical principles of induction motors and develops more detailed models for describing induction motor performance.

In Chap. 4, we showed how stator flux is produced, how the flux is rotated at synchronous speed, and how torque is developed between stator flux and rotor flux. We also studied the characteristics of the synchronous machine, in which the rotor poles are produced by an electromagnet. Now we first investigate the induction of rotor flux (poles) by relative motion between stator flux and rotor conductors. We then derive an equivalent circuit for the induction motor based on Ampère's circuital law, Faraday's law, and conservation of energy. From this circuit, we determine motor characteristics such as efficiency, starting current, and torque as a function of speed. We conclude with a discussion of the various types of single-phase induction motors.

5.1 INDUCTION MOTOR PRINCIPLES

Induction Principles

LEARNING OBJECTIVE 1.

To understand the induction principle that produces torque in induction motors

Stator construction. The stator of an induction motor is identical in principle to that of a synchronous machine: three phases, P poles, sinusoidal mmf and flux distribution, synchronous speed given by Eq. (4.24). The field of the induction motor is the stator; the armature is the rotor.

squirrel-cage rotor

Rotor construction. The rotor of the induction motor shown in Fig. 5.1(a) is an iron cylinder with large embedded conductors. The conductors of the squirrel-cage rotor, shown in Fig. 5.1(b), run the width of the rotor and are shorted at both ends by large conducting rings. Figure 5.2(a) shows the rotor of a small induction motor with part of the metal removed to show the conductors. The conductors are slanted slightly to eliminate resonance effects. The protrusions on the end rings enhance heat transfer. Figure 5.2(b) shows the construction of a small three-phase induction motor.

slip speed

A mental experiment. We investigate how the rotor poles are induced through a mental experiment. We begin with nonrotating stator poles; thus, we excite the stator windings with dc currents of appropriate magnitudes to create a field in the horizontal direction, as shown in Fig. 5.1(a). We now, in our imagination, take hold of the rotor and slowly rotate it in the clockwise direction with an angular velocity ω_Δ. This angular velocity, called the *slip speed*, is the angular velocity in the *negative* direction of the rotor conductors relative to the stator flux, which is stationary in our mental experiment. Thus, in the mental experiment, the mechanical speed of the rotor is $\omega_m = -\omega_\Delta$.

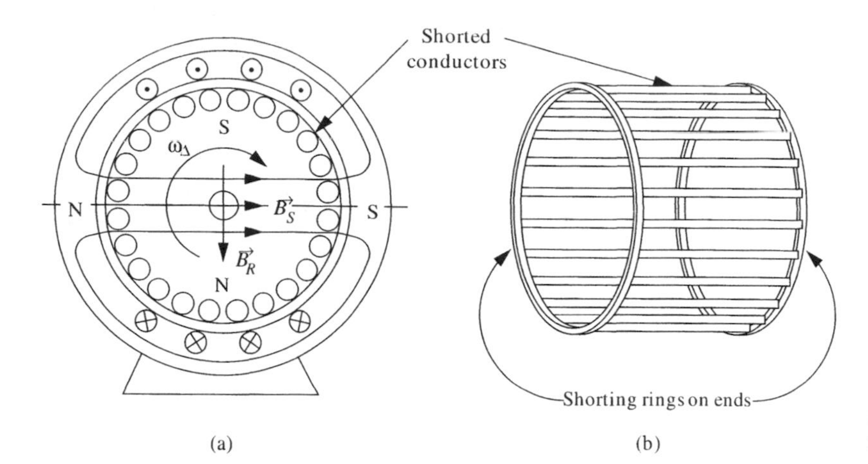

Figure 5.1 (a) Squirrel-cage induction motor; (b) conductors in the rotor.

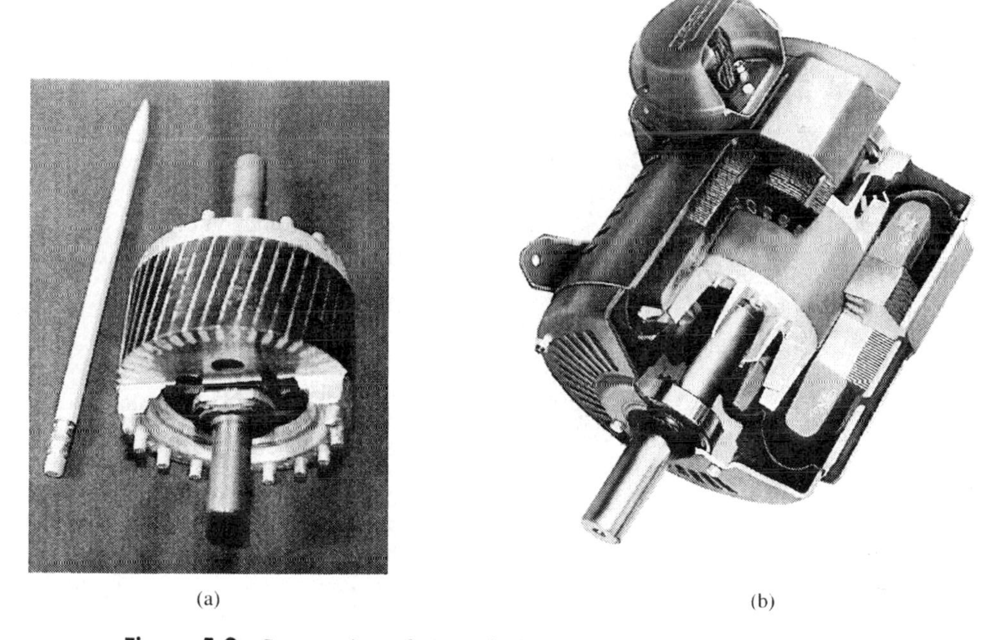

Figure 5.2 Cutaway views of (a) a squirrel-cage rotor and (b) a three-phase induction motor. (Photo (b) courtesy of the Lincoln Electric Company.)

Induced currents. Figure 5.3 diagrams the effects of this rotation. The rotation induces a $\vec{\ell} \cdot \vec{u} \times \vec{B}$ emf in the conductors, Eq. (2.43), and currents flow in the shorted conductors. Applying the right-hand rule, you determine that rotor currents flow out of the page on the right and into the page on the left, with the maximum current in the horizontal plane, under the stationary stator poles. If you consider a single conductor as it rotates under first one pole and then the other, you will realize that an ac current is

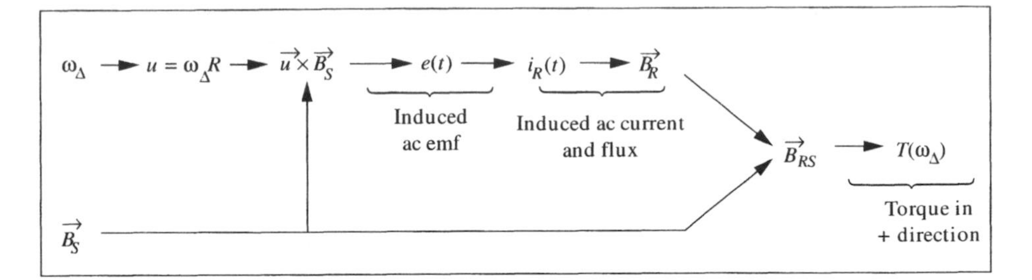

Figure 5.3 Cause and effect in an induction motor.

induced in each conductor. From the viewpoint of the rotor, an ac voltage is induced in the conductor to drive these currents. For this reason, we show in Fig. 5.3 that the steady motion produces a time-varying emf, $e(t)$.

Induced rotor flux. The induced rotor currents produce a magnetic flux, $\vec{B}_R$. Applying the right-hand rule, you determine that this magnetic flux is directed downward, as shown in Fig. 5.1(a). If we associate magnetic poles with $\vec{B}_R$, we require a north pole at the bottom to act as source for the induced flux in the gap, and a south at top, as shown in Fig. 5.1(a). The stator flux and induced rotor flux combine to a rotor–stator flux in the gap, $\vec{B}_{RS}$, and a torque is produced, as described in Sec. 4.1. The torque is counterclockwise, in the positive θ_m direction, because the rotor magnetic poles seek to align with their opposites on the stator. From the viewpoint of Eq. (4.14), the power angle δ_{RS} is $-90°$ and the developed torque is thus in the positive θ_m direction, opposing the torque we apply to cause the rotation. Here are some observations and consequences based on these effects:

- The induced currents and the resulting flux are caused by the stator flux, $\vec{B}_S$. The stator flux is created by dc currents and is not rotating. Although the rotor turns physically, the pattern of currents and induced magnetic flux does not rotate. Thus, the rotor–stator flux pattern, $\vec{B}_{RS}$, is constant in time and space for a given rotation speed (ω_Δ).
- The power angle δ_{RS} is $-90°$ for small slip speed.
- The developed torque, T_{dev}, is proportional to the induced currents and the sine of the power angle and hence varies with slip speed

$$T_{dev}(\omega_\Delta) \propto -i_R(\omega_\Delta)\sin\delta_{RS} = -i_R(\omega_\Delta)\sin(\theta_R) \tag{5.1}$$

  where $i_R(\omega_\Delta)$ represents the rotor current, θ_R the position of the rotor north pole, because $\delta_{RS} = \theta_R - \theta_S$ and $\theta_S = 0°$.
- The mechanical input power that we must supply to turn the rotor against the developed torque is

$$P_{in} = P_R = \omega_\Delta T_{dev}(\omega_\Delta) \tag{5.2}$$

  Conservation of Energy

  where P_R is the rotor copper loss. This power is converted into rotor-copper losses because resistive loss in the rotor conductors is the only energy-conversion mechanism at work if we ignore mechanical losses.

- For the two-pole structure shown in Fig. 5.1(a), the currents induced in each rotor conductor are sinusoidal at an electrical radian frequency of $\omega_e = \omega_\Delta$. If there were more than $P = 2$ poles, the rotor electrical frequency would be $\omega_e = (P/2)\omega_\Delta$.

EXAMPLE 5.1

Rotor current

The rotor in Fig. 5.1(a) has a resistance of 0.0005 Ω/conductor and is turning 15 rad/s relative to a maximum flux density of 1T. The length is 3.4 cm and the radius is 3.8 cm. Find the current in the conductor directly under a stator pole, assuming perfect short circuits at the end.

SOLUTION:
The emf is given by Eq. (2.43)

$$\text{emf} = \ell\omega_m RB_S = 0.034 \times 15 \times 0.038 \times 1 = 0.0194\ \text{V} \tag{5.3}$$

and hence the current is the emf divided by the resistance, or 38.8 A.

WHAT IF?

What if there are 24 slots on a side, and the currents are tapered sinusoidally? What is the total rotor current?[1]

Developed Torque, $T_{dev}(\omega_\Delta)$

Factors influencing torque characteristics. The developed torque, which counters the applied torque causing the rotation, depends on the slip speed, ω_Δ, through three factors:

- The induced voltage increases in proportion to the slip speed because the speed of the conducting bars relative to the magnetic flux increases. This increase in ac voltage causes the induced current to increase as slip speed increases.
- The impedance of the rotor circuit increases with slip speed because of the increase in the frequency of the induced voltage. At very small slip speeds, the impedance is largely resistive, but at larger slip speeds, the inductance of the rotor dominates. Thus, the current tends to approach a maximum value, where the increase in ac voltage is offset by the corresponding increase in impedance.
- The rotor–stator power angle between rotor and stator poles (fluxes) increases beyond the optimum value of $-90°$ because of the phase delay of the rotor currents associated with the inductance. This increase in power angle causes a lessening of the torque at high slip speeds, as shown by Eq. (5.1).

[1] Approximately 590 A.

Combined effect. If we examine the dependence of torque given in Eq. (5.1) and then consider the dependence of induced current and power angle on slip speed, we predict the torque versus mechanical speed, $\omega_m = -\omega_\Delta$, characteristic shown in Fig. 5.4. The developed torque, being a countertorque, has a sign opposite to that of the mechanical speed. For small slip speeds, the torque magnitude increases with slip speed, reflecting the linear increase of emf and current. However, the curve levels off and eventually decreases as slip speed increases because the induced current levels off and because the power angle increases beyond $-90°$. Hence, the torque reaches a maximum and then decreases due to the increasing power angle.

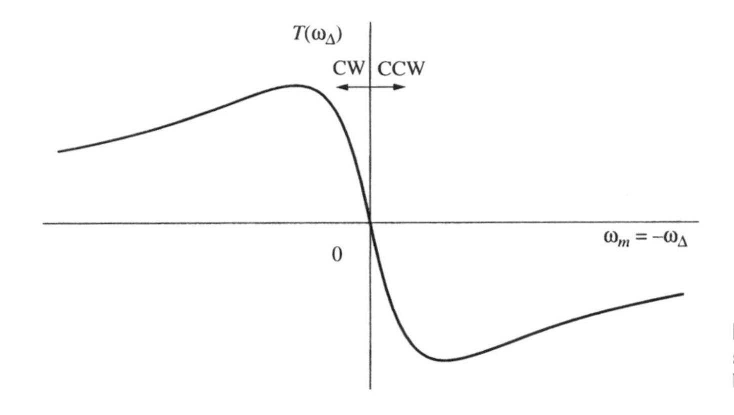

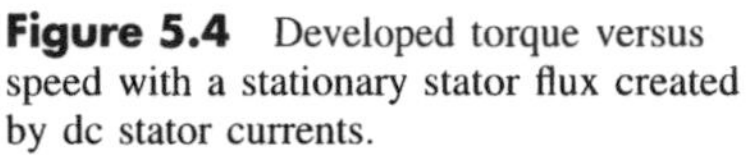

Figure 5.4 Developed torque versus speed with a stationary stator flux created by dc stator currents.

Summary. This completes our mental experiment with dc stator currents. We examined the physical principles at work in the induction motor. We looked at the emf and the resulting current and magnetic flux of the rotor. We examined the frequency of the emf and currents in the rotor. We showed how torque is developed. We developed a formula for the rotor-copper losses, Eq. (5.2). Finally, we anticipated how the developed torque depends on slip speed. In the next section, we excite the stator with three-phase ac currents and the rotating flux creates motor action.

Three-Phase Induction-Motor Characteristics

Effect of rotation. We now apply three-phase currents to the stator field windings. As shown in Chap. 4, the stator flux rotates at synchronous speed. In comparison with our mental experiment, the stator currents change from dc to ac, and the synchronous speed changes from $\omega_s = 0$ to $\omega_s = (2/P)\omega$, where ω is the electrical angular frequency of the three-phase stator currents.

Slip. We assume now that the rotor is turning slower in the positive direction than synchronous speed, such that it is slipping backward by ω_Δ relative to the synchronous speed; thus, the rotor speed changes from $\omega_m = -\omega_\Delta$ to

$$\omega_m = \omega_s - \omega_\Delta = \omega_s - s\omega_s = (1 - s)\omega_s \quad \text{rad/s} \tag{5.4}$$

slip

where s is the normalized slip speed, or more simply the *slip*:

$$s = \frac{\omega_\Delta}{\omega_s} = \frac{\omega_s - \omega_m}{\omega_s} \tag{5.5}$$

Although the rotor is now turning at a high speed, the electrical voltage and currents in the rotor are still generated by the *relative* motion between the rotor conductors and the stator flux; hence, the rotor voltage and current have an electrical frequency of

$$\omega_e = \omega_\Delta \times \frac{P}{2} = s\omega \tag{5.6}$$

where ω is the stator electrical frequency, normally $2\pi \times 60$ rad/s. Even though the rotor is not turning at synchronous speed, the rotor flux is still locked to the stator flux, as in our mental experiment, and hence the rotor *flux* rotates at synchronous speed.

Torque generation. From the viewpoint of the rotor, there is electrically no difference between stationary stator flux and rotating stator flux, provided the slip speed is the same. For the same slip speed, the emf, currents, and rotor magnetic flux are the same in both cases, and hence the developed torque is also the same. The torque characteristic of the motor with ac current, therefore, is identical to Fig. 5.4, except that it now is shifted to the right by the synchronous speed, as shown in Fig. 5.5. Figure 5.5 gives the speed in units of mechanical speed, ω_m rad/s or n rpm, and also in slip, s. Zero slip corresponds to rotation at synchronous speed and a slip of unity corresponds to a stationary rotor, $\omega_m = 0$.[2]

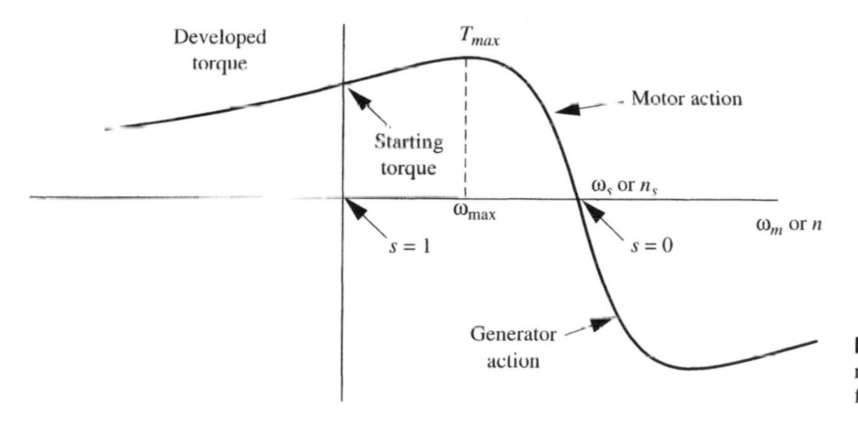

Figure 5.5 Developed torque versus rotational speed and slip with rotating stator flux.

blocked rotor

Torque characteristic. From Fig. 5.5, we note the following motor characteristics:

- Positive mechanical speeds below synchronous speed, $0 < \omega_m < \omega_s$, produce motor action because torque is developed in the same direction as the rotation.

induction generator

- Mechanical speeds exceeding synchronous speed, $\omega_m > \omega_s$, produce generator action because the developed torque is opposite the direction of rotation. External mechanical torque is required to drive the rotor against this countertorque. The electrical power generated has the same frequency as the currents in the stator field windings. This generator produces power when connected to a three-phase power source, but has poor characteristics when connected only

[2] Also called "blocked or locked rotor."

to a load. Such *induction generators* find application in wind-power generators and small power plants.

- When the mechanical speed is equal to the synchronous speed, $\omega_m = \omega_s$, neither motor nor generator action results; the system is merely idling. In this condition, an external mechanical drive is required to supply the mechanical losses.
- For $\omega_m \approx \omega_s$, the torque is proportional to slip. This is the region where the motor is normally operated.
- A point of maximum torque, $T_{\max}$, exists at a speed of $\omega_{\max}$. This speed is slightly below the speed for maximum output power.
- The starting condition corresponds to $s = 1$. The starting torque is positive and hence this motor is self-starting.

developed torque and power, output torque and power, windage

Power relationships. We can derive several useful power relationships from basic principles and from the results of our mental experiment, which gave an expression for the rotor-copper losses in Eq. (5.2). At this point, we must distinguish between the *developed* torque and power and the *output* torque and power. The *developed torque* is the magnetically generated torque and is the torque we have been discussing. The *output torque* is smaller due to the torque required to overcome bearing and *windage* losses, the loss due to air movement. Similarly, the *developed power* is $P_{dev} = \omega_m T_{dev}$, and represents the power converted from electrical to mechanical form. The *output power* is $P_{out} = \omega_m T_{out}$ and is smaller than the developed power because of the mechanical losses.

The developed power is

$$P_{dev} = \omega_m \times T_{dev} = (1 - s)\omega_s T_{dev} \tag{5.7}$$

where T_{dev} is the developed torque and we introduced the synchronous speed through the relationship $\omega_m = (1 - s)\omega_s$. As shown in Eq. (5.2), the power lost in the rotor-copper losses is

$$P_R = \omega_\Delta \times T_{dev} = s\omega_s T_{dev} \tag{5.8}$$

air-gap power

The power crossing the air gap from the stator to the rotor, the *air-gap power*, P_{ag}, by conservation of energy must be the sum of the developed power and the rotor copper losses

$$P_{ag} = P_{dev} + P_R = (1 - s)\omega_s T_{dev} + s\omega_s T_{dev} = \omega_s T_{dev} \tag{5.9}$$

Conservation of Energy

Equation (5.9) shows how air-gap power divides between the rotor-copper losses and the developed power, and it also shows that developed torque is proportional to air-gap power because the synchronous speed is constant.

Conservation of Energy

Power flow. The meaning of the air-gap power is illustrated by Fig. 5.6, which shows the flow of real power in the motor. A portion of the input power becomes copper and iron loss in the stator windings. We have yet to focus on this loss, but it will be represented in the equivalent circuit developed in the next section. The remainder of the input power crosses the air gap into the rotor, similar to power moving from primary to secondary in a transformer. The air-gap power splits into two terms: the rotor-copper loss, which heats the rotor, and the developed power, which turns the rotor and load.

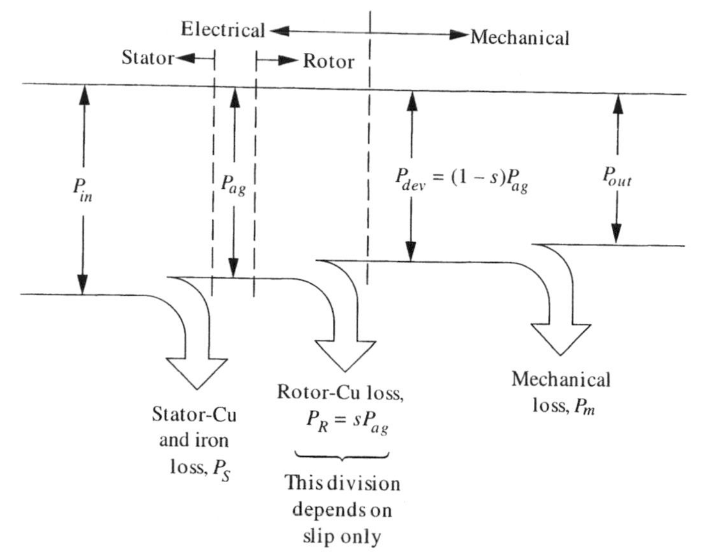

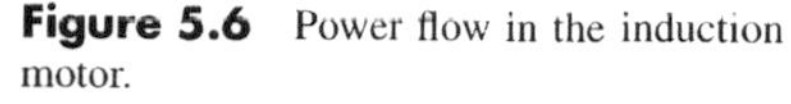
Figure 5.6 Power flow in the induction motor.

From Eqs. (5.7)–(5.9), it follows that the air-gap power divides between rotor-copper loss and developed power in a ratio that depends only on the slip speed

$$P_{ag} = \underbrace{(1-s)P_{ag}}_{= P_{dev}} + \underbrace{sP_{ag}}_{= P_R} \Rightarrow \frac{P_{dev}}{P_R} = \frac{1-s}{s} \tag{5.10}$$

EXAMPLE 5.2 Accounting for the power

A three-phase, four-pole, 60-Hz induction motor has a full-load output power of 5 hp at 1740 rpm. The motor efficiency is 87.5% at full load. The mechanical losses account for 5% of the total losses. Determine all the power quantities in Fig. 5.6.

SOLUTION:
The input power is $5 \times 746/0.875 = 4263$ W, and hence the total losses are $4263 - 5 \times 746 = 533$ W. Of these losses, 5% are mechanical, so the mechanical losses are 26.6 W and electrical losses are 506 W. The developed power is

$$P_{dev} = P_{out} + P_m = 5 \times 746 + 26.6 = 3760 \text{ W} \tag{5.11}$$

where P_m is the mechanical loss. To determine the air-gap power from the developed power, we need the slip. The synchronous speed, from Table 4.1, is 1800 rpm; hence, the slip is

$$s = \frac{n_s - n}{n_s} = \frac{1800 - 1740}{1800} = 0.0333 \tag{5.12}$$

From Eq. (5.10), the air-gap power is

$$P_{ag} = \frac{P_{dev}}{1-s} = \frac{3760}{1 - 0.0333} = 3890 \text{ V} \tag{5.13}$$

Again, from Eq. (5.10), the rotor-copper loss is

$$P_R = sP_{ag} = 0.0333 \times 3890 = 130 \text{ W} \tag{5.14}$$

and hence the stator loss is 506 − 130 = 376 W.

WHAT IF? What if there were no load? What would be the developed power?[3]

Summary. Table 5.1 compares the mental experiment with motor action. All torques are developed torques.

TABLE 5.1 Comparison of Mental Experiment (DC Stator Currents) with Motor Action (Three-Phase AC Stator Currents)

Aspect	*Mental experiment (dc stator currents)*	*Motor action (ac stator currents)*
Synchronous speed	0	$\omega_s = \omega/(P/2)$
Slip speed	ω_Δ	$\omega_\Delta = s\omega_s$
Mechanical speed	$-\omega_\Delta$	$\omega_s - \omega_\Delta = (1-s)\omega_s$
Developed torque	$T(\omega_\Delta)$	$T(\omega_\Delta) = T(s) =$ same
Rotor-copper loss	$\omega_\Delta T(\omega_\Delta)$	$\omega_\Delta T(\omega_\Delta) = s\omega_s T(s) =$ same
Air-gap power	0	$\omega_s T(s)$
Developed power	$-\omega_\Delta T(\omega_\Delta)$	$\omega_m T(\omega_\Delta) = (1-s)\omega_s T(s)$

Check Your Understanding

1. For maximum torque, rotor and stator poles should be aligned, 0°, or spatially orthogonal, 90°. Which?
2. In the mental experiment, all input mechanical power goes into stator loss, rotor-copper loss, or the dc source supplying the stator flux. Which?
3. The slip speed is defined as positive in the negative angular direction. True or false?
4. The effect of rotor inductance is to increase or decrease developed torque as slip speed increases. Which?
5. What is the slip if the rotor is turning at synchronous speed for a six-pole machine?
6. Determine the output speed in rpm of a four-pole, three-phase induction motor operating at 60 Hz and having a slip of 3%.
7. The more poles an induction motor has, the faster it turns. True or false?

[3] 26.6 W, assuming the mechanical losses are constant. The slip is extremely small and hence the rotor copper loss is negligible.

8. In a two-pole, three-phase induction motor, the frequency is 60 Hz and the slip is 5%. Find the following:
   (a) Rotational speed of the rotor (rpm).
   (b) Rotational speed of the rotor flux (rpm).
   (c) Frequency of the stator currents (Hz).
   (d) Frequency of the rotor currents (Hz).
9. If the developed power in a three-phase induction motor is 15 times the rotor-copper loss, what is the slip?

Answers. **(1)** 90°; **(2)** rotor-copper loss; **(3)** true; **(4)** decrease; **(5)** zero; **(6)** 1746 rpm; **(7)** false; **(8)** **(a)** 3420 rpm, **(b)** 3600 rpm, **(c)** 60 Hz, **(d)** 3 Hz; **(9)** 6.25%.

5.2 EQUIVALENT CIRCUITS FOR THREE-PHASE INDUCTION MOTORS

Physical Basis for Equivalent Circuit

LEARNING OBJECTIVE 2.

To understand the physical origin of all elements in the equivalent circuit of an induction motor, and be able to use the circuit to predict motor performance

Energy processes. In this section, we develop an equivalent circuit to account for every major energy process in the motor. As for a transformer, we need resistors to represent rotor-copper, stator-copper, and iron losses; and we need inductors to represent magnetic energy stored in the stray rotor and stator fields and in the flux coupling rotor and stator across the air gap. We also use a resistor to account for the energy leaving the electric circuit as developed mechanical power.

Stator circuit. The per-phase stator circuit in Fig. 5.7 accounts for stator-copper losses, R_S, stray magnetic fields, X_S, and the electromotive force, $\underline{V}_S$, induced in the stator coils by the rotating air-gap flux. The electrical frequency of stator voltage and currents is that of the ac power connecting to the stator, normally 60 Hz. Three times the complex power $\underline{V}_S\underline{I}_S^*$ into the stator emf represents the power and magnetic energy leaving the stator and passing into the air gap. The real power accounts for rotor-copper losses and developed mechanical power. The reactive power accounts for stored energy in the air gap and stray fields in the rotor.

blocked rotor reactance

Rotor circuit. The per-phase rotor circuit shown in Fig. 5.7 shows an emf, $\underline{V}_R$, an inductor, jsX_R, a resistor, R_R, and a short circuit. The electrical frequency of the rotor

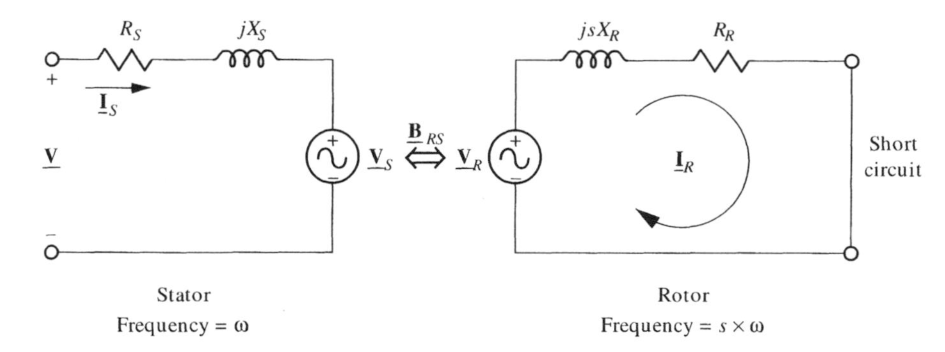

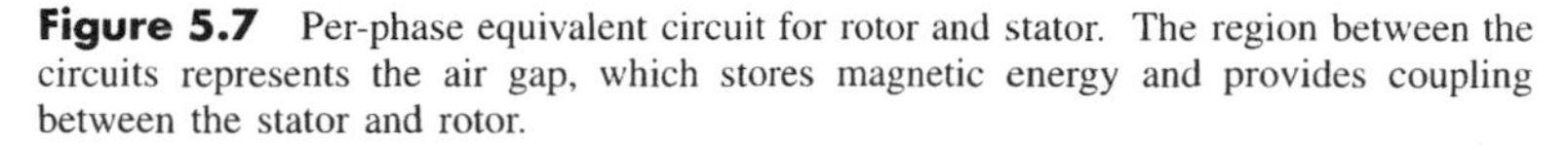
Figure 5.7 Per-phase equivalent circuit for rotor and stator. The region between the circuits represents the air gap, which stores magnetic energy and provides coupling between the stator and rotor.

emf and current is s times the stator frequency, Eq. (5.6), and hence the reactance of the rotor is shown to be proportional to slip. The reactance X_R is the *blocked rotor reactance* because the slip is unity with the rotor stationary. The rotor-copper loss is

$$P_R = 3\ |\underline{\mathbf{I}}_R|^2\ R_R \qquad (5.15)$$

where the factor of 3 accounts for the three phases. Rotor iron losses will be discussed later.

Mechanical output? The equivalent circuit shown in Fig. 5.7 is at present unable to account for the stored energy in the air gap or for the developed mechanical power. Application of Ampère's circuital law introduces a term to account for stored magnetic energy in the air gap. Application of Faraday's law *plus* the conservation of energy introduces a resistor into the rotor circuit to account for the developed mechanical power.

Ampère's circuital law. Ampère's circuital law may be applied to the induction motor in the same way it was applied to electrical transformers, Eq. (3.22). We introduce n_S and n_R as the equivalent turns for the stator and rotor, respectively, and integrate the magnetic field around a suitable path. The results can be put into the form of Eq. (3.25)

$$\underline{\mathbf{I}}_S = \frac{n_R}{n_S}\underline{\mathbf{I}}_R + \frac{\underline{\mathbf{V}}_S}{jX_{ag}} \qquad (5.16)$$

where X_{ag} is a reactance accounting for the stored energy in the air gap. Thus, *for the currents only*, the coupling between rotor and stator are represented by the ideal transformer shown in Fig. 5.8(a).

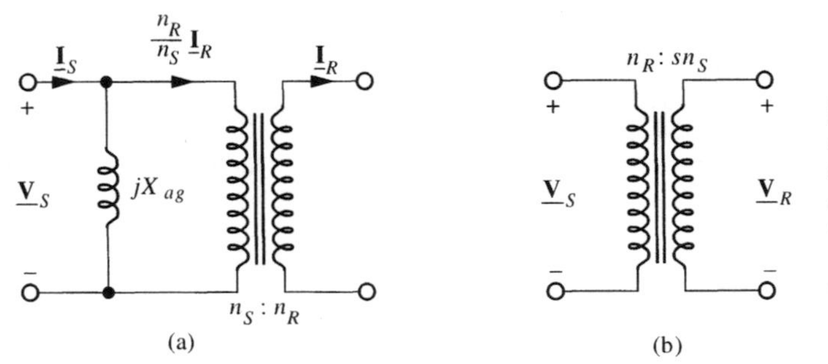

Figure 5.8 (a) Per-phase equivalent circuit showing currents in the stator and rotor. The reactance represents stored energy in the air gap. (b) Per-phase equivalent circuit showing voltage induced in the stator and rotor windings by the rotating flux. The rotor voltage is reduced by the factor s because the rotor is moving in the same direction as the flux.

Faraday's law. Again in analogy with the transformer, Eq. (3.19), the emfs in stator and rotor are related by Faraday's law. However, we must reduce the emf in the rotor by the slip because the electrical frequency in the rotor is $s\omega$. Hence, by Eq. (4.41)

$$\underline{\mathbf{V}}_S = n_S\omega R\ell\underline{\mathbf{B}}_{RS} \qquad (5.17)$$

where R and ℓ are the radius and length of the air gap, respectively, but

$$\underline{\mathbf{V}}_R = n_R s\omega R\ell\underline{\mathbf{B}}_{RS} \qquad (5.18)$$

where $\underline{\mathbf{B}}_{RS}$ is the air-gap flux density. We may eliminate the flux density to obtain

$$\frac{\underline{\mathbf{V}}_S}{n_S} = \frac{\underline{\mathbf{V}}_R}{s n_R} \tag{5.19}$$

which suggests the ideal transformer in Fig. 5.8(b). Note that the equivalent turns ratio of the voltage transformation is lower than that shown in Fig. 5.8(a) by the factor s.

Rotor impedance. The rotor voltage and current are related through

$$\underline{\mathbf{I}}_R = \frac{\underline{\mathbf{V}}_R}{R_R + jsX_R} \tag{5.20}$$

where R_R is the equivalent rotor resistance per phase and X_R is the per-phase reactance of the blocked rotor, at the frequency of the stator voltage.

Nonconservation of electric energy. We now have satisfied all circuit requirements except conservation of electric energy. Equations (5.16) and (5.19) require an ideal transformer with a voltage turns ratio $n_S{:}sn_R$ and a current turns ratio of $n_S{:}n_R$. Such a transformer would not obey conservation of energy, and of course it *should* not because *electrical* energy is not conserved in this device due to the mechanical output.

Conservation of Energy

The Steinmetz transformation. We can eliminate this strange transformer by scaling up the rotor voltage to "impose" conservation of *electrical* energy on the circuit. We divide the numerator and denominator of the right side of Eq. (5.20) by s

$$\underline{\mathbf{I}}_R = \frac{\underline{\mathbf{V}}_R / s}{(R_R/s) + jX_R} \tag{5.21}$$

where $\underline{\mathbf{V}}_R/s$ is a scaled-up rotor voltage. The transformation in Eq. (5.21) was first proposed by Carl Steinmetz (1865–1923). We now may use a normal ideal transformer to express voltage and current relationships between the stator and scaled rotor circuits. The secondary circuit in Fig. 5.9 now accounts for all the power passing from stator to rotor, *including that transformed to mechanical power.* We make the following observations:

- In transforming Eq. (5.20) into Eq. (5.21), rotor voltages and impedance values were scaled upward by $1/s$, but rotor current was unchanged. Thus, the rotor current in Fig. 5.9 is the true rotor current.
- The rotor reactance, X_R, is the per-phase reactance of the blocked rotor, as if the frequency in the rotor circuit were the same as the frequency in the stator circuit. Of course, the frequencies differ in stator and rotor circuits, but the impedance scaling compensates for the different frequencies.

Conservation of Energy

- The rotor part of Fig. 5.9 now accounts for all the power crossing the air gap into the rotor. This air-gap power divides between rotor-copper loss and developed mechanical power. The scaled-up resistor R_R/s must therefore account for both powers. Because the rotor current appears at its true value, we may calculate and subtract the rotor-copper loss from the air-gap power to find the

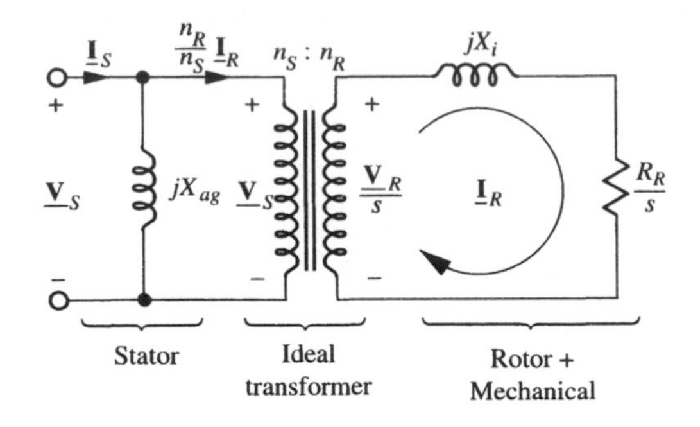

Figure 5.9 Equivalent circuit with rotor voltage and impedance scaled by a factor of $1/s$. This circuit accounts for electrical energy and energy converted to mechanical form.

developed mechanical power

$$P_{dev} = 3\left[|\underline{\mathbf{I}}_R|^2\frac{R_R}{s} - |\underline{\mathbf{I}}_R|^2 R_R\right] = 3|\underline{\mathbf{I}}_R|^2 R_R \times \frac{1-s}{s} \tag{5.22}$$

Equivalent Circuits

Hence, $[(1-s)/s] \times R_R$ is an equivalent resistance accounting for the developed mechanical power per phase. We have thus modeled the developed mechanical power in our equivalent electrical circuit, as identified in Fig. 5.10. The first resistor represents power leaving the electrical circuit and entering the mechanical world as *disordered* mechanical energy, as heat. The second resistor in the rotor circuit represents power leaving the electrical circuit and entering the mechanical world as *ordered* mechanical energy, as work. Figure 5.10 also shows resistors and reactances in the stator circuit to account for stator-copper, R_S, and iron loss, R_i, and magnetic energy as storage due to leakage flux, X_S, and air-gap flux, X_{ag}.

- The kinship between the equivalent circuit in Fig. 5.10 and that of a transformer, Fig. 3.11, is evident. The only difference is that the load representing the mechanical work is always resistive and is derived from the electrical (R_R), and mechanical (s) conditions in this electromechanical device.

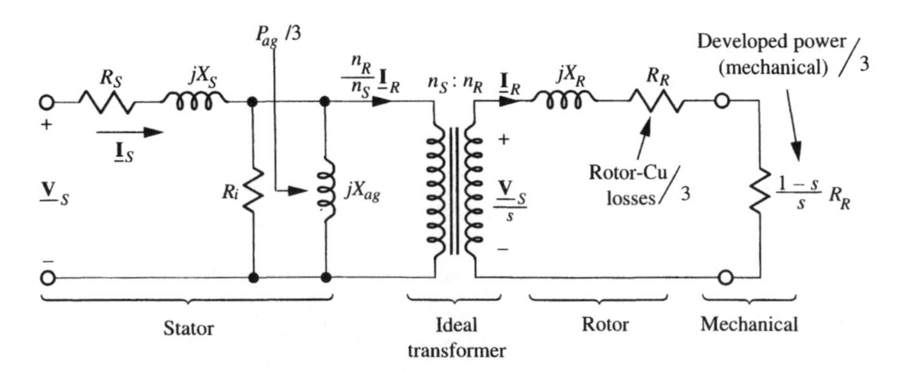

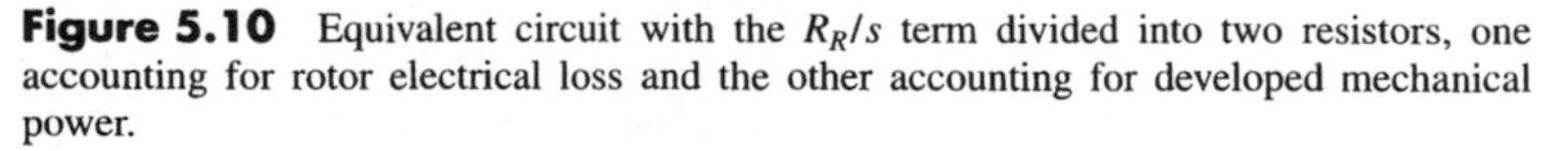

Figure 5.10 Equivalent circuit with the R_R/s term divided into two resistors, one accounting for rotor electrical loss and the other accounting for developed mechanical power.

- The air-gap power in the rotor circuit divides into developed mechanical power and rotor-copper losses in the ratio $(1 - s) : s$, as we established earlier, Eq. (5.10), from our mental experiment and conservation of energy.

Impedance Level

- We may simplify Fig. 5.10 through the impedance-transforming properties of an ideal transformer. Specifically, we multiply secondary (rotor) impedances by $(n_S/n_R)^2$ to obtain the primed values in Fig. 5.11, which represent rotor impedances referred to the stator circuit and frequency. Because energy-related quantities are unchanged by this transformation, the equivalent circuit in Fig. 5.11 can be used to determine many quantities of interest, such as input current, motor efficiency, and output torque as a function of speed.
- The development of the per-phase equivalent circuit in Fig. 5.11 has been inexact in a number of ways. We have been vague in defining rotor and stator turns and in specifying means for determining rotor resistance and reactance. With a more refined analysis, these factors could be defined and calculated, but we do not require such detail. We are interested in the performance and overall characteristics of the motor. Our purposes are served by knowing the nature of the equivalent circuit and identifying the critical parameters. These machine parameters, such as the rotor resistance transformed into the stator, can be measured or deduced from the external characteristics of the machine. From the equivalent circuit, we will be able to predict how motor characteristics depend on the various circuit model parameters, and thus we can gain an understanding of some of the design decisions that must be made in developing an induction motor.

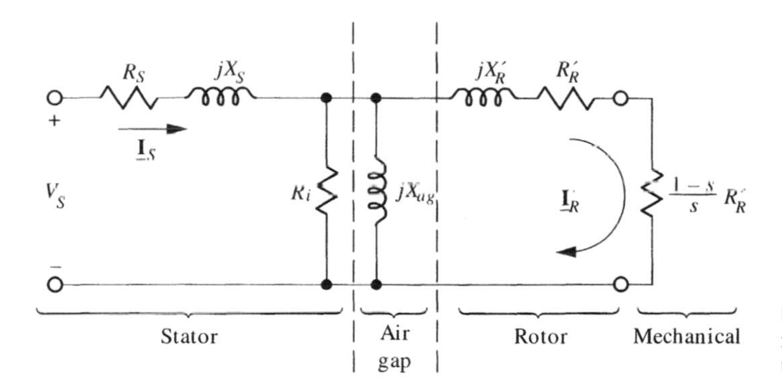

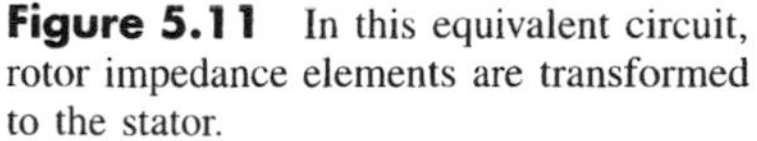

Figure 5.11 In this equivalent circuit, rotor impedance elements are transformed to the stator.

Applications of Equivalent Circuit

Motor analysis. We begin by analyzing a motor in steady state.

EXAMPLE 5.3 Induction-motor analysis

A 60-Hz, 230-V, three-phase, 5-hp, 1740-rpm induction motor is described by the per-phase equivalent circuit in Fig. 5.12. Analyze the circuit at nameplate speed to determine the input current, developed power, and efficiency.

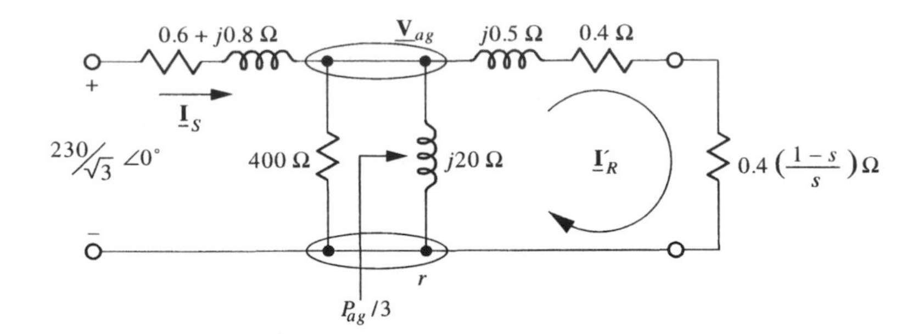

Figure 5.12 Per-phase equivalent circuit for the motor.

SOLUTION:
From the nameplate speed, we learn that this is a four-pole motor; hence the synchronous speed is 1800 rpm and the slip is

$$s = \frac{n_s - n}{n_s} = \frac{1800 - 1740}{1800} = \frac{1}{30} = 0.0333 \tag{5.23}$$

The resistor representing the developed power is

$$R'_R \times \frac{1 - s}{s} = 0.4 \times 29 = 11.6\ \Omega \tag{5.24}$$

and the rotor impedance is $12 + j0.5\ \Omega$.

The analysis of the equivalent circuit is a straightforward but nontrivial task. We performed a nodal analysis, as indicated in Fig. 5.12, with $\underline{V}_{ag}$ the unknown. We skip the details; the results are

$$\underline{V}_{ag} = 121.3 \angle -1.91^\circ \text{ V};\ \underline{I}_S = 12.3 \angle -33.9^\circ \text{ A};\ \underline{I}'_R = 10.1 \angle -4.3^\circ \text{ A} \tag{5.25}$$

From the results of Eq. (5.25), we can determine the motor performance. The air-gap power is

$$P_{ag} = 3 \times (I'_R)^2 \times \left(\frac{R'_R}{s}\right) = 3 \times (10.1)^2 (12\ \Omega) = 3670 \text{ W} \tag{5.26}$$

and the developed power is

$$P_{dev} = (1 - s)P_{ag} = (1 - 0.0333) \times 3670 = 3550 \text{ W} \tag{5.27}$$

The mechanical losses are very small, so this is approximately the output power: $P_{out} = 3550/746 = 4.76$ hp. The developed torque, which is approximately the output torque, is

$$T_{dev} = \frac{P_{dev}}{\omega_m} = \frac{P_{ag}}{\omega_s} = \frac{3670}{60\pi} = 19.5 \text{ N-m} \tag{5.28}$$

The power factor is derived from the angle of the stator current: $PF = \cos 33.9^\circ = 0.830$, lagging. The input power is

$$P_{in} = 3VI \times PF = 3 \times \frac{230}{\sqrt{3}} \times 12.3 \times 0.830 = 4050 \qquad (5.29)$$

Thus, the efficiency and total losses are approximately

$$\eta = \frac{P_{out}}{P_{in}} = \frac{3550}{4050} = 87.6\% \qquad \text{and} \qquad P_{loss} = P_{in} - P_{out} = 403 \text{ V} \qquad (5.30)$$

WHAT IF? What if you want the rotor-Cu, stator-Cu, and iron losses?[4]

Starting current and torque. To calculate the starting torque, we set $s = 1$ and calculate the developed torque. A stationary rotor has no mechanical losses; hence, the developed torque is the starting torque. We can determine the developed torque from the air-gap power and the synchronous speed, Eq. (5.9), and specifically the starting torque,[5] T_{st}, is

$$T_{st} = \frac{P_{ag}(s = 1)}{\omega_s} \qquad (5.31)$$

EXAMPLE 5.4 Starting torque

Calculate the starting torque for the motor described in Exam. 5.3.

SOLUTION:

We show the equivalent circuit for $s = 1$ in Fig. 5.13. The starting current is

$$\mathbf{I}_{st} = \frac{230/\sqrt{3} \angle 0^\circ}{0.6 + j0.8 + 400 \| j20 \| (0.4 + j0.5)} = 81.8 \angle -52.8^\circ \text{ A} \qquad (5.32)$$

The input power is

$$P_{in} = 3 \times \frac{230}{\sqrt{3}} \times 81.8 \times \cos 52.8^\circ = 19{,}700 \text{ W} \qquad (5.33)$$

After we subtract stator-copper losses, we have an air-gap power of 7630 W; hence, the torque is $7630/60\pi = 40.5$ N-m, approximately twice the run torque of 19.5 N-m.

WHAT IF? What if we ignore the $400 \| j20\ \Omega$ parallel impedance?[6]

[4] $P_R = 3 \times (10.1)^2(0.4) = 122$ W; $P_{S(\text{Cu})} = 3 \times (12.3)^2(0.6) = 271$ W; and $P_i = 3 \times (121.3)^2/400 = 110$ W.

[5] Also called "blocked rotor torque."

[6] $I_{st} = 81.0$ A, $T_{st} = 41.7$ N-m.

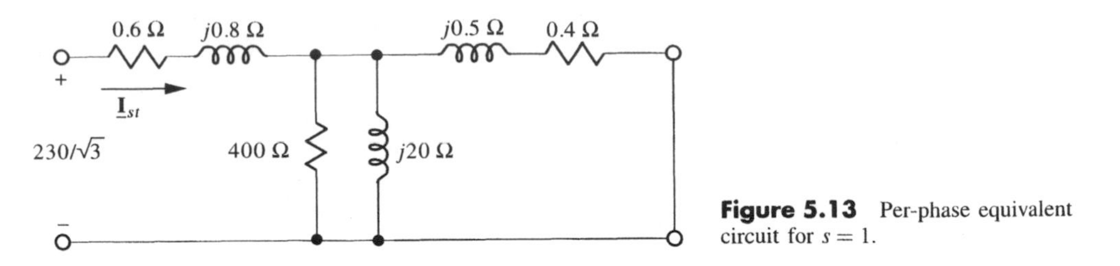

Figure 5.13 Per-phase equivalent circuit for $s = 1$.

Improving starting characteristics. We should not, however, take too seriously the details of these calculations of conditions at starting. Soon we show that the rotor conductors are shaped to lower starting current and increase starting torque; hence, our model is not accurate for starting performance calculations.

The rotor is the load. We demonstrated that the motor performance can be calculated from the equivalent circuit in Fig. 5.11. However, we may simplify the equivalent circuit for purposes of deriving the output-torque characteristics. Figure 5.14 shows the results of representing the power system, stator, and air gap by a Thèvenin equivalent circuit, considering the rotor equivalent circuit to be the load.

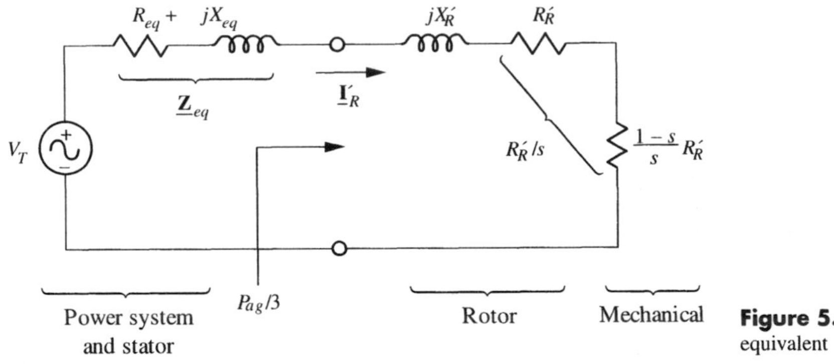

Figure 5.14 The per-phase Thèvenin equivalent circuit presented to the rotor.

EXAMPLE 5.5 Thèvenin equivalent circuit

Derive the per-phase Thèvenin equivalent circuit for the 5-hp motor described by the equivalent circuit in Fig. 5.12.

SOLUTION:

The open-circuit voltage with the load (rotor) removed is

$$\underline{V}_T = \frac{R_i \| jX_{ag}}{R_S + jX_S + R_i \| jX_{ag}} \times \underline{V} \tag{5.34}$$

$$= \frac{400 \| j20}{0.6 + j0.8 + 400 \| j20} \times \frac{230}{\sqrt{3}} \angle 0^\circ = 127.5 \angle 1.54^\circ \text{ V}$$

The output impedance seen by the rotor looking back into the air gap is

$$Z_{eq} = R_{eq} + jX_{eq} = R_i \| jX_{ag} \| (R_S + jX_S) \tag{5.35}$$

$$= 400 \| j20 \| (0.6 + j0.8) = 0.555 + j0.783 \ \Omega$$

Torque characteristic. Equation (5.9) shows that the developed torque may be derived from the synchronous speed and the air-gap power, which we may derive from the equivalent circuit in Fig. 5.14

$$P_{ag} = 3 \times (I_R')^2 \times \frac{R_R'}{s} = \frac{3V_T^2 R_R'}{s[(R_{eq} + R_R'/s)^2 + X_T^2]} \tag{5.36}$$

where V_T is the rms magnitude of the Thèvenin voltage and $X_T = X_{eq} + X_R'$ is the total reactance. Thus, the developed torque is

$$T_{dev} = \frac{P_{ag}}{\omega_s} = \frac{3V_T^2 R_R'}{s\omega_s[(R_{eq} + R_R'/s)^2 + X_T^2]} \tag{5.37}$$

Features of the torque characteristic. Figure 5.15 shows the developed torque characteristic of an induction motor for several values of R_R'. We note the following features:

- The maximum torque[7] is independent of rotor resistance, but the speed at which the maximum torque occurs decreases with increasing rotor resistance.

small slip region

- Near synchronous speed, the torque is a linear function of slip. This *small-slip region*, to be discussed presently, is the normal operating region of the motor.[8]
- The slope of the torque characteristic in the small-slip region decreases as the rotor resistance increases. For a given load-torque requirement, the motor runs slower and has more rotor losses, Eq. (5.8), for larger rotor resistance. Thus, small values of R_R' are desirable to increase efficiency.
- The starting torque is larger for large rotor resistance. Thus, larger values of R_R' are desirable and a compromise is indicated.

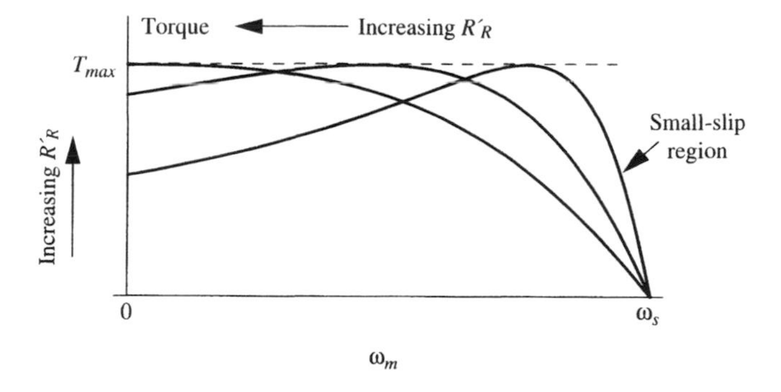

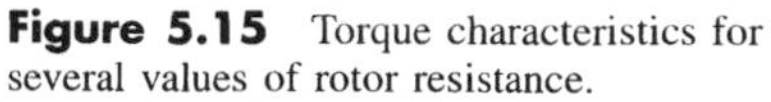
Figure 5.15 Torque characteristics for several values of rotor resistance.

[7] Also called "breakover torque."

[8] See page 37 for a previous discussion of this region.

Small-slip region. In the normal operating region near synchronous speed, the denominator of Eq. (5.37) is dominated by the R'_R/s term; hence, we may approximate Eq. (5.37) in the small-slip region by

$$T_{dev}(s) \approx s \times \frac{3V_T^2}{\omega_s R'_R} \tag{5.38}$$

Thus, the torque in the small-slip region is proportional to slip and inverse to rotor resistance, as suggested in Fig. 5.15. The developed mechanical power in the small-slip region is

$$P_{dev}(s) = \omega_m T_{dev}(s) = (1 - s)\omega_s T_{dev}(s) = 3(1 - s)s\frac{V_T^2}{R'_R} \tag{5.39}$$

The dependence of developed torque on slip allows the prediction of output power and speed throughout the small-slip region.

EXAMPLE 5.6 **Speed for 2 hp**

Find the speed at which the 5-hp motor used in the previous examples has an output power of 2 hp, assuming mechanical losses of 15 W.

SOLUTION:

The developed power for 2 hp would be $2 \times 746 + 15 = 1507$ W. Using Eq. (5.39) and the results of the previous example, we find the required slip

$$1507 = 3(1 - s)s\frac{(127.5)^2}{0.4} \Rightarrow s^2 - s - 0.0124 = 0 \tag{5.40}$$

Thus, $s = 0.0125$ (1.25%) and $\omega_m = (1 - 0.0125)60\pi = 186.1$ rad/s (1777 rpm).

WHAT IF? What if you remove the load? Find the no-load speed in rpm.[9]

wound rotor

Effect of rotor resistance. High efficiency and relatively constant operating speed require small values of the rotor resistance, but high starting torque and moderate starting currents require relatively high values of rotor resistance. This dilemma for the designer can be resolved in two ways. The first requires a different type of rotor from the squirrel-cage type. For a *wound rotor*, slots are milled in the rotor and three-phase coils are wound in the slots. These coils are connected to external resistors through a brush slip-ring assembly. With a wound rotor, we can maximize starting torque with external resistors and then remove all resistance for running. We can also achieve limited

[9] 1799.8 rpm, assuming the 15-W mechanical loss is the developed power.

control of motor speed with a wound-rotor induction motor. However, the wound-rotor machine is more expensive and requires more maintenance than the squirrel-cage motor.

Shaped-rotor conductors. The alternative to the wound rotor is a squirrel-cage rotor with shaped conductors. Figure 5.16 shows two extremes. With small conductors near the surface of the rotor, we have relatively large resistance and small leakage inductance because the nearby air gap limits the leakage flux that encircles the currents in the rotor conductors. Thus, the conductors in Fig. 5.16 (a) give good starting characteristics but poor run characteristics. On the other hand, the larger, deeper conductors in Fig. 5.16 (b) have relatively small resistance and large leakage inductance. Such a rotor gives excellent run characteristics but produces a low starting torque because the high inductance dominates during starting.

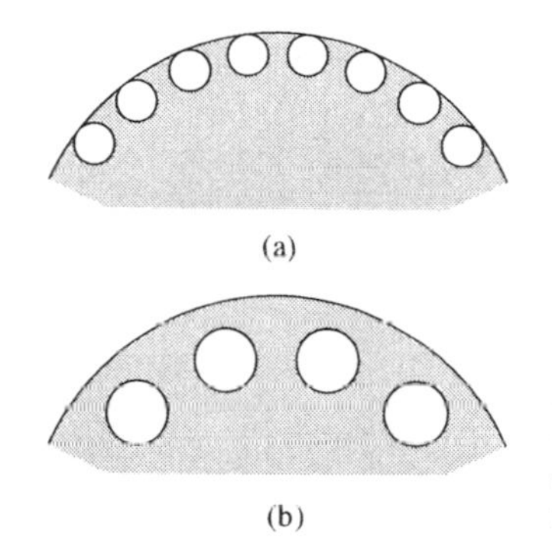

Figure 5.16 (a) Small, shallow rotor conductors give good starting but poor running characteristics; (b) large, deep rotor conductors give poor starting but good running characteristics.

However, we can have both types at once; or rather we can shape the conductors to have a high-resistance, low-inductance portion near the surface and a low-resistance, high-inductance portion deeper in the rotor. At starting, the rotor currents have a high frequency, and the large inductance therefore shields the deeper conductors. The starting currents flow mostly in the surface conductors, which have relatively large resistance. As the motor comes up to speed, however, the rotor frequency decreases, and most of the rotor current shifts to the deeper, low-resistance conductors. Thus, for starting, we have relatively high resistance and for running relatively low resistance in the rotor.

NEMA[10] design classes. Figure 5.17 (a) shows some of the common classes of shaped conductors, with typical torque characteristics resulting from each shown in Fig. 5.17(b). Clearly, the designer can, within limits, control the characteristics of the squirrel-cage motor by shaping the rotor conductors. The design classes in Fig. 5.17 refer to standard designations of squirrel-cage, three-phase induction motors to meet typical run and starting requirements:

- Designs A, not shown in Fig. 5.17 (b), and B have normal starting torque, but design B has lower starting current. Both have low slip, less than 5%, at rated output power. Typical applications are fans, blowers, rotary pumps, unloaded compressors, some conveyors, metal-cutting machine tools, and miscellaneous machinery.

[10] See pages 39 and 41 for information about NEMA.

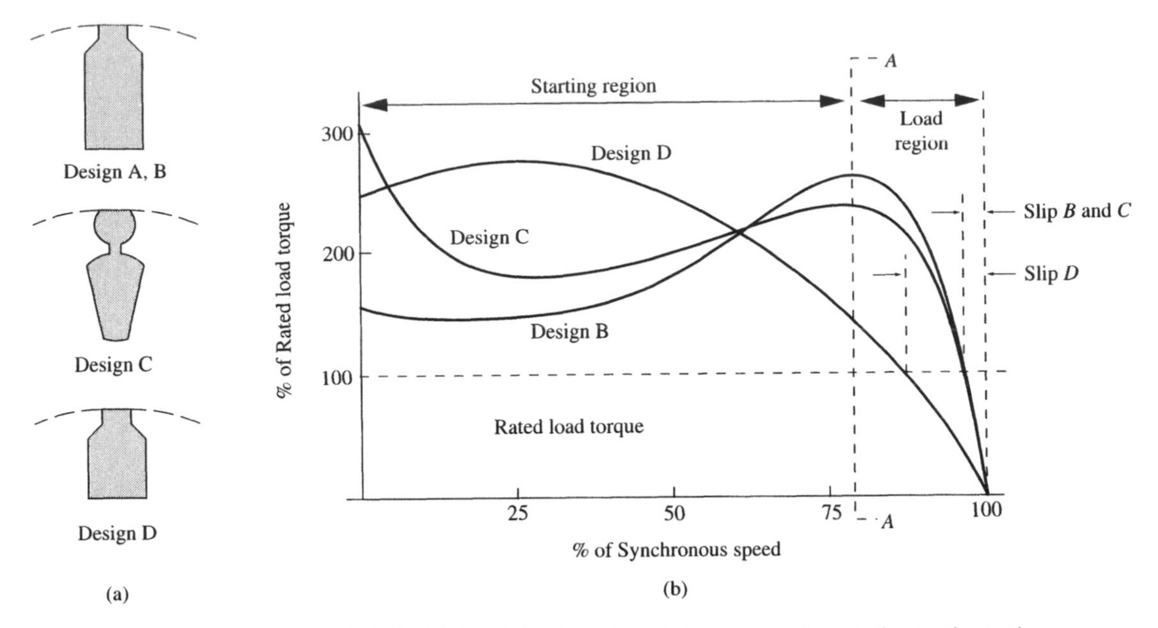

Figure 5.17 (a) Typical slots shaped to compromise starting and running characteristics; (b) typical torque characteristics corresponding to the shaped slots in part (a). (Figure courtesy of the Lincoln Electric Company.)

- Design C has high starting torque and relatively low starting current, and runs with low slip at rated output power. Typical applications are starting of high-inertia loads such as large centrifugal blowers, flywheels, and crusher drums. Loaded starting of piston pumps, compressors, and conveyors also require this design class.
- Design D has high starting torque and low starting current, but runs with relatively high slip, up to 11%, at rated output power. This design is required for very high inertia and loaded starts and also for loads having considerable variation in load torque throughout a load cycle. Typical applications are punch presses, shears and forming machine tools, cranes, hoists, elevators, and oil-well pumping jacks.
- Design E, not shown, is a new NEMA design for high-efficiency motors. The motor has a large starting current.

Dynamic Response of Induction Motors

run-up time

Introduction. Up to this point, we have considered motor performance in steady state. In Chap. 1, we had a brief introduction to dynamic characteristics of motors in general and considered the *run-up time*, the time required for the motor to reach steady state.

In this section, we analyze the dynamics of loaded induction motors in two cases: where the motor drive is constant but the load torque changes; and where the motor drive frequency and voltage are changed but the load torque characteristic is constant.

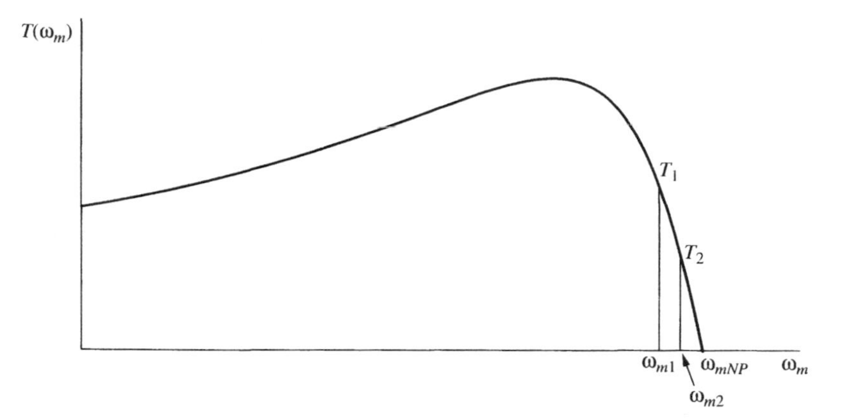

Figure 5.18 The motor characteristic does not change, but the load-torque requirement changes suddenly from T_1 to T_2.

Response to torque changes. Figure 5.18 shows the first response we will analyze. The motor is supplying torque T_1 and running at ω_{m1}, and the torque suddenly changes to T_2. The speed of the motor/load eventually changes to ω_{m2}; we now analyze the dynamics of the change, assuming that the motor operates in its small-slip region and that mechanical losses are negligible. The motor torque curve is

$$T_M(\omega_m) = \frac{\omega_s - \omega_m}{\omega_s - \omega_{mNP}} \times T_{NP} \tag{5.41}$$

where T_M is the motor developed torque and NP signifies a nameplate quantity. The dynamics of the motor/load system is governed by

$$J\frac{d\omega_m}{dt} = T_M - T_L \tag{5.42}$$

where T_L is the load torque and J is the total moment of the system. When we substitute Eq. (5.41) into Eq. (5.42), we have

$$J\frac{d\omega_m}{dt} + \frac{T_{NP}}{\omega_s - \omega_{mNP}} \times \omega_m = \frac{\omega_s}{\omega_s - \omega_{mNP}} T_{NP} - T_L \tag{5.43}$$

Sudden change in torque. Equation (5.43) is valid for arbitrary changes in the load torque, but here we consider only a sudden change. In this case, the right-hand side of Eq. (5.43) changes from one constant to another; thus, we have a simple first-order transient. Therefore, we may determine the motor response from the initial and final values plus the time constant. The initial and final values are shown in Fig. 5.18 and may be determined from the required load torques and Eq. (5.41). When we solve Eq. (5.43), we find the time constant to be

$$\tau = \frac{J}{T_{NP}} \times (\omega_s - \omega_{mNP}) \tag{5.44}$$

Thus, the response of the motor to the sudden change in torque shown in Fig. 5.18 is

$$\omega_m(t) = \omega_{m2} + (\omega_{m1} - \omega_{m2})e^{-t/\tau} \quad (5.45)$$

EXAMPLE 5.7 Change in load

Find the change in speed of the 50-hp motor analyzed on p. 37ff if the load torque changes from half the nameplate torque to full nameplate torque. Assume the moment of inertia of motor and load is 1.1 kg-m^2, and that the no-load speed is the synchronous speed.

SOLUTION:
The analysis of the motor nameplate information is given on pages 37–41. The nameplate torque is found in Eq. (1.52) as 202 N-m and the nameplate speed is 1765 rpm for the 60-Hz motor. The final speed in this example is the nameplate speed, and the initial speed for the half-nameplate torque is 1782.5 rpm, halfway between the nameplate speed and the no-load speed of 1800 rpm. The time constant given by Eq. (5.44) is

$$\tau = \frac{1.1}{202} \times (1800 - 1765) \times \frac{2\pi}{60} = 0.0200 \text{ s} \quad (5.46)$$

Expressing Eq. (5.45) in rpm, we find the motor response to the change in load torque

$$n(t) = 1765 - 17.5e^{-t/20 \text{ ms}} \text{ rpm} \quad (5.47)$$

which is shown in Fig. 5.19.

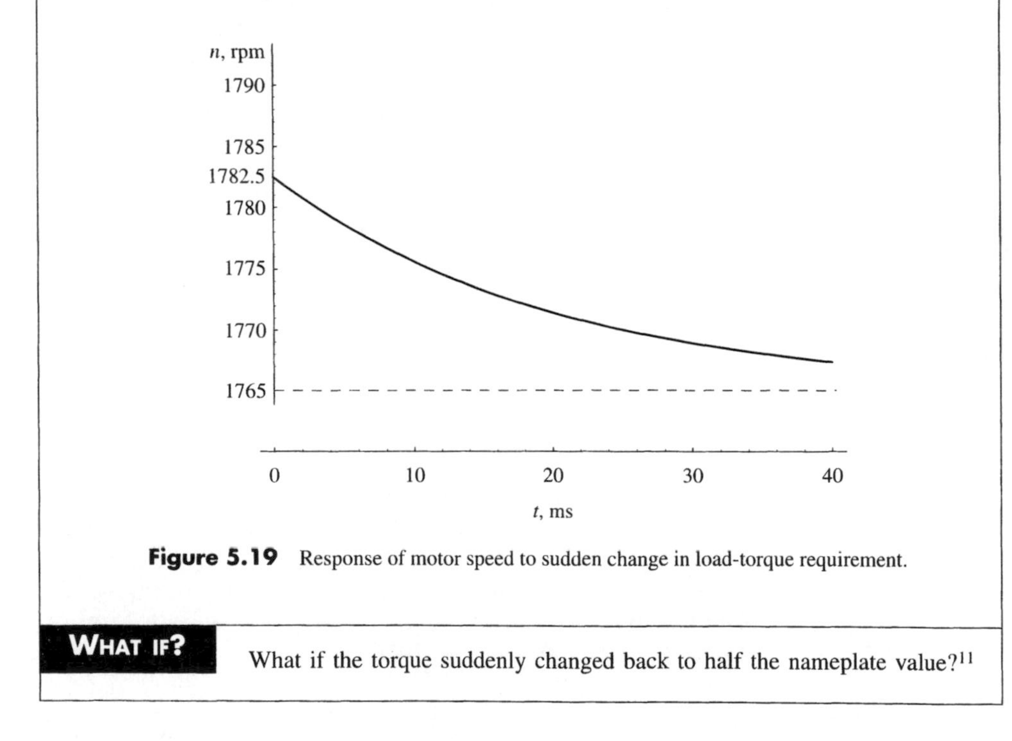

Figure 5.19 Response of motor speed to sudden change in load-torque requirement.

WHAT IF? What if the torque suddenly changed back to half the nameplate value?[11]

[11] $n(t) = 1782.5 - 17.5e^{-t/20 \text{ ms}}$ rpm.

System function. The preceding analysis can be translated into the system notation of Appendix A. If we divide Eq. (5.43) by the moment of inertia and use Eq. (5.44), we have

$$\frac{d\omega_m(t)}{dt} + \frac{\omega_m(t)}{\tau} = \frac{\omega_s}{\tau} - \frac{T_L(t)}{J} \tag{5.48}$$

If we now, in the manner introduced in Appendix A, consider all time variation of the form $e^{\underline{s}t}$, Eq. (5.48) takes the form

$$\left(\underline{s} + \frac{1}{\tau}\right)\underline{\Omega}_m(\underline{s}) = \frac{\omega_s}{\tau} - \frac{\mathbf{T}_L(\underline{s})}{J} \quad \Rightarrow \quad \underline{\Omega}_m(\underline{s}) = \frac{\omega_s/\tau - \mathbf{T}_L(\underline{s})/J}{\underline{s} + 1/\tau} \tag{5.49}$$

where in the last form we solved for the transform of the output speed, $\underline{\Omega}_m(\underline{s})$, in terms of the transform of the input torque variations, $\mathbf{T}_L(\underline{s})$, and the constants of the system.

System diagram. Equation (5.49) is represented by the system diagram in Fig. 5.20, which shows a simple first-order system.

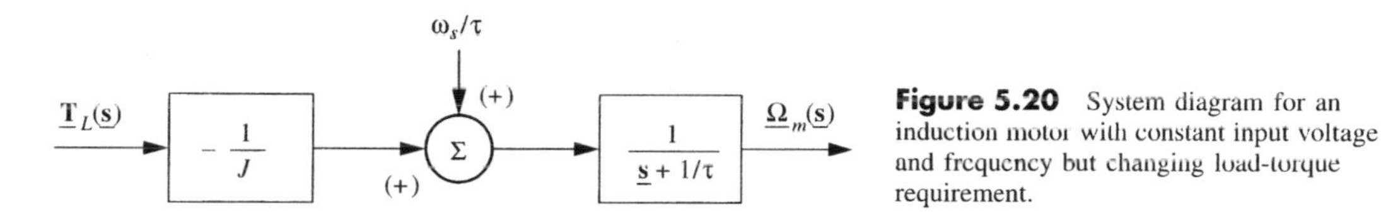

Figure 5.20 System diagram for an induction motor with constant input voltage and frequency but changing load-torque requirement.

Changing motor-drive frequency. In contrast to the passive system just analyzed, active control of the system speed can be accomplished through changing the drive frequency.[12] Such motor drives normally vary the applied voltage in proportion to the frequency to keep the motor flux, controlled by Faraday's law, Eq. (3.19), roughly constant.[13] Thus, the per-phase voltage to the motor varies as

$$V(\omega) = \frac{\omega}{\omega_{NP}} \times V_{NP} \tag{5.50}$$

where ω, ω_{NP}, and V_{NP} are drive frequency, nameplate frequency, and nameplate voltage, respectively. In the small-slip region, the torque of the motor is given by Eq. (5.38)

$$T_M(\omega) = s \times \frac{3V^2}{\omega_s R_R'} = \frac{\omega_s - \omega_m}{\omega_s^2} \frac{3(\omega/\omega_{NP})^2 V_{NP}^2}{R_R'} \tag{5.51}$$

But the synchronous speed, ω_s, also depends on the electrical frequency, ω, Eq. (4.24):

$$\omega_s = \frac{2}{P}\omega \tag{5.52}$$

---

[12] Variable-frequency motor drives are discussed in Chapter 7.

[13] This is called a "constant volts/hertz" drive.

There is, therefore, a cancellation of the square of the electrical frequency in the numerator and denominator of Eq. (5.51), and the result takes the form

$$T_M(\omega) = \left(\frac{2}{P}\omega - \omega_m\right) \times K_T \tag{5.53}$$

where K_T is a torque constant that can be evaluated from either Eq. (5.51) or, better, from the nameplate torque

$$T_{NP} = \left(\frac{2}{P}\omega_{NP} - \omega_{mNP}\right) K_T \tag{5.54}$$

EXAMPLE 5.8 **Torque constant**

Find the torque constant, K_T, in Eq. (5.53) for the 50-hp motor used in Example 5.7.

SOLUTION:
The motor torque constant comes from Eq. (5.54) and the motor nameplate torque, 202 N-m, and speed, 1765 rpm:

$$K_T = \frac{202}{(1800 - 1765) \times 2\pi/60} = 55.1 \tag{5.55}$$

System diagram. Equation (5.53) suggests the system diagram shown in Fig. 5.21, in which $\underline{\Omega}(\underline{s})$ and $\underline{\Omega}_m(\underline{s})$ represent the transforms of electrical frequency and mechanical speed, respectively.

Torque model. To proceed with the analysis, we must assume a model for the torque. We assume the linear model shown in Fig. 5.22

$$T_L(\omega_m) = K_0 + K_1\omega_m \tag{5.56}$$

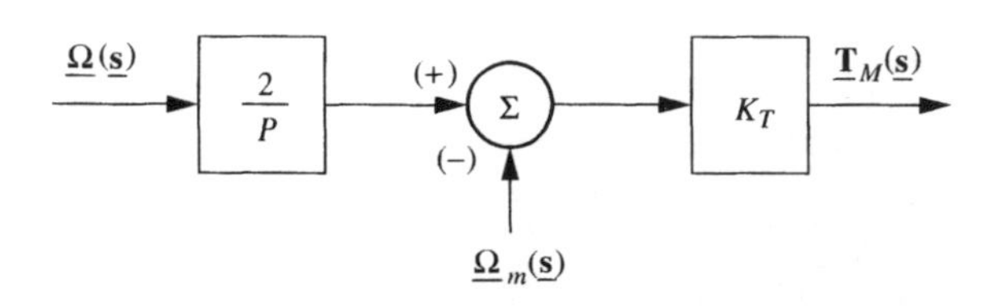

Figure 5.21 System diagram for induction motor with constant volts/hertz drive. The two inputs are the frequency-domain transform of electrical frequency and mechanical speed.

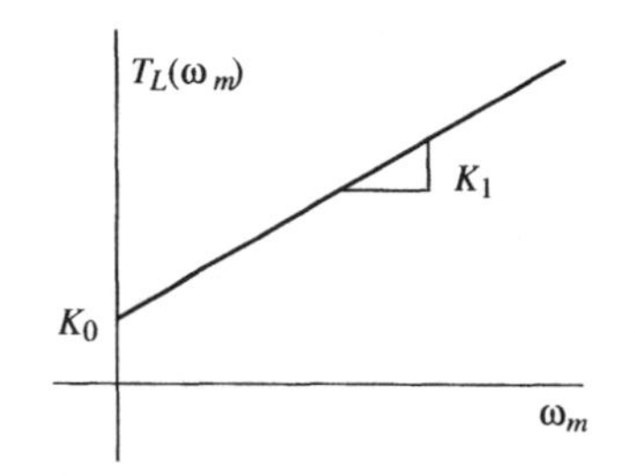

Figure 5.22 Linear torque model.

where K_0 and K_1 are constants. If the load-torque requirements were not linear, Eq. (5.56) could represent a linearization of the exact characteristic in the vicinity of the operating point.

Dynamic analysis. Newton's law for motion, Eq. (5.42), becomes

$$J\frac{d\omega_m}{dt} = T_M(t) - (K_0 + K_1\omega_m) \tag{5.57}$$

When we place the term containing the output speed on the left-hand side, we have

$$J\frac{d\omega_m}{dt} + K_1\omega_m = T_M(t) - K_0 \tag{5.58}$$

Transforming Eq. (5.58) to the frequency domain, we have

$$(J\underline{s} + K_1)\underline{\Omega}_m(\underline{s}) = \underline{T}_M(\underline{s}) - K_0 \tag{5.59}$$

When we solve for the output speed and combine with the motor function in Fig. 5.21, we get the system diagram for the motor/load shown in Fig. 5.23.

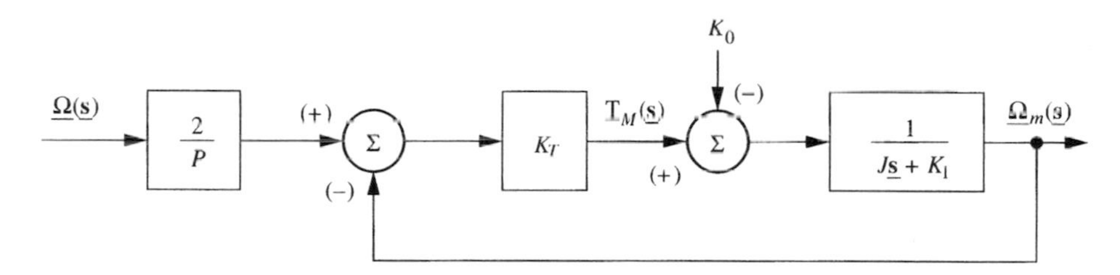

Figure 5.23 System diagram for an induction motor with constant volts/hertz drive with a linear model for load torque.

System time constant. Figure 5.23 shows the motor/load system to be first-order with a feedback loop. We may determine the time constant of the system by finding the pole of the system, as discussed in Appendix A. This requires

$$1 - \underline{L}(\underline{s}) = 0 \Rightarrow 1 - (-1)\frac{K_T}{J\underline{s} + K_1} = 0 \tag{5.60}$$

where $\underline{L}(\underline{s})$ is the loop gain. Equation (5.60) yields

$$\underline{s} = -\frac{K_T + K_1}{J} \Rightarrow \tau = \frac{J}{K_T + K_1} \tag{5.61}$$

Thus, the time constant depends on the moment of inertia and the combined torque constants from the motor and load.

EXAMPLE 5.9 **System time constant**

Calculate the time constant for the 50-hp motor driving a load at nameplate torque. Assume that the load torque is proportional to speed and that the combined moment of the system is 1.1 kg-m^2.

SOLUTION:
For the time constant, we must calculate the two torque constants. The motor-torque constant was determined in Example 5.8. Because the load-torque is proportional to speed, K_0 is zero in Eq. (5.56) and the load-torque constant is

$$K_1 = \frac{202}{1765 \times 2\pi/60} = 1.09 \tag{5.62}$$

Using Eq. (5.61), we find the time constant of the motor/load system

$$\tau = \frac{1.1}{55.1 + 1.09} = 0.0196 \text{ s} \tag{5.63}$$

Thus, if the motor input frequency were suddenly changed from 60 to 45 Hz, the system would slow down approximately 25% with a time constant of 19.6 ms.

WHAT IF? What would be the time constant if the load torque were constant at nameplate value?[14]

Summary. We analyzed an induction motor with a load in two cases: where the motor drive is constant but the load changes, and where the motor-drive frequency changes while the load-torque model is constant. System models were developed describing both cases, and time constants were determined. In the next section, we consider how to select a motor to drive a load that is varying.

Motor-Size Choice for Fluctuating Load Torque

Load requirements. We wish to select a motor to drive the periodic torque shown in Fig. 5.24. The required speed is about 1800 rpm, so we will select a four-pole, 60-Hz, three-phase induction motor. Table 5.2 shows information for a wide selection of four-pole motors from a single manufacturer; our job is to select the appropriate motor for this load.

Power requirements. Because motor powers are listed, we may convert the torques in Fig. 5.24 to power by multiplying by a nominal speed, say, 1750 rpm

$$P_1 = 1750 \times \frac{2\pi}{60} \times 90 = 16{,}490 \text{ W } (= 22.1 \text{ hp}) \tag{5.64}$$

and similarly we get 9.83 hp for the low-torque level. These are the powers that the

[14] 20.0 ms.

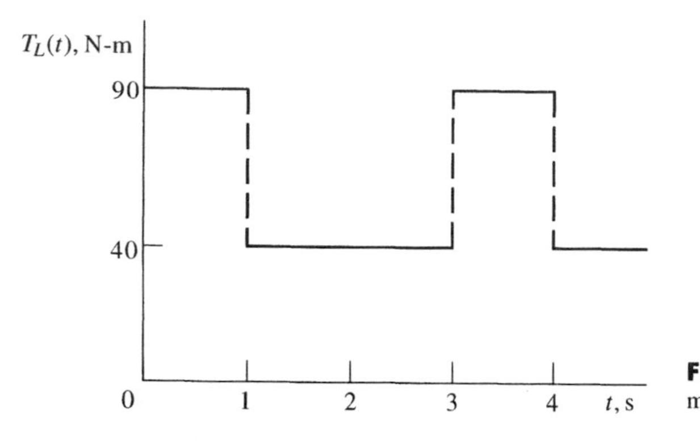

Figure 5.24 Time-varying torque. We must select a motor to drive this load.

TABLE 5.2 Induction-Motor Nameplate Information

hp	*rpm*	*Volts**	*Amps*	*Eff.*	*SF*	*Weight, lb*
1/3	1725	230	1.5	66.0	1.15	16
1/2	1725	230	2.2	69.0	1.15	18
3/4	1725	230	2.8	76.0	1.15	23
1	1725	230	3.8	72.0	1.15	27
1.5	1725	230	5.2	80.0	1.15	41
2	1715	230	5.8	80.0	1.0	50
3	1735	230	8.9	84.0	1.0	75
5	1725	230	13.2	85.5	1.0	97
7.5	1750	230	19.6	87.5	1.0	133
10	1750	230	25.0	88.5	1.0	172
15	1760	230	38.4	88.5	1.0	271
20	1755	230	49.2	89.5	1.0	317
25	1760	230	60.6	91.0	1.0	415
30	1750	230	73.8	91.0	1.0	477
40	1775	230	103.8	90.2	1.0	351
50	1775	230	119.6	91.7	1.0	420
60	1780	230	148.4	91.7	1.0	528
75	1780	230	176.6	91.7	1.0	627
100	1775	230	232.0	92.4	1.0	759

**These motors also operate at 460 V with half the listed current.*

load requires on a cyclical basis. We require that the motor operate with an adequate torque reserve and that the motor not overheat.

Torque reserve. Because motors in this power range are required by NEMA to have a breakover torque that is 200% of the nameplate torque, we conservatively require that the motor selected have a 60% reserve. Any motor rated 22.1 hp/1.6 = 13.8 hp or higher possesses adequate torque reserve by this criterion.

Maximum-horsepower choice. Clearly, we may choose a 25-hp motor and be assured that the motor will not overheat, because a 25-hp motor can produce 22.1 hp continuously. However, this is almost certainly an overdesign, and will increase costs and motor size and weight. Furthermore, motors are designed to have maximum efficiency near their rated values; hence operating costs might also increase.

Average-horsepower choice. The average power required by the load is

$$HP_{avg} = \frac{22.1 \times 1 + 9.83 \times 3}{4} = 12.9 \text{ hp} \tag{5.65}$$

which suggests that a 15-hp motor might be adequate. But average power is a bad criterion for the following reason: Motor current is roughly proportional to load, but motor copper losses increase in proportion to current squared and hence are roughly proportional to output power squared. Thus, a motor chosen on the basis of average power might overheat, and lifetime and efficiency would suffer.

RMS-horsepower choice. The loss considerations suggest that the rms horsepower is a better choice:[15]

$$HP_{rms} = \sqrt{\frac{(22.1)^2 \times 1 + (9.83)^2 \times 3}{4}} = 14.0 \text{ hp} \tag{5.66}$$

This result shows that a 15-hp motor would be adequate. We may confirm this choice with a more careful analysis of motor losses.

Motor losses. The three motor losses are mechanical losses, iron losses, and copper losses.

- Mechanical losses are small and constant as long as the motor is running. Mechanical losses may be determined from the no-load speed of the motor, but should be at most 100 W for a motor of the size required for this load.
- Iron losses are also constant as long as the motor is operating. The iron losses are proportional to motor weight for a given maximum flux. We assume 1.5 watts/pound of weight in this analysis.
- Copper losses are proportional to current squared but only the in-phase component of the current increases with the load; by Faraday's law, the out-of-phase current is roughly constant.

Loss analysis. We estimate the losses of the 15-hp motor for the fluctuating load in Fig. 5.24. The allowed losses in the motor may be determined from the nameplate efficiency

$$Loss = P_{NP}\left(\frac{1}{\eta} - 1\right) = 15 \times 746 \left(\frac{1}{0.885} - 1\right) = 1454 \text{ W} \tag{5.67}$$

Of these, we assume 100 W of mechanical loss and 410 W of iron loss based on the weight, and, hence, we have 944 W of copper loss. Assuming that copper losses are proportional to current squared, the constant of proportionality is

[15] For a more general definition of rms horsepower, see *Electrical Machines,* 5th ed., by A. E. Fitzgerald, Charles Kingsley, Jr., and Stephen D. Umans. New York: McGraw-Hill, p. 377.

$$944 = K_{Cu}(38.4)^2 \quad \Rightarrow \quad K_{Cu} = 0.640 \tag{5.68}$$

To divide the current between the out-of-phase component, which is roughly constant, and the in-phase component, which is proportional to power, we must determine the power factor at the nameplate

$$PF = \frac{15 \times 746/0.885}{\sqrt{3} \times 230 \times 38.4} = 0.8265\,(\theta = 34.3°) \tag{5.69}$$

Thus, the in-phase current is 31.7 A and the out-of-phase component is 21.6 A. Consequently, a model of the motor loss as a function of output power is

$$\text{Loss (hp)} = 100 + 410 + 0.640\left[(21.6)^2 + \left(31.7 \times \frac{\text{hp}}{15}\right)^2\right] \text{ W} \tag{5.70}$$

Using Eq. (5.70), we find that the loss during the high-torque period is 2205 W and during the low-torque period is 1085 W for a weighted average of 1365 W. This is less than the allowed loss of 1454 watts and indicates that the 15-hp motor is adequate. Other factors such as reliability, cost, lifetime, weight, and size should be considered in the final decision.

Summary. We analyzed a motor with a fluctuating load and considered several criteria for choosing a motor. The peak power gives a safe choice, but is an overdesign. Average power is not a valid approach, but rms power has more validity. A careful analysis of the motor losses is better yet, but many assumptions must be made in the absence of detailed test data.

Check Your Understanding

1. For the per-phase equivalent circuit shown in Fig. 5.11, identify what element(s) account for the specified physical effects. If nothing on the circuit is a suitable answer, indicate "none."
   (a) Developed mechanical power.
   (b) Leakage flux in the stator.
   (c) Rotor-copper loss.
   (d) Mechanical loss.
   (e) Stored energy in the air-gap flux.
   (f) Iron loss.
   (g) Stator-copper loss.
   (h) Input three-phase voltage.
2. To operate with small slip, an induction motor should have rotor resistance that is large, small, or does not matter. Which?
3. For high starting torque, an induction motor should have rotor resistance that is large, small, or does not matter. Which?
4. As the load-torque requirements vary, the induction motor has approximately constant speed, constant torque, constant power, or constant rotor losses. Give one or more answers.

5. Choosing a motor to meet average power requirement often leads to an overdesign. True or false?

Answers. **(1) (a)** $R_R(1-s)/s$, **(b)** X_S, **(c)** R'_R, **(d)** none, **(e)** X_{ag}, **(f)** R_i, **(g)** R_S, **(h)** none directly, although V is the input voltage divided by $\sqrt{3}$; **(2)** small; **(3)** large; **(4)** constant speed; **(5)** false.

5.3 SINGLE-PHASE INDUCTION MOTORS

Introduction. The three-phase induction motor is rugged, reliable, long-lived, self-starting, smooth-running, and relatively inexpensive. But three-phase power is not available everywhere, so if possible we need single-phase motors with the same characteristics. In fact, single-phase induction motors have excellent characteristics and outnumber the three-phase variety. Most of the small electric motors used in home, farm, or office are single-phase induction motors of one type or another.

In this section, we show first that the single-phase induction motor runs if started. We look then into the ways that single-phase motors are started, and finally we survey various types of single-phase induction motors.

Forward and reverse slip. As we showed in Sec. 4.2, Eq. (4.33), a single-phase flux can be resolved into two equal counterrotating flux waves. If the rotor is stationary, it will have a slip of unity relative to both waves and receive equal but opposite torques from each. Hence, no starting torque is produced.

forward slip, reverse slip

Forward and reverse slip. Let us assume that the rotor is rotating in the forward direction with an angular velocity ω_m. The *forward slip* is

$$s_f = \frac{\omega_s - \omega_m}{\omega_s} = 1 - \frac{\omega_m}{\omega_s} \tag{5.71}$$

where s_f is the slip relative to the forward flux wave. The synchronous speed of the reverse flux wave is $-\omega_s$, and hence the *reverse slip* of the rotor is

$$s_r = \frac{-\omega_s - \omega_m}{-\omega_s} = 1 + \frac{\omega_m}{\omega_s} = 2 - s_f \tag{5.72}$$

where s_r is the slip relative to the reverse flux wave. However, because both slips can be expressed in terms of the forward slip, we drop the subscripts and use s for the forward slip and $2 - s$ for the reverse slip.

Torque characteristic. We may estimate the torque characteristic of the single-phase induction motor from our results for the three-phase motor. Equation (4.33) shows that an oscillating flux can be divided into two counterrotating fluxes of half strength. Therefore, if $T_{3\phi}$ is the torque characteristic of the corresponding three-phase motor, then the torque as a single-phase motor, $T_{1\phi}$, for the same maximum flux would be

$$T_{1\phi}(s) = (\tfrac{1}{2})^2\ [T_{3\phi}(s) - T_{3\phi}(2-s)] \tag{5.73}$$

where the $(\frac{1}{2})^2$ comes from the equal division of the flux, because torque is proportional

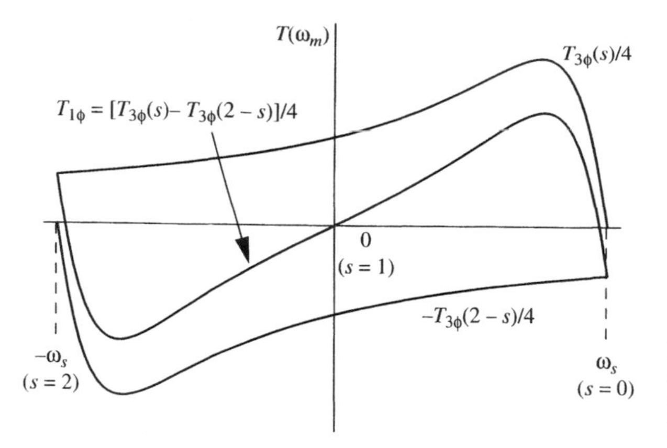

Figure 5.25 The single-phase run characteristic can be derived from the three-phase characteristic. These curves assume that the motor is excited at constant current.

to the square of the flux, and the two terms represent coupling to forward and reverse fluxes, respectively. The torque characteristic of the single-phase motor is therefore as shown in Fig. 5.25. We note that the motor has no starting torque, but, once started, produces a torque in the direction in which it is rotating. Thus, it will run if started.

EXAMPLE 5.10 Single-phase motor slips

Find the forward and reverse slip of a 60-Hz, four-pole, single-phase motor running at 1725 rpm.

SOLUTION:

The synchronous speed is 1800 rpm. From Eqs. (5.71) and (5.72), we find the slip to be $(1800 - 1725)/1800 = 0.0417$ relative to the forward flux and $2 - 0.0417 = 1.9583$ relative to the reverse flux.

WHAT IF? What if you want the electrical frequencies in the rotor?[16]

Constant current supply. The argument supporting the use of Eq. (4.33) in explaining single-phase torque depends on having rotating fluxes that are not influenced by the currents in the rotor. This requires exciting the motor from a constant current source. The characteristic shown in Fig. 5.25 is therefore unrealistic for normal ac sources, which supply constant voltage.

Constant voltage supply. In practice, a single-phase induction motor is driven by a constant voltage supply. By Faraday's law, the applied voltage determines the flux in the stator winding, but in this case, the flux is composed of two counterrotating fluxes, which are not necessarily equal due to the contribution of rotor flux.

[16] The rotor frequencies would be $s \times 60 = 2.5$ Hz and $(2 - s) \times 60 = 117.5$ Hz.

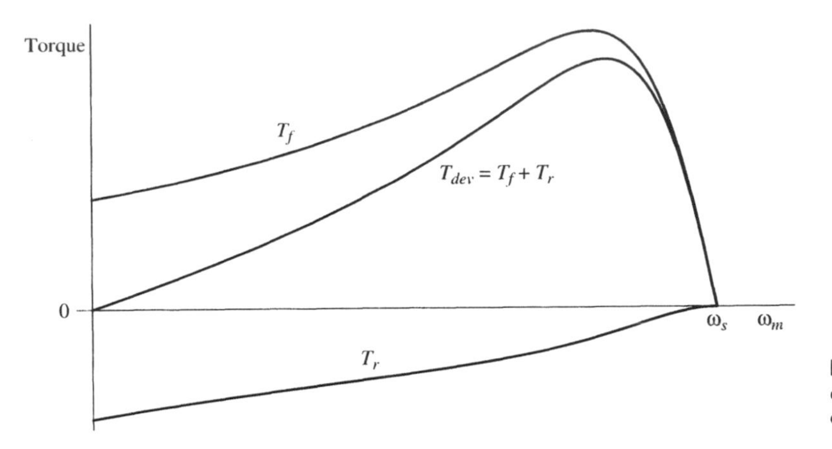

Figure 5.26 Run-torque characteristic of a single-phase induction motor for a constant voltage supply.

An analysis of the single-phase motor driven by a constant voltage supply is beyond the scope of this text. Figure 5.26 shows the torque characteristic of a representative single-phase motor calculated for a constant voltage supply. We show coupling to forward and reverse waves separately. Clearly, the reverse wave has a slight effect in the small-slip region. We conclude that the single-phase motor runs well; our problem is to start it.

LEARNING OBJECTIVE 3.

To understand the characteristics and applications of the various types of single-phase induction motors

Starting methods for single-phase motors. In addition to the main, or run, winding, the single-phase induction motor requires an auxiliary, or starting, winding. The starter winding is physically separated by $90°/(P/2)$ in space from the main winding and thus the stator has a two-phase geometry. In effect, the single-phase motor is started as a two-phase motor. The motor can be started in either direction by reversing the polarity of the auxiliary winding. The auxiliary winding may or may not be designed for continuous operation. We distinguish between several types of single-phase induction motors by how the phase shift is created between the currents in the main and auxiliary windings, and by whether the auxiliary windings are used continuously or only for starting.

starting or auxiliary winding

Normally, the starting winding is disconnected with a centrifugal switch that opens at about 75% of the rated speed. Occasionally, the auxiliary winding is switched with a relay that is activated by the current in the main winding. Figure 5.27(b) shows the current in the main winding and the region where the relay would engage the auxiliary winding. Such arrangements are used in compressor motors, in which the motor is enclosed in a refrigerant system.

Split-phase motors. The split-phase motor, whose circuit is shown in Fig. 5.27(a), has an auxiliary winding that would burn out if used continuously. The phase shift between the main and auxiliary windings in a split-phase motor is created by their differing ratios of resistance to inductance. The auxiliary winding uses a small wire and hence has a higher resistance than the main winding. Thus, the auxiliary current is smaller and more in phase with the line voltage, as shown in Fig. 5.28. This method yields a relatively small phase shift between main and auxiliary currents, and hence gives low starting torque. Split-phase motors serve applications requiring low starting torque, such as fans and bench grinders. Typical power ratings are 1/20 to 1/3 hp.

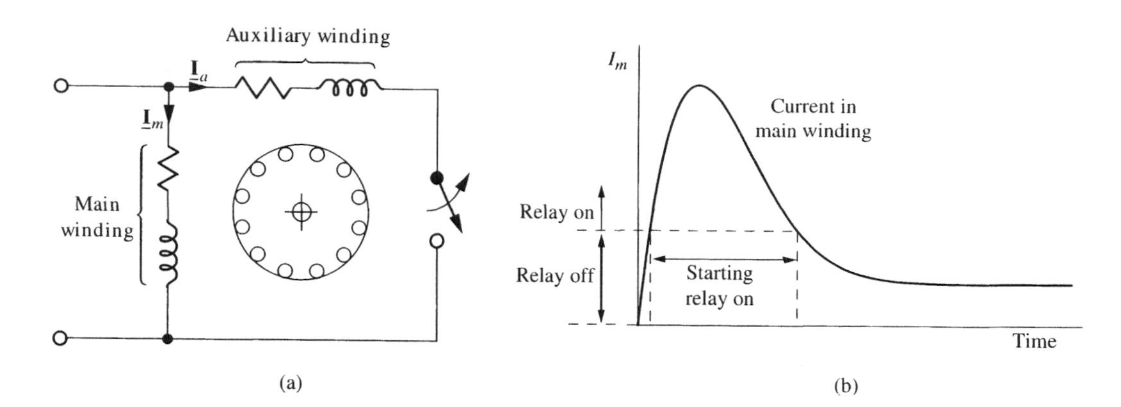

Figure 5.27 (a) For the split-phase motor, the auxiliary, or starting, winding is disconnected after the motor starts; (b) a current-sensing relay can be used to connect and disconnect the auxiliary winding.

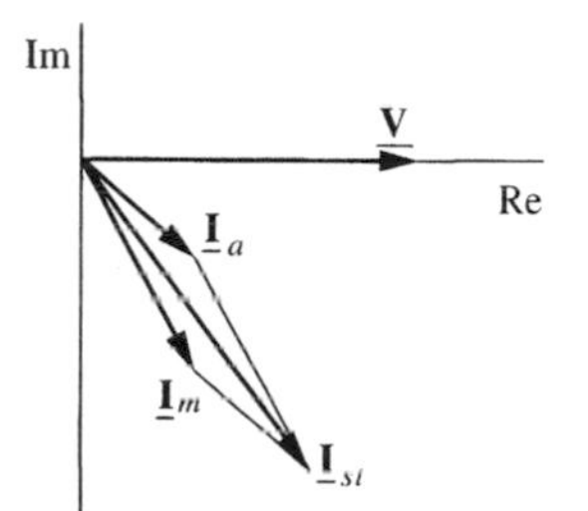

Figure 5.28 The starting current for the split-phase motor. The small phase difference yields low starting torque.

EXAMPLE 5.11 Split-phase motor

Find the phase shift in the starting current of a 115-V, single-phase induction motor if the main and auxiliary winding have impedances of $\underline{Z}_m = 3.3 + j4.0\ \Omega$ and $\underline{Z}_a = 6.0 + j3.8\ \Omega$, respectively.

SOLUTION:

The currents are

$$\underline{I}_m = \frac{115 \angle 0^\circ}{3.3 + j4.0} = 22.2 \angle -50.5^\circ \text{ A}$$

and (5.74)

$$\underline{I}_a = \frac{115 \angle 0^\circ}{6.0 + j3.8} = 16.2 \angle -32.5^\circ \text{ A}$$

The phase shift is $50.5 - 32.5 = 18.1^\circ$.[17]

WHAT IF? What if you want the starting current?[18]

[17] To three-place accuracy, based on the exact angles.

[18] 37.9 A.

Capacitor-start/induction-run motors. Figure 5.29(a) shows the circuit of the capacitor-start motor. A capacitor in series with the auxiliary winding produces a leading current. Ideally, the auxiliary current would lead the main current by 90° of phase for maximum starting torque, as shown in Fig. 5.29(b). The capacitor would be of the electrolytic type, designed for short-time operation, and the auxiliary winding would be disconnected after the motor is started. Such motors have good starting torque and are used for large appliances, some power tools, and large fans. Typical power ratings are 1/4 to 10 hp.

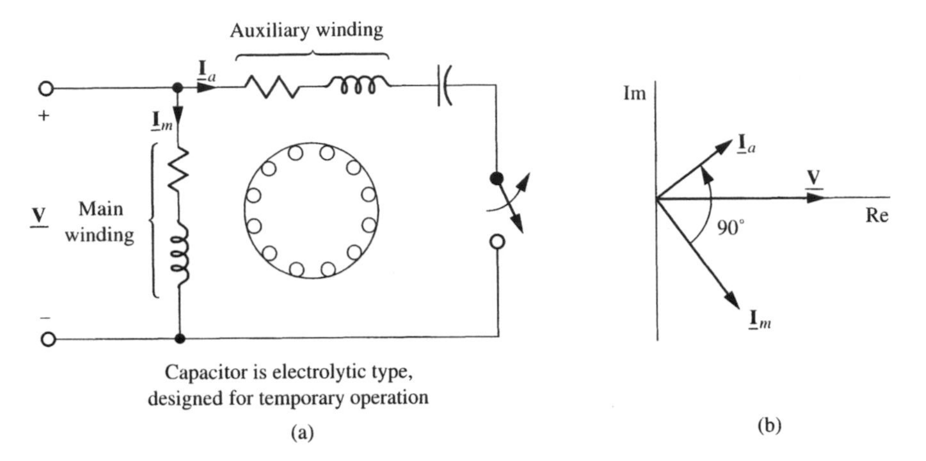

Figure 5.29 (a) A capacitor may be used to increase the phase shift between the currents in the main and auxiliary windings; (b) a phasor diagram showing the 90° shift due to the capacitor.

EXAMPLE 5.12 Phase-shift capacitor

What capacitor, placed in series with the auxiliary winding in Example 5.11, gives a 90° phase shift between the $\underline{I}_m$ and $\underline{I}_a$ at starting?

SOLUTION:

We require

$$\underline{I}_a = \frac{115 \angle 0^\circ}{6.0 + j(3.8 + X_C)} = (\text{magnitude}) \angle -50.5 + 90^\circ \tag{5.75}$$

therefore

$$\frac{3.8 + X_C}{6.0} = \tan(-39.5^\circ) \Rightarrow X_C = -3.8 + 6.0 \times \tan(-39.5^\circ) = -8.75\ \Omega \tag{5.76}$$

This reactance requires 303 μF at 60 Hz.

WHAT IF? What if you use 300 μF and want the starting current?[19]

[19] 26.5 A.

Capacitor-start/capacitor-run motors. The capacitor-start/capacitor-run motor shown in Fig. 5.30 has two capacitors. One capacitor is rated for short-time operation, for starting the motor; the other is rated for long-term operation, for improving the run characteristics of the motor. By giving a rotating wave that is more pure, the run capacitor improves the torque and efficiency characteristics of the motor and reduces vibration. Capacitor-start/capacitor-run motors are available in sizes from 1/4 to 10 hp and are used for conveyors, air compressors, and other heavy intermittent loads.

Permanent-split-capacitor (PSC) motors. In the PSC motor, shown in Fig. 5.31, the auxiliary winding is designed for continuous operation. This motor has a continuous-rated capacitor for start and run, and no switch. The auxiliary winding gives some help in starting torque and some help in run characteristics. The main virtues of the PSC motor are high efficiency and smoother operation. Power ratings of 1/6 to 3/4 hp are typical of PSC motors, which are often used for fans and direct-drive blowers.

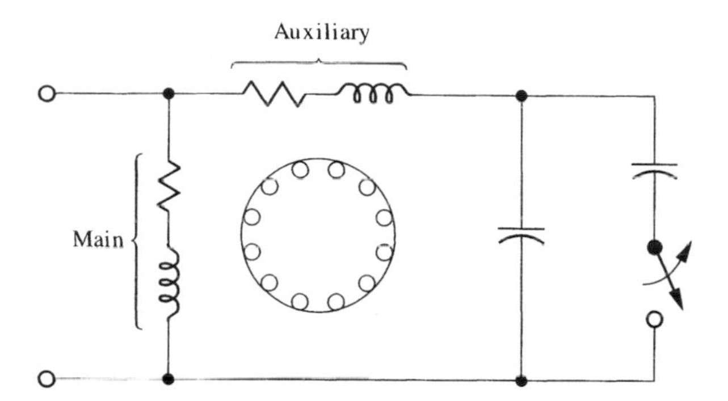

Figure 5.30 Improved run characteristics result if the auxiliary winding is used continuously.

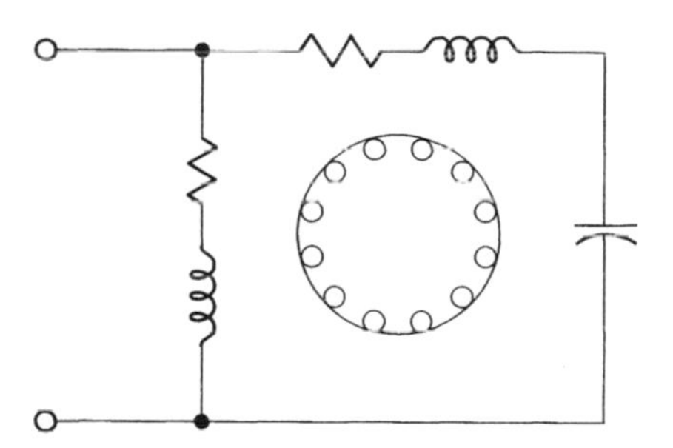

Figure 5.31 The PSC motor has no switch.

Shaded-pole motors. Figure 5.32 shows two types of shaded-pole motor. This motor is characterized by having salient poles, partly encircled by conducting rings. As flux increases, current flows in the "shading ring" to delay flux buildup within the shaded portion of the pole. Later, as flux decreases, the current induced in the ring delays the flux decrease in the shaded portion of the pole. Hence, the flux maximum moves $1 \rightarrow 2 \rightarrow 3 \rightarrow 4 \rightarrow 1$, and so on. The motors shown in Fig. 5.32 turn in one direction only, but shaded-pole motors can be made reversible if both sides of the pole are shaded and the shading coils are brought out for external switching. As shown, the motors are fixed-speed, but multipole motors can be made to run at different speeds by exciting poles in various patterns. Shaded-pole motors are widely used in small fans, pumps, and small household appliances.

Check Your Understanding

1. The single-phase induction motor runs in either direction, depending on which way it is started. True or false?
2. In a single-phase induction motor, the slip of the rotor relative to one rotating wave of stator flux is 6%. What is its slip relative to the other wave of stator flux?

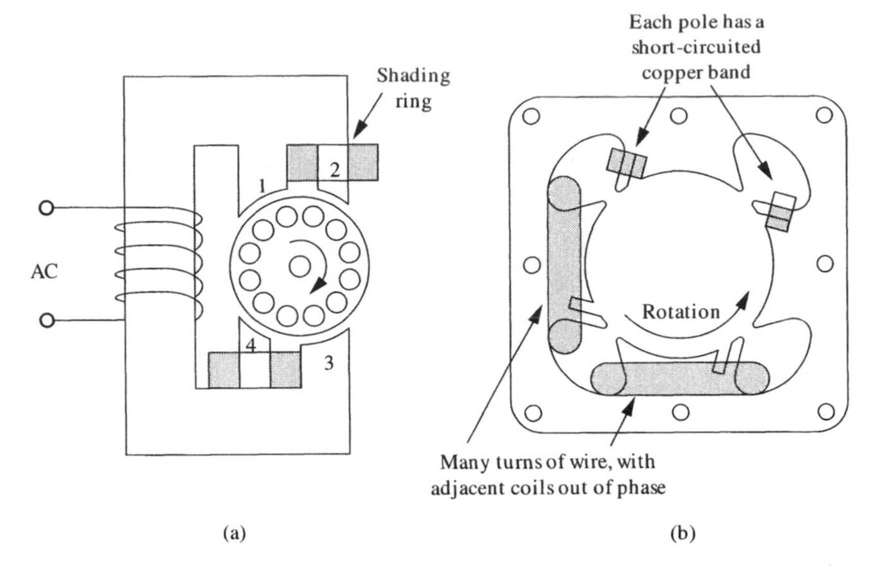

Figure 5.32 (a) A simple two-pole, shaded-pole motor; (b) the magnetic structure for a larger, four-pole, shaded-pole motor. Shading rings and coils are shown on half of the poles.

3. A 60-Hz, single-phase induction motor turns 1740 rpm.
   (a) How many poles does the motor have?
   (b) What is (are) the frequency (frequencies) of the stator current(s)?
   (c) What is (are) the frequency (frequencies) of the rotor current(s)?
4. The unloaded ideal (no mechanical loss) single-phase induction motor will run slower than synchronous speed. True or false?
5. Why does a capacitor-start, single-phase induction motor have greater starting torque than a split-phase motor?
6. A permanent split-capacitor (PSC) motor has less vibration at 120 Hz than a capacitor-start/induction-run motor. True or false?
7. Why does the standard shaded-pole motor turn only one way?

Answers: **(1)** True; **(2)** 194%; **(3)** **(a)** four poles, **(b)** 60 Hz, **(c)** 2 Hz and 118 Hz; **(4)** true, because of the reverse torque; **(5)** the current in the start winding is more out of phase with the current in the main winding; **(6)** true, because the rotating flux has a smaller counterrotating flux; **(7)** because the mechanism to make the flux rotate is built into the iron of the poles.

CHAPTER SUMMARY

We began by investigating the induction principle used to produce torque in an induction motor. The three-phase induction motor was examined thoroughly, both at an intuitive level from basic principles and quantitatively through a per-phase equivalent circuit. The various types of single-phase induction motors were described.

Objective 1: To understand the induction principle that produces torque in induction motors. The rotor of an induction motor has embedded conductors that are shorted to allow the free flow of current. Voltage is induced by the motion of the physical rotor relative to the rotating stator flux. The resulting current produces rotor torque that depends on the slip speed between the physical rotor and the rotating flux.

Objective 2: To understand the physical origin of all elements in the equivalent circuit of an induction motor, and be able to use the circuit to predict motor performance. A per-phase equivalent circuit for the induction motor was developed by considering the various energy processes in the motor. The energy converted to mechanical form was represented by a resistance that depends on the slip. The equivalent circuit permits calculation of output torque and input current and power factor, all as functions of motor speed. Starting current and torque can also be predicted with the equivalent circuit.

Objective 3: To understand the characteristics and applications of the various types of single-phase induction motors. The run principle of the single-phase induction motor was described. The various means for starting such motors were described. Applications of the different types of motors were indicated.

The electronic control of three-phase induction motors is described in Chap. 7. There we discuss the characteristics of three-phase induction motors as ac voltage and frequency are changed.

GLOSSARY

Air-gap power, p. 208, power that crosses from stator to rotor in a motor, or rotor to stator in a generator.

Blocked rotor, p. 207, conditions pertaining to the stationary rotor, such as startup.

Breakover torque, p. 207, the maximum torque of a motor. Any load demanding higher torque will stall the motor.

Developed torque, p. 208, the magnetically generated torque before losses are subtracted.

Forward slip, p. 232, the slip relative to the synchronous speed in the direction of rotation.

Induction generator, p. 207, an induction machine that is driven faster than synchronous speed.

Induction motor, p. 202, a motor in which rotor voltage, and hence current and torque, are produced by relative motion of the rotor relative to a rotating magnetic flux.

NEMA design class, p. 221, a classification of induction motors with regard to starting torque and current, efficiency, and speed regulation.

Output torque, p. 208, the torque on the output shaft after all motor losses are subtracted.

Reverse slip, p. 232, the slip relative to the synchronous speed opposite the direction or rotation.

Run-up time, p. 222, the time required for the motor to reach steady state with a given load.

Slip, p. 206, the slip speed divided by the synchronous speed.

Slip speed, p. 202, the angular velocity in the negative direction of the rotor conductors relative to the stator flux.

Small-slip region, p. 219, the region near the synchronous speed where developed torque is proportional to slip.

Squirrel-cage rotor, p. 203, a rotor in which conductors are embedded in the surface of the rotor with shorting rings at each end.

Wound rotor, p. 220, a rotor with coils of insulated wire that are brought out through slip rings for external connections.

PROBLEMS

Section 5.1: Induction Motor Principles

5.1. What relative values of dc current in coils a, b, and c in Fig 4.11 give a horizontal flux, as required for the mental experiment?

5.2. A three-phase, 60-Hz induction motor has a nameplate speed of 1740 rpm. Find the following:
- **(a)** Slip.
- **(b)** Electrical frequency of the stator currents.
- **(c)** Electrical frequency of the rotor currents.
- **(d)** Mechanical rotational speed of the stator fields.
- **(e)** Mechanical rotational speed of the rotor fields.
- **(f)** Ratio of the developed power to the rotor-copper losses.

5.3. In a three-phase, 60-Hz induction motor turning 1720 rpm, the following losses are known: stator copper loss = 50 W; rotor copper loss = 65 W; iron loss = 80 W; and mechanical losses = 5 W. Find the following:
- **(a)** Number of poles.
- **(b)** Slip.
- **(c)** Air-gap power.
- **(d)** Output power.
- **(e)** Input power.
- **(f)** Efficiency.
- **(g)** Output torque.
- **(h)** Apparent power if the power factor is 0.85.

5.4. A three-phase, 60-Hz induction motor has the speed–torque characteristic shown in Fig. P5.4. The load characteristic is also shown. Find the following:
- **(a)** Number of motor poles.
- **(b)** Operating speed.
- **(c)** Slip at maximum torque.
- **(d)** Motor starting torque
- **(e)** Air-gap power at a slip of 1.0.
- **(f)** Output power at the operating speed (hp).
- **(g)** Approximate rotor-copper losses at the operating speed, neglecting mechanical losses.

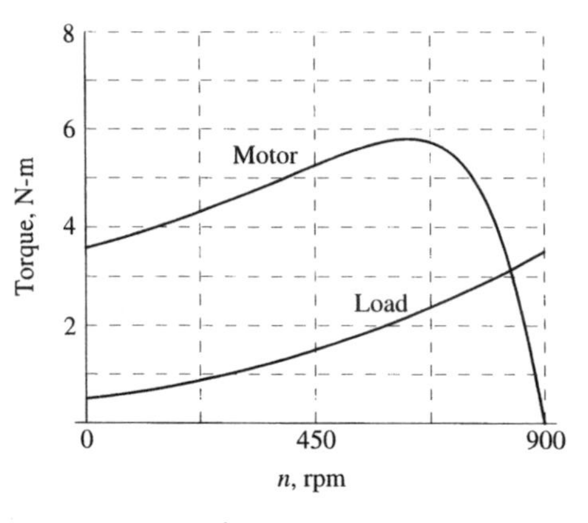

Figure P5.4

5.5. A three-phase, 60-Hz induction motor has the following specifications: 5 hp, 1740 rpm, 230/460 V, 12.8/6.4 A, efficiency 87.5%, and service factor 1.15.
- **(a)** Find the slip of the motor at nameplate speed.
- **(b)** Find the power factor under nameplate conditions.
- **(c)** What is the air-gap power under nameplate conditions, assuming that 95% of the losses are electrical and 5% mechanical?
- **(d)** If the motor were excited by 220 V, would the current be less than, equal to, or greater than 12.8 A with the same ouput power?

5.6. A four-pole, 60-Hz, three-phase induction motor operates from 460 V and draws 2.85 A at 0.76 PF (lagging). Of the input electrical power, 95% crosses into the rotor; of that power, 96% is converted to mechanical form; and of that, 99% is output power. Find the following:
- **(a)** Motor efficiency.
- **(b)** Motor speed in rpm.

(c) Output power.
(d) Output torque.

5.7. A three-phase, 60-Hz induction motor has the following nameplate information: 2 hp, 1725 rpm; 230/460 V, 5.7/2.85 A, 87% efficiency; service factor 1.15. If the machine is operating under nameplate conditions, determine the stator- and rotor-copper losses, assuming the mechanical losses are negligible but iron losses are 60 W.

5.8. The nameplate information on a three-phase, 60-Hz induction motor is 15 hp, 1755 rpm, 230/460 V, 36.0/18.0 A, 88.5% eff., service factor 1.15. Assume that at nameplate speed, the mechanical losses are 0.5% of the output power.

(a) Find the number of poles.
(b) What is the slip at nameplate conditions?
(c) What is the power factor at rared output power?
(d) What is the output torque at nameplate output power?
(e) Find the sum of the stator-iron and stator-copper losses at rated ouput power.

5.9. A three-phase, 60-Hz induction motor has the following nameplate information: 2 hp, 3450 rpm; 230/460 V, 5.2/2.6 A, SF 1.15. At nameplate conditions, the stator-copper, stator-iron, rotor-copper, and mechanical losses divide in the ratios 4:2:3:1. Find the efficiency of the motor.

Section 5.2: Equivalent Circuits for Three-Phase Induction Motors

5.10 Figure P5.10 shows the per-phase equivalent circuit for a three-phase induction motor. The slip is represented by s. At a slip of 4%, the current in the stator is 9.928 A and the current in the rotor part of the circuits is 7.063. Iron losses are ignored.

(a) Find the developed power.
(b) Find the rotor-copper loss.
(c) Find the stator-copper loss.
(d) What is the power factor?

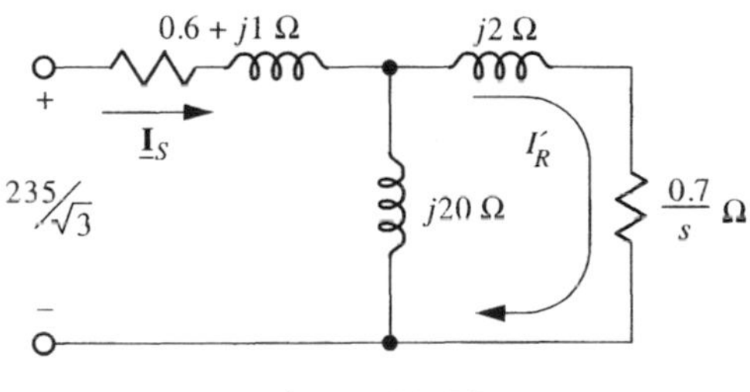

Figure P5.10

5.11 A three-phase, four-pole, 60-Hz induction motor is represented at one speed by the per-phase circuit shown in Fig. P5.11. The 8-Ω resistor represents the power being converted to mechanical form in the motor.

(a) What is the motor speed in rpm?
(b) Find the developed mechanical torque. The current in the rotor part of the circuit is 5.0 A.
(c) Estimate the starting torque. (Ignore $100 \parallel j16\ \Omega$.)

5.12 A three-phase, 230-V, 60–Hz induction motor has the following nameplate information: 10 hp, 90% efficiency, 0.82 power factor (lagging), 1740 rpm.

(a) Determine the apparent power rating of the machine.
(b) Find the reactive power required by the motor at nameplate operation.
(c) What is the output torque at nameplate operation?
(d) Find the total stator losses at nameplate operation, assuming 10 W of mechanical loss.

5.13. The per-phase equivalent circuit of a two-pole, three-phase, 60-Hz induction motor is shown as a function of slip in Fig. P5.13. Ignore mechanical losses.

(a) For an ouput power of 1 hp, find the speed of the two-pole motor. *Hint:* Use a Thèvenin equivalent circuit.

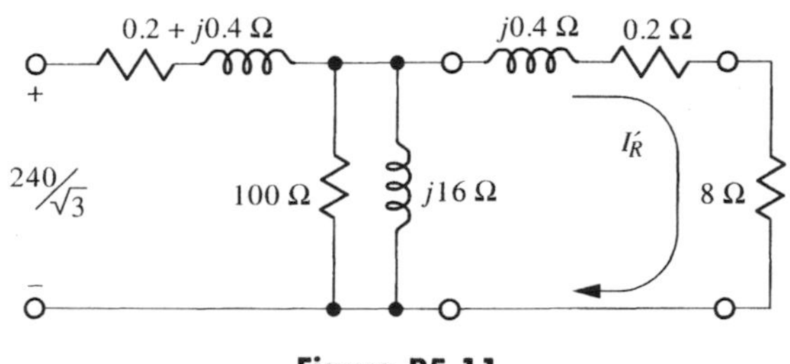

Figure P5.11

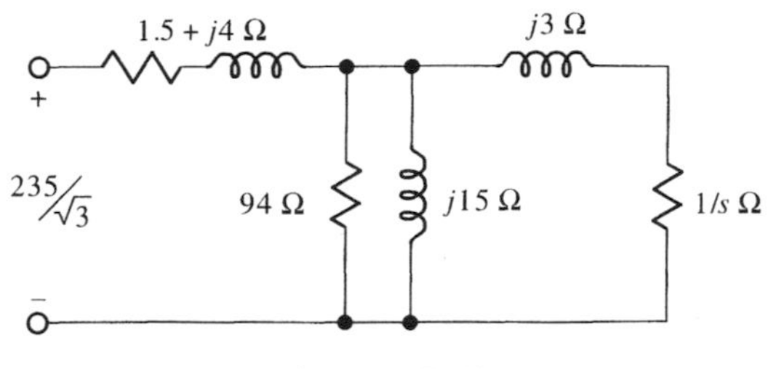

Figure P5.13

(b) For an output power of 1 hp, determine the input current and the efficiency of the motor.
(c) Estimate the starting torque for this motor.

5.14. A 60-Hz, three-phase induction motor runs at 3599.5 rpm at no load and at 3500 rpm with 5 hp out. Assume operation in the small-slip region.
(a) Find the slip at no load and at 5 hp out.
(b) Find the mechanical losses, assumed independent of speed. Assume operation in the small-slip region.
(c) Find the air-gap power at 3500 rpm.

5.15. The per-phase equivalent circuit for a 60-Hz, three-phase, four-pole induction motor is shown in Fig. P5.15. At a slip of 4%, the input current is 49.75 A and the power factor is 0.9391.
(a) What is the input power to the motor at this slip?
(b) Find the air-gap power at this slip.
(c) What is the developed torque for the motor at this slip?
(d) *Estimate* the input current for starting the motor. (Ignore $j10 \parallel 44\ \Omega$.)
(e) *Estimate* the starting torque for the motor.

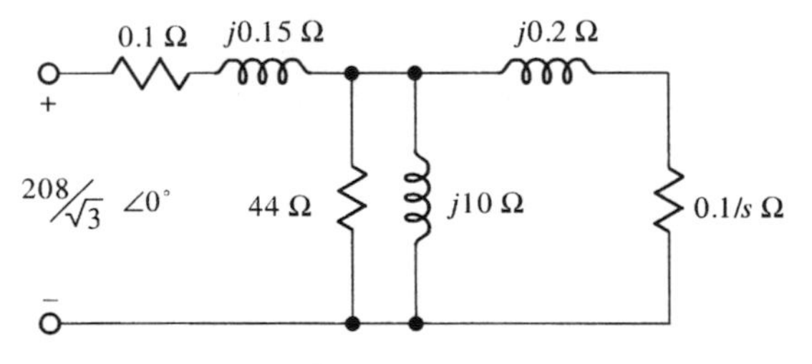

Figure P5.15

5.16. The equivalent per-phase circuit shown in Fig. P5.16 represents a 60-Hz, three-phase induction motor with six poles.
(a) Estimate the starting current. (Ignore $j20 \parallel 60\ \Omega$.)
(b) Estimate the slip for 25-hp output power, ignoring mechanical losses.

5.17. A 60-Hz, three-phase induction motor has the following nameplate information: 100 hp, 1780 rpm, 230/460 V, 244.0/122.0 A, 91.7% efficiency, service factor 1.15, and 732 lb. Assume that mechanical losses are 100 W and that iron losses are 1.8 W/lb.
(a) Find the number of poles and the slip at nameplate operation.
(b) Find the three electrical loss components.
(c) Estimate the motor speed if the motor is operated at 115 hp continuously.

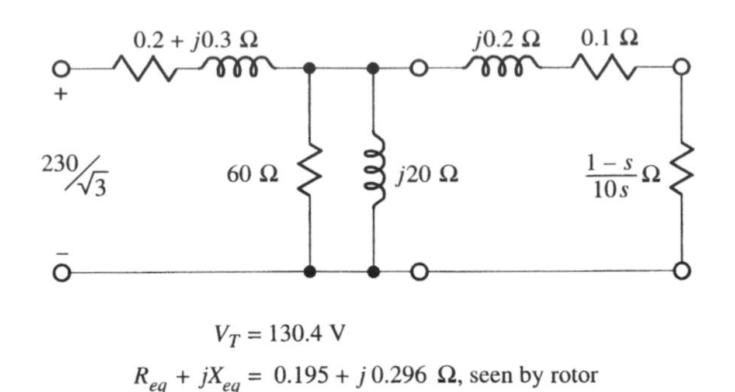

$V_T = 130.4$ V

$R_{eq} + jX_{eq} = 0.195 + j0.296\ \Omega$, seen by rotor

Figure P5.16

5.18. The per-phase equivalent circuit of a three-phase induction motor is shown in Fig. P5.18. The rotor current at a slip of 5% is 10 A rms.
(a) Find the developed power of the rotor at this slip.
(b) Estimate the efficiency of the motor at this slip.

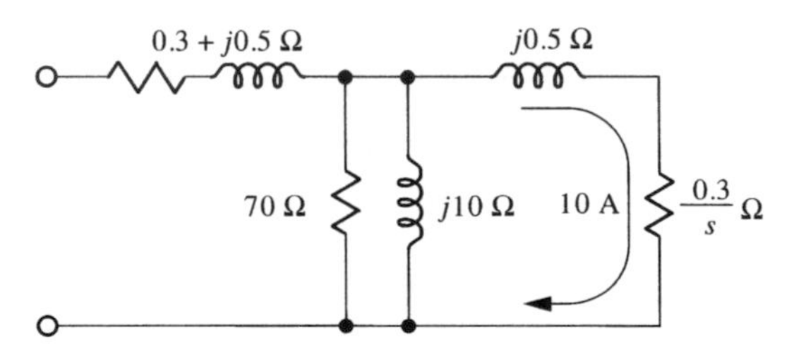

Figure P5.18

5.19. A six-pole, 60-Hz, three-phase induction motor has the per-phase equivalent circuit shown in Fig. P5.19.
(a) Determine the developed power at 1140 rpm.
(b) At what speed does the motor produce 4 hp? Assume 10 W mechanical loss and a small-slip region.

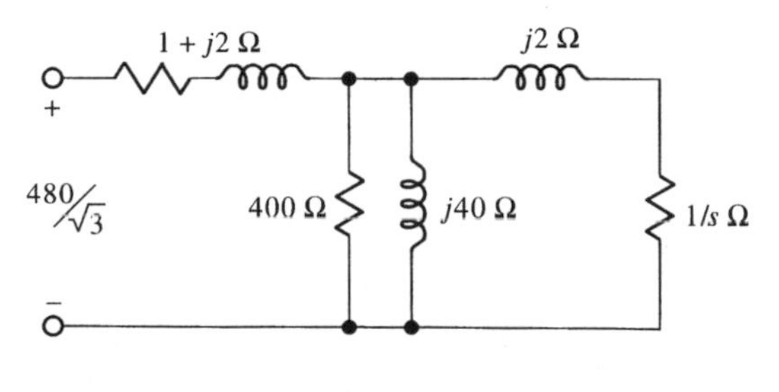

$V_T = 257.7$ V

$\underline{Z}_{eq} = 0.914 + j1.918\ \Omega$

Figure P5.19

5.20. A three-phase, 60-Hz induction motor has the following nameplate information: 25 hp, 1750 rpm, 230/460 V, 64.0/32.0 A, 91.0% eff., and 1.15 service factor. Assume negligible mechanical losses but a 500-W iron loss.

(a) Find the reactive power requirement of the motor at nameplate conditions.

(b) Find the stator-copper loss of the motor at nameplate conditions.

(c) Estimate the output torque at a speed of 1760 rpm.

5.21. An industrial-duty 60-Hz, three-phase induction motor has the following nameplate information: 20 hp, 3515 rpm, 230/460 V, 50.0/25.0 A, efficiency = 86.5%, and 320 lb. The mechanical losses are 100 W. Assume iron losses are 1.5 W/lb of weight. At nameplate conditions, find the following:

(a) Air-gap power.

(b) Stator losses.

(c) Power factor.

(d) Developed torque.

(e) No-load speed.

(f) The per-phase stator resistance when the motor is connected for 460-V operation.

5.22. The nameplate specifications of a 60-Hz, three-phase induction motor are as follows: 15 hp, 284 T frame, 1170 rpm, 67.3 lb-ft torque, 94.2 lb-ft locked-rotor torque, 135 lb-ft breakdown torque, 42 A, 230 V, 232 A locked-rotor current, 89.5% efficiency, and 1.15 service factor. (1 ft-lb = 1.356 N-m.) Assuming no mechanical losses but 300-W iron loss, find the following:

(a) The power factor under nameplate conditions.

(b) The rotor-copper loss under nameplate conditions.

(c) The per-phase stator resistance.

(d) The speed at which the motor output torque is 60% of the nameplate torque.

(e) The air-gap power for $s = 1$.

5.23. Figure P5.23 shows the per-phase equivalent circuit for a six-pole, 60-Hz, three-phase induction motor. The stator circuit has been converted to a Thèvenin equivalent circuit.

(a) Find the speed (rpm) for an output power of 10 hp. Assume the small-slip region.

(b) The nameplate speed is 1135 rpm. Find the nameplate output power, assuming 1% of the developed power is mechanical loss.

(c) From the previous information, estimate the no-load speed, assuming constant mechanical loss.

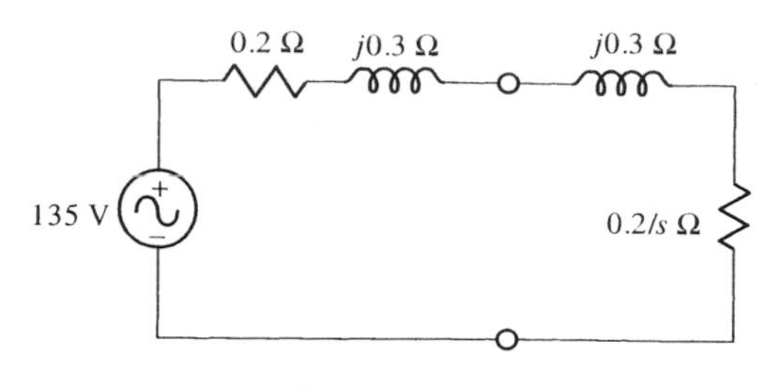

Figure P5.23

5.24. A three-phase, 60-Hz induction motor has the following nameplate information: 5 hp, 1740 rpm, 230/460 V, 12.8/6.4 A, efficiency 87.5%, service factor 1.15, blocked-rotor current 93 A at 230 V, blocked-rotor torque 180% of rated torque, and weight 95 lb.

(a) Find the power factor under nameplate conditions.

(b) Find the air-gap power under blocked rotor conditions.

(c) What happens to the air-gap power under blocked rotor conditions?

(d) Assuming that the iron losses are 1.5 W/lb of weight, find the stator per-phase resistance when connected for 230-V operation. Ignore mechanical losses.

5.25. A three-phase, 60-Hz induction motor has the following nameplate information: 2 hp, 3500 rpm, 230/460 V, 6.0/30. A, 80.0% eff., and 50 lb. Tests at 230 V show that the motor has a no-load speed and input current of 3599.2 rpm and 3.2 A, respectively.

(a) Find the mechanical loss, assumed independent of speed.

(b) Find the rotor loss at nameplate operation.

(c) Find the stator loss at nameplate operation, assuming 2-W/lb iron loss.

(d) Find the power factor at nameplate operation.

(e) Estimate the power factor at no load.

5.26. A three-phase, 10-hp motor has the following specifications: 60-Hz, 1750 rpm, 230/460 V, 88.5% eff., 26.2/13.1 A, and service factor 1.35. Assume that the mechanical losses are 0.5% of the output power at nameplate conditions and are constant throughout the small-slip region. Assume iron losses of 250 W.

(a) What is the output torque at nameplate conditions?

(b) Give the power factor at nameplate conditions.

(c) Find the per-phase stator resistance of the motor for 230-V excitation.

(d) Estimate the per-phase rotor resistance referred to the stator for 230-V excitation.

5.27. The (simplified) per-phase equivalent circuit in Fig. P5.27 is for a 460-V, four-pole, three-phase, 60-Hz induction motor. The motor has 200 W of iron and the mechanical loss, and the total losses must be less than 900 W to avoid overheating.

(a) What is the approximate nameplate power rating for this motor?

(b) Estimate the slip at rated load.

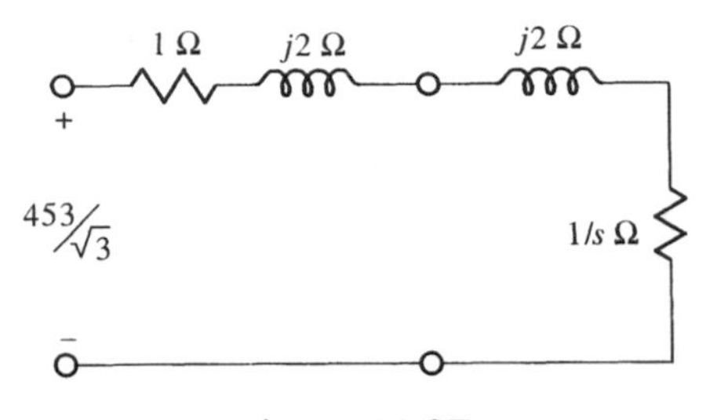

Figure P5.27

5.28 Figure P5.28 shows the per-phase equivalent circuit for a 50-Hz, three-phase induction motor. Estimate the speed for 4.5 hp out, ignoring mechanical losses.

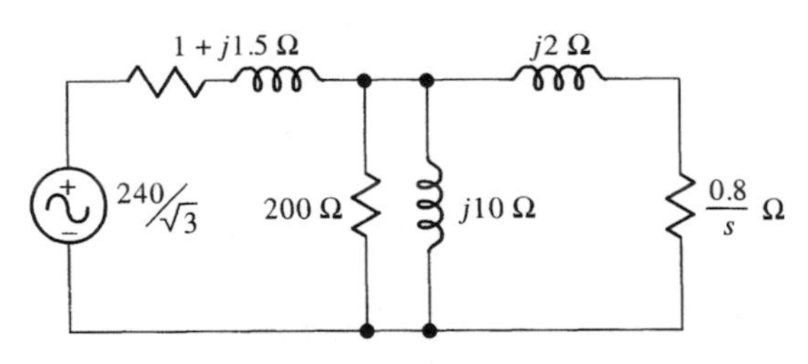

Figure P5.28

5.29. A three-phase, 60-Hz induction motor has the following nameplate information: 7.5 hp, 1755 rpm, 230/460 V, 19.4/9.7 A, 88.5% eff., and service factor 1.15. Neglect mechanical losses but assume 200 W of iron loss.

(a) Find the power factor at nameplate operation.

(b) Find the output torque at nameplate operation.

(c) Find the slip speed in radians/second at nameplate operation.

(d) Find the rotor losses at nameplate operation.

(e) Find the stator losses at nameplate operation.

(f) A load requiring $T_L(n) = 10 + 12.8\,(n/1800)^2$ N-m is connected to the motor. Find the speed at which the motor operates.

5.30. A three-phase, 60-Hz induction motor has the following nameplate information: 1/4 hp, 1725 rpm, 230/440 V, 1.1/0.55 A, 66.8% eff., and 1.35 service factor. Consider mechanical losses to be negligible.

(a) Find the output torque at nameplate operation.

(b) Find the power factor at nameplate operation.

(c) If the load torque requirement is $T_L = 0.004\omega_m$ N-m, find the speed in rpm at which the motor/load will operate.

(d) Under the conditions in part (c), estimate the rotor-copper losses.

5.31. A 60-Hz, three-phase induction motor has the following nameplate information: 40 hp, 1775 rpm, 230/460 V, 91.7% eff., 94.8/47.4 A, $1269.97 wholesale, and 548 lb. Neglect mechanical losses but assume 1 W/lb of iron loss.

(a) Find the slip speed in radians/second at nameplate conditions.

(b) Find the apparent power in kVA used by the motor at nameplate conditions.

(c) Find the power factor of the motor at nameplate conditions.

(d) Find the rotor-copper losses at nameplate conditions.

(e) Find the stator losses at nameplate conditions.

5.32 A junior engineer was given the task of evaluating the motor in the previous problem through a series of measurements, so as to produce a per-phase equivalent circuit. The engineer measured the mechanical losses as 10 watts and presented to the boss the equivalence circuit shown in Fig. P5.32.

(a) Does the $j13\ \Omega$ look reasonable? Explain why or why not.

(b) Does the $143\ \Omega$ look reasonable? Explain why or why not.

(c) Does the $0.28 + j0.2\ \Omega$ look reasonable? Explain why or why not.

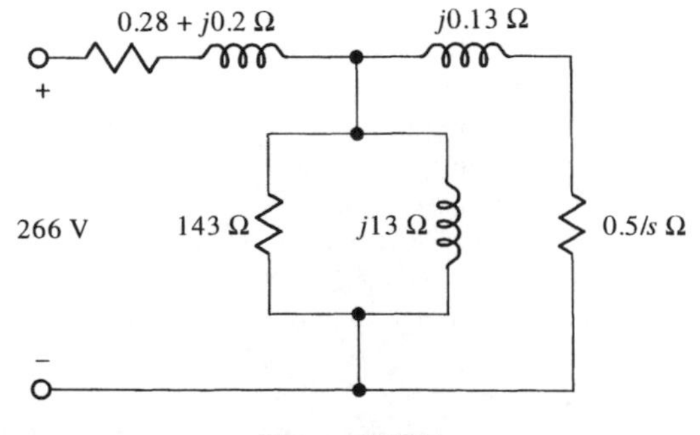

Figure P5.32

(d) Does the $0.5/s\ \Omega$ look reasonable? Explain why or why not.

N.B.: These questions should be answered without a detailed analysis of the equivalent circuit. Your "reasons" should include some simple calculations.

5.33. A 60-Hz, three-phase induction motor has the following nameplate specifications: 20.5 A at 230 V; runs 1725 rpm with an efficiency of 88.0%. The no-load speed and current are 1799.2 rpm and 11.1 A, respectively. We ignore iron loss. A simplified per-phase equivalent circuit for the motor is shown in Fig. P5.33. From the given information, estimate as many of the equivalent circuit resistors and reactances as you can.

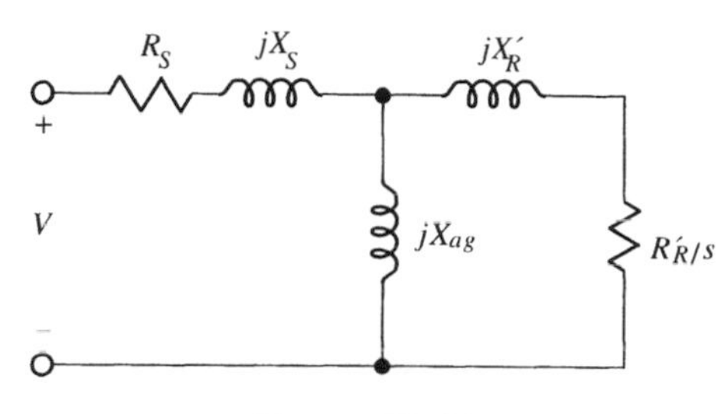

Figure P5.33

5.34. A three-phase, 60-Hz induction motor has the following nameplate information: 30 hp, 1750 rpm, 39.0 A, 460 V 87.5% eff., and 1.15 SF.
A simplified per-phase equivalent circuit for the motor is shown in Fig. P5.34.

(a) From the given information, estimate the three impendance values in the circuit. Ignore iron and mechanical losses.

(b) The motor is connected to a load that requires a torque given by the equation

$$T_L(n) = 10 + 12\left(\frac{n}{1800}\right) + 30\left(\frac{n}{1800}\right)^2 \text{ N-m}$$

Assuming operation in the small-slip region, find the operating speed of the system. Do not use the equivalent circuit from part (a), but rather work directly from the nameplate information.

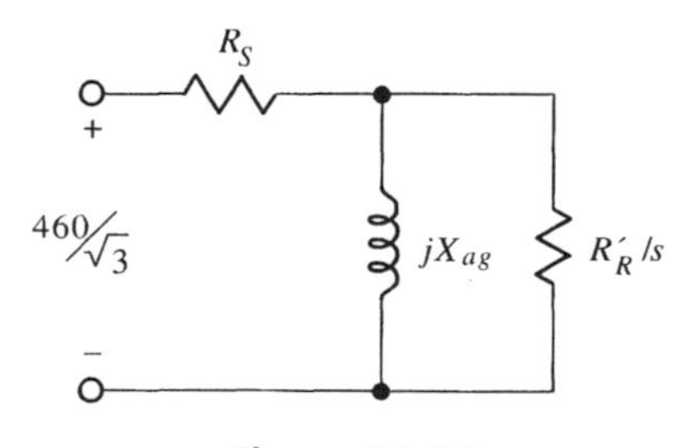

Figure P5.34

5.35 Design a motor-drive system for a load with the following characteristics: The motor is OFF for 2 seconds and ON for 1 second, and then the pattern repeats. While ON, the power required is 1.5 hp. The current required to start is 400% the rated current, but lasts only for 0.1 second. Neglect mechanical losses and assume 2 watts/pound of weight for the iron losses. Pick a suitable motor drive from Table 5.2 and confirm that your choice meets the specifications for the job through an analysis of the losses.

5.36. A high-slip, 30-hp, three-phase motor is used on an unbalanced pumping jack. The nameplate information on the motor is 460 V, 36 A, 1115 rpm, 86.4% eff., service factor = 1.0. Ignore mechanical losses. The equivalent circuit for the motor is shown in Fig. P5.36. The load is unbalanced, with a torque requirement of 250 N-m on the upstroke and a load requirement of −100 N-m on the down stroke, meaning that the motor acts as an induction generator and returns energy to the electrical power system. The gearing is such that the shaft turns 100 revolutions on the up-stroke and 100 revolutions on the downstroke.

(a) Find the total energy used by the motor on one up–down cycle.

(b) Is the motor overloaded in this application? Explain.

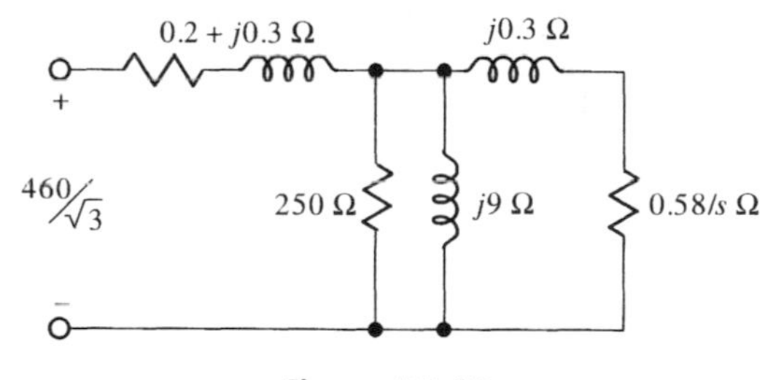

Figure P5.36

5.37. A load with the time-varying torque shown in Fig. P5.37 turns about 1750 rpm. We need to specify a motor-drive this load, to satisfy two criteria: (1) the load torque never exceeds 175% of the nameplate torque (avoid breakover); and (2) the rms horsepower required by the load does not exceed the nameplate rating of the motor (avoid overheating). To simplify, consider the speed to be exactly 1750 rpm. Determine the rating of a three-phase induction motor to meet these criteria, using the motor information in Table 5.2.

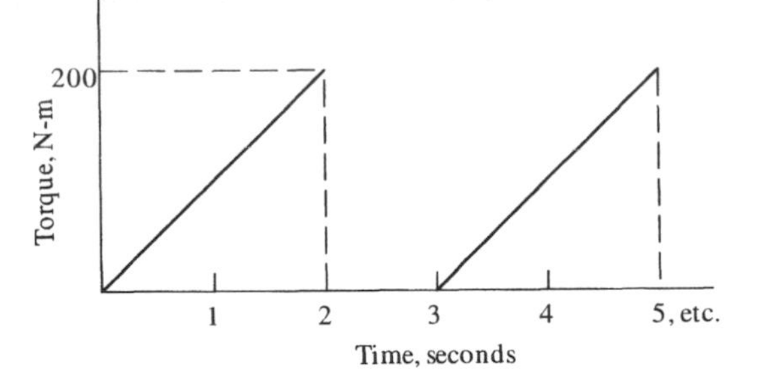

Figure P5.37

Section 5.3: Single-Phase Induction Motors

5.38. A single-phase, 120-V induction motor takes a current of 6 A. The electrical loss is 5% of the input power and the mechanical loss is 1% of the output power. If the total losses are 30 W, find the power factor of the motor.

5.39. A 115-V, 60-Hz, single-phase induction motor produces 2 hp at 3500 rpm with an input current of 18.8 A. Assume losses at this power to be 20% of the output power, divided between mechanical and electrical losses in the ratio 1:20. Find the power factor and the efficiency.

5.40. A 60-Hz, single-phase induction motor has the following nameplate information: 115/230 V, 8.2/4.1 A. 1725 rpm and 3/4 hp. Assume a power factor of 0.78 lagging.

(a) Determine the efficiency of the motor at nameplate operation.
(b) What is the output torque at nameplate operation.
(c) What are the electrical frequencies in the rotor at nameplate operation?
(d) Explain how the motor can be connected for either 115- or 230-V operation.
(e) Would you expect the motor to have better performance when connected for 115- or 230-V operation? Explain.

5.41. A single-phase, 60-Hz induction motor has the following specifications: 3/4 hp, 1725 rpm, 115/230 V, 11.6/5.8 A, ball bearings, and reversible.

(a) Assuming a power factor of 0.77, determine the efficiency of the motor.
(b) Find the output torque under nameplate conditions.
(c) Explain how to reverse the motor.
(d) If the motor were excited at 113 V, would the current be less than, the same as, or greater than 11.6 A for the same output power?

5.42. A high-efficiency, single-phase, 60-Hz induction motors has the following nameplate information: 1/2 hp, 1725 rpm, 115 V, and 5.5 A. This is capacitor-start/capacitor-run motor.

(a) How many poles does this motor have?
(b) Assuming an efficiency of 75%, determine the power factor.
(c) What electrical frequencies exist in the rotor?
(d) What is the output torque?
(e) Draw the motor circuit that exists during starting.

5.43. A 60-Hz, 120-V, single-phase induction motor has a linear region described by the equation $P_{dev} = a + bs$. The motor runs at 1798 rpm at no load and 1740 rpm with 1-hp output power. Find the output power at 1785 rpm, assuming the same mechanical losses at all speeds in the problem.

5.44. A capacitor-start/induction-run, 60-Hz, single-phase induction motor has the following nameplate information: 2 hp, 1725 rpm, 60 Hz, 230 V, 10.3 A, and efficiency = 84.0 %. Find the following:

(a) The number of poles.
(b) The slip at nameplate load.
(c) The power factor at nameplate load.
(d) The rated torque at nameplate load.
(e) The yearly operating cost based on 6 cents/kW-hour, 40 h/week, 50 weeks/year.

5.45. The circuit diagram for a 120-V, 60-Hz, single-phase capacitor-start/induction-run motor is shown in Fig. P5.45. The circuit applies only with the rotor at rest. After the motor comes to full speed, the current in the main winding is one-third of its value at the instant of starting. The auxiliary winding is connected only during starting. Find the ratio of the maximum starting current to the run current.

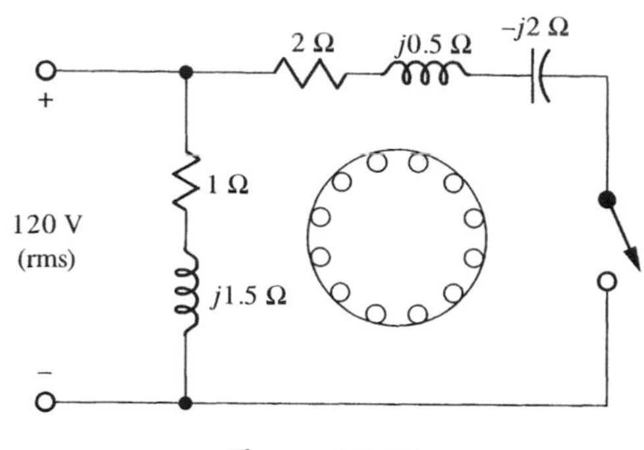

Figure P5.45

5.46 A single-phase, 60-Hz capacitor-start/induction-run motor has the equivalent circuit (for starting) shown in Fig. P5.46.

(a) Find C for a 90° phase shift between starting currents in run and auxiliary windings at starting.

(b) Determine the starting current for this value of C.

(c) Assuming the nameplate run current is 3.9 A, the speed is 1740 rpm, and the power is 1/4 hp, estimate the power factor and output torque of the motor under nameplate conditions. Ignore iron and mechanical losses.

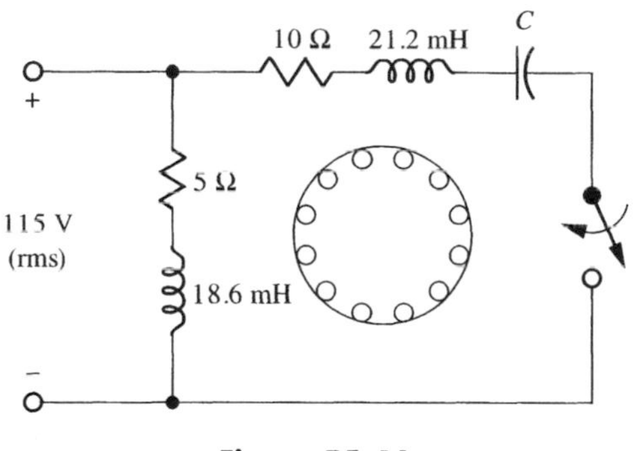

Figure P5.46

5.47. A motor catalog lists a 60-Hz, high-efficiency (capacitor-start/capacitor-run) single-phase motor with the following nameplate information: 1 hp, 1725 rpm, 115/230 V, and 9.2/4.6 A. The catalog also claims that this motor saves 100 W compared to the normal capacitor-start/induction-run motor, which has the following nameplate information: 1 hp, 1725 rpm, 115/230 V, and 14.8/7.4 A. Assume that electrical losses are $K(I_{in})^2$, where K is the same for both motors. Also assume that mechanical losses are 2% of the output power for both motors. Determine the power factors of the two motors.

5.48. A 1/4 hp, 120-V, 60-Hz, single-phase induction motor runs at 3599 rpm no-load speed and 3450 rpm at rated output power. *Hint:* Assume $P_{out} = a + bs$.

(a) Find the two electrical frequencies in the rotor at 1/4 hp out.

(b) Estimate the slip for 1/8 hp out.

(c) Find the mechanical power required to drive the motor at 3600 rpm.

5.49. A capacitor-start/induction-run, 115-V, 60-Hz single-phase induction motor has stator field parameters as shown in Fig. P5.49.

(a) What value of capacitor is required to produce a 90° phase shift between main and auxiliary currents for starting?

(b) Find the starting current with the value of capacitance determined in part (a).

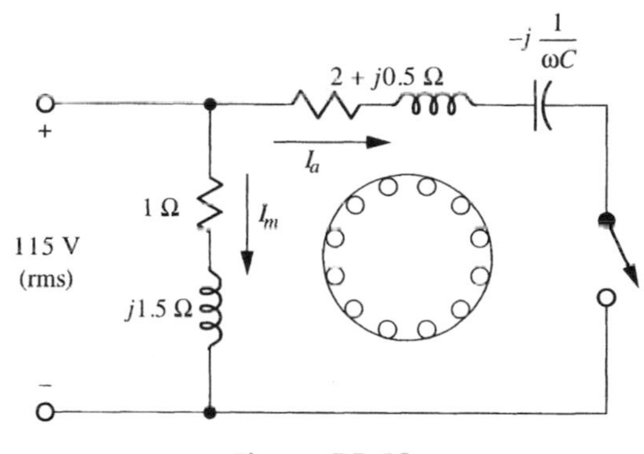

Figure P5.49

5.50. Figure P5.50 shows a shaded-pole induction motor. The conducting bands "shade" a portion of the poles to create rotating flux.

(a) Estimate the flux in the iron.

(b) Estimate the input current to the motor if its output power is 25 W. Ignore iron losses.

(c) Mark the direction that the motor will turn.

5.51. A 1/4-hp, 120-V, capacitor-start/induction-run motor has the current, PF, and speed characteristics given in Table P5.51.

(a) Find the efficiency at 60% and 100% of nameplate power.

(b) Is this motor designed to have maximum efficiency at nameplate conditions? Yes or no?

(c) Estimate the stator losses at no load. (*N.B.*: Assume no mechanical losses. But there are copper losses at no-load because the rotor is in

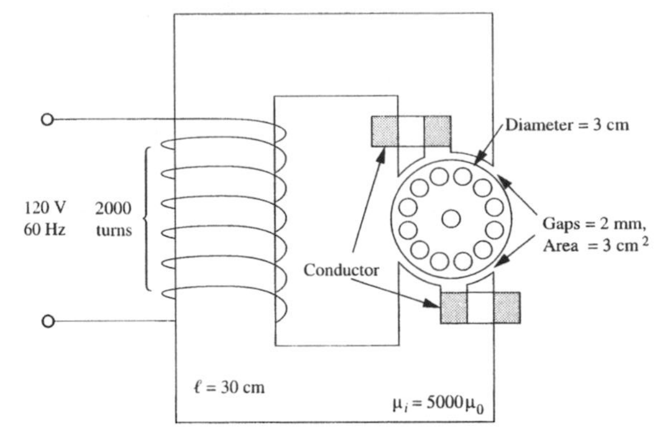

Figure P5.50

TABLE P5.51

Power, hp	*rpm*	*PF*	*Current*
0	1787	0.246	6.25
0.05	1777	0.292	6.21
0.10	1767	0.341	6.24
0.15	1756	0.392	6.28
0.20	1745	0.433	6.34
0.25	1731	0.476	6.41
0.30	1716	0.520	6.55
0.35	1702	0.561	6.78
0.40	1686	0.594	7.11
0.45	1669	0.625	7.55
0.50	1650	0.652	8.07

equilibrium between forward and reverse components of the rotating fields and losses due to each torque appears as rotor-copper loss. Consider the reverse-torque to be constant over the operating range of the motor.)

5.52. A single-phase induction motor has the following nameplate information: 1/3 hp, 60 Hz, 3450 rpm, 115/230 V, 9.8/4.9 A, and power factor 0.72, lagging. The equivalent circuit, connected for 230-V operation, is shown in Fig. P5.52. The resistance values are valid for starting, but change as the rotor moves to account for power transformed into mechanical work.

(a) Find the starting current for the motor.

(b) Find the stator-copper losses at nameplate conditions.

(c) If the no-load speed were 3585 rpm, estimate the speed for 0.25-hp output power.

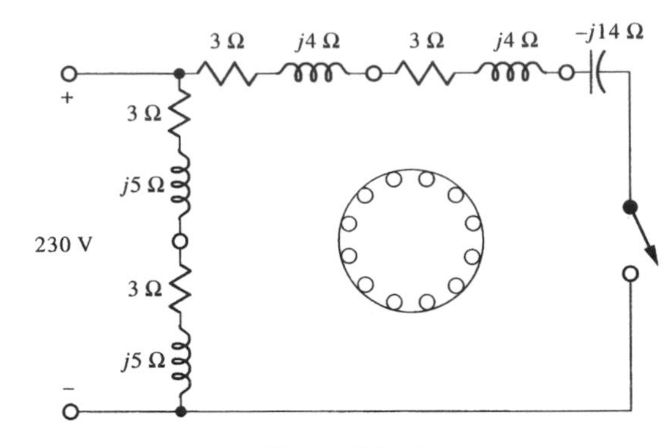

Figure P5.52

5.53. The equivalent circuit for a single-phase motor is shown in Fig. P5.53. Under nameplate conditions, the currents in the two windings are equal in magnitude and 90° out of phase.

(a) What type of motor is it? Your choices are (1) capacitor-start/induction-run, (2) split-phase, (3) capacitor-start/capacitor-run, (4) permanent-split-capacitor, and (5) shaded-pole.

(b) The rotating field in this motor depends on one phase, two phases, or three phases? Which?

(c) The input run current to the motor is 10 A. What is the magnitude of the currents in the windings?

(d) The run current in the winding without the capacitor lags input voltage by 45°. Find the value of the capacitor.

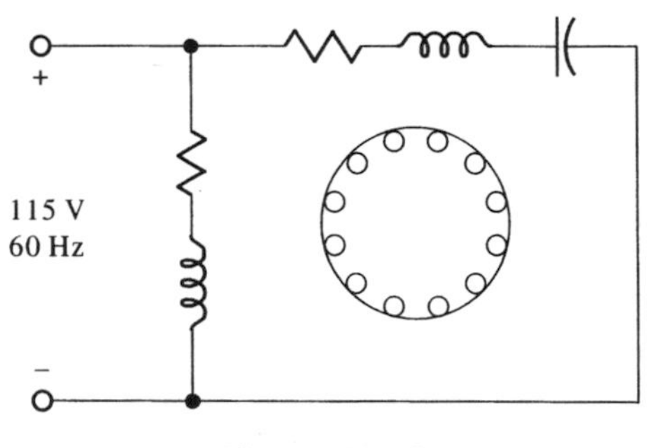

Figure P5.53

5.54. A single-phase motor drives a machine that has a duty cycle requiring 2 N-m for 15 s and 4 N-m for 5 s, and then repeats with a 20-s cycle time. The speed should be in the range 1650–1750 rpm, so a 60-Hz, four-pole motor is indicated. A section out of a motor catalog is given in Table P5.54. Assume that all motors have an efficiency of 80% when operating

TABLE P5.54

hp	*rpm*	*Amps*	*SF*	*Cost, $*
1/4	1725	5.6	1.35	83.16
1/3	1725	8.0	1.0	85.14
1/2	1725	10.0	1.0	101.97
3/4	1725	10.8	1.25	141.57
1	1725	15.0	1.0	156.42

near their nameplate conditions, total losses are proportional to current squared, and that current is proportional to torque. Select the best motor for the job.

5.55. A 60-Hz, single-phase, capacitor-start/induction-run induction motor has the following nameplate information: 3 hp, 3450 rpm, 115/230 V, 30.0/15.0 A, service factor 1.15, $347.34, and 63.0 pounds. Measurements show the motor losses to be 600 W.

(a) Find the slip at nameplate conditions.
(b) Find the number of poles.
(c) What is the efficiency of the motor at nameplate conditions?
(d) What is the in-phase component of the current at nameplate conditions for 230-V operation?
(e) What is the out-of-phase component of the current at nameplate conditions for 230-V operation?

5.56. By accident, the run windings for the motor described in the previous problem are connected in series for 230-V operation, but the auxiliary windings are connected in parallel. The motor has 230 V applied. Explain the following:

(a) Will this affect the starting torque?
(b) Will this affect the starting current?
(c) Will this affect the run torque?
(d) Will this affect the run current?
(e) Will this affect the direction of rotation?

General Problems

5.57. A motor catalog lists the following motor: 5 hp, 230/460 V, 12.8/6.4 A, 1740 rpm, 60 Hz, 85.5% eff., 1.15 service factor, 60 lb. Assume operation at nameplate conditions and negligible mechanical losses. Assume further that the electrical losses divide between rotor and stator in the ratio 1:3.

(a) What type of motor is this: induction or synchronous? Explain how you know.
(b) In this a three-phase or single-phase motor? Explain how you know.
(c) What is the power factor of the motor?
(d) Find the rotor- and stator-copper losses and the iron losses.
(e) Find the per-phase rotor resistance, assuming 230-V operation.

5.58. Observe the comparison in Table P5.58 between similar 60-Hz, single-phase and three-phase induction motors.

TABLE P5.58

Phases	*hp*	*rpm*	*Volts*	*Amps*	*lb*	*Cost, $*
3	1	1750	230	3.5	32	120.06
1	1	1740	230	7.2	40.7	174.75

Indicate for the following statements: true, false, or uncertain, and explain.

(a) The single-phase motor is heavier because of the starting capacitor.
(b) The single-phase motor runs slower because it has a counter-rotating flux that opposes the principal rotation.
(c) The single-phase has higher current because it has a worse efficiency.
(d) Both motors have four poles.

5.59. Here are some short-answer questions:

(a) Two input power wires of an operating three-phase induction motor are exchanged suddenly. The slip at that instant would be about -1, 0, $+1$, or $+2$?
(b) Assuming voltages are OK, a single-phase induction motor can be operated from two wires of a three-phase system, but a three-phase induction motor cannot be operated off the two wires of a single-phase power system. True or false?
(c) A two-phase induction motor would have a torque/speed characteristic similar to that of a three-phase induction motor except it would have no starting torque. True or false?
(d) If a 60-Hz, single-phase induction motor were run at 50 Hz, the motor speed for the same torque would be approximately 5/6 of the 60-Hz speed. True or false?
(e) If a three-phase induction motor were operating and one of the input wires became disconnected, the motor would stop immediately. True or false?

Answers to Odd-Numbered Problems

5.1. $I_a = I$, $I_b = I_c = -I/2$.

5.3. (**a**) 4; (**b**) 4.44%; (**c**) 1460 W; (**d**) 1390 W; (**e**) 1590 W; (**f**) 87.4%; (**g**) 7.73 N-m; (**h**) 1870 VA.

5.5. (**a**) 3.33%; (**b**) 0.836; (**c**) 3890; (**d**) greater than.

5.7. Rotor = 64.9 W and stator = 98.1 W.

5.9. 87.2%.

5.11. (**a**) 1800 rpm; (**b**) 33.4 N-m; (**c**) 76.4 N-m.

5.13. (**a**) 3518 rpm; (**b**) 7.86 A, 77.8%; (**c**) 2.65 N-m.

5.15. (**a**) 16,800 W; (**b**) 15,100 W; (**c**) 80.1 N-m; (**d**) 298 A; (**e**) 141 N-m.

5.17. (**a**) 1.11%; (**b**) rotor = 839 W, iron = 1320 W, stator = 4500 W; (**c**) 1777 rpm.

5.19. (**a**) 8370 W; (**b**) 1182 rpm.

5.21. (**a**) 15,400 W; (**b**) 1390 W; (**c**) 0.866; (**d**) 40.8 W; (**e**) 3599.4 rpm; (**f**) 0.741 Ω.

5.23. (**a**) 1166 rpm; (**b**) 12,200 W; (**c**) 1199.35 rpm.

5.25. (**a**) 12.4 W; (**b**) 43.0 W; (**c**) 218 W; (**d**) 0.780; (**e**) 0.341, lagging.

5.27. (**a**) 7660 W, about 10 hp; (**b**) 4.37%.

5.29. (**a**) 0.818; (**b**) 30.4 N-m; (**c**) 4.71 rad/s; (**d**) 143 W; (**e**) 384 W; (**f**) 1767 rpm.

5.31. (**a**) 2.62 rad/s; (**b**) 37.8 kVA; (**c**) 0.862; (**d**) 420 W; (**e**) 1730 W.

5.33. $R'_R \approx 0.4\ \Omega$, $X_{ag} \approx 11\ \Omega$, $R_s \approx 0.37\ \Omega$; reactances are small and not easily estimated except by careful measurement.

5.35. RMS hp suggests 1 hp; average losses estimate to about 220 W, well below the allowed 290 W.

5.37. 30 hp, mainly on the maximum torque requirement.

5.39. 0.828, 83.3%.

5.41. (**a**) 54.5%; (**b**) 3.10 N-m; (**c**) reverse the auxiliary winding connections; (**d**) greater than 11.6 A; approximately 11.8 A.

5.43. 167 W.

5.45. 3.60.

5.47. $PF_1 = 0.779$, $PF_2 = 0.543$.

5.49. (**a**) 1450 μF; (**b**) 79.8 A.

5.51. (**a**) $\eta_{60\%} = 37.9\%$, $\eta_{100\%} = 50.9\%$; (**b**) no; (**c**) 88 W.

5.53. (**a**) PSC; (**b**) two phases; (**c**) 7.07 A; (**d**) 115 μF.

5.55. (**a**) 4.17%; (**b**) 2; (**c**) 78.9%; (**d**) 12.3 A; (**e**) 8.53 A, lagging.

5.57. (**a**) induction motor: (**b**) three phase, because its power factor would exceed unity if it were a single-phase motor; (**c**) 0.856; (**d**) rotor = 129 W, stator = 388 W, iron = 118 W; (**e**) 0.789 Ω.

5.59. (**a**) +2, because the synchronous speed reverses sign; (**b**) true: two wires give single-phase power; (**c**) false: it would have the same characteristics; (**d**) true: the synchronous speed is 5/6 of 60-hz synchronous speed; (**e**) false: it becomes a single-phase motor, which will run if started.

CHAPTER

Direct-Current Motors

1. **To understand the physical basis for the equivalent circuit of a dc motor**
2. **To understand the speed–torque characteristic of the shunt-connected dc motor**
3. **To understand the speed-torque characteristic of the series-connected dc motor**
4. **To understand how to develop system models for separately excited dc motors from nameplate and load characteristics**

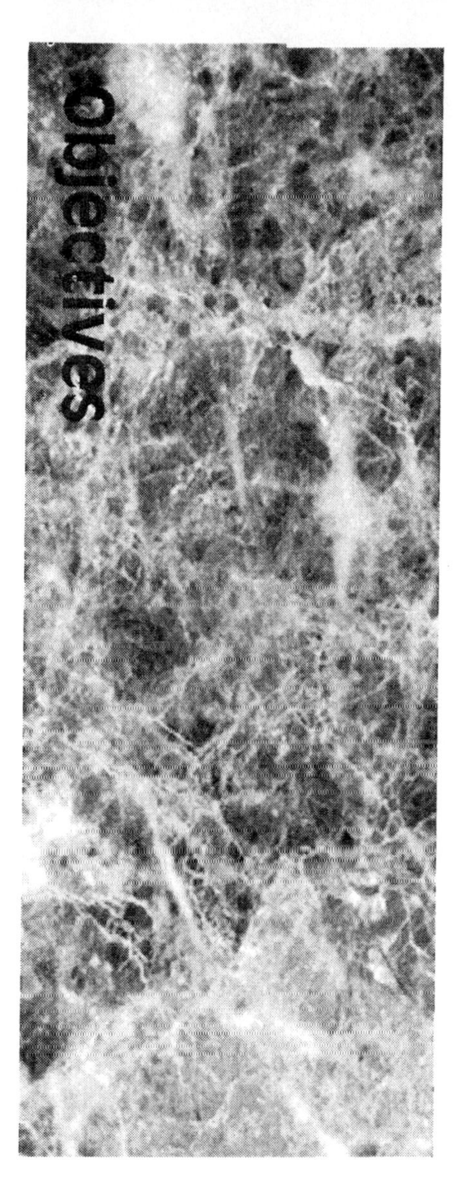

We still need lots of dc **motors for portable operation such as in automobiles, for control applications, and for household tools and appliances that, ironically, run on ac power. In this chapter, we consider operating principles and external characteristics of several types of dc motors.**

6.1 PRINCIPLES OF DC MACHINES

alternator

Importance. The dc machine can be used either as a motor or a generator. However, because semiconductor rectifiers can generate dc voltage from ac with electronic power supplies, dc generators are unneeded except for remote operations. Even in the automobile, the dc generator has been replaced by the *alternator*, a synchronous generator plus diodes for rectification. On the other hand, generator operation must still be considered because motors operate as generators in braking and reversing.

Portable devices operating from battery power require dc motors, such as auto starters, window lifts, and portable tape players. Equally important, the dc machine is readily controlled in speed and torque and hence is useful for control systems. Examples are robots, elevators, machine tools, rolling mills, and large power shovels.

In this chapter, we examine the fundamental principles of dc motors and explore the wide range of characteristics that can be achieved through various motor configurations. DC generators are mentioned briefly, primarily because of the role of the back-generated voltage in the operation of dc motors. The universal motor, a type of dc motor that runs on ac power, is also discussed.

LEARNING OBJECTIVE 1.

To understand the physical basis for the equivalent circuit of a dc motor

Stator Magnetic Structure

Salient-pole structures. The stator magnetic structure for a dc machine is shown in Fig. 6.1(a). The dc machine uses salient poles,[1] which are extended in width to leave as little interpole space as practical. The field coils are wrapped around these poles. The rotor is cylindrical, with slots for wires, as shown in Fig. 6.1(b). The flux-density distribution approximates a square wave, as shown in Fig. 6.2 for the four-pole structure in Fig. 6.1(a).

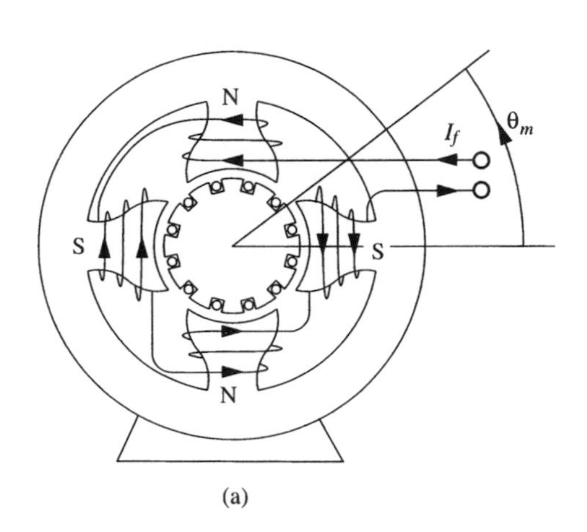

(a)

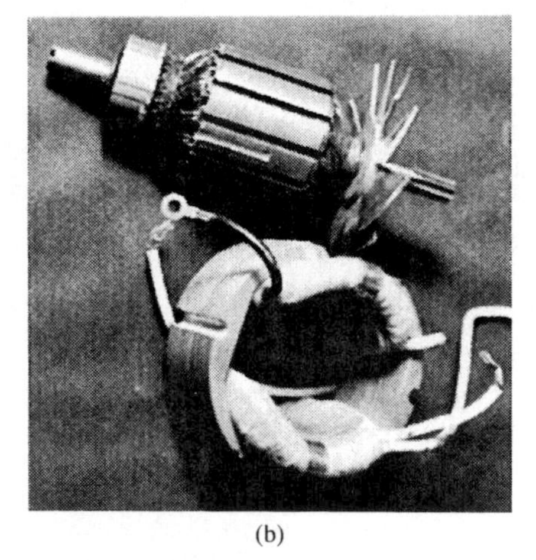

(b)

Figure 6.1 (a) Salient-pole magnetic structure with $P = 4$ poles; (b) a small dc motor with two poles.

Analysis. The maximum flux density can be determined from Ampère's circuital law around a path passing through two adjacent poles.

[1] In this context, "pole" refers to the mechanical protrusion as well as the magnetic pole associated with it.

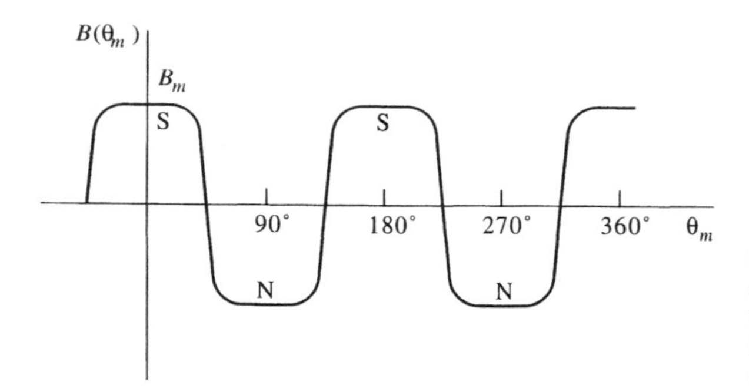

Figure 6.2 Flux density for four-pole, dc magnetic structure. Flux is positive in the radially outward direction. This is also the form of the voltage generated in a single wire of a rotating armature.

$$\oint \vec{H} \cdot \vec{dl} = ni \quad \Rightarrow \quad B_m = \frac{\mu_0 2nI_f}{2g} \tag{6.1}$$

where B_m is the maximum flux density, nI_f is the mmf per pole, and g is the width of the air gap. The factor 2 in the denominator arises because the air gap is traversed twice by the path of integration, and the factor 2 appears in the numerator because n is the turns on one pole and the path of integration passes through two poles. The mmf loss in the iron has been neglected.

EXAMPLE 6.1 Field mmf

A small dc motor has a rotor OD of 3.80 ± 0.01 cm and a stator ID of 3.90 ± 0.01 cm. Find the mmf/pole for a nominal flux density of 0.8 tesla.

SOLUTION:

By using the nominal diameters, the value of $2g$ in Eq. (6.1) is 3.90 − 3.80 = 0.10 cm. From Eq. (6.1), the required mmf is

$$0.8 = \frac{4\pi \times 10^{-7} \times 2nI_f}{0.10 \times 10^{-2}} \quad \Rightarrow \quad nI_f = 318 \text{ A-t} \tag{6.2}$$

WHAT IF?

What if 318 ampere-turns are used, and the *maximum* possible flux density is desired?[2]

Alnico

Permanent magnet fields for dc machines. Figure 6.3 shows an automotive air-conditioner/heater-blower motor, which uses a field structure of permanent magnets made from molded ceramic ferrite. Larger structures might use *Alnico*[3] permanent magnets.

[2] 1.0 tesla.

[3] *Alnico* is a tradename of Alcoa for aluminum–nickel–cobalt, from which many permanent magnets are fashioned.

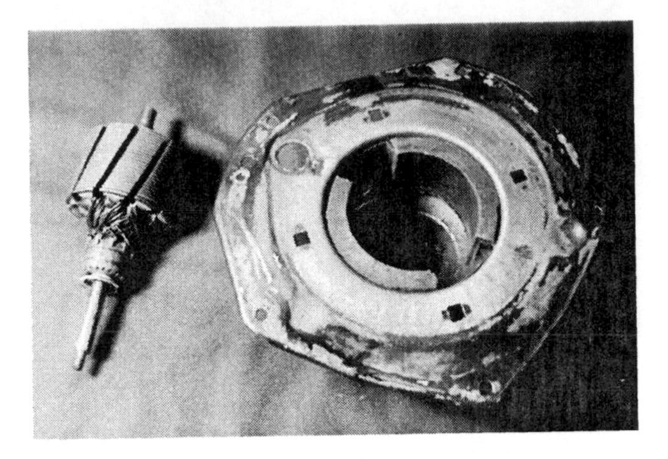

Figure 6.3 Automotive fan motor. Permanent magnets on the stator produce the magnetic field. This stator has four poles.

Rotor Construction

Rotor currents and flux. The dc machine requires a brush-commutator system, indicated in Fig. 6.4, to produce currents out of the paper on the right side and into the paper on the left. The right-hand rule shows that the rotor currents produce a downward flux, a north pole at the bottom and a south pole at the top. These poles are attracted to their opposites on the stator and a counterclockwise torque is produced on the rotor from these currents.

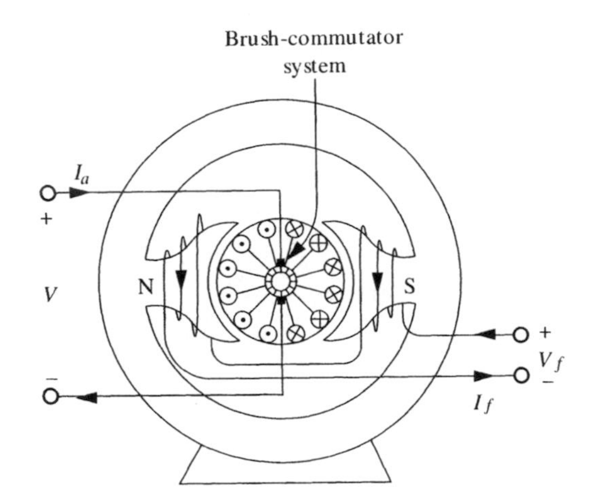

Figure 6.4 A two-pole dc machine with two salient stator poles and a brush-commutator system.

commutator, brushes

Brush–commutator system. Figure 6.5 shows a brush and the commutator from a dc machine. The commutator has a cylindrical surface of wedge-shaped segments connected to the rotor conductors. The commutator is part of the rotor and participates in its rotation. The brushes are stationary and rub against the commutator as the rotor rotates. Figure 6.6 shows a schematic view of how the currents reverse due to commutation. The radial lines represent the active lengths of the rotor conductors, and the current returns and internal connections are shown. The currents reverse on opposite sides of the rotor, as indicated in Fig. 6.6.

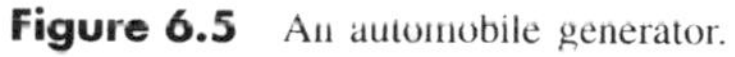

Figure 6.5 An automobile generator.

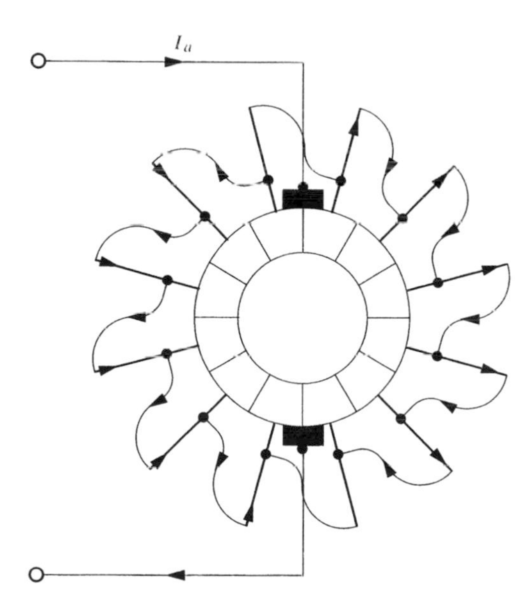

Figure 6.6 Schematic diagram of the armature conductors. The heavy lines represent the active lengths of the armature conductors. The currents are in opposite directions on opposite sides.

The brush–commutator system provides, therefore, two related functions:

- Electrical connection is made with the moving rotor.
- Switching of the rotor currents is accomplished mechanically in a way that automatically synchronizes switching with rotor motion.

Due to the brush–commutator system, the *spatial pattern* of the rotor currents is always the same, independent of the physical position of the rotor.

Many of the problems with dc machines arise due to commutation. Not only must the brush–commutator system carry large currents across a sliding contact, but the switching of currents in the individual coils causes an inductive effect that limits performance. These problems have been solved in measure, with the result that we have available dc machines with excellent characteristics, although they require occasional maintenance.

In dc motors, both stator and rotor magnetic poles remain fixed in space, and the mechanical rotor rotates relative to the rotor magnetic poles. The rotor conductors carry "ac" currents[4] under the control of the brush–commutator system.

IDEA 6 **Equivalent Circuits**

Circuit Model

separately excited motor

Field circuit. The field circuit is modeled in Fig. 6.7 as a series connection of resistance and inductance, with a field-voltage supply, V_f. The motor is *separately excited* because the field circuit is independent of the armature circuit. In steady-state operation, the field current follows from Ohm's law

$$I_f = \frac{V_f}{R_f} \tag{6.3}$$

where I_f is the field current and R_f the resistance of the field windings. Neglecting magnetic saturation and residual magnetism, we may relate the flux to the field current through the reluctance of the magnetic structure, $\mathfrak{R}$

$$\Phi = \frac{2nI_f}{\mathfrak{R}} \tag{6.4}$$

where Φ is the flux and n is the equivalent number of turns/pole. Normally, magnetic saturation and residual magnetism are significant, however, and the relationship between field current and flux is nonlinear.

The number of poles runs from 2 or 4 for a small motor to as large as 30 for a large motor. As we will see, the number of poles has no effect on the speed of the motor.

Armature circuit. The armature circuit is located on the rotor. The circuit model for the rotor consists of a resistance and inductance in series with an emf, E. Figure 6.7 shows the armature resistance and inductance outside the brush–commutator system, which introduces the emf, E, into the armature circuit. The armature resistance, R_a, and inductance, L_a, belong physically between the brushes but customarily are shown outside for artistic reasons. In either event, it does not matter for a series connection.

Mechanical output. We indicate the mechanical output with symbols indicating rotation and developed torque. As a consequence of its rotation in a magnetic field, the rotor also has iron losses; however, these customarily are not represented in the electrical equivalent circuit but are combined with the mechanical losses because they depend on mechanical speed and not on the electrical variables in the armature.

[4] Actually, chopped dc currents.

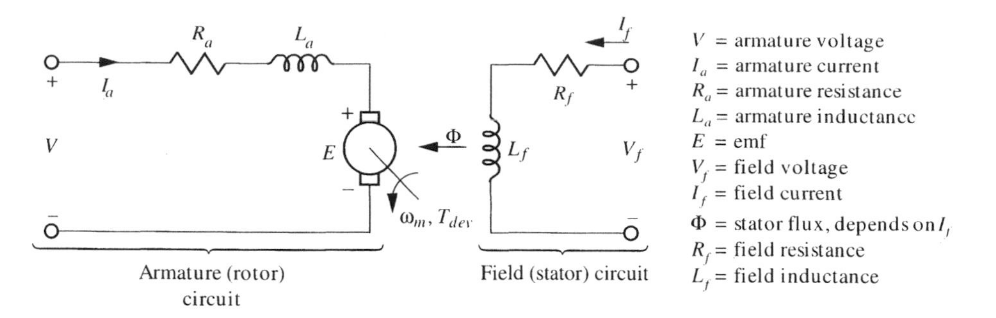

Figure 6.7 Circuit model for a separately excited dc machine.

back emf

Electromotive force. When the rotor is rotated in the flux produced by the field, an ac voltage is produced in each rotor conductor. These voltages are rectified and summed by the brush–commutator system to produce a dc emf.[5] By Faraday's law, this emf is proportional to flux and rotation speed, and hence may be expressed by the relationship

$$E = K_E \Phi \omega_m \tag{6.5}$$

where K_E is a constant that depends on rotor size, the number of rotor turns, and details of how these turns are interconnected.

Developed torque. If an armature current flows through the brush–commutator system, this current passes through the rotor conductors and a torque is developed. By Ampère's force law, this developed torque is proportional to flux and armature current and hence may be expressed as

$$T_{dev} = K_T \Phi I_a \tag{6.6}$$

where K_T is a constant that also depends on rotor size, the number of rotor turns, and details of how these turns are interconnected. We soon demonstrate that conservation of energy requires that the constants in Eqs. (6.5) and (6.6) be the same.

Power Flow in DC Machines

Conservation of energy. In steady state, KVL applied to the armature circuit yields

$$V = R_a I_a + E \quad \Rightarrow \quad I_a = \frac{V - E}{R_a} \tag{6.7}$$

where V is the armature voltage, I_a is the armature current, R_a is the armature resistance, and E is the armature emf. We may convert Eq. (6.7) into a power equation by multiplying by the armature current

[5] In a motor called the *back emf.*

$$\underbrace{VI_a}_{P_{in}} = \underbrace{R_a I_a^2}_{\substack{\text{armature} \\ \text{Cu loss}}} + \underbrace{EI_a}_{\substack{\text{developed} \\ \text{power}}} \tag{6.8}$$

Conservation of Energy

Equation (6.8) shows that the input power divides between armature-copper losses and EI_a, which represents power leaving the electrical circuit as mechanical power. Thus, conservation of energy requires the developed power to be

$$P_{dev} = \omega_m T_{dev} = EI_a \tag{6.9}$$

The output power, P, is less than the developed power by the rotational losses, P_{rot}

$$P = P_{dev} - P_{rot} \tag{6.10}$$

The machine constant. If we substitute Eqs. (6.5) and (6.6) for E and T_{dev}, respectively, Eq. (6.9) takes the form

$$K_T \Phi I_a \omega_m = K_E \Phi \omega_m I_a \tag{6.11}$$

and thus $K_E = K_T = K$, the machine constant,[6] as asserted earlier.

Motor action. Combining Eq. (6.7) with Eq. (6.9), we may express the developed power as

$$P_{dev} = EI_a = \frac{E(V - E)}{R_a} \tag{6.12}$$

When the armature voltage exceeds the emf, the armature current is positive relative to the reference direction shown in Fig. 6.7, electrical power is delivered to the armature, and the machine acts as a motor. In this case, the developed torque has the same direction as the direction of rotation, and mechanical power is delivered to the mechanical load.

Generator action. When the emf exceeds the armature voltage due to external mechanical drive, the current becomes negative relative to the reference direction in Fig. 6.7 and flows out of the + polarity mark on the armature. The machine then acts as a generator, and the developed torque is opposite to the direction of rotation. In many applications, the machine alternatively acts as a motor or generator, depending on changes in the mechanical load or armature voltage.

Causality. Figure 6.8 shows the causal factors governing dc machine behavior. Let us consider an at-rest dc machine with no mechanical load. If we apply voltage to the armature circuit, the resulting current will produce torque, which accelerates the rotor. As the rotor speed increases, so does the back emf, E, and the current decreases. The no-load equilibrium is reached when $E \approx V$ and a small armature current flows to supply rotational losses.

[6] The product of this machine constant and the magnetic flux, $K\Phi$, is also called "the machine constant." Context will make it clear which is meant.

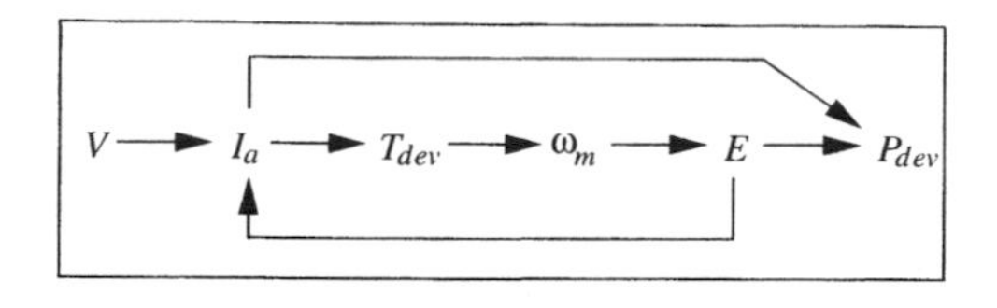

Figure 6.8 Causality in a dc motor.

Effect of load. Continuing the previous discussion, we now apply a mechanical load. This slows down the rotor and the back emf is reduced proportionally. The lower emf causes an increase in current and torque until a new equilibrium is reached. Thus, we expect the motor to slow down with increasing mechanical load.

EXAMPLE 6.2 Maximum power

What is the maximum power that can be developed in an armature if armature voltage is constant?

SOLUTION:

For fixed V and R_a, Eq. (6.12) describes a parabola with a maximum at $E = V/2$. Thus, the maximum power that can be developed is

$$P_{\max} = \frac{V}{2} \times \frac{V - V/2}{R_a} = \frac{V^2}{4R_a} \tag{6.13}$$

WHAT IF? What if the voltage source in not ideal but has an output impedance, R_{eq}?[7]

magnetization curve

Magnetization curve. Equation (6.5) allows the relationship between stator flux and field current to be represented as a *magnetization curve* of emf versus field current for a constant speed, as shown in Fig. 6.9. From the magnetization curve, we can determine the machine constant and the emf at other speeds, because the emf is strictly proportional to rotational speed by Eq. (6.5). Furthermore, we can also derive torque information because Eq. (6.9) gives

$$T_{dev} = \frac{E}{\omega_m} \times I_a \tag{6.14}$$

where E/ω_m depends on field current and can be determined from the magnetization curve.

[7] $P_{max} = V^2/4(R_{eq} + R_a)$.

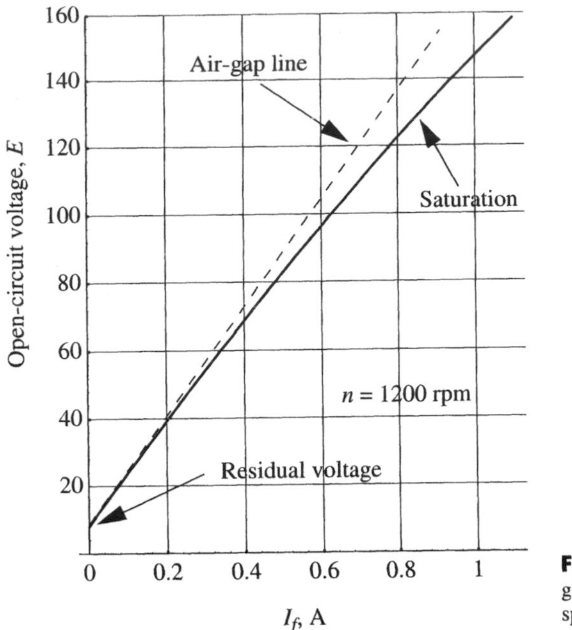

Figure 6.9 The magnetization curve gives emf versus field current at fixed speed.[8]

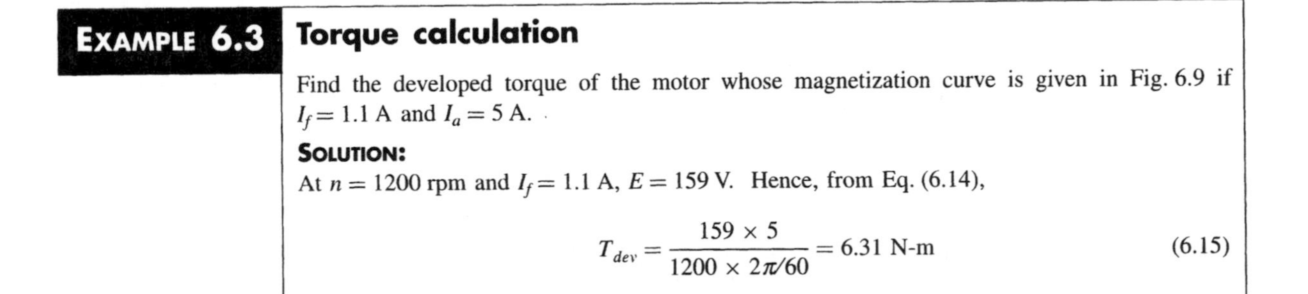

EXAMPLE 6.3 **Torque calculation**

Find the developed torque of the motor whose magnetization curve is given in Fig. 6.9 if $I_f = 1.1$ A and $I_a = 5$ A.

SOLUTION:

At $n = 1200$ rpm and $I_f = 1.1$ A, $E = 159$ V. Hence, from Eq. (6.14),

$$T_{dev} = \frac{159 \times 5}{1200 \times 2\pi/60} = 6.31 \text{ N-m} \tag{6.15}$$

Residual magnetism. At zero field current, Fig. 6.9 indicates a small voltage. This effect from residual (permanent) magnetism in the stator magnetic structure plays an important role in the buildup of voltage if the machine is operated as a generator.

air-gap line

Air-gap line. Figure 6.9 shows the effect of saturation in the iron of the magnetic structure. The linear approximation, the *air-gap line*, is important in the following sections because, to derive approximate motor characteristics with the various means of field excitation, we often assume that the motor is operating with the stator flux proportional to field current. The results of such an analysis suggest the general features of the motor characteristics, but any resulting calculations are approximate.

[8] To aid in working examples and problems based on this magnetization curve, we confess that Fig. 6.9 was generated by the following parabola: $E(I_f) = 160I_f - 21I_f^2 + 8$. In practice, it has to be measured.

EXAMPLE 6.4 Output power calculation

A 120-V dc motor has an armature resistance of 0.70 Ω. At no-load, it requires 1.1 A armature current and runs at 1000 rpm. Find the output power and torque at 952 rpm output speed. Assume constant flux.

SOLUTION:

From the no-load condition, we can calculate the machine constant and the rotational losses. The input power at no load is 120 V × 1.1 A = 132 W, and the armature loss is $0.70(1.1)^2 =$ 0.85 W; hence, the rotational losses at 1000 rpm are 131.2 W. The emf at this speed can be calculated from Eq. (6.7) as 120 − 0.70(1.1) = 119.2 V. Hence, the product of the machine constant and the stator flux is $K\Phi = 119.2/1000$.[9]

At 952 rpm, the emf is reduced to $E' = K\Phi \times 952 = 119.2 \times 952/1000 = 113.5$ V. This reduced voltage implies an input current of $I'_a = (120 - 113.5)/0.70 = 9.28$ A; and hence the developed power is $P_{dev} = EI'_a = 113.5 \times 9.28 = 1052.9$ W. Mechanical losses at 952 rpm are approximately the same as at 1000 rpm, and therefore the output power is 1052.9 − 131.2 = 921.7 W (1.23 hp). The output torque is

$$T_{out} = \frac{P_{out}}{\omega_m} = \frac{921.7}{952(2\pi/60)} = 9.25 \text{ N-m} \tag{6.16}$$

WHAT IF?

What if the no-load current were 1.4 A and the rotational losses were proportional to speed? Find the output torque.[10]

Check Your Understanding

1. On a dc machine, the magnitude of the rotor-stator power angle is about 90°. True or false?
2. DC motors are still used because they (a) can be portable, (b) have a good power factor, (c) can be controlled for speed, (d) can be controlled for torque, and (e) have low maintenance. Which? (May be more than one.)
3. DC motor torque is proportional to (a) armature current, (b) field current, (c) speed, or (d) developed power. Which? (May be more than one.)
4. DC motor emf is proportional to (a) armature current, (b) field current, (c) speed, or (d) developed power. Which? (May be more than one.)
5. To increase the speed of a separately excited dc motor, we would increase the (a) input voltage, (b) field current, (c) output torque, or (d) number of poles. Which?
6. The armature circuit of a separately excited dc motor is shown in Fig. 6.10. Determine the input and developed power of the motor.

[9] In this expression, the units for K include the conversion between rpm and rad/s. Because we employ scaling principles, such conversion factors cancel.

[10] 9.31 N-m.

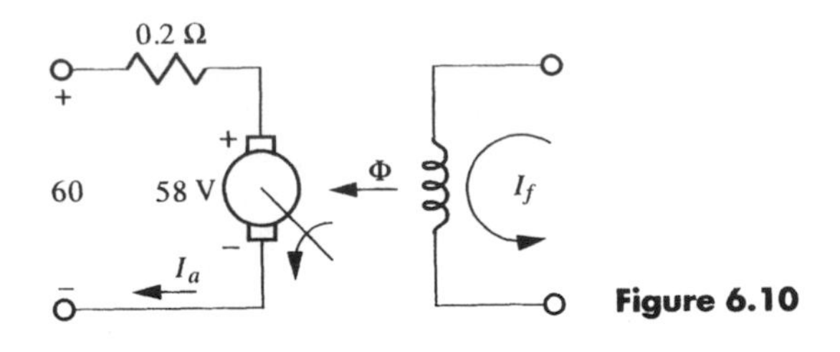

Figure 6.10

Answers. **(1)** True; **(2)** (a), (c), and (d); **(3)** (a) and (b); **(4)** (b) and (c); **(5)** (a); **(6)** 600 and 580 W, respectively.

6.2 CHARACTERISTICS OF DC MOTORS

LEARNING OBJECTIVE 2.

To understand the speed-torque characteristic of the shunt-connected dc motor

Shunt-Connected Field

shunt-connected, shunt-excited, parallel-excited, rheostat

Circuit. Figure 6.11 shows a *shunt-connected*[11] motor. Here the field circuit is connected in parallel, or shunt, with the armature circuit. Normally, the field has a large resistance, so the field current is small compared with the armature current. The shunt-connected motor is similar to the separately excited motor, except that here the field current must be controlled by a field *rheostat*, R_F.[12]

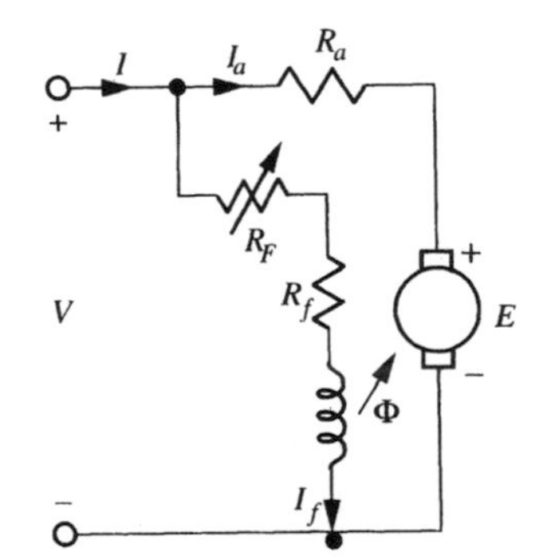

Figure 6.11 Circuit model for a shunt-connected motor.

Analysis. We now derive the torque as a function of speed with fixed input voltage and fixed field current. The nonlinear behavior of the magnetic structure is not a factor because the field current is constant. We begin with KVL in the armature circuit, Eq. (6.7), eliminate I_a through Eq. (6.6), and eliminate E through Eq. (6.5).[13] The results are

$$V = R_a I_a + E = R_a \frac{T_{dev}}{K\Phi} + K\Phi\omega_m \tag{6.17}$$

[11] Also called *shunt-excited* and *parallel-excited.*

[12] A *rheostat* is a variable resistor used to control a current.

[13] Recall that $K_T = K_E = K$.

Solving for developed torque, we obtain

$$T_{dev} = \frac{K\Phi}{R_a}[V - K\Phi\omega_m] \tag{6.18}$$

If we assume the rotational-loss torque is constant or varies linearly with speed, the output torque will have the form of a straight line

$$T_{out}(\omega_m) = C_1 - C_2\omega_m \tag{6.19}$$

where C_1 and C_2 are constants.

EXAMPLE 6.5 **No-load speed**

Consider the machine whose characteristic is shown in Fig. 6.9. Assume the line voltage to be 120 V, the armature resistance to be 0.5 Ω, and the total field resistance to be 120 Ω, such that the field current is 1.0 A. The rotational losses at 1200 rpm are 25 W, assumed constant. Find the speed for no load.

SOLUTION:

With no external load, the power into the armature must supply armature resistive loss and rotational losses. For the moment, we ignore the electrical loss; hence, the input power to the armature is 25 W, and the required current is 25 W/120 V = 0.208 A. From KVL, we calculate the emf to be 120 − 0.208 × (0.5 Ω) = 119.9 V. From Fig. 6.9, we have, for a field current of 1.0 A, a generated voltage of 147 V at 1200 rpm. Hence, we scale the speed proportional to the generated voltage and estimate the no-load speed as

$$n_{NL} = \frac{119.9}{147} \times 1200 = 978.7 \text{ rpm} \tag{6.20}$$

WHAT IF? What if we assume rotational losses are proportional to speed? Find the corresponding no-load speed.[14]

Finding the speed generally. Because back emf, E, and speed are strictly proportional, we may determine speed from developed power with Eq. (6.12), which is quadratic in E

$$E^2 - VE + R_a P_{dev} = 0 \tag{6.21}$$

Equation (6.21) gives two real and positive roots; the larger value of E is the realistic solution. From the resulting E, the speed can be found, and from the speed and power, the torque can be calculated.

[14] 978.9 rpm.

EXAMPLE 6.6 1 hp output

Find the speed and torque for an output power of 1 hp for the motor in the previous example.

SOLUTION:
The developed power in the armature must now be 746 + 25 = 771 W. From Eq. (6.21):

$$E^2 - 120E + 0.5 \times 771 = 0 \Rightarrow E = 3.30 \text{ and } 116.7 \text{ V} \tag{6.22}$$

Thus, for 1 hp out, the speed must drop to

$$n = \frac{116.7}{147} \times 1200 = 952.6 \text{ rpm} \tag{6.23}$$

The output torque for 1 hp and 952.6 rpm is

$$T_{out} = \frac{746}{952.6 \times 2\pi/60} = 7.48 \text{ N-m} \tag{6.24}$$

WHAT IF?

What if we assume rotational losses are proportional to speed? What then would be the speed at 1 hp output power?[15]

Torque–speed characteristic. Equation (6.19) shows that the torque–speed characteristic for the shunt-connected dc motor is a straight line. Thus, the full-torque characteristic may be derived from two points. The speed is approximately constant because $E \approx V$ and $n \propto E$. The motor torque is limited by the ability of the armature to dissipate the armature-copper loss without damage, or perhaps by the ability of the brush–commutator system to handle the required current.

EXAMPLE 6.7 Torque characteristic

Find the output torque, $T_M(n)$, for the motor used in the previous two examples.

SOLUTION:
From Eq. (6.19) and the results of the previous two examples:

$$\begin{aligned} 0 &= C_1 - C_2 \times 978.7 \\ 7.48 &= C_1 - C_2 \times 952.6 \end{aligned} \tag{6.25}$$

Thus, $C_1 = 280$ N-m and $C_2 = 0.286$ N-m/rpm, and the torque characteristic is shown in Fig. 6.12.

[15] 952.8 rpm.

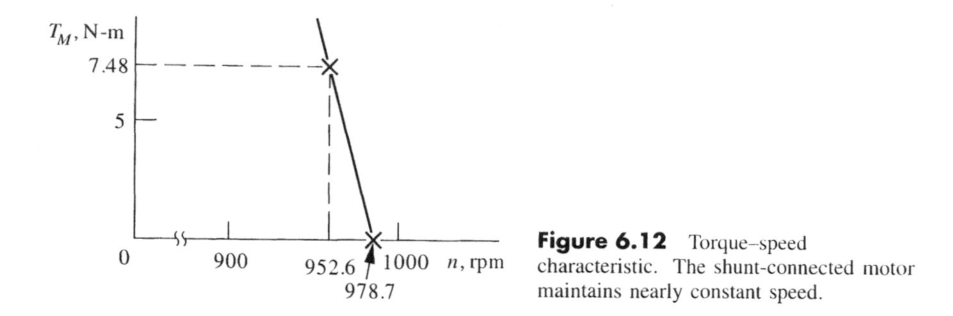

Figure 6.12 Torque–speed characteristic. The shunt-connected motor maintains nearly constant speed.

Rotational losses. The rotational losses of a dc motor consist principally of iron losses, with a small component of mechanical loss. The iron losses occur in the rotor, which rotates in a stationary magnetic flux. The ac frequency is proportional to motor speed, and hence rotational power losses are approximately proportional to speed and the loss torque is approximately constant.

Speed control. We note in Fig. 6.12 that for the dc motor, the speed is nearly constant over a wide range of torques. We can control the speed of the dc motor by varying the armature voltage while keeping the field current constant, or by varying the field current while keeping the armature voltage constant. Both these methods are effective, but the control of the armature voltage offers broader range and more desirable dynamic properties. The armature voltage may be varied by several methods, such as the electronic means discussed in Chap. 7.

Efficiency. In calculating the efficiency of the dc motor, the losses of the field circuit and rheostat much be charged against the motor.

EXAMPLE 6.8 Efficiency calculation

Find the efficiency of the motor in the previous three examples at 1 hp output power.

SOLUTION:

The armature current is

$$I_a = \frac{V - E}{R_a} = \frac{120 - 116.7}{0.5} = 6.61 \text{ A} \tag{6.26}$$

Thus, the input current to the motor is

$$I = I_a + I_f = 6.61 + 1.00 = 7.61 \text{ A} \tag{6.27}$$

and the motor efficiency is

$$\eta = \frac{P_{out}}{P_{in}} = \frac{746}{120 \times 7.61} = 81.7\% \tag{6.28}$$

WHAT IF? What if you wish to know the fraction of motor losses caused by the field circuit?[16]

[16] 71.9%.

Permanent-magnet (PM) DC motors. Large numbers of dc machines are manufactured with fields provided by permanent magnets. Applications include fan and window-lift motors on automobiles, small appliances such as electric toothbrushes, tape recorders, and hair dryers, instruments like tachometers, and novelties such as toy trains. Some large machines, up to 200 hp, are designed with permanent-magnetic fields to meet special requirements for size, weight, or efficiency. Machines with permanent-magnet fields have characteristics similar to separately and shunt-excited machines, except that field flux cannot be varied.

EXAMPLE 6.9

PM dc motor

A permanent-magnet dc motor has the following nameplate information: 50 hp, 200 V, 200 A, 1200 rpm, and armature resistance of 0.05 Ω. Determine the speed and output power if the voltage is lowered to 150 V and the current is 200 A. Assume that rotational losses are proportional to speed.

SOLUTION:

First, we analyze the nameplate information to determine the machine constant and rotational losses. At the nameplate voltage and current, the emf is

$$E = V - I_a R_a = 200 - 200(0.05) = 190\text{ V} \tag{6.29}$$

and the machine constant in volts/rpm is

$$K\Phi = \frac{190\text{ V}}{1200\text{ rpm}} = 0.158\text{ V/rpm} \tag{6.30}$$

Thus, the developed power and rotational losses are

$$P_{dev} = 190 \times 200 = 38{,}000\text{ W} \Rightarrow P_m = 38{,}000 - 50 \times 746 = 700\text{ W} \tag{6.31}$$

With the input voltage at 150 V, and the armature current unchanged, the new emf is

$$E' = 150 - 200(0.05) = 140\text{ V} \tag{6.32}$$

We may determine the new speed by scaling

$$n' = n \times \frac{E'}{E} = 1200 \times \frac{140}{190} = 884\text{ rpm} \tag{6.33}$$

This same result could be obtained from the machine constant in Eq. (6.30). The new developed power is

$$P'_{dev} = E' \times I_a = 28{,}000\text{ W} \tag{6.34}$$

We have assumed rotational losses proportional to speed, so the new losses are 516 W. Hence, the new output power is 27,500 W (36.8 hp). This is the rated output power of the machine at 150 V.

WHAT IF? What if the input voltage is 160 V? Find the new rated output power.[17]

[17] 39.5 hp.

Characteristics of shunt-connected and permanent-magnet-field dc motors. The characteristics of the shunt-connected and permanent-magnet dc motors are as follows:

- For fixed field current and input voltage, the speed is nearly constant. This is true because the emf is approximately equal to the input voltage.
- For fixed field current, the speed is approximately proportional to armature voltage. Because the emf is proportional to the product of the rotational speed and the stator magnetic flux, for fixed armature voltage, the speed is inversely affected by field current. To increase motor speed, we must decrease field current.
- Reversing the input voltage to the shunt-connected motor does not reverse the direction of rotation because both stator flux and armature current reverse. To reverse the motor directions, we must reverse the polarity of either field or armature. The permanent-magnet motor reverses when input voltage is reversed.

LEARNING OBJECTIVE 3.

To understand the speed–torque characteristic of the series-connected dc motor

Series-Connected Field

Circuit. Figure 6.13 shows a series-connected dc motor, where the armature and field carry the same current.

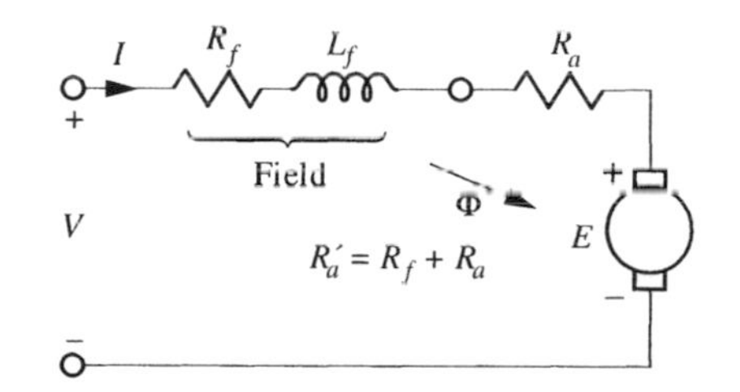

Figure 6.13 Circuit model for series-connected motor.

Speed–torque characteristic. We now derive the speed–torque characteristic of the series-connected motor. We ignore magnetic saturation by assuming that the stator flux is proportional to field current; thus, we combine Eqs. (6.4) and Eq. (6.6):

$$T_{dev} = K\Phi I = \frac{K2n}{\Re} I^2 = K' I^2 \tag{6.35}$$

where I is the current, n is the number for turns/pole in the field, $\Re$ is the reluctance of the magnetic structure, and K' replaces all the other constants. We let $R_a' = R_f + R_a$ represent the combined field and armature resistance; hence, KVL in the armature circuit becomes, after using Eq. (6.5)

$$V = R_a' I + K\Phi\omega_m = (R_a' + K'\omega_m) I \tag{6.36}$$

To obtain the torque–speed characteristic, we solve Eq. (6.36) for the current and substitute into Eq. (6.35)

$$T_{dev} = K' \left[\frac{V}{R_a' + K'\omega_m} \right]^2 \tag{6.37}$$

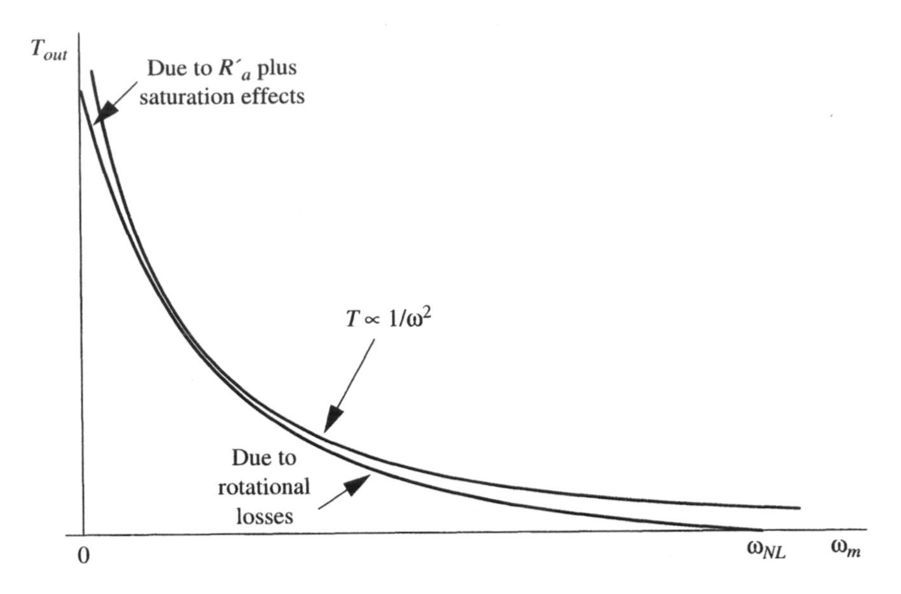

Figure 6.14 Output speed–torque characteristic for a series-excited motor.

The speed–torque characteristic in Eq. (6.37), modified to include rotational losses and magnetic saturation, is shown in Fig. 6.14.

EXAMPLE 6.10 3 hp output

A 50-V series-connected motor has $R'_a = 0.05\ \Omega$ and an output power of 1 hp at 500 rpm. Find the speed and motor current for an output power of 3 hp, ignoring rotational losses.

SOLUTION:

An approximate solution is easy. If we assume we are in that part of the characteristic where the effects of resistance and rotational losses are negligible, the torque is inversely proportional to the square of the speed. In this region, the power, $P = \omega_m T$, is inversely proportional to speed; hence

$$P = \omega_m T \approx \frac{C}{n} \tag{6.38}$$

where C is a constant and n is the motor speed in rpm. From the given information, we know that $C = 500$ when power is expressed in hp; hence, for 3 hp, the speed, n' , must be

$$3 = \frac{C}{n'} \quad \Rightarrow \quad n' = \frac{500 \times 1}{3} = 167 \text{ rpm} \tag{6.39}$$

The current, I' , can be determined from conservation of energy, Eq. (6.8)

$$50I' = 0.05(I')^2 + 3 \times 746 \tag{6.40}$$

The quadratic has two solutions, and we pick the smaller root as the realistic current, $I' = 47.0$ A. Although the speed was derived from an approximate analysis, the current was derived from conservation of energy and thus is exact.

We begin a more accurate analysis by calculating the current, I, for 1 hp output

$$50I = 0.05I^2 + 746 \tag{6.41}$$

Hence, $I = 15.1$ A. Because the torque for 1 hp is

$$T = \frac{746}{500(2\pi/60)} = 14.2 \text{ N-m} \tag{6.42}$$

Eq. (6.35) gives K' as

$$K' = \frac{T}{I^2} = \frac{14.2}{(15.1)^2} = 0.0621 \text{ N-m/A}^2 \tag{6.43}$$

This value is used in Eq. (6.37), which we convert to a power equation by multiplying by ω_m

$$P = \omega_m T = K'\left[\frac{V}{R'_a + K'\omega_m}\right]^2 \times \omega_m \tag{6.44}$$

With the known values of power in watts (3 × 746), input voltage (50 V), total resistance (0.05 Ω), and K' (0.0621), Eq. (6.44) is quadratic in ω_m. The two roots are $\omega_m = 0.0397$ and 16.3 rad/s. In this case, the larger root, 156 rpm, is the realistic answer. The current for 3 hp was calculated earlier to be 47.0 A.

WHAT IF? What if we want the speed for 2 hp?[18]

Features of the series-connected motor. Reversing the polarity of the input voltage does not reverse the direction of rotation because both field and armature currents are reversed. To reverse the motor, we must reverse field or armature polarity separately.

The sloping speed–torque characteristic of the series-connected motor offers benefits for many applications. The motor gives good torque for starting without excessive current. For this reason, series-connected motors are used as starter motors for automobiles.

Universal (AC/DC) Motors

Principle of operation. The series-connected dc motor operates on alternating current. With the field in series with the armature, as shown in Fig. 6.15(a), the flux is proportional to the current, and a time-average torque is produced by ac current, as shown in Fig. 6.15(b). Thus, we may time average Eq. (6.35):

$$\langle T(t)\rangle = K'\langle i^2(t)\rangle = K'I^2_{\text{rms}} \tag{6.45}$$

where the angle brackets indicate time averaging. The speed–torque characteristic of the universal motor is similar to that for the series-connected dc motor in Fig. 6.14.

[18] 242 rpm.

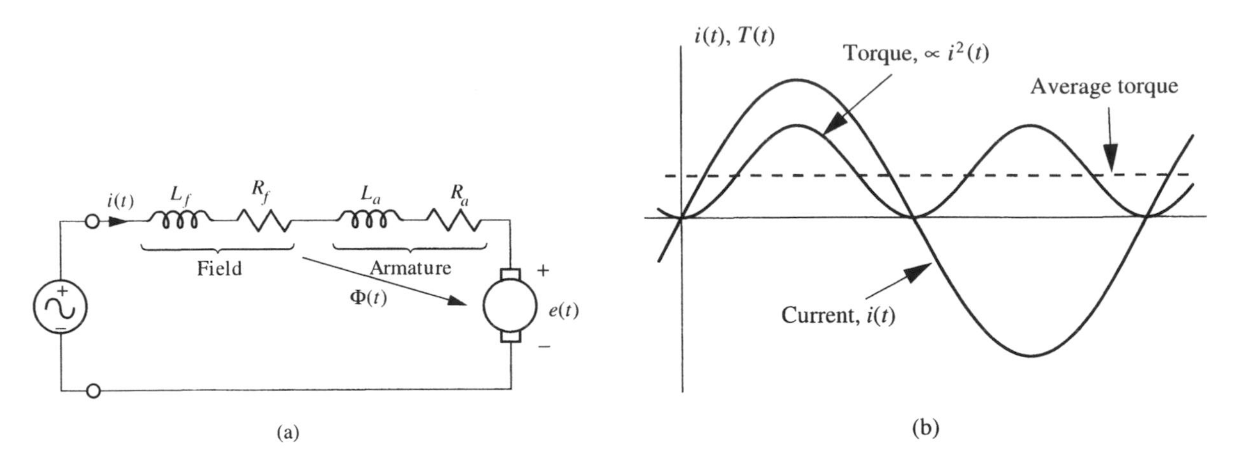

Figure 6.15 (a) Circuit model for a universal motor; (b) the ac current produces a time-average torque.

Applications. Although the machine operates on ac or dc, most universal motors are designed to operate only on ac. When you see an ac motor with a brush–commutator system, you recognize a universal motor. When an application requires speeds higher than 3600 rpm, a universal motor is required. Also important is the ease with which the speed of a universal motor can be controlled with a "dimmer" circuit.[19] For these reasons, universal motors are used for tools such as drills and routers and for household appliances such as mixers, blenders, and vacuum cleaners.

The speed–torque characteristic of the series motor serves many applications. For example, in a hand drill, we want a high speed for a small drill bit, but with the heavier load of a large bit, we would like a slower speed. The overall characteristic of the universal motor is to slow down and increase torque as the mechanical load increases and to allow stall at a moderate torque. The motor is, therefore, tolerant of a wide variety of load conditions. Finally, the universal motor gives more horsepower per pound than other ac motors because of its high speed, and thus universal motors are used for handheld tools such as drills, sanders, and saws of various types.

Check Your Understanding

1. A shunt-connected dc motor will reverse if the polarity of the input voltage is reversed. True or false?
2. A shunt-connected dc motor runs at 800 rpm. The field current is increased by 10% but the armature voltage remains constant. What is the new speed approximately?
3. A shunt-connected dc motor draws 10 A at 24 V. The motor torque drops to half its former value. What is the new input power?
4. A series-connected dc motor with constant input voltage will run with essentially constant (a) speed, (b) current, (c) power, (d) torque, or (e) none of these. Which?
5. A series-connected dc motor will reverse if the polarity of the input voltage is reversed. True or false?

[19] Discussed in Sec. 7.1.

6. Name the motor most likely to be used in an electric chain saw.
7. The starter motor in an automobile would be (a) shunt-connected, (b) series-connected, or (c) have a permanent magnet field. Which?

Answers. **(1)** False; **(2)** 727 rpm; **(3)** 120 W; **(4)** e; **(5)** false; **(6)** universal motor; **(7)** series-connected dc motor.

6.3 DYNAMIC RESPONSE OF DC MOTORS

System component. DC motors are frequently used in control systems. In this section, we analyze the performance of a dc motor as a system component, whether driven by a current source or a voltage source, using the methods and notation summarized in Appendix A. We treat only the separately excited motor with constant field current; hence, the armature circuit is analyzed for constant field flux. Figure 6.16 shows the armature circuit in the time domain with all time variables explicit.

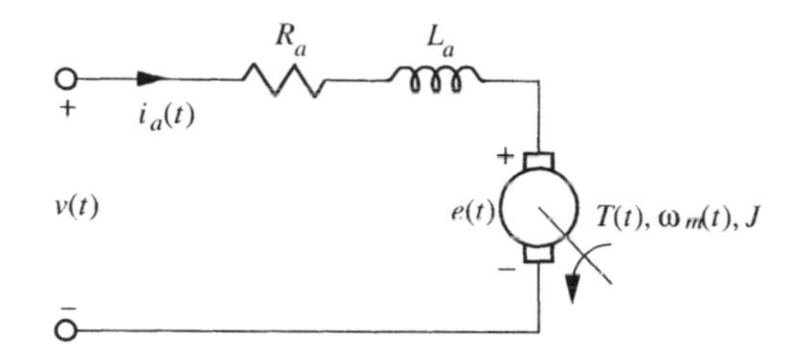

Figure 6.16 Armature circuit of a dc motor. The voltage, current, developed torque, and mechanical speed are functions of time.

Analysis of motor nameplate. The system model involves motor parameters that can be deduced from the nameplate information plus the armature inductance and total moment of the rotating system. We begin by analyzing the nameplate of a specific motor that we use in the numerical examples: 1 hp, 180 V, 4.9 A, 1.78 Ω armature resistance, 30-mH armature inductance, and 1750 rpm (183.3 rad/s). The armature moment of inertia is 5.9×10^{-4} kg-m²; we assume a total moment of twice this value, $J_T = 1.18 \times 10^{-3}$ kg-m².

Machine constant. The machine constant can be determined from the nameplate, *NP*, conditions

$$K\Phi = \frac{E_{NP}}{\omega_{mNP}} = \frac{V_{NP} - R_a I_{NP}}{\omega_{mNP}} \tag{6.46}$$

$$= \frac{180 - 1.78 \times 4.9}{183.3} = 0.935 \ \text{V}/(\text{rad/s})$$

Rotational losses. The rotational power losses are approximately proportional to speed, which means the loss torque is approximately constant. We may determine this loss torque from the nameplate conditions

$$P_{rot} = V_{NP} I_{NP} - R_a I^2_{NP} - P_{NP} \tag{6.47}$$

$$= 180 \times 4.9 - 1.78 \times (4.9)^2 - 746 = 93.3 \text{ W}$$

and thus the loss torque is

$$T_{loss} = \frac{P_{rot}}{\omega_{mNP}} = \frac{93.3}{183.3} = 0.509 \text{ N-m} \tag{6.48}$$

Load model. The dynamics of the system depend also on the load characteristics. We assume a load with a moment of inertia equal to that of the motor armature and a torque requirement proportional to speed. We assume further that the load requires nameplate power at nameplate speed. Thus

$$T_L(\omega_m) = K_L\omega_m \tag{6.49}$$

where the torque constant, K_L, can be determined from the nameplate conditions

$$K_L = \frac{T_{NP}}{\omega_{mNP}} = \frac{P_{NP}}{(\omega_{mNP})^2} = \frac{746}{(183.3)^2} = 2.22 \times 10^{-2} \tag{6.50}$$

Current-driven Motor Dynamics

LEARNING OBJECTIVE 4.

To understand how to develop system models for separately excited dc motors from nameplate and load characteristics

The Frequency Domain

System model. The system model for the current-driven dc motor results from combining motor equations with Newton's law. The developed torque is

$$T_{dev}(t) = K\Phi i_a(t) \quad \Rightarrow \quad \underline{\mathbf{T}}_{dev}(\underline{\mathbf{s}}) = K\Phi\underline{\mathbf{I}}_a(\underline{\mathbf{s}}) \tag{6.51}$$

The first equation is a time-domain equation; the second is the frequency-domain version. The transform of Eq. (1.48) is

$$\underline{\mathbf{s}}J_T\underline{\Omega}_m(\underline{\mathbf{s}}) + K_L\underline{\Omega}_m(\underline{\mathbf{s}}) = \underline{\mathbf{T}}_{dev}(\underline{\mathbf{s}}) - T_{loss} \tag{6.52}$$

where $\underline{\Omega}_m(\underline{\mathbf{s}})$ is the transform of the speed, and J_T is the moment of the rotating system. Equations (6.51) and (6.52) may be combined

$$(\underline{\mathbf{s}}J_T + K_L)\underline{\Omega}_m(\underline{\mathbf{s}}) = K\Phi\underline{\mathbf{I}}_a(\underline{\mathbf{s}}) - T_{loss} \quad \Rightarrow \quad \underline{\Omega}_m(\underline{\mathbf{s}}) = \frac{K\Phi\underline{\mathbf{I}}_a(\underline{\mathbf{s}}) - T_{loss}}{\underline{\mathbf{s}}J_T + K_L} \tag{6.53}$$

Equation (6.53) may be represented by the system diagram of Fig. 6.17. The input from mechanical losses is present if the motor is turning but would go away if the motor were at rest.

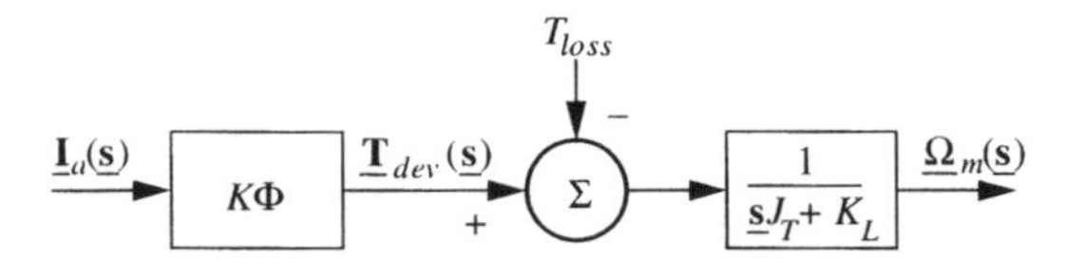

Figure 6.17 System diagram for armature rotational response. The input variable is the transform of armature current.

Time constant. Equation (6.53) and Fig. 6.17 show the current-driven motor to be a simple first-order system. The natural frequency, $\underline{\mathbf{s}}_n$, comes from the pole of the system function:

$$\underline{\mathbf{s}}_nJ_T + K_L = 0 \quad \Rightarrow \quad \underline{\mathbf{s}}_n = -\frac{K_L}{J_T} \quad \text{and} \quad \tau = \frac{J_T}{K_L} \tag{6.54}$$

where τ is the time constant.

EXAMPLE 6.11 Startup transient

Calculate the motor response from rest if nameplate current were suddenly applied to the motor just described.

SOLUTION:

Because this is a first-order system, we need only initial and final values plus the time constant. The initial speed is zero, and the final speed is the nameplate speed because the load requires nameplate torque at nameplate speed. The time constant from Eq. (6.54) is

$$\tau = \frac{1.18 \times 10^{-3}}{2.22 \times 10^{-2}} = 53.2 \text{ ms} \tag{6.55}$$

Thus, the start-up transient is

$$\omega_m = \omega_m(\infty) + [\omega_m(0) - \omega_m(\infty)]e^{-t/53.2 \text{ ms}} = 183.3(1 - e^{-18.8t}) \text{ rad/s} \tag{6.56}$$

which is shown in Fig. 6.18.

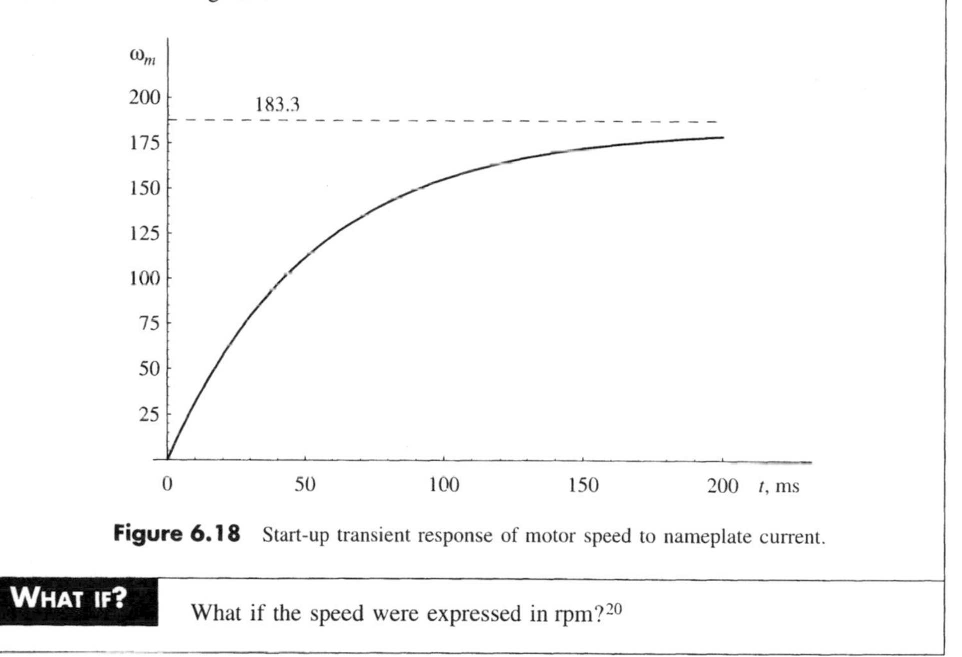

Figure 6.18 Start-up transient response of motor speed to nameplate current.

WHAT IF? What if the speed were expressed in rpm?[20]

Summary. With a current-driven motor, the armature circuit is open-circuited and hence the energy stored in the inductance of the armature is not a factor in motor dynamics.

Voltage-Driven Motor Dynamics

IDEA 7 The Frequency Domain

System function. We now consider a voltage-driven motor, in which stored electrical and mechanical energies interact. Kirchhoff's voltage law in the armature circuit of Fig. 6.16 is

[20] $1750(1 - e^{-18.8t})$ rpm.

$$v(t) = L_a \frac{di_a}{dt} + R_a i_a + e(t) \quad \Rightarrow \quad \underline{\mathbf{I}}_a(\underline{\mathbf{s}}) = \frac{\underline{\mathbf{V}}(\underline{\mathbf{s}}) - \underline{\mathbf{E}}(\underline{\mathbf{s}})}{\underline{\mathbf{s}}L_a + R_a} \tag{6.57}$$

where $e(t)$ is the emf and $v(t)$ the armature voltage. The frequency-domain version of KVL is solved for the armature current, which drives the motor response as shown in Fig. 6.17. Equation (6.57) may be modeled with a summer and a block representing the armature impedance, as shown in Fig. 6.19. Because the emf is proportional to motor speed, Eq. (6.5), we have

$$e(t) = K\Phi\omega_m(t) \quad \Rightarrow \quad \underline{\mathbf{E}}(\underline{\mathbf{s}}) = K\Phi\underline{\Omega}_m(\underline{\mathbf{s}}) \tag{6.58}$$

We may thus complete the system diagram of the voltage-driven motor with a feedback loop,[21] as shown in Fig. 6.19.

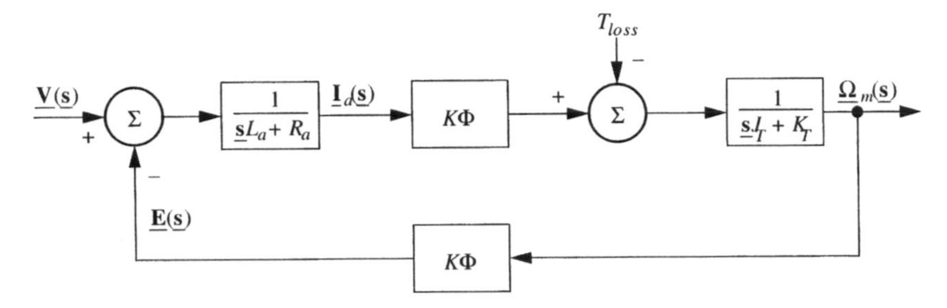

Figure 6.19 System diagram for a dc motor that is voltage-driven. The motor contains an inherent feedback loop.

Dynamic response. The system now is second-order. We may determine the natural frequencies of the system from the loop gain, which is

$$\underline{\mathbf{L}}(\underline{\mathbf{s}}) = -\frac{(K\Phi)^2}{(\underline{\mathbf{s}}L_a + R_a)(\underline{\mathbf{s}}J_T + K_L)} \tag{6.59}$$

According to feedback theory the natural frequencies of the system come from

$$1 - \underline{\mathbf{L}}(\underline{\mathbf{s}}) = 0 \quad \Rightarrow \quad (\underline{\mathbf{s}}L_a + R_a)(\underline{\mathbf{s}}J_T + K_L) + (K\Phi)^2 = 0 \tag{6.60}$$

The nature of the response depends on the parameters of the motor and load.

EXAMPLE 6.12 System response

Determine the nature of the response of the motor/load system in Example 6.11 to the sudden application of nameplate voltage.

SOLUTION:

Equation (6.60) is

$$(0.03\,\underline{\mathbf{s}} + 1.78)(1.18 \times 10^{-3}\,\underline{\mathbf{s}} + 2.22 \times 10^{-2}) + (0.935)^2 = 0 \tag{6.61}$$

which has the solutions $\underline{\mathbf{s}}_n = -39.1 \pm j156 \text{ s}^{-1}$. Thus, the response is a damped sinusoid.

[21] This feedback loop is inherent to the motor and is not something we have added to the motor for control purposes.

Dynamic response. We now know that the response is underdamped and can work out the total solution. The form of the answer is

$$\omega_m(t) = \omega_m(\infty) + Ae^{-\sigma t}\cos(\omega t + \theta) \tag{6.62}$$

where $\underline{s}_n = \sigma \pm j\omega$ are the natural frequencies of the underdamped system, given in Example 6.12. The initial condition is zero speed, and the initial derivative of the speed must also be zero because the current, which generates the torque, is zero at $t = 0^+$.

$$\omega_m(0) = 0 = \omega_m(\infty) + A\cos\theta \tag{6.63}$$

and

$$\left.\frac{d\omega_m(t)}{dt}\right|_{t=0} = 0 = A(-\sigma\cos\theta - \omega\sin\theta) \tag{6.64}$$

which are readily solved for the amplitude and phase of the start-up transient.

EXAMPLE 6.13 Transient response

Determine the response of the voltage-driven motor if nameplate voltage is applied suddenly to the at-rest motor/load system.[22]

SOLUTION:

The final value is nameplate speed, 183.3 rad/s. From Eq. (6.64), the phase is

$$\tan\theta = -\frac{\sigma}{\omega} = -\frac{39.1}{156} \Rightarrow \theta = -14.1^\circ \tag{6.65}$$

and from Eq. (6.63), the amplitude is

$$A = -\frac{183.3}{\cos 14.1^\circ} = -188.9 \text{ V} \tag{6.66}$$

The transient response is

$$\omega_m(t) = 183.3 - 188.9e^{-39.1t}\cos(156t - 14.1^\circ) \text{ V} \tag{6.67}$$

which is shown in Fig. 6.20. The motor overshoots and oscillates around its final speed before settling down to steady state.

[22] Applying nameplate voltage to an at-rest system is not generally recommended, as it causes excessive currents and can damage the motor. We use nameplate voltage for convenience because we know the final speed to be nameplate speed. The nature of the response is independent of the magnitude of the input change.

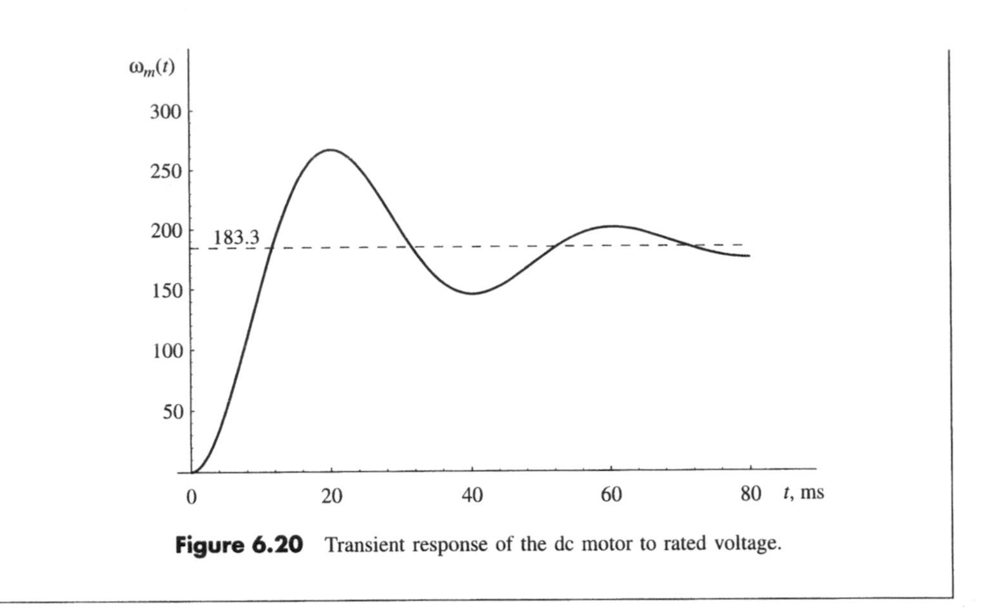

Figure 6.20 Transient response of the dc motor to rated voltage.

Check Your Understanding

1. If the load required a constant torque, the current-driven motor would reach steady-state immediately. True or false?
2. If the load required a constant torque, the voltage-driven motor would reach steady-state faster. True or false?
3. Changing the motor field current would have no effect on the dynamic response of the motor. True or false?

Answers. (1) False; (2) true; (3) false.

CHAPTER SUMMARY

The dc motor is important for portable and control applications. In the dc motor, the motor flux is stationary in space and the rotor rotates physically with respect to its flux by means of a synchronizing switch, the commutator. The characteristics of the motor can be changed markedly by connecting the field in series or parallel with the armature. The voltage-driven armature exhibits a second-order behavior in control applications.

Objective 1: To understand the physical basis for the equivalent circuit of a dc motor. The field of the dc motor is on the stator and thus the magnetic flux of the machine is stationary in space. The rotor is the armature, and currents are brought through a brush-commutator switch that synchronizes flux direction with rotation.

Objective 2: To understand the speed–torque characteristic of the shunt-connected dc motor. The shunt-connected and permanent-magnet-field dc motors maintain fairly constant speed as output torque varies widely. If the field is separately connected, the speed can be controlled by the armature voltage, or the torque can be controlled by the armature current.

Objective 3: To understand the speed–torque characteristic of the series-connected dc motor. With the field in series with the armature, the dc motor speed depends on output torque. The series-connected motor is called a universal motor when operated with ac input power. Series-connected motors find applications in automotive starters, hand-held tools, and household appliances.

Objective 4: To understand how to develop system models for separately excited dc motors from nameplate and load characteristics. System models for the separately excited dc motor are derived from the analysis of the nameplate information, supplemented by knowledge of the rotating moment of inertia and the load-torque characteristics. We examine the dynamic responses of current-driven and voltage-driven armatures.

Electronic drive of dc motors is described in Chap. 7. We show that unfiltered rectifiers can drive dc motors, but motor characteristics differ from motors driven by pure dc sources.

GLOSSARY

Air-gap line, p. 262, a linear approximation to the magnetization curve, used to derive approximate motor characteristics.

Alnico, p. 255, a tradename of Alcoa for aluminum–nickel–cobalt, from which many permanent magnets are fashioned.

Brushes, p. 256, stationary conductors that rub against the commutator as the rotor rotates and makes electrical connection with the rotor.

Commutator, p. 256, a cylindrical surface of wedge-shaped conductors connected to the rotor windings. The commutator is part of the rotor and participates in its rotation.

Electromotive force, p. 259, the voltage induced in the rotor by its motion in the stator flux.

Magnetization curve, p. 261, the open-circuit voltage (emf) versus field current for a dc machine driven externally at a fixed speed.

Parallel-excited motor, p. 264, a dc motor in which the field power is derived from the armature circuit in a parallel connection, also called shunt excited or shunt connected.

Rheostat, p. 264, a variable resistor used to control current.

Salient poles, p. 254, iron protrusions from the stator with windings to produce the stator field.

Separately excited motor, p. 258, a dc motor in which the field power is derived from a source independent of the armature circuit.

Series excited motor, p. 269, a dc motor in which the field power is derived from the armature circuit in a series connection.

Universal motor, p. 271, a series-connected motor that will run off dc power but normally is run off ac power.

PROBLEMS

Section 6.1: Principles of DC Machines

6.1. A dc motor is shown in Fig. P6.1. The stator coil has 200 turns per pole and carries 7 A dc in the direction shown. The rotor is turning counterclockwise, as shown, and the rotor currents come out on the right and go in on the left, as shown. The width of the air gap is 1.5 mm on each side and the iron has $\mu_i = \infty$.

(a) Determine the direction of the stator flux and mark the poles N and S accordingly on the stator poles.

(b) Find the flux density in the air gap due to the field current. Assume a linear magnetic structure.

(c) What is the direction of the magnetically generated torque on the rotor?

(d) Is the machine acting as a generator or a motor?

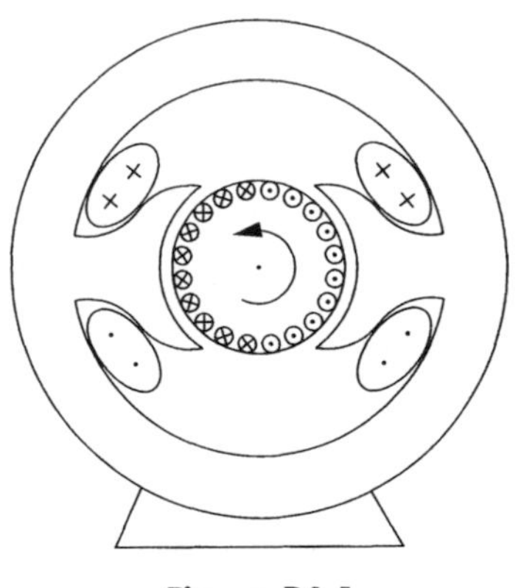

Figure P6.1

6.2. The magnetic structure for a dc motor is shown in Fig. P6.2. Each field pole has 1000 turns, the rotor radius and length are 10 cm, and the gap between rotor and stator is 2 mm wide. The rotor is turning at 5000 rad/s in the direction shown. The pole faces cover 75% of the rotor circumference. The rotor current is 10 A, and there are 18 rotor conductors, as shown. The air-gap flux density is 1.6 T. Assume a linear magnetic structure.

(a) What is the field current?

(b) Estimate the developed torque.

(c) Is the machine acting as a motor or generator? How much power is being converted between electrical and mechanical form?

(d) Estimate the armature voltage.

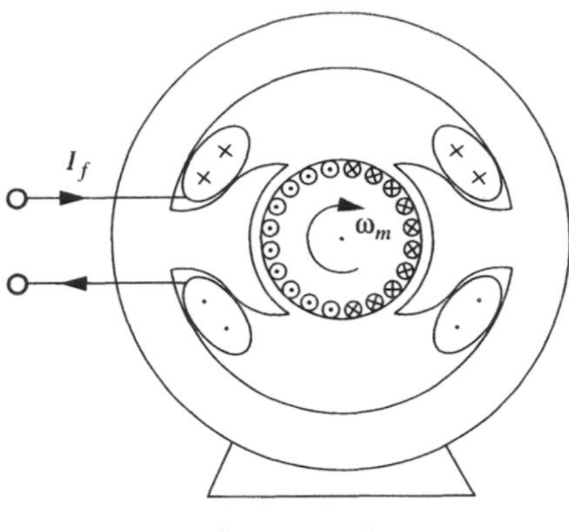

Figure P6.2

6.3. A dc motor is separately excited, that is, has independent control over armature and field. The ratings are P watts, n rpm, I_a amperes, V volts on the armature, T newton-meters, and I_f amperes of field current. Assuming a linear magnetic structure and ignoring losses, fill in Table P6.3.

TABLE P6.3

Field Current	*Voltage*	*Current*	*Speed*	*Torque*	*Power*
I_f	V	I_a	n	T	P
I_f	V			$\frac{1}{2}T$	
$\frac{1}{2}I_f$	V	I_a			
$2I_f$	$\frac{1}{2}V$	$\frac{1}{2}I_a$			
I_f	$2V$				P

6.4. For the motor model in Fig. 6.7, the input voltage and current are 24 V and 10 A, respectively, the armature resistance is 0.10 Ω, the field voltage is 24 V, the field current is 1.0 A, the rotational losses are 10 W, and the motor speed is 1200 rpm. Find the following:

(a) The input power, including that required for the field current.

(b) The emf.

(c) Developed power.

(d) The output power.

(e) The output torque.

(f) The efficiency.

(g) What is the machine constant in volts/(radian per second)?

6.5. An 80-V, 25-A dc motor is separately excited from a constant voltage, and thus has constant field current. The armature resistance is 0.100 Ω. The motor is used to hoist an elevator weighing 500 lb (Fig. P6.5). Neglect rotational losses throughout this problem.

(a) What is the lifting rate in feet per second with the motor drawing nameplate current?

(b) The motor voltage is then reversed in polarity to lower the load. The voltage is adjusted to lower at the same rate as in part (a), again with nameplate current. What is the required voltage?

(c) What percentage of the energy used to raise the load is then returned to the electrical supply in lowering the load? Neglect rotational loss and the effects of acceleration and deceleration.

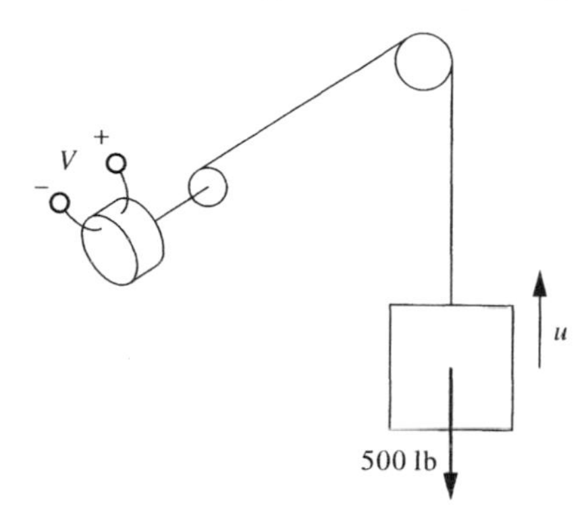

Figure P6.5

6.6. A separately excited dc motor has the following nameplate information: 180 V, 9.5-A armature current, 1150 rpm, 1.03-Ω armature resistance, and 2 hp. Determine the no-load speed of the motor for the same field current, assuming the rotational losses are proportional to the motor speed.

6.7. An 80-V dc motor has a constant field flux (separately excited) and a nameplate speed of 1150 rpm with 710-W output power. The nameplate armature current is 10 A and the no-load current is 0.5 A. Assume constant rotational loss.

(a) Estimate the rotational loss.

(b) Determine the armature resistance.

(c) What is the no-load speed in rpm?

(d) Find the machine constant in V/(rad/s).

(e) Find the efficiency at nameplate load, including field-circuit losses of 30 W.

Section 6.2: Characteristics of DC Motors

6.8. A separately excited dc motor runs 1000 rpm for a line voltage of 120 V dc, a field current of 2 A, and an output power of 2 hp. Give the effect on motor speed (speed up or slowdown, and whether the change is large or small) of the following changes, made one at a time and then restored before the next change is made. Assume negligible armature resistance and field flux proportional to field current.

(a) The voltage changes to 60 V.

(b) The output power doubles.

(c) The output torque doubles.

(d) The field current doubles.

6.9. The circuit model for a shunt-excited dc motor is shown in Fig. P6.9. The machine constant is 0.2 V/(rad/s). Find the speed in rpm for a developed power of 500 W.

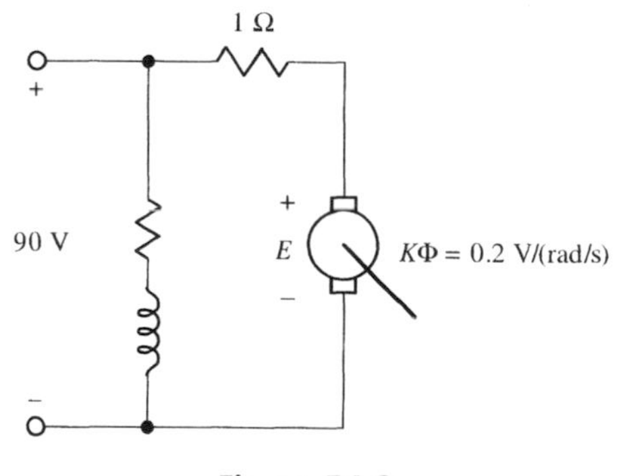

Figure P6.9

6.10. A dc motor has the following nameplate information: 5 hp, 180 V, 25 A, 1200 rpm, and 0.25 Ω armature resistance. Assume rotational losses are proportional to speed. The motor is separately excited at nameplate field and must drive a load having a requirement $T_L(n) = 15 + (n/1200)^2$ N-m, where n is the speed in rpm. Find the speed at which the system operates and the input current under this load.

6.11. A 1.5-hp shunt-excited dc motor has the following nameplate information: 180 V, 2500 rpm, 7.5 A, $R_a = 0.563\ \Omega$, $L_a = 12$ mH, $I_f = 0.56$ A, and $R_f = 282\ \Omega$. Assume rotational losses are constant. Consider the field current constant throughout the entire problem.

(a) Find the total losses of the motor at nameplate operation. These are losses in the physical motor and do not include losses associated with external circuitry.
(b) Find the rotational losses at nameplate operation.
(c) Find the required current for a developed power of 1.2 hp with $V = 180$ V.
(d) Find the output power if the developed power is 1.2 hp with $V = 180$ V.
(e) Find the required input voltage for a no-load speed of 2800 rpm.

6.12. For the motor in Problem 6.8, the armature resistance is 0.8 Ω and the rotational loss is 30 W, assumed constant.
(a) Find the armature current for 2 hp out.
(b) Find the no-load speed.

6.13. A shunt-connected dc motor operates from 24 V and has an armature resistance of 0.30 Ω. The rotational losses are 5% of the output power. The armature current is 10 A and the speed is 1200 rpm.
(a) Find the input power. Ignore field losses.
(b) Find the output power in horsepower.
(c) Find the machine constant $K\Phi$.
(d) Approximate the no-load speed in rpm.

6.14. The nameplate information on a dc motor is the following: armature 2 hp, 180 V, 9.5 A, 1150/1380 rpm (with reduced field), 1.03 Ω, 28 mH; field 120 V, 0.76 A for 1150 rpm. The motor is externally excited for operation at nameplate conditions, that is, with 0.76 A field current.
(a) Estimate the field current at 1380-rpm operation with nameplate conditions (hp and armature voltage).
(b) Find the efficiency at nameplate conditions, including the field loss.
(c) Determine the rotational losses at nameplate conditions.
(d) Find the no-load speed assuming constant loss torque.
(e) Find the no-load armature current, assuming constant loss torque.
(f) Find the speed at which the motor output is 1 hp, assuming constant loss torque.

6.15. A shunt-excited dc motor has an armature resistance of 0.5 Ω, a field resistance of 100 Ω, and an applied voltage of 90 V. The magnetization curve is shown in Fig. P6.15. Neglect rotational losses.
(a) Find the no-load speed of the motor.
(b) What input current is required for 10 N-m of developed torque?

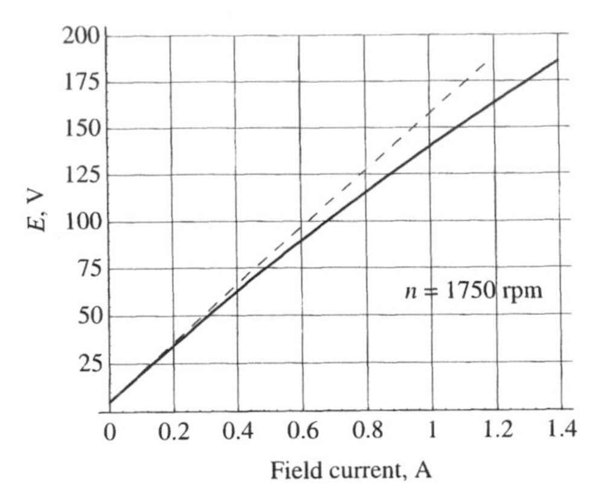

Figure P6.15

(c) Determine the efficiency of the motor with 2 hp out, including field losses.

6.16. A shunt-connected, 1-hp, 180-V dc motor puts out nameplate power at 1800 rpm but runs at 1850 rpm at no load. The rotational loss is a constant 50 W and field resistance is 810 Ω.
(a) Draw a circuit of the motor.
(b) What is the field current?
(c) Determine the armature resistance.
(d) Find the input current to the armature at nameplate load.
(e) Find the efficiency, including field losses.
(f) Find the torque out at nameplate load.
(g) What is the machine constant in N-m/ampere?

6.17. A dc motor has the following nameplate information: 1.5 hp, 1750 rpm, 180 V in the armature, 7.3 A in the armature, 1.05-Ω armature resistance, 180 V for the field, and 0.55 A for the field. The motor is shunt-connected. Assume constant rotational losses in this problem.
(a) Find the rotational losses at 1750 rpm.
(b) Find the developed torque at 1750 rpm.
(c) Determine the no-load speed.

6.18. A 180-V, shunt-excited dc motor runs 1150 rpm at nameplate load with 2 hp out and 1190 rpm at no load. Estimate the armature resistance. State assumptions.

6.19. For the motor with the magnetization curve shown in Fig. P6.19, find the voltage required for 5 hp out at 1000 rpm in shunt-connected operation with $I_f = 1.5$ A. The rotational losses at 1000 rpm are 100 W, and $R_a = 0.8$ Ω.

6.20. For the motor with the magnetization curve shown in Fig. P6.19, determine the developed torque at the following currents:

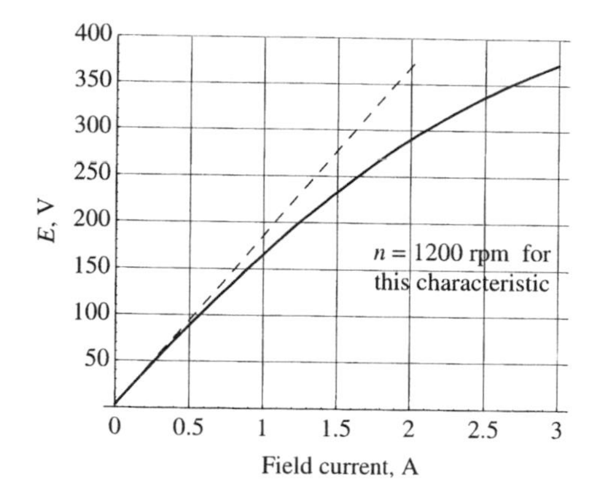

Figure P6.19

(a) $I_f = 2$ A, $I_a = 15$ A.
(b) $I_f = 2$ A, $I_a = 10$ A.
(c) $I_f = 1$ A, $I_a = 10$ A.

6.21. A 5-hp, shunt-connected, 180-V dc motor has 0.25-Ω armature resistance and 50-W rotational loss at the nameplate speed of 600 rpm. The field current is 0.5 A.

(a) What is the developed torque at the nameplate output power of 5 hp?
(b) What is the efficiency at 5 hp output, including field losses?
(c) What is the no-load speed? Assume the same rotational losses.

6.22. A dc motor has the following nameplate information: armature 180 V, 7.3 A, 1.5 hp, 1750 rpm, and resistance 1.05 Ω; field 100 V, and 0.5 A.

(a) Draw the circuit for parallel excitation, including the field rheostat.
(b) What value of the field rheostat resistance gives the nameplate field current?
(c) Determine the rotational losses at nameplate speed.
(d) Determine the efficiency at nameplate conditions.
(e) The motor has a load described by $T_L(n) = 2 + 0.002n + 10^{-6}n^2$ N-m. Find the operating speed assuming rotational losses are proportional to speed.

6.23. A dc motor has the following nameplate information: armature 50 hp, 50 V, 80 A, 0.18 Ω, 8.2 mH, and 1200 – 1350 rpm; field 200 V, 10 A (for 1200 rpm), and 15-H inductance. Assume rotational losses proportional to speed.

(a) Draw a circuit of the motor, connected for parallel excitation. Determine the range of the field rheostat for field speed control, assuming a linear magnetic field.
(b) Find the efficiency at nameplate conditions, 1200 rpm.
(c) Find the formula for the rotational losses as a function of speed in rpm.
(d) Find the current to the armature for 20 hp output at 300 rpm with a 10-A field current. The voltage is reduced for this operation.
(e) Would the motor operate successfully for long periods of time under the conditions of part (d)? Explain.
(f) If the machine's speed is up to 1350 rpm by reduction of the field current, what is the maximum output power possible at this speed?

6.24. Consider a dc motor with the following nameplate information: armature 2500 rpm, 1.5 hp, 180 V, 7.5 A, 0.563 Ω, and 12 mH; field 0.56 A (2500 rpm), 0.43 A (2750 rpm), 282 Ω, and 85 H. Assume the motor is separately excited at the nameplate field current (2500 rpm) and has a true 180 V (dc) on the armature throughout the problem.

(a) Determine the rotational loss at nameplate conditions.
(b) Find the motor current for 1 hp output, assuming constant rotational loss.
(c) Find the motor speed in rpm for the condition in part (b).
(d) Suppose you wanted the motor to be running 2000 rpm at 1 hp output, and you add a resistor in the armature circuit to slow down the motor. What value of resistance should you use?

6.25. A dc motor has the following nameplate armature information: 500 V, 40 hp, 66.2 A, and 1750 rpm. Field nameplate information is 240 V and 3.0 A. The motor is to be shunt-connected. Assume that armature-copper losses are equal to rotational losses under nameplate condition.

(a) Draw a circuit diagram for the motor and put in a field rheostat resistance for nameplate condition.
(b) Determine the total efficiency of the motor under nameplate conditions.
(c) Find the no-load speed, assuming rotational losses are proportional to the speed.
(d) Determine the output torque if the output power is 20 hp, assuming rotational losses are proportional to the speed.

6.26. A dc motor has the following nameplate: armature 1 hp, 1150/1380 rpm (with reduced field), 5.0 A, 180 V, 2.43 Ω, and 0.049 H; field: base-speed field current = 0.54 A; top-speed field current = 0.33 A, 200 V, 282 Ω, and 86 H.

(a) If the field were really operated from 200 V dc, what range (max. and min.) of series resistance would have to be added to give the stated range of field current?

(b) Find the rotational losses of the armature.

(c) Find the no-load speed of the motor.

(d) Assuming that rotational power losses are proportional to speed, find the output torque as a function of speed in rpm.

(e) The field current is adjusted for 0.54 A and the armature voltage is 180 V. Find the input current if the motor drives a load with the torque requirement $T_L(n) = 2 + n/240$ N-m.

6.27. We have two motors. Motor A: three-phase induction motor, 60 Hz, 1 hp, 230/460 V, 3.4/1.7 A, 1735 rpm, 13-lb-ft locked rotor torque, 18-lb-ft breakdown (maximum) torque, 25/12.5-A locked rotor current, 0.968-lb-ft^2 rotor moment, 82% efficient, and service factor 1.15. A no-load test was performed on the motor and the speed of the motor was observed to be 1799.1 rpm, the input current to be 2.8 A, and the input power 95 W. Assume that mechanical losses are constant. Motor B: dc motor, 1 hp, 1750 rpm, 180 V, 4.9 A, 1.78-Ω armature resistance, 30-mH armature inductance, 0.14-lb-ft^2 rotor moment, 200-V field voltage, and 0.78-A field current. Motor A and motor B, both turning in the same direction, have their shafts connected together, so that they must turn at the same speed. Both are operated at nameplate voltage, and the dc motor is operated with nameplate field. Assume rotational losses in the dc motor are proportional to speed.

(a) Find the resulting speed.

(b) Find the power going down the shaft and the direction it goes, that is, which motor is the source and which the load.

6.28. A shunt-connected dc motor has an armature resistance of 0.2 Ω, a line voltage of 300 V, and a field resistance of 100 Ω. The magnetization curve is shown in Fig. P6.19. Ignore rotational losses.

(a) At what speed would the motor run with no load?

(b) A load is applied and the speed drops 5%. Find the input current and output power in horsepower.

(c) With the same torque requirement as in part (b), the input voltage is dropped to 250 V. Find the new speed.

6.29. A 12.6-V permanent-magnet-field dc motor has negligible rotational losses. The stall ($n = 0$ rpm) current is 30 A. Find the maximum output power (watts) this motor can produce when supplied with nameplate voltage.

6.30. A dc motor with a permanent-magnet-field structure has the characteristics shown in Table P6.30.

TABLE P6.30

Condition	*Armature V*	*Armature I (A)*	*Speed (rpm)*
No load	12.6	0.37	1210
Full load	12.6	5.85	627

(a) Find the armature resistance and the rotational losses at no load. Assume rotational losses are proportional to speed.

(b) Find the full-load power.

6.31. A 12.6-V, permanent-magnet-field dc motor is used to power a window lift in an automobile. The motor requires 10.2 A and runs at 1180 rpm when lifting the window, but requires only 7.6 A and turns at 1220 rpm when lowering the window (with reversed voltage, current, and direction of rotation). Assume friction and rotational losses are proportional to speed and hence can be represented by a constant loss torque.

(a) Determine the armature resistance.

(b) Determine the torque required to lift the window, excluding the effects of friction.

6.32. A 90-V, permanent-magnet-field dc motor has an armature resistance of 1.0 Ω and draws 10 A at 5000 rpm. Rotational loss at this speed is 100 W and is proportional to the speed.

(a) Find the output power at 5000 rpm.

(b) Find the no-load speed if the input voltage is decreased to 85 V.

(c) Find the speed at which the output power of the motor is maximum with 90-V input.

6.33. A 12.6-V window-lift motor on an automobile has a permanent-magnet field and runs 800 rpm raising the window with a current of 8.1 A. The stall current (motor is stopped, with the window fully up) is 22 A. The torque required to lower the window is 80% of that required to lift the window because the weight of the window is aiding the descent. Find the current drawn in lowering the window and the corresponding speed. Ignore rotational losses in the armature.

6.34. A permanent-magnet-field, window-lift motor for an automobile runs at 2000 rpm and draws 10 A from a 12.6-V battery source. The stall current of the motor is 25 A.

(a) Find the developed power at 2000 rpm.

(b) What resistance must be placed in series with the motor to give 1000 rpm if the power required by the load, including rotational losses, is 61% that for 2000 rpm?

6.35. An engineer purchased a dc motor with a permanent-magnet field. The nameplate gives 90 V and 9.2 A but does not give the rated power. The engineer applied 90 V to the unloaded motor and measured the input current (0.5 A) and speed (1843 rpm). The engineer then loaded the motor mechanically until the current reached 9.2 A and measured the speed (1750 rpm). The engineer assumed that rotational losses were constant, and determined from these data the output power at nameplate load. What was the result?

6.36. Assume a series-wound dc motor to have negligible rotational losses. The nameplate voltage is 24 V, the nameplate current 20 A, and the efficiency and speed at nameplate load are 78% and 3000 rpm, respectively.

(a) Find the combined armature and field resistance.

(b) Find the speed for an output power of 1/4 hp.

(c) Find the stall torque of the motor, assuming a 50% decrease in the magnetic flux (from the ideal value) due to magnetic saturation.

6.37. The circuit of Fig. P6.37 shows a series-excited motor. Ignore rotational losses.

(a) Find the current for 1 hp out.

(b) If the load torque is decreased by a factor of 2, what is the new current, assuming no magnetic saturation?

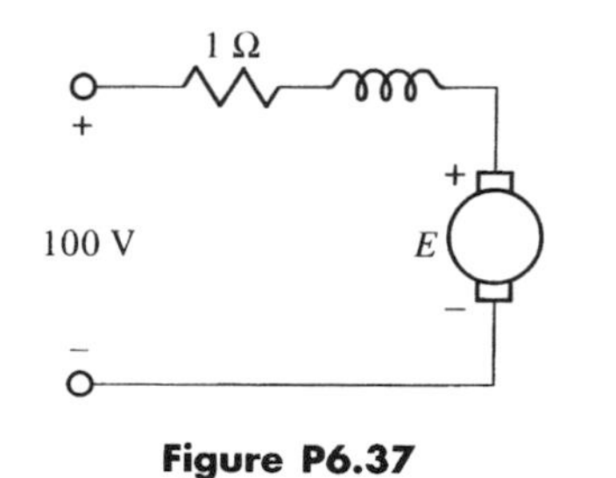

Figure P6.37

6.38. A series motor operates on 12.6 V and has a combined field and armature resistance of 0.4 Ω. At 2000 rpm, the input current is 6.8 A. Ignore rotational losses.

(a) Find the output torque at 2000 rpm.

(b) Find the speed at which the motor produces twice the torque found in part (a).

(c) Find the input current for an output power of 30 watts.

6.39. A series-connected dc motor runs at 1200 rpm with an input voltage of 180 V and an output power of 1 hp. Ignore all losses in this problem. Fill in Table P6.39.

TABLE P6.39

Voltage (V)	*Power (hp)*	*Speed (rpm)*
180	1	1200
120	$\frac{1}{2}$	
120		1800
	1	800

6.40. An 80-V series-excited dc motor has an armature resistance of 1 Ω and a field resistance of 1 Ω. The motor turns 10,000 rpm with a load of 3/4 hp. Ignore rotational losses.

(a) Find the motor current for 3/4 hp out.

(b) Give an equation for the emf of the motor as a function of speed in rpm and current.

(c) At what speed is the power out of the motor a maximum?

6.41. An 80-V, series-excited dc motor draws 8 A at an output torque of 6 N-m. Find the torque at a speed of 1200 rpm. Ignore losses and magnetic nonlinearities.

6.42. A series-connected dc motor has a rotational power loss that is proportional to speed. Assume the armature resistance is negligible. A test of the output torque versus speed produces the data shown in Table P6.42.

TABLE P6.42

Speed (rpm)	*Output Torque (N-m)*
8000	No load
6000	0.86
4000	3.3
3000	6.7
2000	8.2
1000	10.4
0	11.6

(a) Find the loss torque of the motor.
(b) Estimate the maximum power out of the motor.

6.43. A 24-V, series-excited dc motor has a combined armature and field resistance of 1.2 Ω. The stall torque of the motor is 10 N-m, and the no-load speed is 10,000 rpm. Find the rotational losses at the no-load speed, assuming no magnetic saturation.

6.44. A 60-V, series-excited dc motor has a speed–torque characteristic as shown in Fig. P6.44. The motor has a loss torque of 3 N-m, which is constant over the range of speeds shown, except at rest and at extremely low speeds. At speeds below about 80 rad/s, the current is so high in the motor that magnetic saturation is a strong factor in establishing the torque.

(a) Estimate the factor K' based upon the information supplied.
(b) Estimate the stall current, the current that would result if the motor were stopped with the nameplate voltage applied.
(c) Estimate the maximum power out of this motor and the speed at which it occurs.

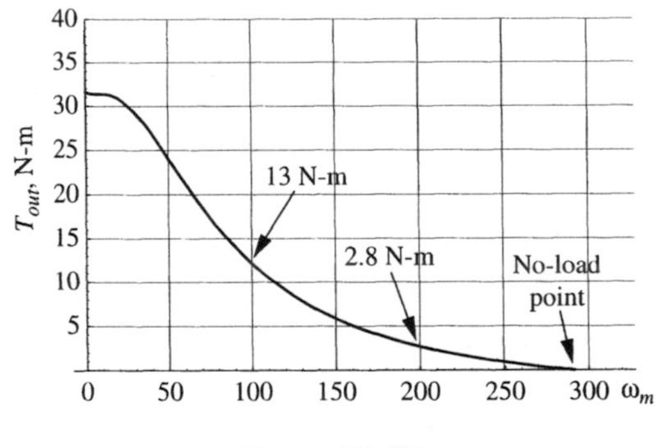

Figure P6.44

6.45. A series-connected dc motor has a combined field and armature resistance of 0.5 Ω, an input voltage of 24 V dc, and runs at 5000 rpm with 1/4-hp output power. Ignore rotational losses in this problem.

(a) Find the output torque.
(b) Find the input current.
(c) At twice the input current for part (b), what is the speed? Assume that magnetic flux is proportional to current.

6.46. A dc series-connected motor drives a load whose input torque requirement is proportional to speed. The combined armature and field resistance is 3 Ω. With an input voltage of 100 V dc, the motor input power is 50 W and the output speed of the load is 100 rpm. Derive the equation for the voltage required for other speeds. Assume that the flux of the field is proportional to the current. Ignore rotational losses.

6.47. Does the output power of a series-excited dc motor increase, decrease, or remain the same as load torque increases? Explain.

6.48. A 12-V, series-connected dc motor has rotational losses that are proportional to speed. At no-load, the motor runs at 10,000 rpm and draws 0.83 A. Find the current and speed for an output power of 50 W. Assume no armature or field resistance.

6.49. A 90-V, series-connected dc motor has a no-load speed of 20,000 rpm. Its nameplate output power of 1.5 hp occurs at a speed of 10,000 rpm. Assume rotational loss is proportional to speed, but assume electrical losses are negligible. Find the no-load current.

6.50. A 24-V, series-excited dc motor has a current-vs.-speed curve given by Fig. P6.50. Find the range of torques developed in the normal operating region. Assume that magnetic saturation reduces the *starting* torque by 20%.

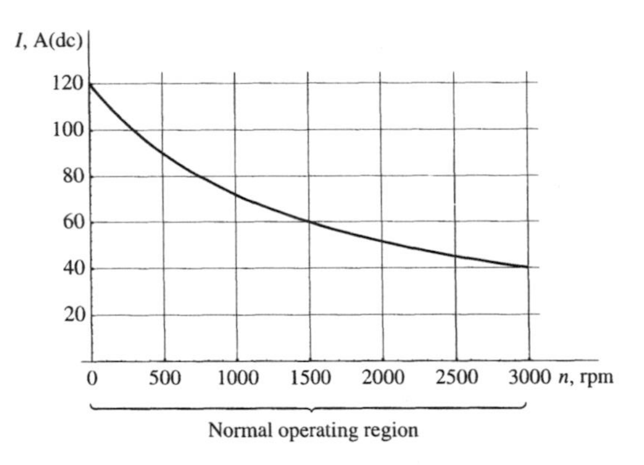

Figure P6.50

6.51. A hand-held drill powered by a universal motor runs at 1200 rpm with no load. With a 3/8-in. drill bit in operation, the drill slows down to 900 rpm. Ignore electrical loss and also the effects of inductance. Mechanical loss is assumed to be proportional to speed. Determine at the slower speed the following ratio: the output power to the bit divided by the rotational loss at the lower speed.

6.52. A 24-V, series-connected dc motor has a combined armature and field resistance of 2 Ω. The motor runs at 5000 rpm with a developed power of 22 W. Find the maximum power that the motor can develop and the associated speed. Assume that flux is proportional to current.

6.53. A 250-V dc motor is equipped for speed control with both a field rheostat in the shunt field (0 to 150 Ω) and a series armature rheostat (0 to 5 Ω). The resistance of the shunt field is 100 Ω and the armature resistance is 0.5 Ω. Figure P6.53 shows the motor circuit. The rotational losses are 300 W at 5000 rpm and are proportional to speed. Assume the field is proportional to field current.

(a) With both rheostats set to zero, the motor draws a total input current of 18.5 A at 5000 rpm. What is the output power?

(b) Find the minimum no-load speed and the combination of field and armature rheostat resistances for minimum speed.

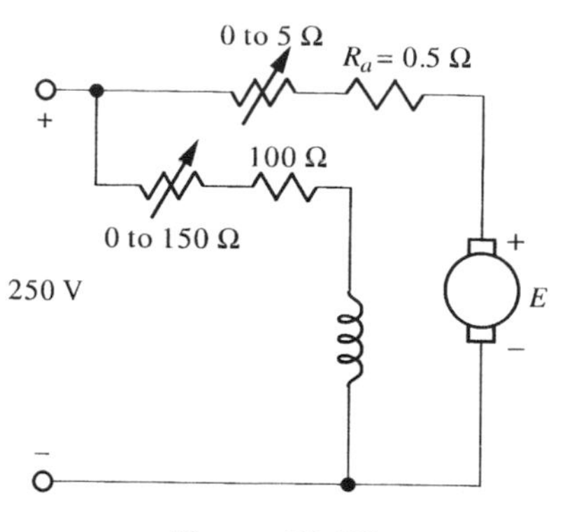

Figure P6.53

(c) Find the maximum no-load speed and the combination of field and armature rheostat resistances for maximum speed.

Section 6.3: Dynamic Response of DC Motors

6.54. A dc permanent magnet motor has 0.8-Ω armature resistance and negligible inductance. The machine constant is $K\Phi = 0.12$ V/rpm. The motor is operated at 24 V with a load, inducing rotational losses of $T_L = 0.4 + 0.1\omega_m$, and the total moment of the system is 0.01 kg-m^2.

(a) Find the steady-state operating speed in radians/second with 24 V applied to the armature.

(b) With the motor running at the speed calculated in part (a), if the 24 V were suddenly removed and the armature circuit open-circuited, how long would it be before the motor/load would come to rest?

(c) With the motor running at the speed calculated in part (a), if the 24 V were suddenly removed and the armature circuit short-circuited, how long would it be before the motor/load would come to rest?

6.55. A dc motor has the following nameplate information: 3500 rpm, 15 hp, 54 A, 230 V, 0.153-Ω armature resistance, 1.1-mH armature inductance, and 0.068-kg-m^2 moment of inertia. The field is separately excited at the nameplate field current. The nameplate voltage is suddenly applied to the armature. Show that the armature-current-response transfer function is second-order and determine the natural frequencies. The motor is unloaded and rotational losses are proportional to speed.

6.56. A separately excited dc motor has the following nameplate information about the armature: 180 V, 1.05 Ω, 22 mH, 1750 rpm, 1.5 hp, and 7.3 A. The field is excited at the nameplate value. The load-torque requirement is $T_L = \omega_m/20$ N-m and the total moment of the system is 0.01 kg-m^2. Assume that rotational losses are proportional to speed.

(a) If the armature is suddenly excited at the nameplate current, find the time constant of the response and the speed as a function of time.

(b) If the armature is suddenly excited at the nameplate voltage, find the natural frequencies of the response and the speed as a function of time.

6.57. A small permanent-magnet dc motor has an armature inductance of 5 mH and negligible resistance. Assume the load torque is proportional to speed. The motor has negligible inertia and runs at 1000 rpm with an input voltage of 12 V and a current of 3 A.

(a) Make a system diagram describing the operation of the motor in the frequency domain. Use numerical values in the boxes where possible.

(b) Using feedback theory, determine the transfer function of the motor, $\underline{\mathbf{M}}(\underline{s}) = \underline{\Omega}_m(\underline{s})/\underline{\mathbf{V}}(\underline{s})$.

(c) If the motor is at standstill and 12 volts is suddenly applied, calculate the speed as a function of time, $n(t)$ rpm.

6.58. A dc motor has the following nameplate information: armature 2 hp, 1150/1380 maximum rpm, 9.5 A, 180 V, 1.03 Ω, 28 mH, and $J = 0.32 \times 4.21 \times 10^{-2}$ kg-m^2; field 189 Ω, 0.76 A for nameplate speed, 0.43 A for max. speed, and 85 H.

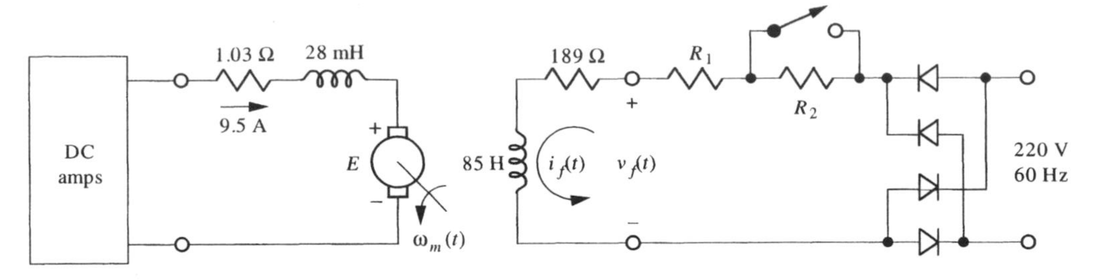

Figure P6.58

Throughout the entire problem, consider that the armature current is kept constant at the nameplate value of 9.5 A dc. The field is excited by a full-wave rectifier, as shown in Fig. P6.58.

(a) The field circuit has two rheostat resistors and a switch to adjust the field current between the nameplate value (0.76 A dc) and the value for maximum speed (0.43 A dc). Find the value of R_1 and R_2. Assume ideal diodes.

(b) The load is of the form $\tau_L = K_L\omega_m$ and is such that at nameplate speed (1150 rpm), the motor produces nameplate power (2 hp). Further, the rotational losses are assumed to vary as the speed, $P_{rot} = D_R\omega_m$. Find K_L and D_R.

(c) Assume that the switch is closed for a long time and then suddenly opened. Using the time-domain equations describing the system operation, derive a system diagram in the frequency domain with $\underline{\mathbf{V}}_f(\underline{\mathbf{s}})$ the input and $\underline{\Omega}_m(\underline{\mathbf{s}})$ the output. Let I_a (= 9.5 A dc) represent the constant armature current. Find the motor system function $\underline{\mathbf{M}}_f(\underline{\mathbf{s}})$ under these conditions. Use symbols ($K\Phi$, J, K_L, n, $\Re$, etc.) on this part. The definition of the motor system function is $\underline{\Omega}_m(\underline{\mathbf{s}}) / \underline{\mathbf{V}}_f(\underline{\mathbf{s}}) = \underline{\mathbf{M}}(s)$.

Note: Although the field current changes are a result of a resistance change via a switch opening, the system diagram works out better if the field voltage is considered as the input variable. We can work out the initial and final speeds from the motor characteristics; after all we do not know the field turns or reluctance. All we need from the system function are the natural frequencies.

(d) Find the response of the system $\omega_m(t)$ caused by the switch opening.

6.59. A dc motor has the following nameplate information: 15 hp, 230 V, 54 A, 3500 rpm, 0.153-Ω armature resistance, 1.1-mH armature inductance, and 0.068-kg-m^2 armature moment. The motor is driving a load requiring a torque proportional to speed and that requires nameplate power at nameplate speed. The load moment is nine times that of the armature; thus, the total moment is 0.68 kg-m^2. The system can be represented by the feedback system shown in Fig. 6.19.

(a) Find the load torque constant.

(b) Analyze the motor for the $K\Phi$ and T_{loss} constants.

(c) Find the loop gain using numerical values.

(d) Determine the motor system function, $\underline{\mathbf{M}}(\underline{\mathbf{s}}) = \underline{\Omega}_m(\underline{\mathbf{s}}) / \underline{\mathbf{V}}(\underline{\mathbf{s}})$ using feedback theory. Ignore the loss-torque effect.

(e) Find the natural frequencies of the system, and describe the response to a sudden change in armature voltage (underdamped? overdamped? etc.).

Answers to Odd-Numbered Problems

6.1. **(a)** N on the right, flux from right to left; **(b)** 1.17 T; **(c)** CW; **(d)** generator.

6.3.

Field Current	*Voltage*	*Current*	*Speed*	*Torque*	*Power*
I_f	V	I_a	n	T	P
I_f	V	$\frac{1}{2}I_a$	n	$\frac{1}{2}T$	$\frac{1}{2}P$
$\frac{1}{2}I_f$	V	I_a	$2n$	$\frac{1}{2}T$	P
$2I_f$	$\frac{1}{2}V$	$\frac{1}{2}I_a$	$\frac{1}{4}n$	T	$\frac{1}{4}P$
I_f	$2V$	$\frac{1}{2}I_a$	$2n$	$\frac{1}{2}T$	P

6.5. **(a)** 2.86 ft/s; **(b)** –75.0 V; **(c)** 93.8%.

6.7. **(a)** 40 W (39.9 exact); **(b)** 0.500 Ω (0.501 exact); **(c)** 1227 rpm (1223 exact); **(d)** 0.623 V/(rad/s); **(e)** 85.5%.

6.9. 4013 rpm.

6.11. **(a)** 319 W; **(b)** 199 W; **(c)** 5.05 A; **(d)** 694 W; **(e)** 198 V.

6.13. **(a)** 240 W; **(b)** 200 W; **(c)** 0.167 V/(rad/s); **(d)** 1371 rpm (1364 exact).

6.15. **(a)** 1250 rpm; **(b)** 14.5 A; **(c)** 85.6%.

6.17. **(a)** 139 W; **(b)** 6.86 N-m; **(c)** 1820 rpm.

6.19. 211 V.

6.21. **(a)** 60.2 N-m; **(b)** 93.5%; **(c)** 618 rpm.

6.23. **(a)** $R_{F\max} = 47.4\ \Omega$ and $R_{F\min} = 30\ \Omega$

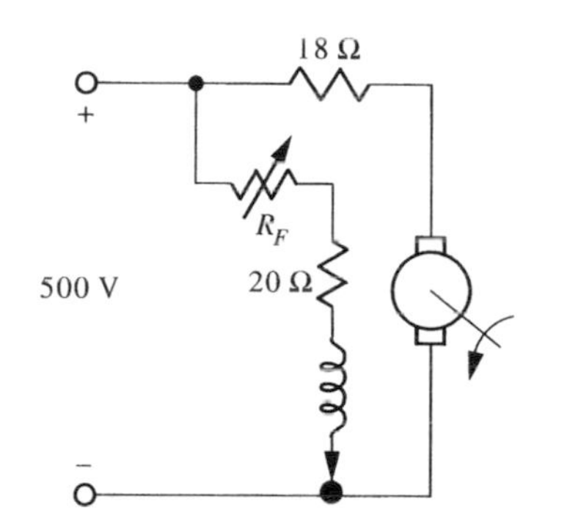

(b) 82.9 %; **(c)** 1550 $(n/1200)$ W; **(d)** 144 V for 126 A; **(e)** losses are 5250 W compared with the allowed 4700 W, so it is overloaded; **(f)** 46,000 W (61.7 hp).

6.25. **(a)**

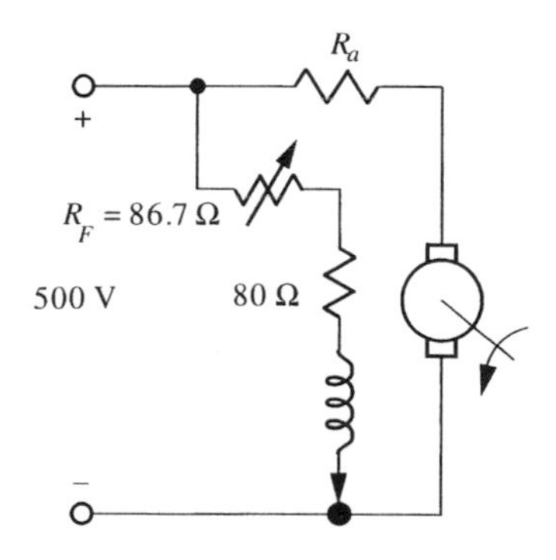

(b) 86.2%; **(c)** 1836 rpm; **(d)** 79.4 N-m.

6.27. **(a)** 1813 rpm; **(b)** 158 W, the dc motor is the motor and the ac motor is the load (generator).

6.29. 94.5 W.

6.31. **(a)** 0.205 Ω; **(b)** 0.105 N-m.

6.33. 6.48 A and 893 rpm.

6.35. 743 W (739 exact).

6.37. **(a)** 8.12 A; **(b)** 5.74 A.

6.39. 120 V for 0.296 hp at 1800 rpm; 120 V for 1/2 hp at 1067 rpm; 147 V for 1 hp at 800 rpm.

6.41. 4.32 N-m.

6.43. 20.1 W.

6.45. **(a)** 0.0391 N-m; **(b)** 9.75 A; **(c)** 1863 rpm.

6.47. P varies as $\sqrt{\text{torque}} \times$ constant $- T \times$ a small constant. Thus, the power at first increases, but eventually levels out and decreases.

6.49. 8.39 A with 74.6 W.

6.51. 0.778.

6.53. **(a)** 3570 W (4.79 hp); **(b)** 5042 rpm for maximum armature resistance and minimum field resistance; **(c)** 12,833 rpm for minimum armature resistance and maximum field resistance.

6.55. The natural frequencies are $\underline{s}_n = -69.5 \pm j7.52\ \text{s}^{-1}$.

6.57. **(a)**

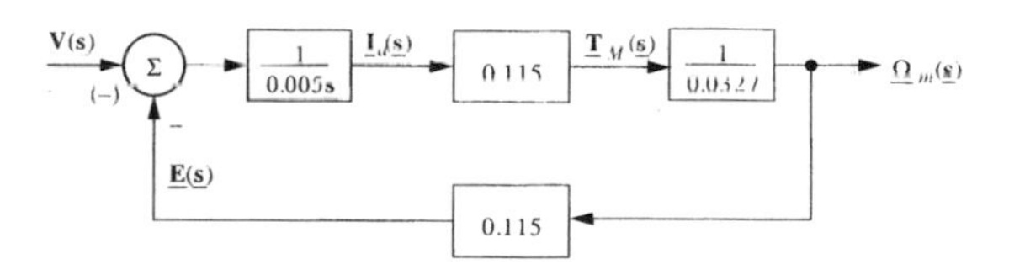

(b) $\underline{\mathbf{M}}(\underline{s}) = 702/(\underline{s} + 80.2)$; **(c)** $n(t) = 1000\,(1 - e^{-80.2t})$ rpm.

6.59. **(a)** 0.0833; **(b)** 0.605 V/(rad/s) and 2.14 N-m;

(c) $\underline{\mathbf{L}}(\underline{s}) = -\dfrac{(0.605)^2}{(0.0011\underline{s} + 0.153)(0.68\underline{s} + 0.0833)}$

(d) $\underline{\mathbf{M}}(\underline{s}) = \dfrac{0.605}{1 + \dfrac{(0.605)^2}{(0.0011\underline{s}+0.153)(0.68\underline{s}+0.0833)}}$

(e) $\underline{s} = -3.79, -90.9\ \text{s}^{-1}$, overdamped.

CHAPTER 7

Power Electronic Systems

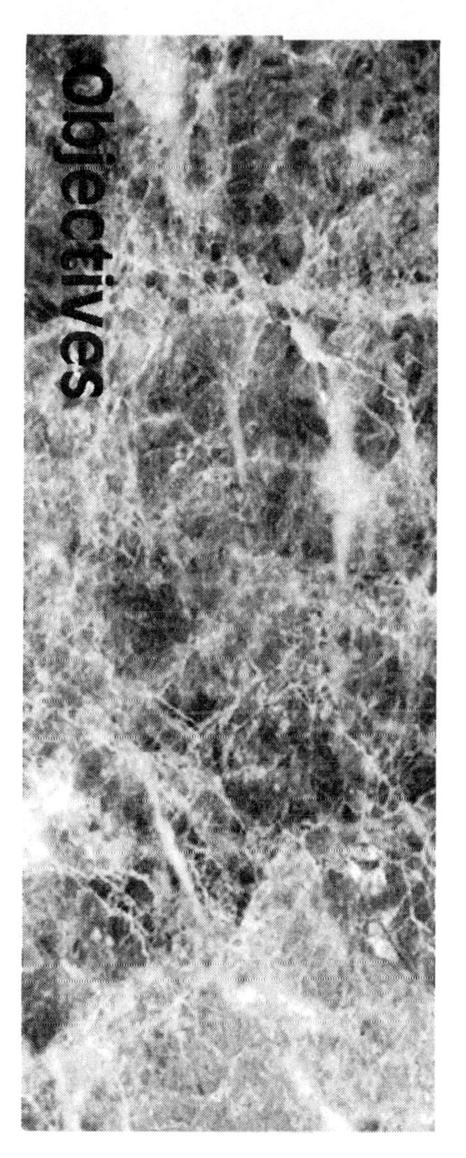

1. To understand the variety and limitations of semiconductor switches
2. To understand in full the operation of the common light-dimmer circuit
3. To understand the operation of a controlled single-phase, full-wave rectifier driving a dc motor
4. To understand the methods and complexities of ac motor controllers

"Power" and "electronics" do not seem to go together, but increasingly electronic techniques are used to control large power equipment such as motors and transmission systems. The recent appearance of semiconductor devices capable of switching large amounts of power has made possible these new techniques. We consider these switches and some basic applications in dc and ac motors controllers.

7.1 INTRODUCTION TO POWER ELECTRONICS

power electronics

Introduction. *Power electronics* applies electronic techniques to the control of electric power. The field is not new because vacuum tubes have long been used for this purpose. But recently, power electronics has grown rapidly as solid-state devices have been developed to control electric power, specifically, power transistors, silicon-controlled rectifiers (SCRs), and gate-turnoff thyristors (GTOs).

HVDC

Ordinary electronics deals with the control of electric power in small quantities, normally for its information content. Power is voltage times current; hence, ordinary electronic devices, such as transistors and diodes, must handle small voltages, or small currents, or both. Power electronic devices must handle large voltages and large currents. How large depends on the context. The variable-speed hand-held drill requires voltages around 200 V (peak) and currents of a few amperes. By contrast 4500-V, 2500-A SCRs can be operated in tandem to handle conversion of 200 MW of ac power to dc power for a high-voltage, direct-current (HVDC) transmission line. The principles are the same for both applications.

This chapter introduces basic concepts of power electronics. We begin by discussing semiconductor switches. We then analyze a simple power controller such as might be used in a light dimmer or hand-held, variable-speed drill. After giving some general principles of motor controllers, we discuss the electronic control of both dc motors and ac motors. Our purpose is to illustrate the principles of power electronic systems through the analysis of representative applications.

Semiconductor Switches

LEARNING OBJECTIVE 1.

To understand the variety and limitations of semiconductor switches

Introduction. A switch is a device that has two stable states, ON and OFF. When ON, the switch has an impedance much smaller than its load, and when OFF, it has an impedance much larger than its load. Semiconductor switches have recently been developed to handle large amounts of power. In this section, we describe principles of operation, terminal characteristics, and limitations of four-level diodes, thyristors, gate-turnoff thyristors, and power transistors.

Four-level diode. Figure 7.1(a) shows the circuit symbol and Fig. 7.1(b) shows the structure of the *pnpn* diode, which consists of four alternating layers of *p*- and *n*-type semiconductor, forming three *pn* junctions. Two of the *pn* junctions face the same direction, and the middle one faces the opposite direction. Thus, you might anticipate that the device will operate as three diodes in series, with the middle one turned around. Hence, no current ought to flow in either direction, because at least one of the diodes will always be reverse-biased.

Characteristics of *pnpn* diodes. The characteristic displayed in Fig. 7.1(c) reveals that no current flows for negative voltages, where two *pn* junctions are reverse- biased; nor does current flow for positive voltages, where the middle *pn* junction is reverse-biased, until a threshold voltage, V_{th}, is reached. After this threshold voltage is exceeded, the four-level diode begins to conduct freely: It "fires," acting as if the middle *pn* junction has disappeared. The threshold phenomenon occurs because the doping levels in the four layers differ greatly. The outside *p* and *n* materials are doped heavily; hence, they have many carrier holes and electrons, respectively, available to diffuse into the middle *n* and *p* regions, which are lightly doped. Once the breakdown occurs in the

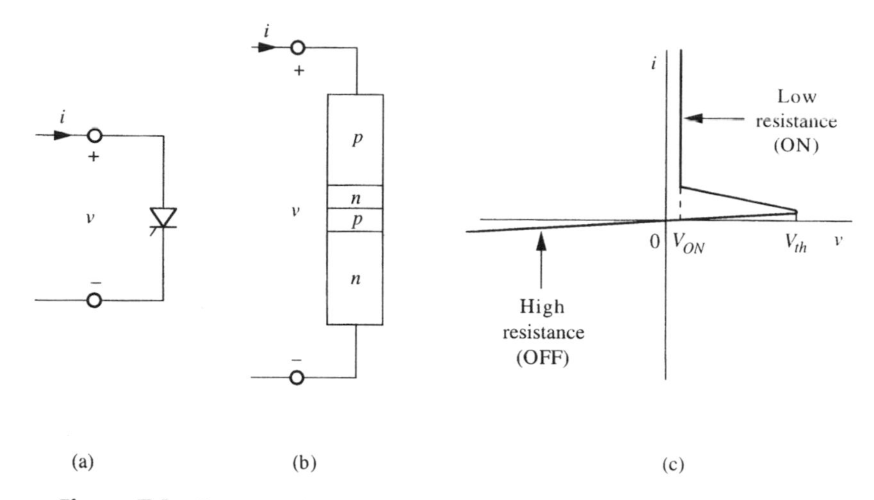

Figure 7.1 Four-level diode: (a) symbol, (b) structure, and (c) voltage–current characteristic.

middle junction at V_{th}, the holes from above and the electrons from below flood into the depletion region of the middle junction, where the electric field due to uncovered charges reinforces their movement across the junction. Thus, this middle depletion region effectively disappears due to the carriers from the forward-biased junctions, and we are left with two forward-biased junctions in series. The small voltage for the *pnpn* diode in the ON state results from the contributions from each ON junction. The turn-ON voltage is typically less than 0.7 V because the excess carriers from each junction help each other.

Voltage-actuated switch. Thus, the *pnpn* has two states. It is OFF for negative voltage, and remains OFF for positive voltage until the threshold voltage is reached, after which it turns ON. It remains ON until the current is reduced to a small value, after which it turns OFF again. The value of the threshold voltage, V_{th}, can be controlled over a modest range by the semiconductor designer. Typical *pnpn* diodes have threshold voltages from 6 to 32 V. Thus, this device, like the *pn*-junction diode, is a voltage-controlled switch, except that the four-level diode requires much more than 0.7 V to turn ON.

gate

Thyristor (SCR). Figure 7.2(a) shows the circuit symbol and Fig. 7.2(b) shows the structure of the thyristor, or silicon-controlled rectifier (SCR). The thyristor has a *pnpn* structure with an external *gate* to turn ON the device. With no gate current, the SCR characteristic is that of a four-level diode, as shown. The important difference, however, is that the breakdown threshold occurs at a much higher voltage, indeed, high enough that the SCR should never conduct because the input voltage exceeds its threshold. On the contrary, the forward-biased SCR should fire only when a pulse of current is delivered to the gate. This is shown by the $i_G > 0$ characteristic in Fig. 7.2(c); in effect, the threshold is reduced to a very small value when the gate conducts. Physically, the gate injects holes into the lightly doped *p* region and floods the depletion region with carriers, thus initiating breakdown.

SCR limitations. From the instant when voltage is applied to the gate, there is a small delay before the SCR turns ON, but this is not a problem at power frequencies.

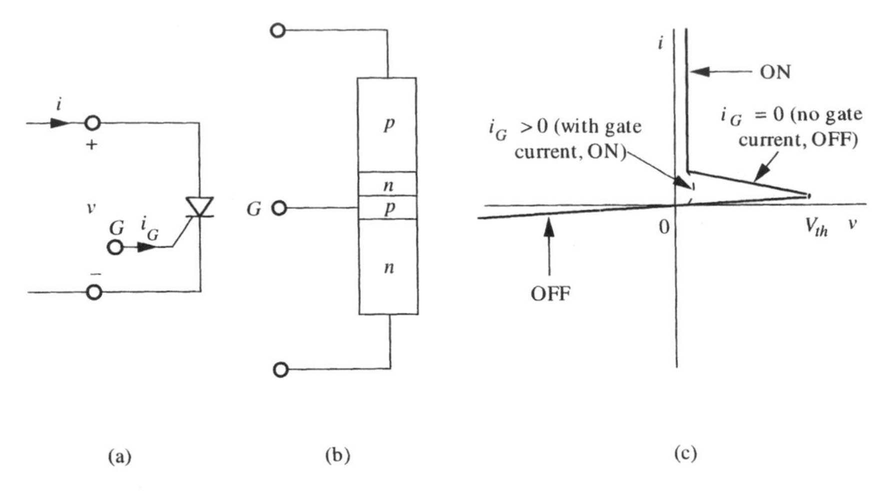

Figure 7.2 Silicon-controlled rectifier (SCR): (a) symbol, (b) structure, and (c) voltage–current characteristic.

More problematic is the rate of rise of SCR current, which can cause device failure due to localized heating in the junction. If not limited by load inductance, an external inductor must be added in series to limit the *di/dt* of the SCR.

snubber

Also critical is the rate of voltage rise of the SCR in its OFF state. In certain applications, a rapid voltage increase can initiate conduction, independent of the gate signal, and cause device malfunction. The *RC snubber* circuit[1] shown in Fig. 7.3 limits the *dv/dt* across the SCR. With the SCR OFF, the small (10- to 100-Ω) resistor in series with C and the inductance in the load circuit limit the *dv/dt* across the SCR, and the capacitor blocks dc current. When the SCR fires, the energy stored in the capacitor is dissipated in R and the SCR.

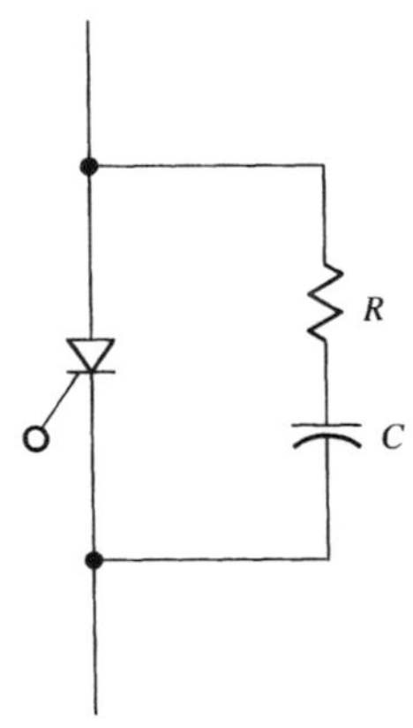

Figure 7.3 The *RC* snubber circuit limits the rate of voltage rise of the SCR in its OFF state.

To turn OFF, the SCR must be reverse-biased for a sufficient period of time for carriers to recombine and reestablish the blocking junction. This time can vary from 20 to 200 μs, depending on SCR size and type.

Line and forced commutation. *Commutation* refers to the switching of a conducting SCR from the ON state to the OFF state. As we have seen, an SCR may be commutated by reducing its current below the value required to sustain conduction. Normally, this is accomplished by reverse biasing the device for a period of time.

When in an ac cycle, the voltage across an SCR changes from forward bias to reverse bias, the SCR ceases to pass current and is *line commutated.* When an auxiliary circuit is used to commutate the SCR independent of the line voltage, the SCR has a *forced commutation.*

commutation, line commutation, forced commutation

Summary. In the previous section, we discussed the use of SCRs as power switches. We gave the conditions for turning SCRs ON and OFF. The need for auxiliary circuits to force the commutation of the SCR is a serious liability, and for this and other reasons alternative electronic switches have been developed.

[1] A *snubber* circuit protects a switch circuit from excessive voltage, current, energy, or power.

Gate-turnoff thyristors (GTOs). Figure 7.4(a) shows the circuit symbol for a GTO, which is like an SCR symbol except for a mark on the gate, and Fig. 7.4(b) shows GTO construction. The forward-biased GTO is turned ON by a pulse of positive current to its gate, but unlike the SCR, it is turned OFF by a pulse of negative gate current. The negative current removes carriers from the cathode region, and the inner *pn* junction blocks the forward current.

The switching characteristics of the GTO are excellent, and this device is appearing in a variety of power controllers. Development in GTOs and power field-effect transistors is very active, and the circuit designer has an increasing number of semiconductor devices to consider in power controllers.

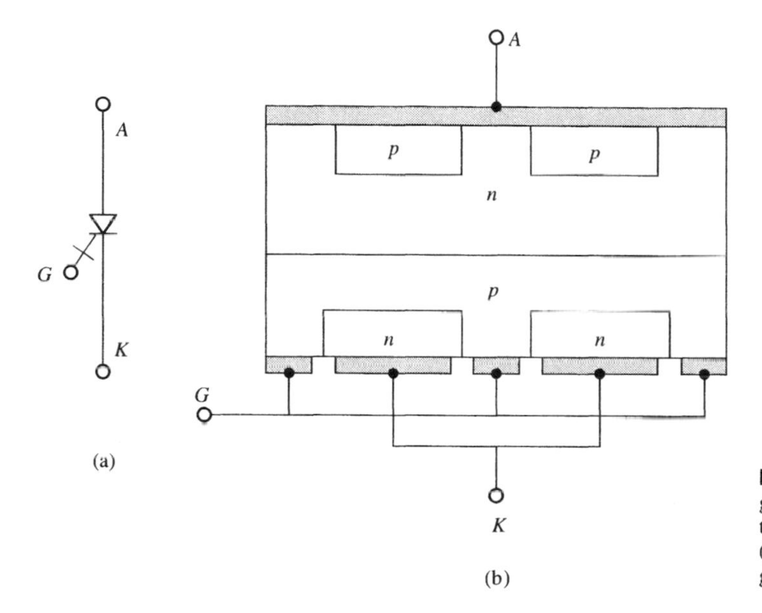

Figure 7.4 (a) Circuit symbol for a gate-turnoff thyristor (GTO); (b) construction of a GTO. The anode (A) and cathode (K) are the power terminals and G is the gate terminal.

Power bipolar-junction transistors. Recent advances in bipolar power transistors (BJTs) have allowed their use in power electronic circuits. Although SCRs are still dominant in applications requiring high voltage and high power, transistors are replacing SCRs in ac motor controllers for voltages below 460 V and power levels below 400 hp. The main advantage of transistor switches is that turn-ON and turn-OFF are controlled by the base current, and no forced-commutation circuit is required. Transistors also offer fast switching times, which is important in many ac motor controller circuits. The disadvantages of transistors are that they require medium-power base drive circuits and that they have higher switching and operating losses than SCRs. In this section, we consider transistor switches with resistive and inductive loads.

rise time

Transistor switch with a resistive load. Figure 7.5(a) shows a BJT switching a resistive load, and Fig. 7.5(b) shows voltage and current during turn-ON. Base current is applied at $t = 0$. After a delay time, t_d, the collector current begins to rise and the collector–emitter voltage begins to drop. The *rise time*, t_r, is defined as the time required for the collector current to rise from 10 to 90% of its final value, I_{Cp}.

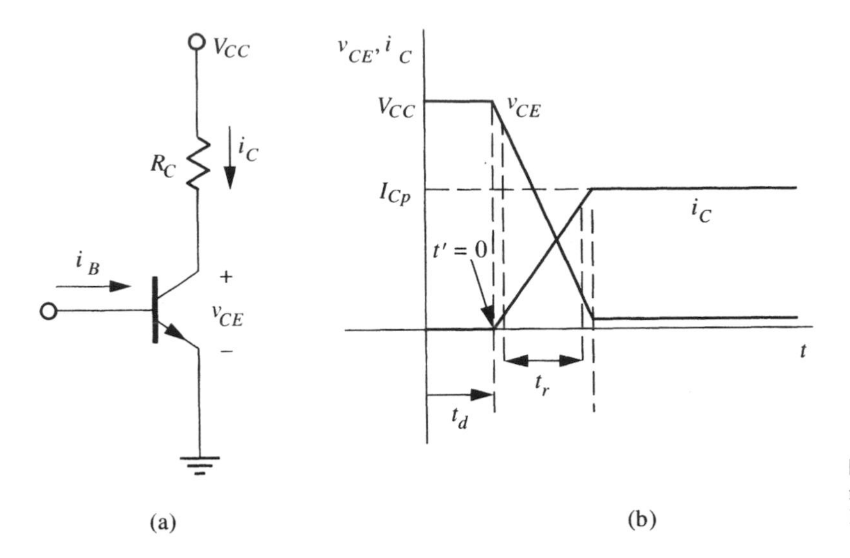

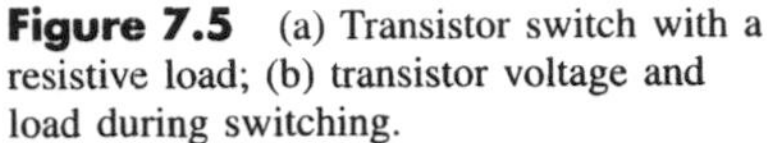

Figure 7.5 (a) Transistor switch with a resistive load; (b) transistor voltage and load during switching.

Switching losses. When the transistor is OFF, it has no collector current and hence no power dissipation. After switching, the transistor is saturated and has a low collector–emitter voltage and hence low dissipation. During switching, however, the transistor has appreciable voltage and current; hence, the power level can become high. We calculate the power during switching and the total energy dissipated by the transistor during switching. We assume linear voltage and current transitions and assume an origin (t') at the instant the current begins to rise. The voltage and current are, therefore

$$i_C(t') = 0.8t' \times \frac{I_{Cp}}{t_r} \quad \text{and} \quad v_{CE}(t') = V_{CC}\left(1 - \frac{0.8t'}{t_r}\right) \tag{7.1}$$

where $t_r/0.8$ is the total time of transition and we have set the saturation voltage to zero. The expressions in Eq. (7.1) are valid only during the time $0 < t' < t_r/0.8$. The power into the transistor is the product of voltage and current

$$p_C(t') = V_{CC}I_{Cp}\frac{0.8t'}{t_r}\left(1 - \frac{0.8t'}{t_r}\right) \tag{7.2}$$

which has a maximum value of $V_{CC}I_{Cp}/4$ at $t' = 0.625t_r$. For a 300-V, 100-A device, the peak power into the transistor during switching is 7500 W. The energy given to the transistor in switching, W, is the integral of the power

$$W = \int_0^{t_r/0.8} v_{CE}(t')i_C(t')\,dt$$

$$= V_{CC}I_{Cp}\int_0^{t_r/0.8} \frac{0.8t'}{t_r}\left(1 - \frac{0.8t'}{t_r}\right)dt' = \frac{V_{CC}I_{Cp}t_r}{4.8} \tag{7.3}$$

which gives an energy of 12.5 mJ for each switching transition, assuming a 2-μs rise time. This appears small, but if the transistor is switching at a 2-kHz rate, this requires dissipation of 50 W from switching alone. During saturation, the collector–emitter voltage drop is about 1.2 V at high current levels, so the collector dissipation is 120 W while the transistor is ON. In the switching mode, the transistor would be ON approximately half of the time, so the average power would be 60 W due to the ON voltage. Therefore, the switching power is an important contribution to the total transistor dissipation.

EXAMPLE 7.1

Resistive switching losses

A BJT with a rise time of 3 μs is used to switch 100 V to a 10-Ω resistor at a 5-kHz rate. Find the switching losses.

SOLUTION:

The peak current is $I_{Cp} = 100\ \text{V}/10\ \Omega = 10\ \text{A}$. Thus, Eq. (7.3) gives

$$W = \frac{100 \times 10 \times 3 \times 10^{-6}}{4.8} = 0.625\ \text{mJ}/\text{transition} \tag{7.4}$$

Because there are 10^4 transitions/second, the average switching power is 6.25 W.

WHAT IF?

What if the transistor saturation voltage is 1.0 V and the maximum allowable power is 8 W. What is the minimum load resistance?[2]

Transistor switching losses with inductive loads. Figure 7.6(a) shows a BJT with an inductive load. The free-wheeling diode is required to protect the transistor from large inductive voltages. Figure 7.6(b) compares the switching trajectory in the v_{CE}–i_C plane for the resistive and inductive loads. To understand the trajectory for an inductive load, we must investigate the role of the transistor in controlling the current through the inductive load, and also investigate the importance of the time-average voltage in controlling the current through the inductive load.

Inductive loads. For the resistive load, the transistor turns the current on and off, and thereby controls the average voltage and power to the load. For an inductive load, by contrast, the transistor is turned ON and OFF to control the average voltage across the load, and this average voltage determines the average current through the load, as shown in the following paragraph. The current through the load does not cease because the inductance keeps it more or less constant, provided the transistor switching period is much smaller than the time constant of the *RL* load.

IDEA 7 **The Frequency Domain**

Average voltages. We now consider why the average current through an *RL* load is controlled by the time-average voltage across the load. This is true for all periodic

[2] 20.8 Ω.

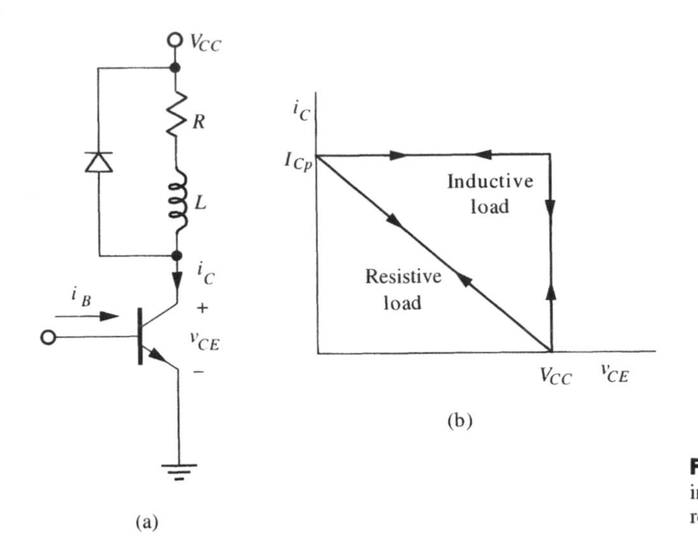

Figure 7.6 (a) Transistor switch with an inductive load; (b) switching trajectories for resistive and inductive loads.

voltages and currents because the time-average voltage across the inductance must be zero. To see that this is true, consider the current through the *RL* load as a discrete spectrum of sinusoidal harmonics, plus a dc component. The voltage across the inductance consists, therefore, of sinusoidal components, and the time average of a sinusoid is zero. Therefore, the time-average voltage across the inductance is zero. It follows also that the time-average voltage across the *RL* load is equal to the dc component of the current times the resistance of the load, because the time average of the resistor voltage due to the harmonics of the current must also be zero. Thus, the transistor controls the current in the *RL* load by controlling the time-average voltage across the load.

Returning to the transistor. The transistor in Fig. 7.6(a) switches ON and OFF, but the current through the inductor continues more or less constant. Therefore, while the transistor is OFF and any time the transistor is not carrying the full-load current, some current passes through the free-wheeling diode, and the collector voltage of the transistor is equal to the supply voltage, assuming an ideal diode. It follows that the switching trajectory for an inductive load is that shown in Fig. 7.6(b), with constant voltage at V_{CC} while the collector current is moving from zero to I_{Cp} and back during the switching cycle.

An analysis of the power and energy shows that the peak power into the transistor is four times that with a resistive load, and the energy per cycle is three times that required for switching a resistive load. The strain on the transistor is greatly increased, and snubber circuits are required to reduce the power requirement on the transistor.

EXAMPLE 7.2 Switching losses

A BJT switches 100 V to an inductive load with $R = 5\ \Omega$ and $L = 20$ mH. The transistor is ON for 0.1 ms and OFF for 0.2 ms. Find the average current and switching loss in the transistor if $t_r = 3\ \mu$s and $V_{ON} = 1$ V.

SOLUTION:

The RL time constant is 4 ms, much longer than the switching cycle, so the current is essentially constant. The voltage across the RL load is as shown in Fig. 7.7. The average voltage is 67.0 V, so the time-average current is $I_{dc} = 67.0\text{ V}/5\ \Omega = 13.4\text{ A}$. As stated, the switch loss/transition is three times that given by Eq. (7.3)

$$W = \frac{V_{cc}I_{cp}t_r}{1.6} = \frac{100 \times 13.4 \times 3 \times 10^{-6}}{1.6} = 2.51 \text{ mJ/transition} \tag{7.5}$$

Hence, the switching losses are $2 \times 2.51 \times 10^{-3}/0.3 \times 10^{-3} = 16.8$ W.

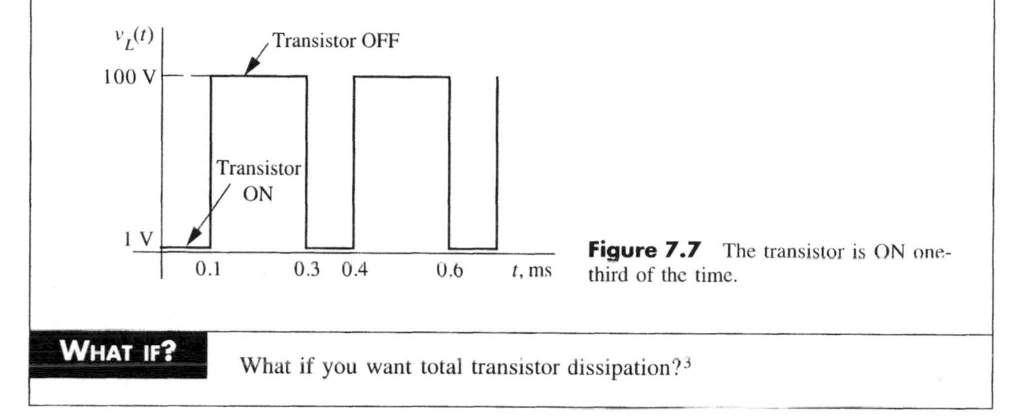

Figure 7.7 The transistor is ON one-third of the time.

WHAT IF? What if you want total transistor dissipation?[3]

Summary. The section introduced SCRs, GTOs, and power transistors as semiconductor switches. We considered how to turn these devices ON and OFF and discussed their limitations. In the next section, we analyze a light-dimmer circuit to show a common application of power electronic techniques.

LEARNING OBJECTIVE 2.

To understand in full the operation of the common light-dimmer circuit

Common Application of Power Electronics

Power electronic circuit. In this section, we examine the operation of the circuit shown in Fig. 7.8. The ac source represents a standard 120-V, 60-Hz supply. The circuit in the box is a power controller, the resistor with the arrow represents a variable resistor, and R_L represents the load receiving the power. The circuit uses a four-level diode and an SCR as semiconductor switches.

Common applications for this circuit. Circuits of this type are used in light dimmers. In this application, the variable resistor is adjusted by a rotary mechanism, and an on–off switch is normally built into the same mechanism. The load is an incandescent light. Variable-speed drills are controlled by this or similar circuits. In that application the variable resistor is adjusted by the squeeze trigger on the drill handle and the load is a universal motor, discussed in Sec. 6.2.

Four-layer diode function. To explain the operation of the circuit in Fig. 7.8, we start with the simpler circuit shown in Fig. 7.9 to see how the SCR-gate current pulses

[3] 21.2 W.

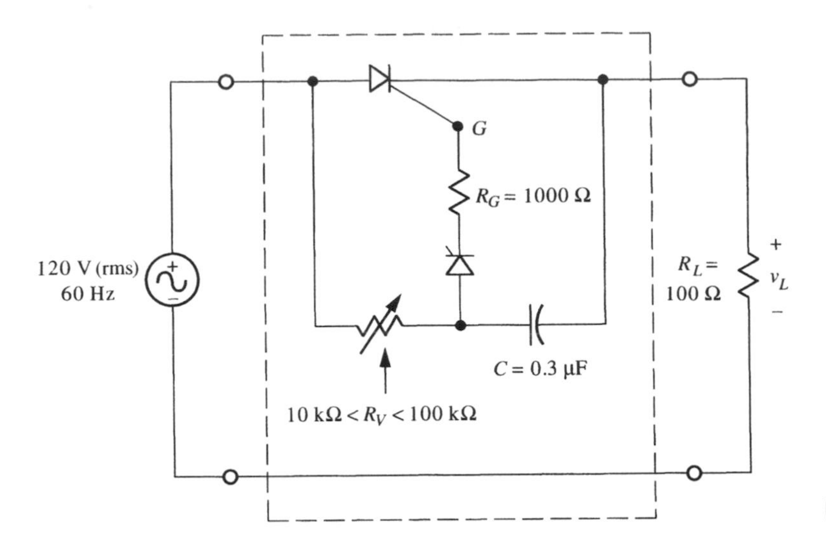

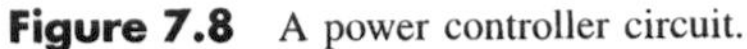

Figure 7.8 A power controller circuit.

are formed. We begin with a discharged capacitor. Closing the switch at $t = 0$ allows current to flow and the capacitor voltage increases. The *pnpn* diode remains OFF and no current flows through R_G until the voltage across the capacitor reaches the threshold voltage, V_{th}. Thus, until the *pnpn* diode fires, we have a simple *RC* transient. The time constant is $R_V C$, the initial value of the voltage across the capacitor is 0 V, and the final value would be V_p if the transient reached completion. Using standard techniques, we determine the voltage across the capacitor to be

$$v_C(t) = V_p(1 - e^{-t/R_V C}) \tag{7.6}$$

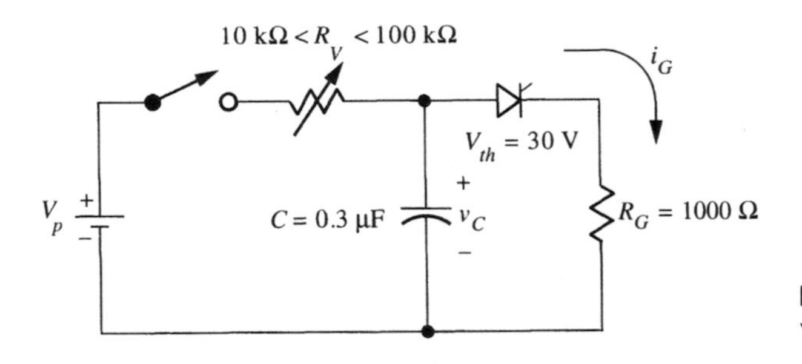

Figure 7.9 The four-level diode fires when $v_C = 30$ V.

The capacitor voltage increases according to Eq. (7.6) until the threshold voltage of the *pnpn* diode is reached. At this point, the *pnpn* diode turns ON and allows current to flow through R_G. Current from the dc source is limited by the large value of R_V, but the current from the capacitor discharge can be quite large. The time at which the discharge occurs, t_α, is

$$V_{th} = V_p(1 - e^{-t_\alpha/R_V C}) \quad \Rightarrow \quad t_\alpha = -R_V C \times \ln\left(1 - \frac{V_{th}}{V_p}\right) \tag{7.7}$$

EXAMPLE 7.3 Firing time

Let $V_{th} = 30$ V, $V_p = 120\sqrt{2}$ V, $C = 0.3$ μF, and 10 kΩ < R_V < 100 kΩ. Find the minimum firing time.

SOLUTION:

Using Eq. (7.7), we calculate

$$t_\alpha = -10^4 \times 0.3 \times 10^{-6} \ln\left(1 - \frac{30}{120\sqrt{2}}\right) = 0.584 \text{ ms} \tag{7.8}$$

WHAT IF?

What if $R_G = 1000\ \Omega$ and $V_{ON} = 0.5$ V? What is the current in R as a function of time?[4]

Pulse cycle. After discharge, the *pnpn* diode turns OFF, there being insufficient current flowing to keep it ON, and the cycle begins again. Hence, the capacitor voltage and the current through the small resistor are as shown in Fig. 7.10, where there is a 5-ms cycle time.

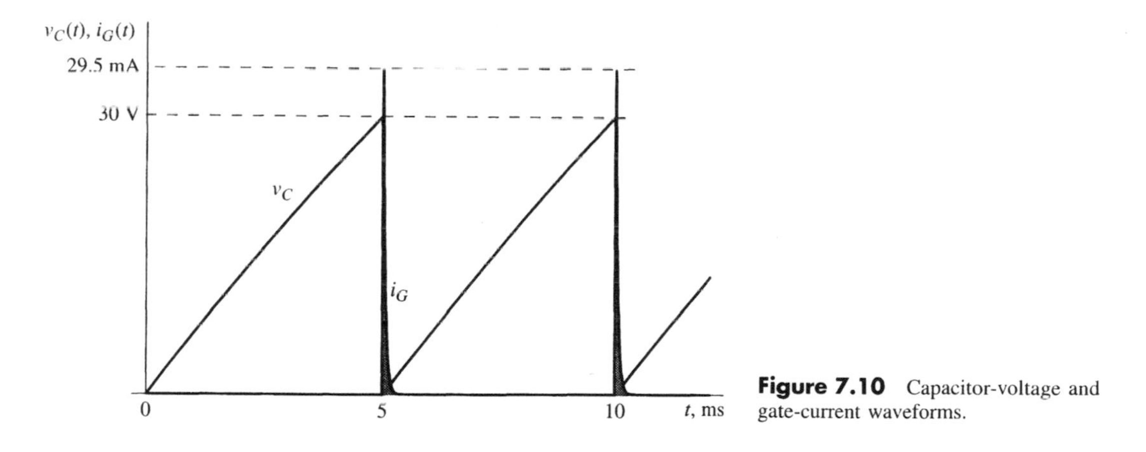

Figure 7.10 Capacitor-voltage and gate-current waveforms.

With an AC source. We now consider what would happen if the dc source were replaced by an ac source, as shown in Fig. 7.11(a). We draw the input voltage as a sine function and show the capacitor voltage, v_C, as it would be if the *pnpn* diode did not conduct. If the capacitor were initially uncharged and if the four-level diode never fired, there would be a start-up transient, and then the voltage of the capacitor would lag the input voltage with a phase shift between 0° and 90°, depending on the value of $R_V C$. However, the four-level diode fires at the time labeled t_α, when the capacitor voltage reaches the threshold voltage, Fig. 7.11(b).

[4] $i_G(t) = 29.5 e^{-(t - t_\alpha)/300\ \mu s}$ mA.

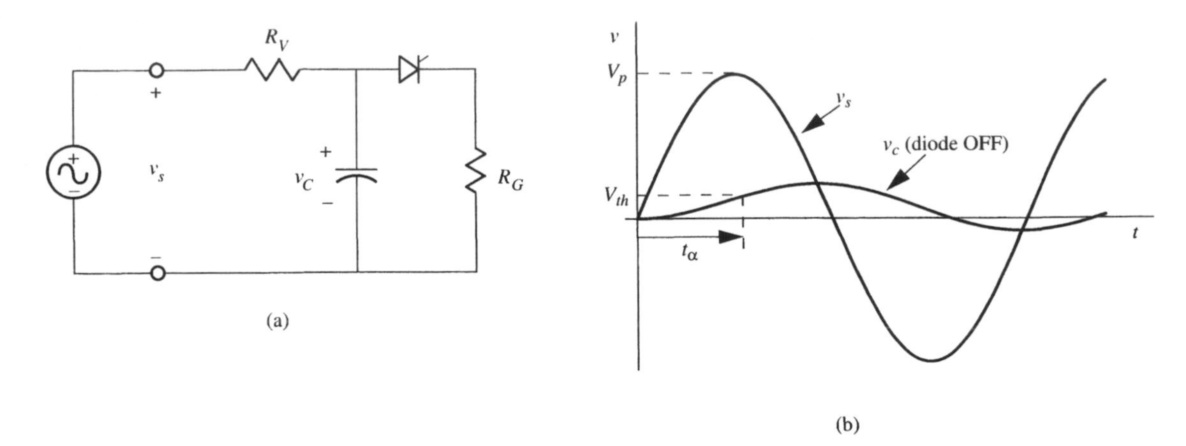

Figure 7.11 Circuit response with a sinusoidal source.

The Time Domain

Analysis. The calculation of t_α combines transient and ac steady-state techniques but is basically a time-domain analysis. The steady-state voltage found by ac circuit techniques is

$$v_C(t) = \frac{V_p}{\sqrt{1+(\omega R_V C)^2}} \sin(\omega t - \phi) \tag{7.9}$$

where V_p is the peak ac voltage, $\phi = \tan^{-1}(\omega R_V C)$, and $t = 0$ when the input voltage is zero. The transient portion of the capacitor voltage is $Ae^{-t/R_V C}$, where A is a constant to be determined from the initial conditions. The capacitor is initially uncharged, so

$$0 = \frac{V_p}{\sqrt{1+(\omega R_V C)^2}} \sin(0 - \phi) + Ae^{-0/R_V C} \Rightarrow \tag{7.10}$$

$$A = \frac{V_p \sin\phi}{\sqrt{1+(\omega R_V C)^2}}$$

Firing time. The firing time, t_α, for the *pnpn* diode occurs when the capacitor voltage equals the threshold voltage of the *pnpn* diode:

$$V_{th} = \frac{V_p}{\sqrt{1+(\omega R_V C)^2}} [\sin(\omega t_\alpha - \phi) + e^{-t_\alpha/R_V C} \sin\phi] \tag{7.11}$$

delay angle, firing angle

The *delay angle*, $\alpha = \omega t_\alpha$, describes the position in the cycle when the *pnpn* diode fires. For the values of R_V and C that we show, the delay angle lies between 30° and 180°. Hence, we can control the firing time, or *firing angle*, with R_V as we did with the dc source. When the four-level diode fires, the capacitor discharges through the gate resistor with a short time constant, as before. After the capacitor discharges, the four-level diode turns OFF, there being insufficient current passing through R_V to keep it conducting. The capacitor voltage again begins building up and may fire the four-level diode again before the input voltage goes negative, but once the input becomes negative, the four-level diode does not fire again until the source voltage again goes positive.

Summary. We can create a pulse of current with the four-level diode. The timing of this pulse of current is controlled by the R_VC time constant and can be varied over a range of delay angles.

IDEA 5 **Impedance Level**

Back to the original circuit. We now can investigate the operation of the circuit in Fig. 7.8, repeated in Fig. 7.12. Figure 7.12 shows the SCR gate circuit to be the "load" to which the four-level diode delivers its current pulses. The load of the SCR, R_L, has presumably a low impedance level and does not affect greatly the charging of the capacitor. Hence, when the input voltage goes positive, the capacitor begins charging through R_V and the load. When the capacitor voltage builds up to the threshold voltage, the four-level diode fires and the current pulse passes through the gate, turning ON the SCR.

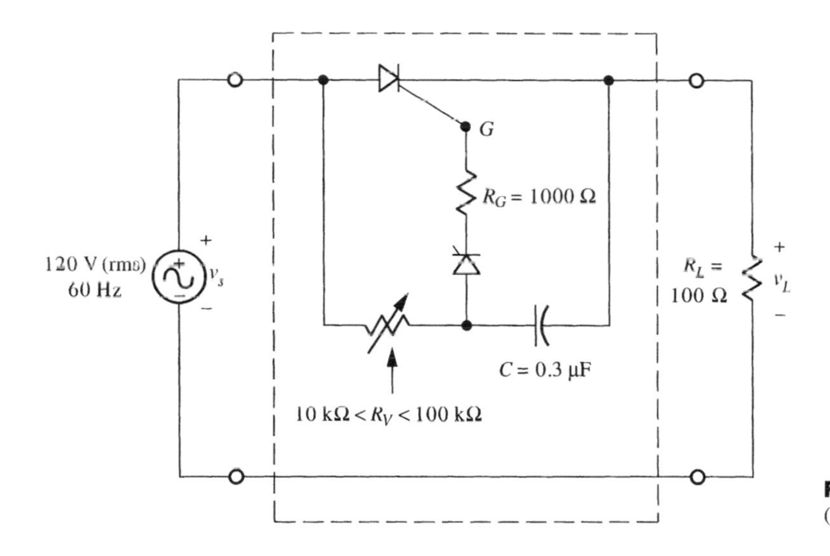

Figure 7.12 Power controller circuit (identical to Fig. 7.8).

Load voltage. The voltage to the load is thus that shown in Fig. 7.13. By adjusting the delay angle with R_V, we can control the conduction angle of the SCR and hence the power to the load. The resistor in series with the four-level diode limits the gate current to an acceptable value. With the SCR gate for a load, repeated firing of the four-level diode does not matter. Once the SCR is turned ON, it remains ON until the input voltage goes negative to turn it OFF by line commutation.

EXAMPLE 7.4 Gate current

What resistance, R_G, should be put in series with the SCR gate to limit the peak gate current at 10 mA if the circuit is that used in the previous examples?

SOLUTION:
The voltage driving the transient is $V_{th} - V_{ON} = 29.5$ V. Thus

$$10 \text{ mA} = \frac{29.5}{R_G} \quad \Rightarrow \quad R_G = 2950 \ \Omega \tag{7.12}$$

WHAT IF? What if you want the new discharge time constant?[5]

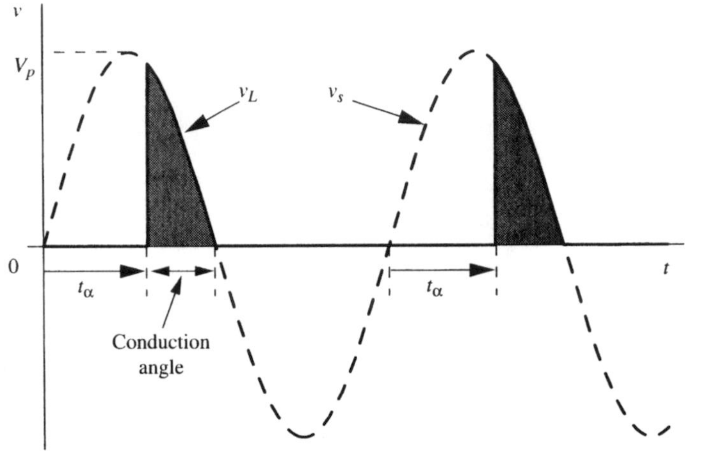

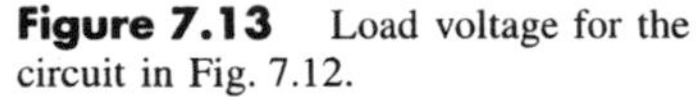
Figure 7.13 Load voltage for the circuit in Fig. 7.12.

diac, triac

Diacs and triacs. Figure 7.14 shows a nonrectifying power controller. The device replacing the four-level diode is called a *diac*. It is like two parallel four-level diodes facing in opposite directions and fires in either direction. Similarly, the device replacing the SCR is called a *triac*, and it functions like two parallel SCRs facing opposite directions. This circuit operates like the circuit in Fig. 7.12, except that it does not rectify. Hence, the load voltage is as shown in Fig. 7.15. This is the preferred circuit for light dimmers and universal-motor tools, which do not require dc voltage.

Check Your Understanding

1. Turning ON an SCR requires (a) forward bias, (b) a pulse of current to the gate, or (c) both. Which?
2. Turning OFF an SCR requires (a) reverse-biasing the SCR, (b) keeping the SCR reverse-biased for a prescribed period of time, and/or (c) removing the gate signal. Which? (May be more than one.)
3. To operate successfully, the SCR must be protected from (a) high dv/dt in the OFF condition, (b) high di/dt in the ON condition, and/or (c) excessive reverse voltage in the OFF condition. Which? (May be more than one.)

[5] 885 μs.

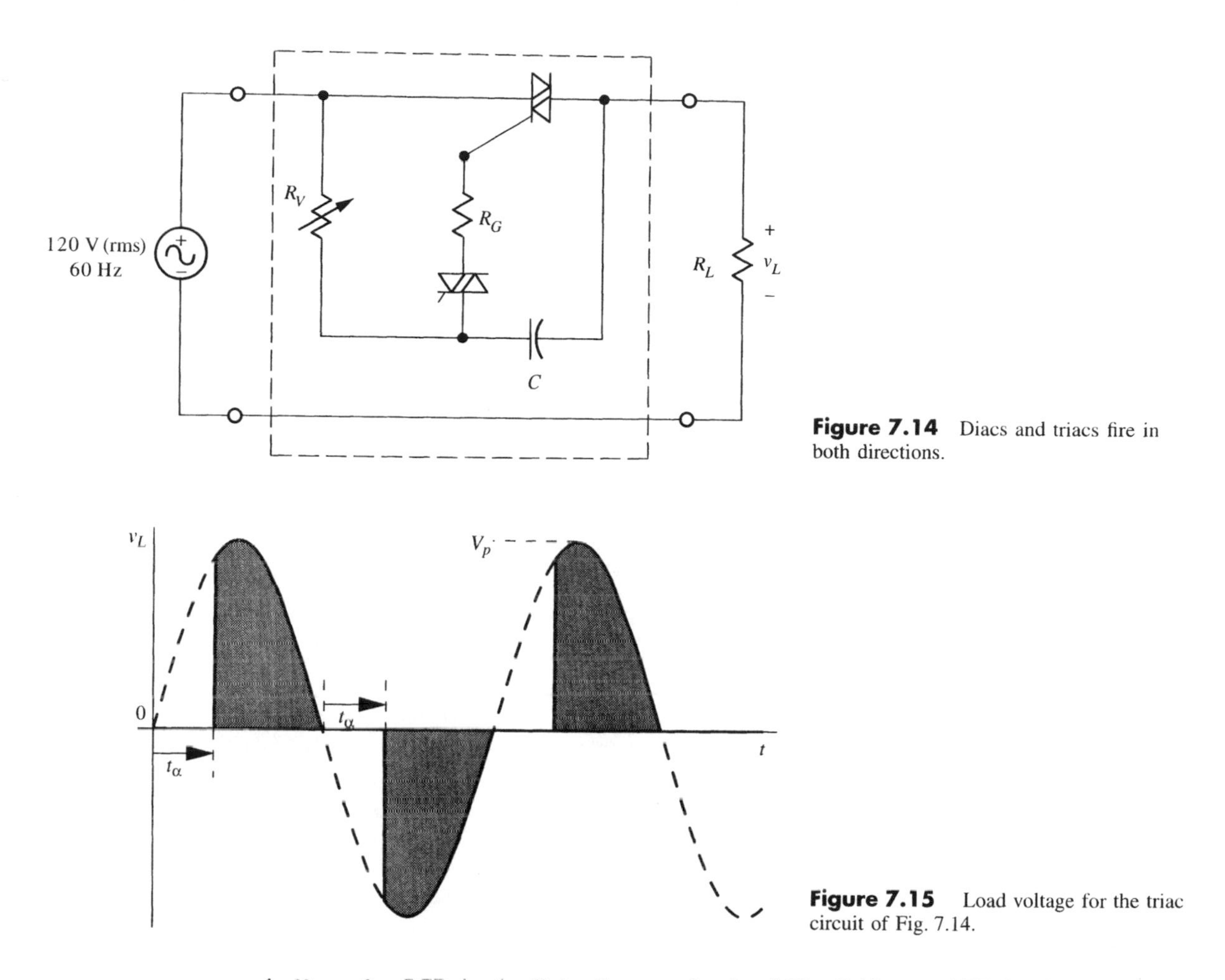

Figure 7.14 Diacs and triacs fire in both directions.

Figure 7.15 Load voltage for the triac circuit of Fig. 7.14.

4. Does the SCR in the light-dimmer circuit of Fig. 7.12 turn OFF by line or forced commutation?
5. The power dissipation limits of a power transistor affect the rate at which it can be cycled as a switch. True or false?
6. If the *pnpn* diode in Fig. 7.9 fired at 20 V, what would be the resistor value required for it to fire in 3 ms?

Answers. **(1)** (c); **(2)** (b) and (c); **(3)** (a), (b), and (c); **(4)** line commutation; **(5)** true; **(6)** 79.7 kΩ.

7.2 DC MOTOR CONTROLLERS

Introduction to Motor Controllers

Power electronic converters. A power electronic converter controls the power exchange between an electrical supply and a load, as shown in Fig. 7.16. The information that controls the power electronics may be a simple manual control, as in the previous section, or may be produced by a control system. The electrical supply may be ac or dc. The load may be a passive load, such as a bank of lights or an induction furnace,

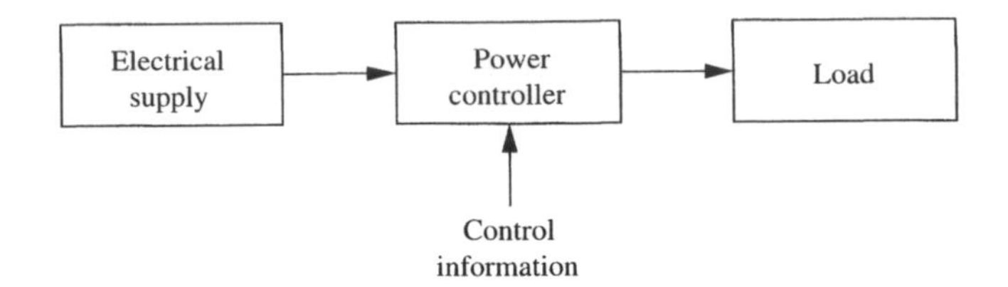

Figure 7.16 General power controller.

or may be an electromechanical system, such as a dc or ac motor. The load may even be another electrical supply. Normally, the power flows from the electrical supply to the load, but for certain loads may flow in both directions. We introduce briefly the various types of systems:

chopper

- **DC–DC Converters.** A converter that takes fixed-voltage dc voltage and produces variable-voltage dc is normally called a *chopper*. We could, for example, control a dc motor armature circuit and hence control motor speed with a dc–dc converter.
- **AC–DC Converters.** An ac–dc converter is a controlled rectifier. In such a converter, the rectifier diodes are replaced by SCRs or similar switches, and the dc voltage is controlled by the switches. We will deal with single-phase converters of this type.

inverter

- **DC–AC and AC–DC–AC Converters.** A dc–ac converter is called an *inverter*. Combined with an ac–dc converter, we can develop an ac–dc–ac converter to make a variable-frequency ac source. For example, if we wished to control the speed of an induction motor, one approach would be to rectify the available 60-Hz ac-to-dc power and then invert the dc to ac at a controlled frequency to drive the motor. Such systems can also link together unsynchronized ac power systems for exchange of ac power between different power grids, sometime over a high-voltage dc, HVDC, transmission line. On a smaller scale, an ac–dc–ac converter can make a noninterruptible power supply if the dc portion maintains a battery bank to supply temporary dc power in case of a power failure.
- **AC–AC Controllers.** In addition to the ac–dc–ac converters described before, there exist ac–ac converters that perform direct conversion from fixed-frequency, fixed-voltage ac power to variable-frequency, variable-voltage ac power.

Quadrants of operation. Our principal concern is motor controllers. Figure 7.17 shows the four quadrants of motor operation. Because power is the product of torque and speed, the mechanical power out of the motor is positive in the first and third quadrants, which correspond to forward and reverse driving of the motor, respectively. In the fourth quadrant, the motor rotates in the forward direction, but the torque is supplied in the reverse direction, tending to slow down the motor. In this forward-braking region, the power output of the motor is negative, and the motor acts as a source of electric power. This power may be delivered back to the electrical supply or may be dissipated in the motor or the converter. Similarly, in the second quadrant, the motor is braked from reverse rotation. Operation in the second and fourth quadrants is sometimes called

plugging, regenerative braking

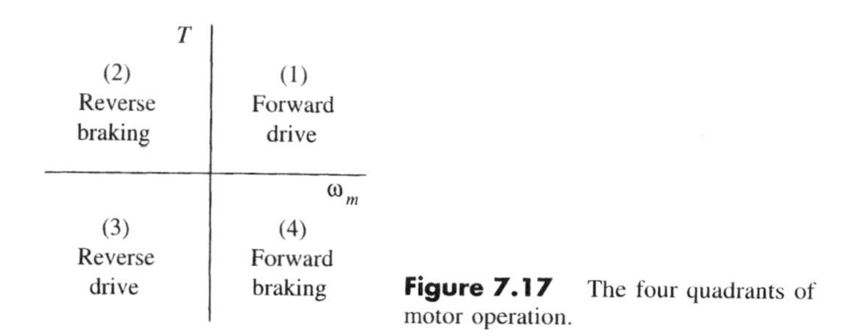

Figure 7.17 The four quadrants of motor operation.

plugging, or *regenerative braking*. As we will see, motor controllers are classified according to the quadrants in which they operate.

DC Motor Model

Equivalent Circuits

Time-average torque and power. We begin with ac–dc power electronic controllers for dc motors. We reexamine the model of a dc motor for the effect of periodic time-varying armature voltage and current. Figure 7.18(a) shows the armature circuit of a separately excited dc motor. The armature emf is

$$E = K\Phi\omega_m \tag{7.13}$$

where $K\Phi$ is the machine constant, assumed constant. Although the armature voltage is periodic, the emf is constant because of the inertia of the armature and mechanical load. The armature current is also periodic, but the time-average torque is

$$T_{dev} = \langle K\Phi i_a(t)\rangle = K\Phi\langle i_a(t)\rangle = K\Phi I_a \tag{7.14}$$

where $< >$ means time average and I_a is the dc component of the armature current. Because the emf is constant, the developed power also depends on the dc component of the armature current:

$$P_{dev} = \langle Ei_a(t)\rangle = E\langle i_a(t)\rangle = EI_a \tag{7.15}$$

Input voltage. Although the armature voltage, $v_a(t)$, is periodic, the dc component of the armature voltage is simply related to the emf and the dc current. Kirchhoff's voltage law in the armature circuit is

$$v_a(t) = Ri_a(t) + v_L(t) + E \tag{7.16}$$

where $v_L(t)$ is the voltage across the inductor. As explained earlier on page 297, the time-average voltage across the inductance must be zero for steady-state operation, so the time average of Eq. (7.16) is

$$V_a = RI_a + E \tag{7.17}$$

where $V_a = \langle v_a(t)\rangle$ is the time-average voltage applied to the armature circuit.

Quadrants of operation. Because torque is proportional to current and emf is proportional to speed, the four quadrants of operation in Fig. 7.17 correspond to identical quadrants in the E–I_a plane shown in Fig. 7.18(b). For example, in the first quadrant, the emf and dc current are positive, which corresponds to positive torque and speed, hence to forward drive. The input voltage to the armature circuit is not identical

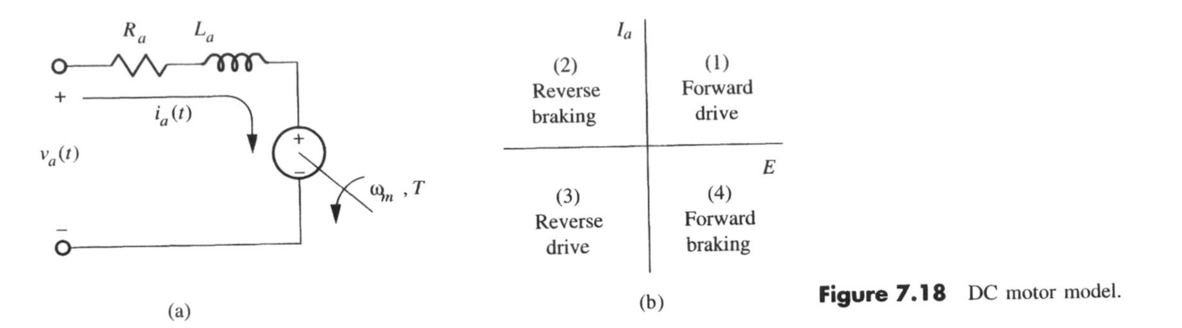

Figure 7.18 DC motor model.

to the emf, but in steady state, the emf follows the input voltage closely. Thus, we may assume that motor operation in the first quadrant corresponds to positive input voltage and current, and so on.

Notation for switches. In the remainder of this chapter, we symbolize a semiconductor switch as a "switch in a box," as shown in Fig. 7.19. Because the SCR, GTO, and BJT switches allow current flow in one direction only and require forward bias in that direction to conduct, we have shown the allowed current direction by the orientation of the open end of the switch in a box, as shown. By this symbol we imply the existence of circuits to switch the device ON and OFF for forced commutation, including necessary protective circuits.

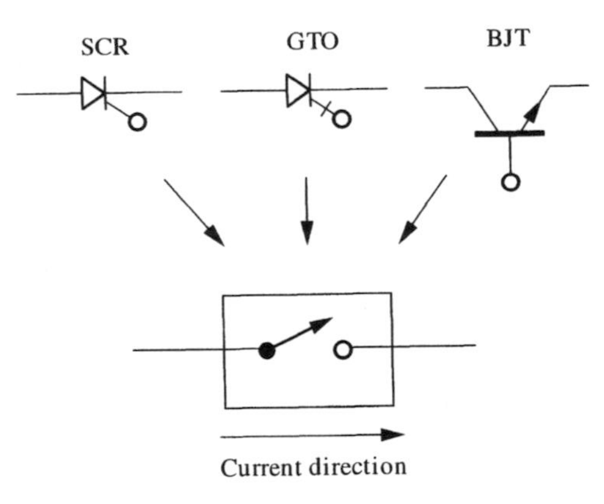

Figure 7.19 The "switch in a box" represents a semiconductor switch. The allowed current direction is shown. All switches require forward bias in this direction to conduct.

Single-Phase Uncontrolled Rectifier Analysis

LEARNING OBJECTIVE 3.

To understand the operation of a controlled single-phase, full-wave rectifier driving a dc motor

Introduction. In this section, we analyze a single-phase rectifier that drives a dc motor from an ac source. A full treatment of this subject would analyze both single- and three-phase rectifiers and deal with dc motors with and without free-wheeling diodes. Due to limitation of space, we consider only two circuits: uncontrolled and controlled full-wave, single-phase rectifiers driving a dc motor with a free-wheeling diode.

Problem description. Figure 7.20 shows the circuit we now analyze. The full-wave rectifier uses a center-tapped transformer, but a bridge rectifier also works. The free-wheeling diode prevents the armature voltage from going negative, as can happen due to the inductance in the armature. We assume a separately excited motor with con-

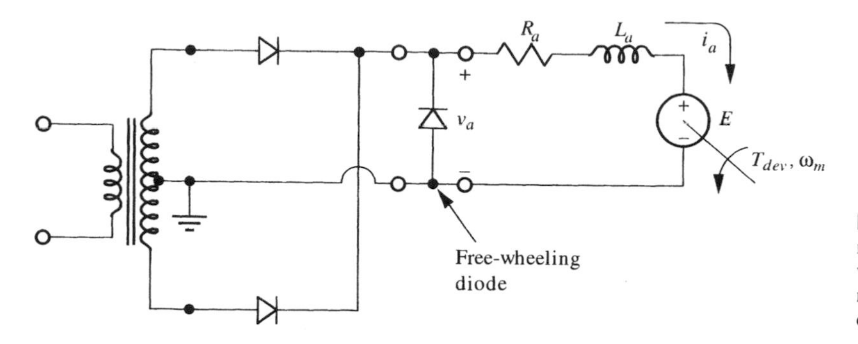

Figure 7.20 Uncontrolled full-wave rectifier driving a dc motor with free-wheeling diode. The rectifier could be replaced with a bridge circuit with four diodes.

stant nameplate field flux. The goal of our analysis is to determine the conditions required to drive the motor at nameplate conditions and to examine the performance of the rectifier–motor at other load conditions.

Motor nameplate conditions. Throughout this section we illustrate with the motor analyzed at the beginning of Sec. 6.3. The nameplate information is 1 hp, 180 V, 4.9 A, 1.78-Ω armature resistance, 30-mH armature inductance, 1750 rpm (183.3 rad/s), and 2050 rpm maximum. The nameplate emf and torque are $E_{NP} = 180 - 4.9\text{ A} \times 1.78\ \Omega = 171.3\text{ V}$, and $T_{NP} = 746\text{ W}/183.3\text{ rad/s} = 4.07\text{ N-m}$, respectively. The brochure describing this motor states the motor should operate with full-wave controlled rectifiers with 230 nominal ac voltage input. However, we are at this stage of the analysis using uncontrolled rectifiers and leaving the ac voltage unspecified.

Continuous-current assumption. The analysis of the circuit in Fig. 7.20 depends on whether the current in the armature is continuous or discontinuous. We assume the current to be continuous, that is, to flow throughout the entire ac cycle. These conditions follow:

- If the current is continuous, then one of the diodes is always ON.
- The diode that is ON must be the diode that is connected to the most positive input voltage. Thus, the armature voltage is always positive.
- The free-wheeling diode never conducts because the armature voltage is always positive.

These conditions lead to a relatively simple analysis. We therefore assume continuous current and nameplate conditions and check afterward to confirm the assumption of continuous current.

The Frequency Domain

Armature voltage. Because the diode connected to the positive end of the transformer is ON, the armature voltage is simply a full-wave rectified sinusoid, as shown in Fig. 7.21. Our analysis of the circuit is based on the Fourier series of the armature voltage

$$v_a(t) = v_{FW}(t) = V_p\left[\frac{2}{\pi} + \frac{4}{3\pi}\cos(2\omega t) - \frac{4}{15\pi}\cos(4\omega t) + \cdots\right] \tag{7.18}$$

where ω is the radian frequency of the ac voltage, 120π for 60 Hz, and V_p is the peak voltage from the centertap to the ends of the transformer. The voltage consists of a dc component plus the even harmonics.

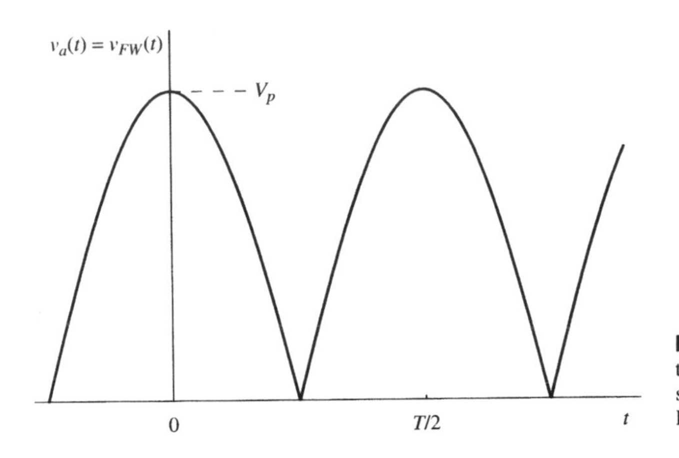

Figure 7.21 With continuous current, the armature voltage is a full-wave rectified sinusoid. The Fourier series is given in Eq. (7.18).

DC current. At dc, the armature inductance is a short circuit, and the armature dc voltage and current are

$$V_a = \frac{2}{\pi} V_p \text{ and } I_a = \frac{(2/\pi)V_p - E}{R_a} \tag{7.19}$$

EXAMPLE 7.5

Required ac voltage

Find the required ac voltage into the full-wave rectifier for nameplate operation of the motor assuming continuous current.

SOLUTION:
For nameplate operation, the input dc current is 4.9 A and the back emf is 171.3 A. With the resistance of the armature at 1.78 Ω, Eq. (7.19) shows the dc voltage to be 180 V, which is, of course, the nameplate input voltage. Therefore

$$\frac{2}{\pi} V_p = 180 \quad \Rightarrow \quad V_p = 282.7 \text{ V, or } 199.9 \text{ V(rms)} \tag{7.20}$$

The transformer turns ratio must be chosen to have approximately 400 V rms on the full secondary, or 200 V rms to the centertap.

Second-harmonic current. We must consider the fluctuations in the current to confirm that the current remains positive. At the second and higher harmonics, the emf, which is a dc source, is treated as a short circuit. The second-harmonic current magnitude is the second-harmonic voltage divided by the magnitude of the circuit impedance at the second-harmonic frequency:

$$I_2 = \frac{(4/3\pi)V_p}{|R_a + j2\omega L_a|} = \frac{4V_p}{3\pi\sqrt{R_a^2 + 4\omega^2 L_a^2}} \tag{7.21}$$

and similarly for the higher harmonics. The approximate condition for continuous current is that the second-harmonic peak current is smaller than the dc current; otherwise, the fluctuations of the current try to go negative. This requires

$$I_2 < I_a \quad \Rightarrow \quad \frac{(2/\pi)V_p - E}{R_a} < \frac{4V_p}{3\pi\sqrt{R_a^2 + 4\omega^2 L_a^2}} \tag{7.22}$$

Equation (7.22) places a limitation on the emf

$$E < \frac{2}{\pi} V_p \left(1 - \frac{2}{3\sqrt{1 + 4\tan^2\phi}}\right) \tag{7.23}$$

where $\tan\phi = \omega L_a / R_a$ is the angle of the armature impedance at the fundamental frequency. Equation (7.23) gives the approximate condition for continuous current in the motor.

EXAMPLE 7.6

Is the previous example valid?

Confirm, if possible, that the motor in Example 7.5 has continuous current.

SOLUTION:

The angle of the armature impedance at the fundamental is

$$\tan\phi = \frac{\omega L_a}{R_a} = \frac{120\pi \times 0.030}{1.78} = 6.354 \rightarrow \phi = 81.1^\circ \tag{7.24}$$

and thus Eq. (7.23) yields

$$E < 180\left[1 - \frac{2}{3\sqrt{1 + 4(6.354)^2}}\right] = 170.6 \tag{7.25}$$

which is less than the value of 171.3 at nameplate conditions. Thus, the current is not continuous at nameplate operation by this criterion.

WHAT IF?

What if the fourth harmonic is considered?[6]

Conclusion. We conclude that this mode of operation is acceptable, but the analysis is not robust. The copper losses in the armature are increased by approximately 50% due to the second-harmonic current. This motor is designed for this type of operation and tolerates the extra losses. If the load demanded less than the nameplate torque, the motor current would become discontinuous. This would not hurt the motor; it would merely hurt our simple analysis. However, if the load were too small, the motor speed would increase dangerously.

[6] The harmonics are out of phase: $I_2 - I_4 = 5.29 - 0.53 = 4.76$ A, so the current is barely continuous.

Controlled Rectifier Operation: Constant Speed Analysis

Introduction. We move on to the analysis of a controlled rectifier rather than analyzing the uncontrolled rectifier circuit for discontinuous current, because the latter is in fact a special case of the analysis we now undertake.

Effect of gating. Figure 7.22 shows a controlled full-wave rectifier circuit driving a dc motor. We replaced the diodes by controlled switches SW1 and SW2. The ac voltage is applied at $t = 0$, and we analyze circuit behavior for $t > 0$. Figure 7.23 shows the

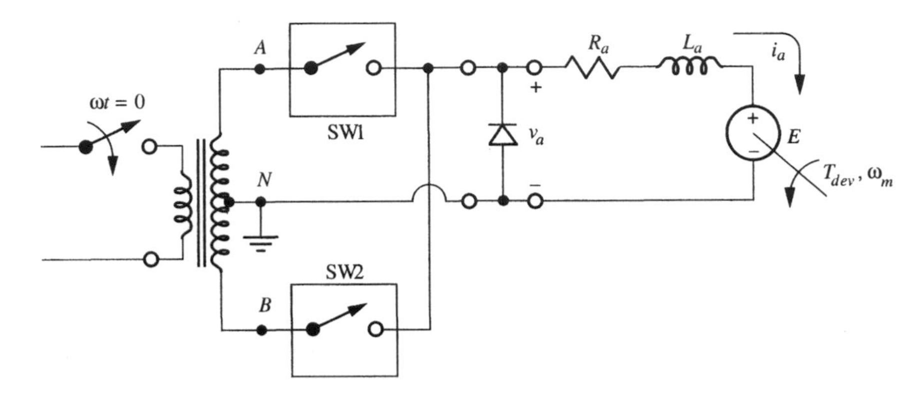

Figure 7.22 A switched full-wave rectifier with a dc motor for a load.

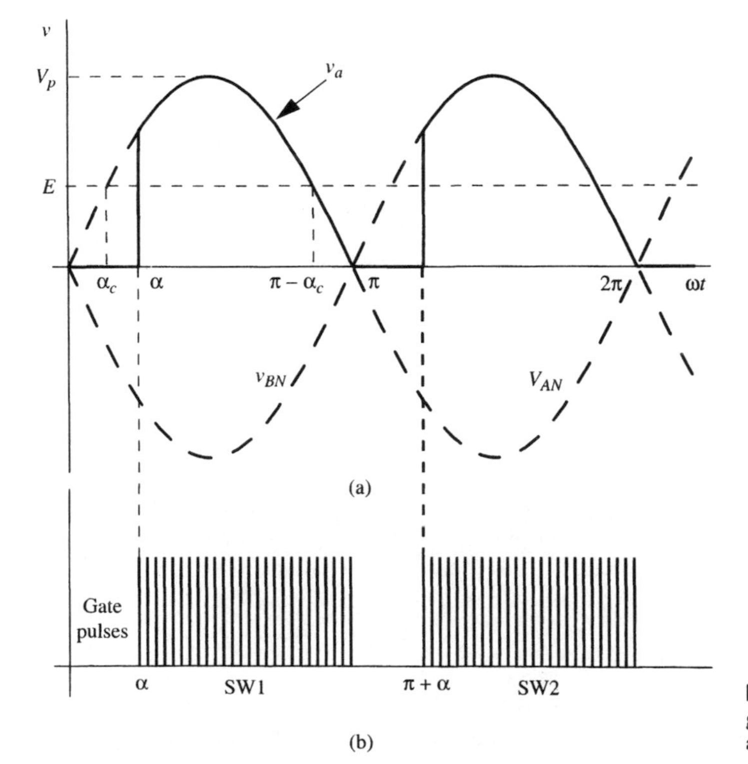

Figure 7.23 (a) Applied voltage and (b) gating signals as a function of electrical angle, ωt. The critical angle is α_c.

ac voltages and the gating signals with the time axis expressed in electrical angle. The gating signals consist of sequences of pulses beginning at $\omega t = \alpha$ for SW1 and $\omega t = \pi + \alpha$ for SW2 and repeated with the ac period. This type of gating signal is required because a switch may not conduct when first gated because it may be reverse-biased due to the emf.

critical delay angle

Critical delay angle. If SW1 were gated before the critical delay angle, α_c, it would come ON at α_c when the input voltage first exceeds the emf, E. This *critical delay angle* is

$$\alpha_c = \sin^{-1} \frac{E}{V_p} \tag{7.26}$$

where V_p is the peak amplitude of the ac voltage. Similarly if α exceeds $\pi - \alpha_c$, neither switch will come ON and the rectifier supplies no current. We analyze for $\alpha_c < \alpha < \pi - \alpha_c$, because this is the region of normal operation. Figure 7.23(a) shows the input ac gated ON at $\omega t = \alpha$ and remaining connected to the armature until $\omega t = \pi$, when the free-wheeling diode prevents the armature voltage from going negative. The voltage marked v_a would be the armature voltage if the current flows continuously.

Three cases. With fixed speed (fixed emf), we find three regimes of operation. When the firing angle is large, current flows through the armature for a short period of time and the free-wheeling diode never conducts. We call this Case A. Then there is a range of firing angles for which the motor current flows for a longer period of time, but not continuously, and the free-wheeling diode conducts for part of the cycle, Case B. Finally, there is a range of firing angles that causes the current to flow continuously, Case C. Each of these cases must be analyzed separately because different equations apply in each.

Circuit operation for Case A. Figure 7.24 shows the waveform applied to the load for $\alpha = 115°$, which is Case A. The input voltage is $230\sqrt{2}$ and the motor speed is $n = 1750$ rpm. Firing the switch begins a buildup of current in the load because the input ac voltage exceeds the back emf. The armature current builds up gradually due to the inductance, and thus the current remains positive after the source voltage falls below the emf. When the load current goes to zero and tries to reverse, the switch turns OFF by line commutation. During the periods when the instantaneous armature current is zero, the motor is not connected to the ac voltage and the load voltage is equal to the emf.

Armature-current analysis. The analysis of the circuit response involves a transient solution with a steady-state condition consisting of a sinusoidal and a dc response. Once SW1 is fired, the circuit is that shown in Fig. 7.25. The DE for the circuit is

$$L_a \frac{di_a}{dt} + R_a i_a = V_p \sin(\omega t) - E \tag{7.27}$$

and its solution consists of a transient and a steady-state response

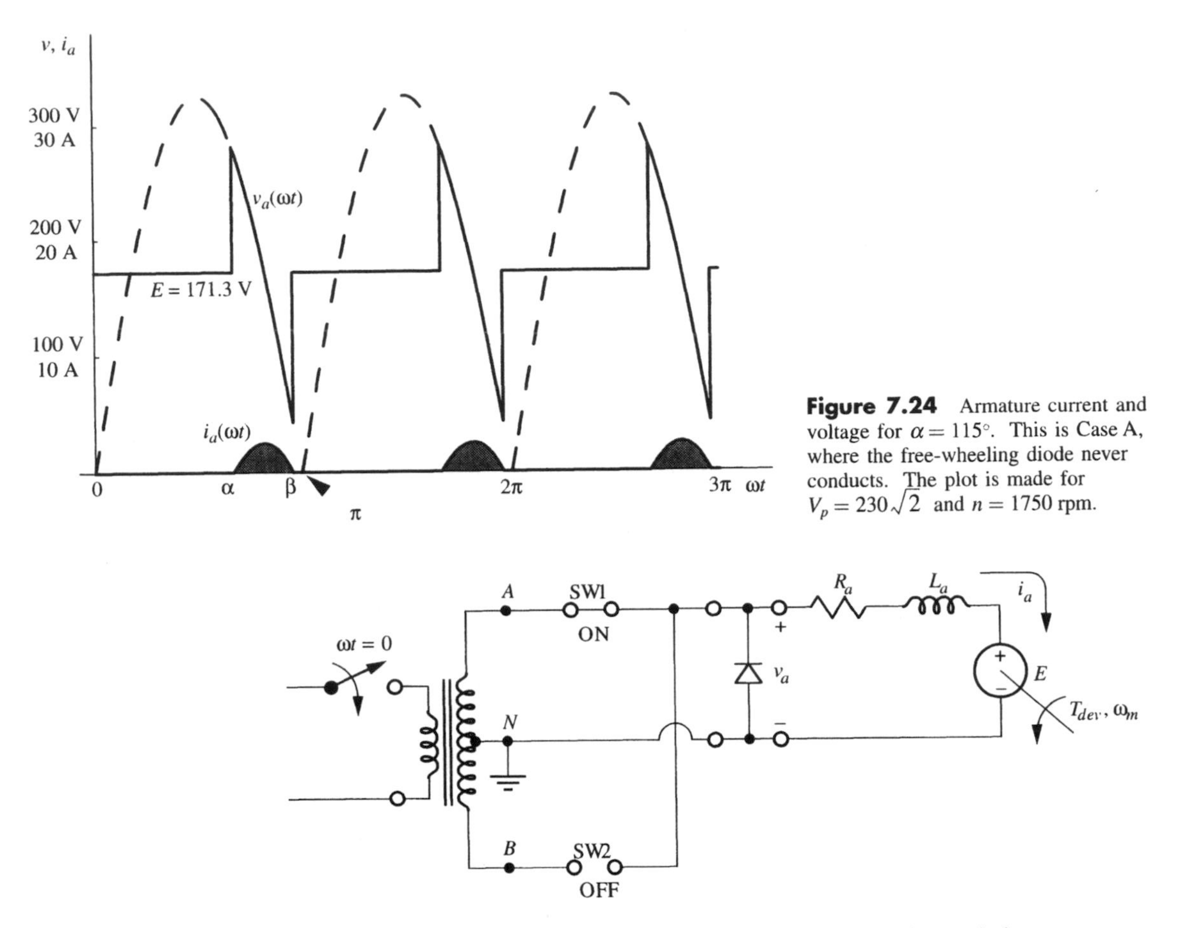

Figure 7.24 Armature current and voltage for $\alpha = 115°$. This is Case A, where the free-wheeling diode never conducts. The plot is made for $V_p = 230\sqrt{2}$ and $n = 1750$ rpm.

Figure 7.25 Circuit after SW1 is fired, for the transient analysis.

$$i_a(t) = Ae^{-t/\tau} + \frac{V_p}{Z}\sin(\omega t - \phi) - \frac{E}{R_a} \tag{7.28}$$

where A is an unknown constant to be determined from the initial condition, $\tau = L_a/R_a$ is the time constant of the transient, $Z = \sqrt{R_a^2 + (\omega L_a)^2}$ is the magnitude of the impedance at the input frequency, ω, of the resistor and inductor in series, and $\phi = \tan^{-1}(\omega L_a/R_a)$ is the phase shift from the inductor. The sinusoidal portion of the current is determined from routine ac analysis, using the sine function for the ac source. The dc portion of the forced response involves only the emf and the resistor because the inductor and the ac source are short circuits at dc.

In Case A, $\alpha > \alpha_c$ such that SW1 fires at $\omega t = \alpha$. Because of the inductor, the initial current must be zero, so the initial condition at $\omega t = \alpha$ is

$$i_a(\alpha) = Ae^{-\alpha/\tan\phi} + \frac{V_p}{Z}\sin(\alpha - \phi) - \frac{E}{R_a} = 0 \tag{7.29}$$

where in the exponential we have replaced t/τ by $\alpha/\omega\tau$, and then replaced $\omega\tau$ by $\omega L_a/R_a = \tan\phi$. Solving for A, we find

$$A = \left[\frac{E}{R} - \frac{V_p}{Z}\sin(\alpha - \phi)\right] e^{\alpha/\tan\phi} \tag{7.30}$$

and so the current is

$$i_a(\omega t) = \frac{V_p}{Z}[\sin(\omega t - \phi) - e^{(\alpha - \omega t)/\tan\phi}\sin(\alpha - \phi)] - \frac{E}{R}[1 - e^{(\alpha - \omega t)/\tan\phi}] \tag{7.31}$$

extinction angle

Equation (7.31) is valid as long as $i_a > 0$. When the current tries to reverse, the switch turns OFF through line commutation, and the current ceases. The *extinction angle*, β, is defined as the angle where $i_a(t)$ crosses zero going negative, so β satisfies Eq. (7.32)

$$i_a(\beta) = 0 = \frac{V_p}{Z}[\sin(\beta - \phi) - e^{(\alpha - \beta)/\tan\phi}\sin(\alpha - \phi)] - \frac{E}{R_a}[1 - e^{(\alpha - \beta)/\tan\phi}] \tag{7.32}$$

Figure 7.24 shows β. For a specific circuit and value of α, Eq. (7.32) may be solved for β with numerical techniques; be sure to seek the first zero crossing after $\omega t = \alpha$.

EXAMPLE 7.7 Find the extinction angle, β

Find the extinction angle, β, for $\alpha = 115°$.

SOLUTION:
Equation (7.32) is nonlinear and may be solved by a "solve" routine on a calculator or computer. Another approach is to use Eq. (7.31) and seek the first zero crossing. For $\alpha = 115°$ and the system parameters given on page 309, we calculate the currents in Table 7.1.

TABLE 7.1 Value of Armature Current for $\alpha = 115°$

ωt (deg)	$i_a(\omega t)$
160	2.675
170	1.273
180	−0.939

The angle for which $i_a(\omega t) = 0$ may be estimated by linear interpolation between the last two values to be 175.8°. Thus, the extinction angle is approximately $\beta = 175.8°$.

WHAT IF? What if you want more accuracy?[7]

Armature voltage for Case A. While the switch is ON and current flows, the armature voltage is equal to the input sinusoidal voltage, and the armature voltage is E for the remainder of the cycle.[8] The time-average load voltage, averaged over one-half the period, is therefore

$$V_a = \frac{1}{\pi}\int_{\alpha}^{\alpha+\pi} v_a(\omega t)\, d(\omega t) = \frac{1}{\pi}\int_{\alpha}^{\beta} V_p \sin(\omega t) d(\omega t) + \frac{1}{\pi}\int_{\beta}^{\alpha+\pi} E d(\omega t) \tag{7.33}$$

$$= \frac{V_p}{\pi}(\cos\alpha - \cos\beta) + E\left(1 - \frac{\gamma}{\pi}\right)$$

conduction angle

where $\gamma = \beta - \alpha$ is the *conduction angle* in radians.

EXAMPLE 7.8 Find voltage, current, and power

Find the time-average armature voltage, current, and power for $V_p = 230\sqrt{2}$, $n = 1750$ rpm, and the motor parameters on page 309.

SOLUTION:

For $\alpha = 115°$ and $\beta = 176.2°$, the average load voltage is from Eq. (7.33)

$$V_a = \frac{230\sqrt{2}}{\pi}(\cos 115° - \cos 176.2°) + 171.3\left(1 - \frac{176.2° - 115°}{180°}\right) = 172.6 \text{ V} \tag{7.34}$$

As we showed in Eq. (7.17), the dc current can be determined from the dc load voltage, the emf, and the resistance:

$$I_a = \frac{V_a - E}{R_a} = \frac{172.6 - 171.3}{1.78} = 0.744 \text{ A} \tag{7.35}$$

The total power given to the emf is given by Eq. (7.15):

$$P_{dev} = EI_a = 171.3 \times 0.744 = 127.5 \text{ W} \tag{7.36}$$

and the total power to the armature circuit is slightly higher because of loss in the armature resistance. The developed power minus the rotational losses gives an output power of 34.2 W for this condition.

---

[7] $i_a(175°) = 0.2868$ A and $i_a(176°) = 0.0432$ A; therefore, $\beta = 176.2°$.

[8] In Case A, the free-wheeling diode never conducts.

Limits for Case A. The maximum delay angle for Case A occurs when the input voltage falls below the emf, $\alpha = \pi - \alpha_c = 148.2°$. For larger values of α, the switches are reverse-biased and never conduct. The minimum delay angle for Case A occurs when $\beta = \pi$, which requires

$$i_a(\pi) = 0 = \frac{V_p}{Z}[\sin(\pi - \phi) - e^{(\alpha - \pi)/\tan\theta} \sin(\alpha - \phi)] - \frac{E}{R_a}[1 - e^{(\alpha - \pi)/\tan\phi}] \tag{7.37}$$

Equation (7.37) can be solved for α by numerical means. For our case, the solution is $\alpha = 109.4°$. The corresponding armature currents for Case A, $109.4° < \alpha < 148.2°$ are $1.12\ \text{A} > I_a > 0\ \text{A}$.

Condition for Case B. For $\alpha < 109.4°$, the extinction angle β exceeds 180° and the armature voltage tries to go negative. For this case, the free-wheeling diode conducts for part of the cycle. Figure 7.26 shows the voltage and current for $\alpha = 75.2°$, which is the value of α that drives the motor at nameplate voltage.

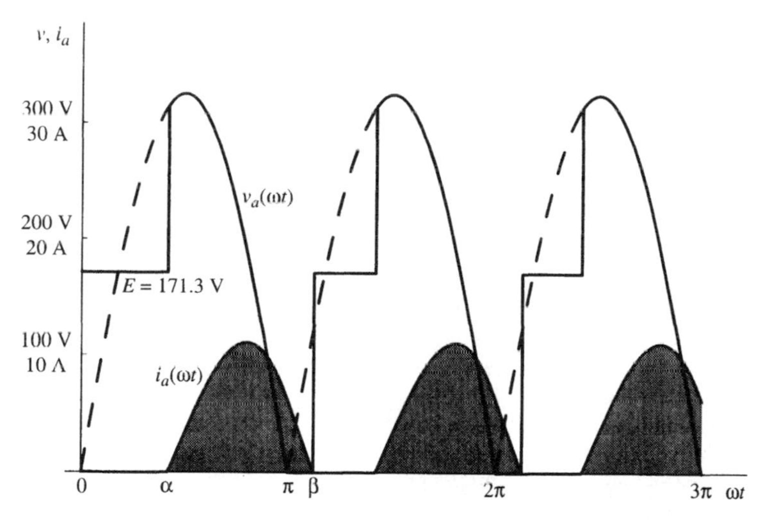

Figure 7.26 Armature current and voltage for $\alpha = 75.2°$. This is Case B, where the free-wheeling diode conducts for part of the cycle.

Role of the free-wheeling diode. The current to the load is increased by the free-wheeling diode across the load. When SW1 is ON and SW2 OFF, the diode is in parallel with the source and is reverse-biased while the source voltage is positive. When the source voltage goes negative, the diode turns ON, and SW1 turns OFF by line commutation. The armature current is free-wheeling through the diode.

Analysis for Case B. The current for the period $\alpha < \omega t < \pi$ is given by Eq. (7.31) if we assume zero current when SW1 is fired. When $\omega t = \pi$, SW1 turns OFF, and the current is

$$I_a(\pi) = \frac{V_p}{Z}[\sin(\pi - \phi) - e^{(\alpha - \pi)/\tan\phi} \sin(\alpha - \phi)] - \frac{E}{R_a}[1 - e^{(\alpha - \pi)/\tan\phi}] \tag{7.38}$$

When $\omega t = \pi$, the free-wheeling diode conducts, SW1 is OFF by line commutation and the current begins to decay in a simple dc transient. The initial value is $i_a(\pi)$, and the final value is $-E/R_a$; hence, the current for $\pi < \omega t < \beta$ is

$$I_a(\omega t') = -\frac{E}{R_a} + \left[i_a(\pi) + \frac{E}{R_a}\right] e^{-t'/\tau} \tag{7.39}$$

where $t' = 0$ at $\omega t = \pi$. Extinction occurs when

$$\beta = \pi + (\omega\tau)\ln\left[1 + \frac{i_a(\pi)R_a}{E}\right] \tag{7.40}$$

At extinction, the free-wheeling diode turns OFF, and the load current remains at zero until a switch is again gated ON in the next cycle. For the present case, we require that $\pi < \beta < \alpha + \pi$, which is true for $109.4° > \alpha > 36.5°$.

Armature voltage for Case B. The armature voltage is equal to the source voltage while the switch is ON, is zero while the diode is ON, and is equal to E while both switch and diode are OFF, as shown in Fig. 7.26. The average load voltage is

$$V_a = \frac{1}{\pi}\int_{\alpha}^{\pi} V_p \sin(\omega t)\, d(\omega t) + \frac{1}{\pi}\int_{\beta}^{\alpha+\pi} E\, d(\omega t) \tag{7.41}$$

$$= \frac{V_p}{\pi}(1 + \cos\alpha) + E\left(1 - \frac{\gamma}{\pi}\right)$$

which yields nameplate conditions of 180.0 V for $E = 171.3$ V, $\alpha = 75.2°$, and the circuit parameters given on page 309. Equation (7.17) gives the average load current as 4.90 A. The full range of currents for Case B is $1.12 < I_a < 8.68$ A. Thus, Case B covers the remainder of the operating region of the motor. However, for the sake of completeness, we discuss Case C, in which the armature current flows continuously.

Analysis for Case C. For $\alpha < 36.5°$, the current becomes continuous and a transient period passes before the current reaches steady state. Figure 7.27 shows the first three half cycles for $\alpha = \alpha_c = 31.8°$. The diode never conducts, and the average voltage in steady state is

$$V_a = \frac{1}{\pi}\int_{\alpha}^{\pi} V_p \sin(\omega t)\, d(\omega t) = \frac{V_p}{\pi}(1 + \cos\alpha) \tag{7.42}$$

which is the same as Eq. (7.41) for a conduction angle of 180°. For this firing angle, Eq. (7.42) yields $V_a = 191.6$ V and the armature current is 11.4 A. As the firing angle continues to increase, the applied waveform approaches that of a conventional full-wave rectifier, and the dc current to the armature approaches that given by Eq. (7.19).

Summary. The controlled full-wave rectifier delivers dc voltage and current to the dc motor armature. The current may be continuous or discontinuous. A free-wheeling

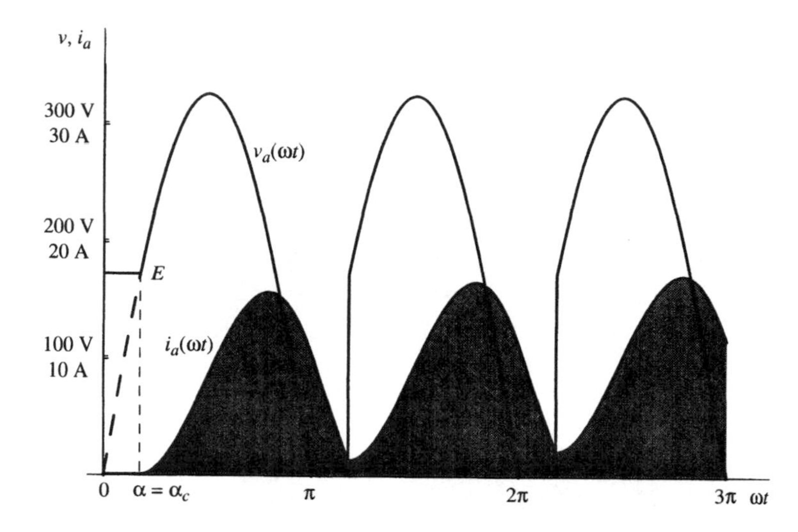

Figure 7.27 For $\beta > \alpha + \pi$, the current becomes continuous and builds up to a steady state. This is Case C. We show the first three cycles for $\alpha = \alpha_c = 31.8°$.

diode across the load prevents the armature voltage from going negative and thus increases the armature current and reduces harmonics. Figure 7.28 shows the delay and extinction angles plotted against armature current for Cases A and B. These curves were calculated by first assuming the delay angle, α, then calculating the extinction angle, β, then the armature voltage, and from that the current. The analysis must distinguish three cases, depending on whether the free-wheeling diode conducts and whether the current is continuous or discontinuous. As we shall see in what follows, there are even other states that can come into play.

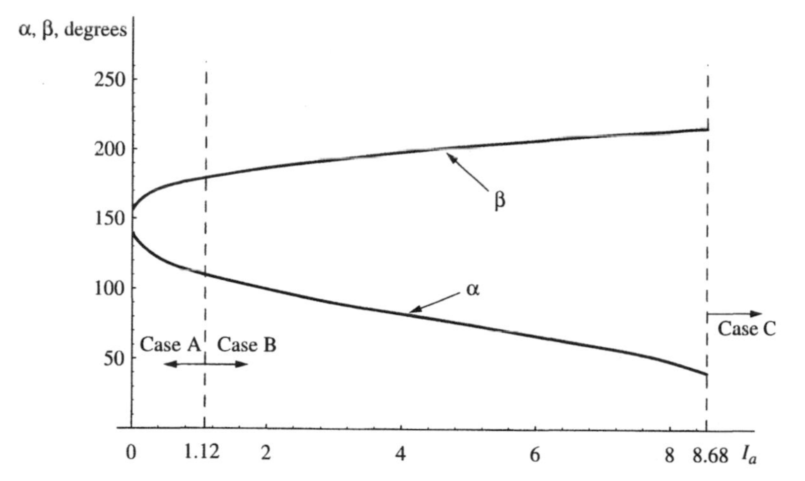

Figure 7.28 Delay and extinction angle as a function of the resulting current. This assumes a fixed motor speed of 1750 rpm. The independent variable is the delay angle, α, and the extinction angle, β, and armature current are calculated. For Cases A and C, the free-wheeling diode never conducts.

Motor Performance with Constant Firing Angle

Nameplate conditions. As asserted earlier, nameplate conditions occur with $\alpha = 75.2°$ provided the load requires nameplate torque at nameplate speed. In this section, we analyze the motor with constant firing angle and determine the output torque as a function of motor speed. The procedure is to assume a speed, determine the back emf from the speed, and then find the extinction angle, β, from either Eq. (7.32) or (7.40), depending on whether Case A or B proves to be consistent.[9] From the firing and extinction angles, we can determine the dc armature voltage from either Eq. (7.33) or (7.41), and then the armature current and developed torque. After subtracting the loss torque determined in Eq. (6.48), we have the output torque.

Case B conditions. We now determine the limits for which Case B conditions exist. The minimum speed is that for which the conduction angle, γ, is 180°, which requires that $\beta = \alpha + 180° = 255.2°$. Thus, we can substitute this angle into Eq. (7.40) and solve for the emf, E. The result is $E = 112.6$ V, and the corresponding speed and armature current are 1151 rpm and 9.98 A, respectively.

The maximum speed for which Case B conditions exist is that at which the extinction angle is 180°. Thus, we can substitute $\alpha = 75.2°$ and $\beta = 180°$ into Eq. (7.32) and solve for the emf. The result is $E = 215.9$ V, which corresponds to a speed of 2206 rpm and current of 2.60 A.

Case A conditions. Case A begins where Case B leaves off and continues until the critical angle equals the delay angle. This occurs when

$$E = V_p \sin \alpha = 230\sqrt{2} \sin 75.2° = 314.5 \text{ V} \tag{7.43}$$

which corresponds to 3213 rpm, $\beta = 119.4°$, and $I_a = 0.03$ A. Because the motor losses require more current than this, we conclude that this condition is not reached in practice. Indeed, this high speed could damage the motor and should be prevented from occurring in practice.

Torque versus speed for $\alpha = 75.2°$. The results of the calculations described before are shown in Fig. 7.29 alongside the torque characteristic for fixed nameplate voltage of 180 V. We have marked the limits imposed by armature current and speed constraints. Clearly, active control of the system is required to prevent exceeding motor limitations as load requirements change.

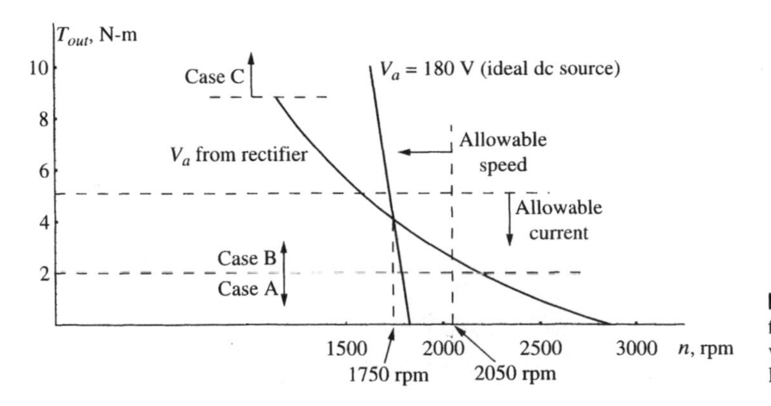

Figure 7.29 Torque characteristic with fixed delay angle and fixed armature voltage. The armature speed and current limits are shown.

[9] Case C exceeds the motor limits and will not be considered.

Harmonics in the ac line. When a motor is driven by an controlled rectifier, the current that flows in the ac power system is nonsinusoidal, containing in general all the even harmonics of 60 Hz. The harmonics constitute noise for other users of the ac power circuit and can disturb sensitive electronic circuits. The issue of power quality is growing in importance to power companies as more power-electronic equipment comes into use.

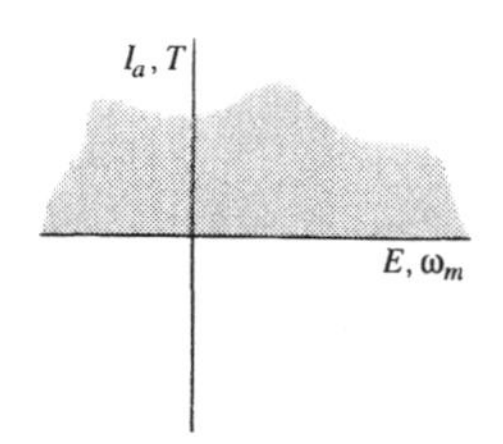

Figure 7.30 The controlled rectifier in Fig. 7.22 operates in the first and second quadrants.

Operation in the second quadrant. Without the free-wheeling diode, the circuit can put out negative armature voltage for $\alpha > 90°$ and small or negative emf. As α is advanced more and more, the switches are gated ON and continue conducting when the input voltage is negative, making the load voltage negative. However, the load current must be kept positive in the indicated direction because the switches in the controlled rectifier in Fig. 7.22 permit only positive load current. This can occur when $V_a > E$. If E is positive, V_a has to be more positive. This can also occur if E is negative and V_a is less negative. This circuit controls motor torque, therefore, in the first and second quadrants, as shown in Fig. 7.30. The controller is incapable of drive in the reverse direction, but could operate as a brake if the motor direction were reversed by the mechanical system.

Inverter operation. In the circumstance where the emf is negative, but the current is positive, the emf acts as a generator and delivers power to the ac source. The system thus can function as an inverter, converting dc power to ac power. The same may be accomplished in the four-quadrant converter described in what follows, in which the current may reverse.

Four-quadrant controller. Figure 7.31(a) shows a controlled rectifier that operates in all four quadrants without a free-wheeling diode, as shown in Fig. 7.31(b). The original switches, SW1 and SW2, are gated for positive load current, and SW3 and SW4 are gated for negative load current. We can envision the rectifier operating with continuous positive current for small α and going into discontinuous positive current as α is increased. Somewhere near $\alpha = 90°$, the current reverses and the motor slows and reverses. Further increases of α drive the motor in the reverse direction. At some point, the current may become continuous in the negative direction.

The circuit in Fig. 7.31 is completely symmetrical and may be analyzed by the methods given before; indeed, the various formulas may be adapted for this type of operation. However, the control of the switches is complicated by the possibility of reversal of the load current, for the direction of the current flows determines which switches to turn ON.

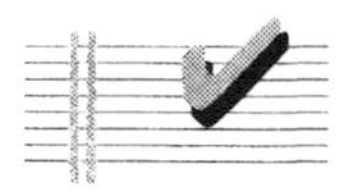

Check Your Understanding

1. A free-wheeling diode increases or decreases the armature current to a dc motor. Which?
2. If a dc motor draws continuous current from a source at a given speed, it will draw continuous current for higher speeds, everything else being held the same. True or false?
3. The peak of the current in Fig. 7.26 corresponds roughly to what condition between input voltage and emf?

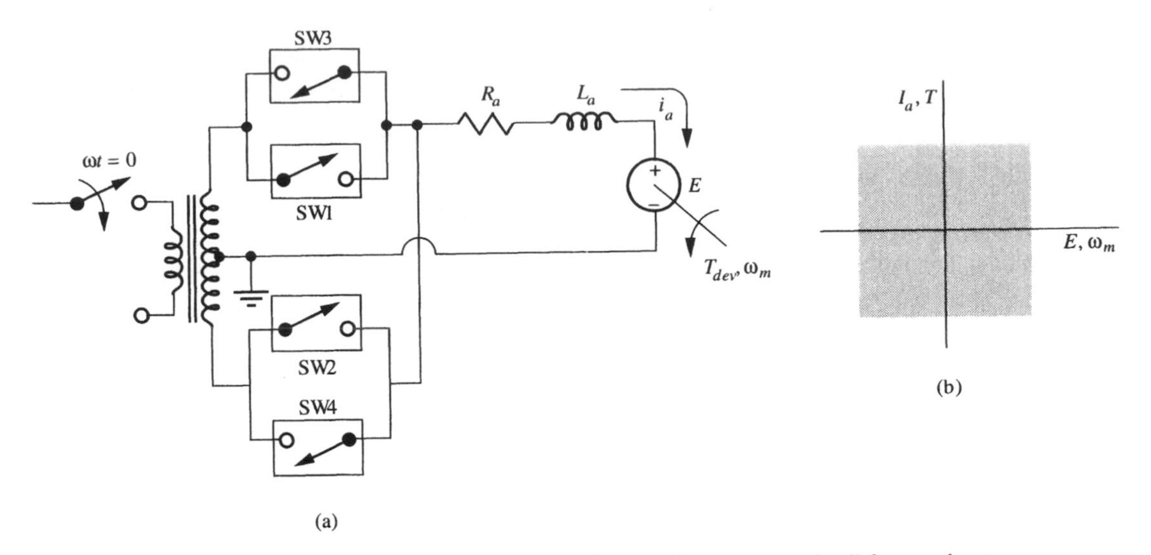

Figure 7.31 A controlled rectifier capable of operation in all four quadrants.

4. For an uncontrolled rectifier drive of a dc motor, the value of α is always the critical value. True or false?
5. For a controlled single-phase rectifier, the extinction angle cannot be greater than the firing angle by more than 180°. True or false?

Answers. **(1)** Increases for Cases B and C; **(2)** false; **(3)** the current maximum occurs roughly when the input voltage and emf are equal because the armature impedance is largely inductive; **(4)** true, if the current is discontinuous; **(5)** true.

7.3 AC MOTOR CONTROLLERS

Many types of ac motor controllers are currently in use, and this is an active field of research and development, with continual innovations in response to improvements in semiconductor devices and control techniques. Controllers are tailored to specific types of motors, such as induction motors and synchronous motors. We limit our discussion to the control of induction motors, and examine the more common types of power controllers.

LEARNING OBJECTIVE 4.

To understand the methods and complexities of ac motor controllers

Speed control of induction motors. Of the various means for controlling the speed of a three-phase induction motor, we consider the two most common: variation of the magnitude of the applied voltage and the variation of frequency of the applied voltage.

Figure 7.32 shows the effect of lowering the voltage applied to an induction motor. The principal effect is to lower the torque, but with certain loads, this leads to a measure of speed control.

Figure 7.33 shows the torque characteristics of a three-phase induction motor as the frequency of the applied voltage is varied but the voltage amplitude is kept constant. Magnetic saturation has been ignored. At low frequencies, the impedance of the motor

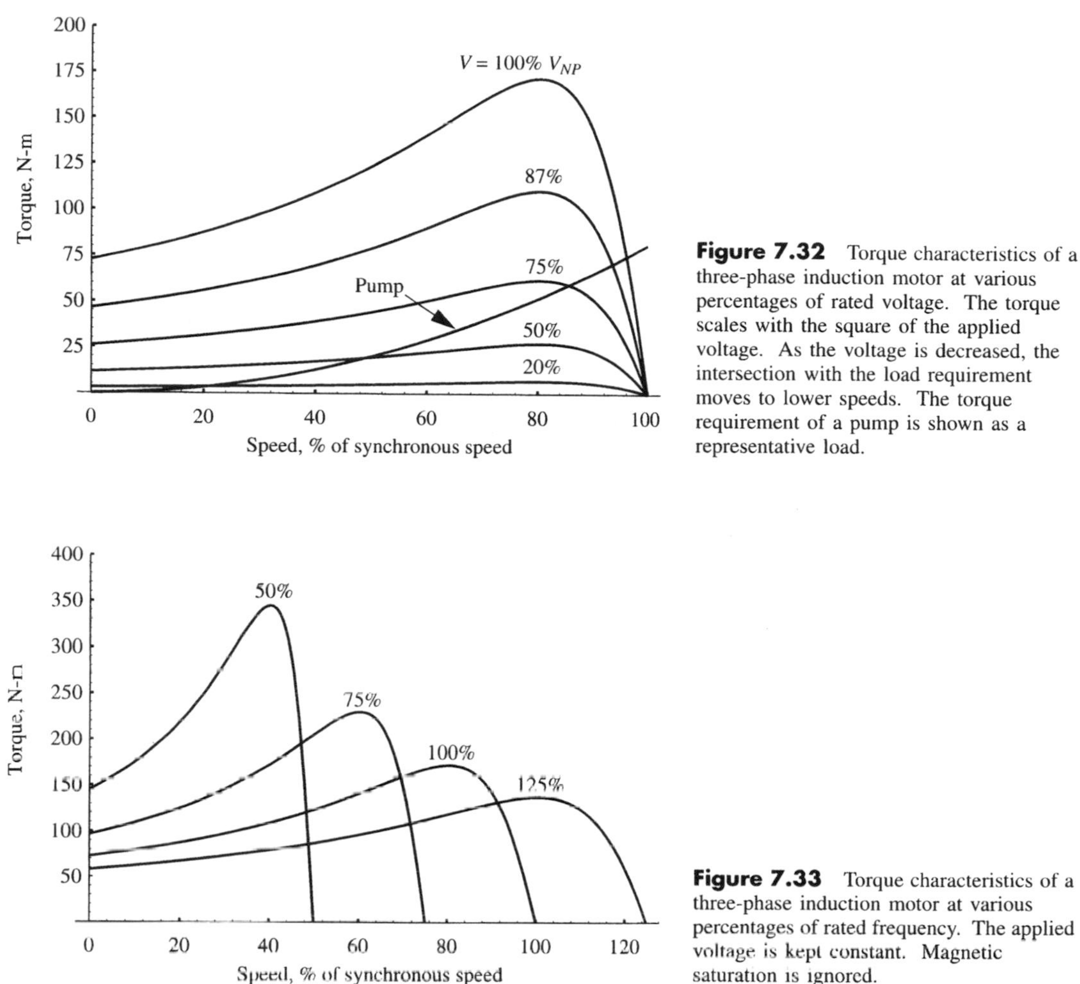

Figure 7.32 Torque characteristics of a three-phase induction motor at various percentages of rated voltage. The torque scales with the square of the applied voltage. As the voltage is decreased, the intersection with the load requirement moves to lower speeds. The torque requirement of a pump is shown as a representative load.

Figure 7.33 Torque characteristics of a three-phase induction motor at various percentages of rated frequency. The applied voltage is kept constant. Magnetic saturation is ignored.

is low, current is increased, and torque is high. However, high currents cause magnetic saturation and increased losses. For this reason, motor controllers vary applied voltage in proportion to frequency to keep the current roughly constant, which is known as constant volts/hertz drive.

Voltage control of induction motors. Figure 7.34 shows a fixed-frequency, variable-voltage controller for a three-phase induction motor. Such a controller is useful for loads whose torque requirements increase strongly with speed, such as fans and pumps. The controller consists of six back-to-back switches that switch the three-phase ac waveform. The output of the converter is an ac waveform that can be varied in amplitude, plus harmonics. The model for the induction motor is that of Fig. 5.14, except that for simplicity, we have treated the Thèvenin voltage as the per-phase voltage and ignored mechanical losses.

IDEA 6 **Equivalent Circuits**

Torque characteristics. Considering only the fundamental in the voltage applied to the induction motor, Eq. (5.37) gives the developed torque as

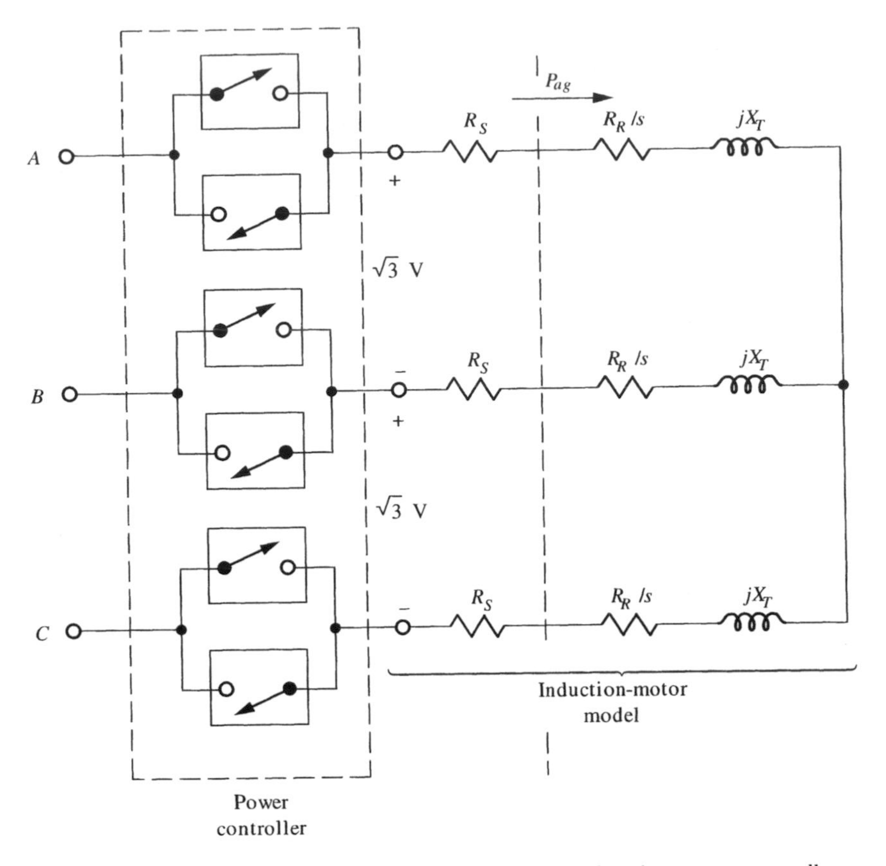

Figure 7.34 Circuit model of a fixed-frequency, variable-voltage motor controller with a model of a three-phase induction motor. The slip, s, represents the motor speed and the air-gap power, P_{ag}, is proportional to the motor torque.

$$T_{dev} = \frac{3V^2}{\omega_s} \times \frac{R_R'/s}{[R_S + R_R'/s]^2 + X_T^2} \tag{7.44}$$

where V is the per-phase voltage, R_S is the stator per-phase resistance, R_R' is the rotor per-phase resistance as seen from the stator, and X_T is the total reactance per phase. Torque is proportional to the voltage squared, and thus the motor/load speed can be controlled if the load torque varies strongly with speed.

EXAMPLE 7.9 Induction motor with a pump as load

Consider a six-pole, 230-V, 60-Hz, three-phase induction motor with $R_S = 0.2\ \Omega$, $R_R' = 0.2\ \Omega$, and $X_T = 1.0\ \Omega$. The load torque varies as the square of speed and operates with a slip of 4% at rated voltage. Calculate the system speed as the ac voltage amplitude is varied. Figure 7.32 shows the output torque of the motor and the load-torque requirements.

SOLUTION:
The developed torque at 4% slip is, from Eq. (7.44),

$$T_{dev}(s = 4\%) = \frac{3(230/\sqrt{3})^2}{40\pi} \times \frac{0.2/0.04}{[0.2 + (0.2/0.04)]^2 + 1^2} = 75.1 \text{ N-m} \quad (7.45)$$

and because speed is proportional to $1 - s$, the load torque is generally

$$T_L(s) = 75.1 \times \left(\frac{1-s}{0.96}\right)^2 \quad (7.46)$$

Combining Eqs. (7.44) and (7.46), we have for the system

$$75.1\left(\frac{1-s}{0.96}\right)^2 = \frac{3V^2}{40\pi} \times \frac{0.2/s}{[0.2 + (0.2/s)]^2 + 1^2} \quad (7.47)$$

which may be solved for slip or voltage if either is assumed. Figure 7.35 shows speed versus voltage from Eq. (7.47), and also the speed characteristics of a high-slip motor ($R_R' = 1.0\,\Omega$) . We include the latter to demonstrate its improved speed-control characteristics, albeit at lower efficiency.

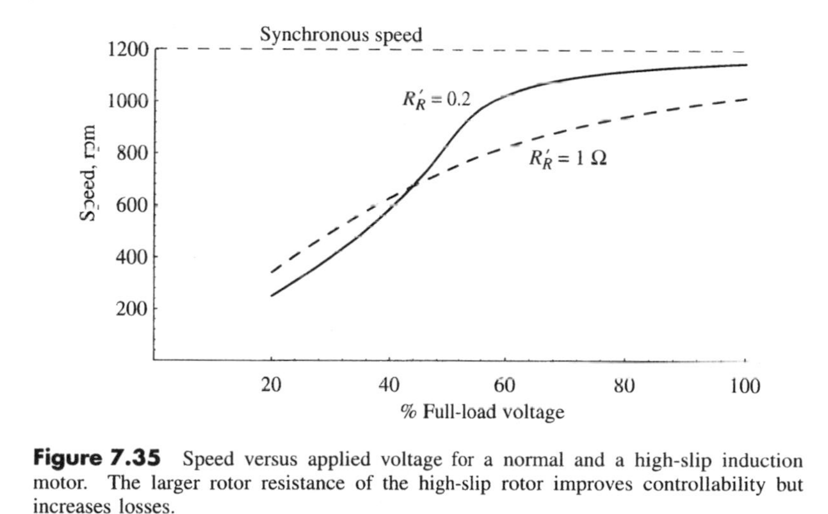

Figure 7.35 Speed versus applied voltage for a normal and a high-slip induction motor. The larger rotor resistance of the high-slip rotor improves controllability but increases losses.

Discussion of fixed-frequency voltage control. This method is limited to smaller loads with favorable speed–torque characteristics. The efficiency is low, and harmonics are generated in the power system. The controller operates in the first and second quadrants, although the latter does not occur with typical loads.

DC-Link Variable-Frequency Controllers

Basic configuration. Figure 7.36 represents a dc-link variable-frequency motor controller. The input ac power is converted to dc, filtered, and then converted to variable-frequency three-phase ac by an inverter. The switches are numbered in the order that they are turned ON, with the complementary switch turned OFF simultaneously; that is,

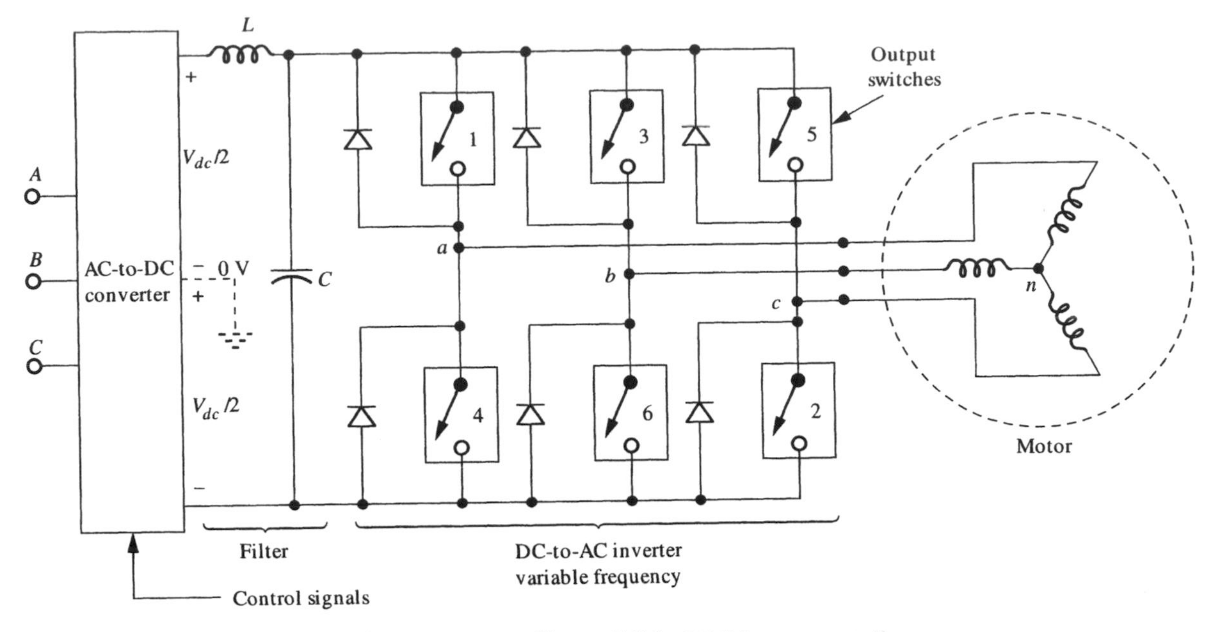

Figure 7.36 DC-link motor controller.

when switch 1 is ON, switch 4 is OFF, and so on. We examine several means by which the switches may convert the dc to three-phase ac power to drive an induction motor. The bypass diodes are required for reactive power flow and to limit the voltage to that of the dc supply. The filter we show supplies the inverter with dc voltage that is largely independent of load current. When the capacitor is omitted, the dc voltage may vary, but the inductor tends to keep the current constant, which offers some advantages. The ac-to-dc converter output may be fixed or variable voltage (or current), depending on the type of inverter scheme and filter used. We show the neutral as a dashed ground because no actual ground is required.

Square-wave inverter. In the square-wave inverter, each phase is connected alternatively to the positive and negative power-supply outputs to give a square-wave approximation to an ac waveform at a frequency of ω_e determined by the gating of the switches. The voltage in each output line is phase shifted by 120° to synthesize a three-phase source. Voltage waveforms for the square-wave inverter are shown in Fig. 7.37(a). The first three waveforms are voltage of phases *a*, *b*, and *c* relative to the neutral on the centertap of the power supply. The bottom waveform gives the voltage of motor terminal *a* relative to the ungrounded neutral on the motor. The circuit in Fig. 7.37(b) shows the configuration when *a* and *c* are connected to the positive dc voltage and *b* is connected to the negative dc voltage. The two phases in parallel give half the impedance of an individual phase and thus get one-third the total dc voltage, whereas the unpaired phase gets two-thirds the dc voltage. Thus, for the first 30°, $v_{an} = \frac{1}{3}V_{dc}$ and for the next 30°, $v_{an} = \frac{2}{3}V_{dc}$, and so on. In this manner, the stair-step voltage for each motor phase is produced. The fundamental component in each waveform is about $0.637 \times V_{dc}$.

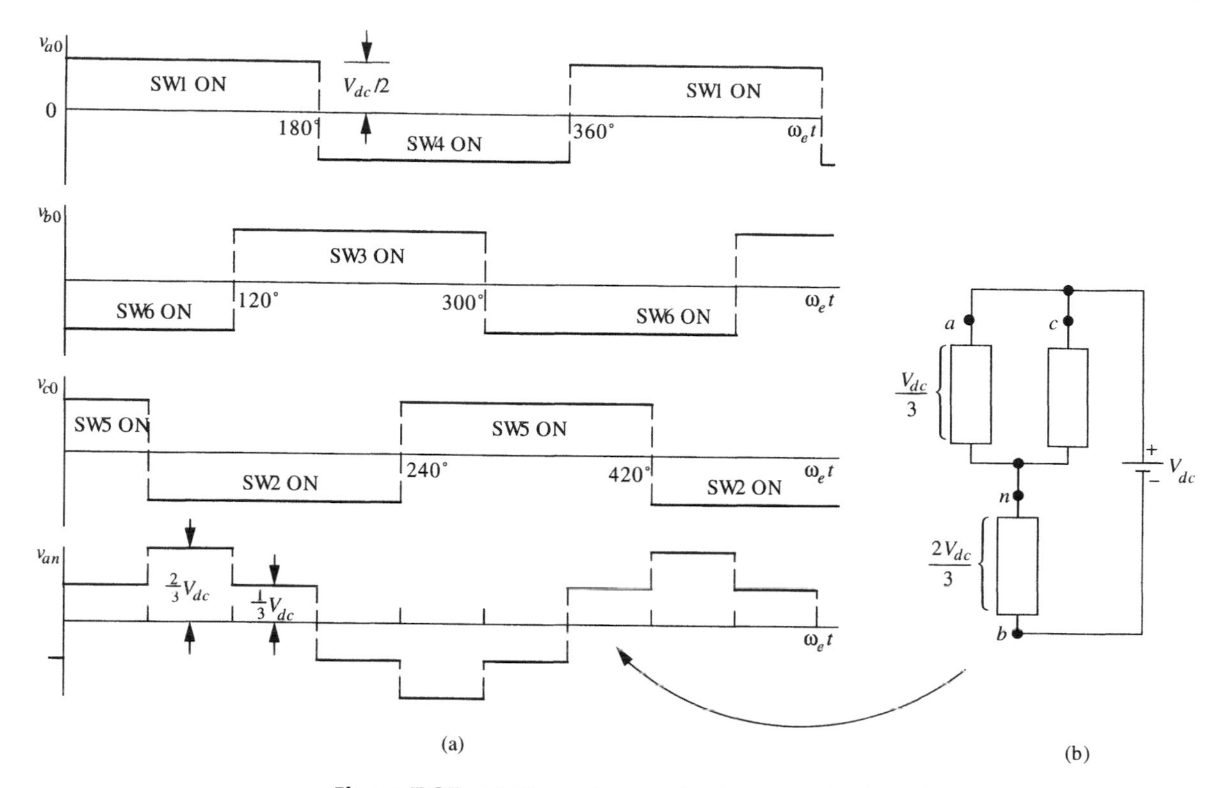

Figure 7.37 (a) The synthesis of the line-to-neutral voltage from a square-wave inverter. The stair-step waveform results from the changing neutral voltage on the motor. (b) Circuit configuration when a and c are connected to the positive and b is connected to the negative dc voltage.

Need for constant voltage/frequency. At frequencies below the rated frequency of the motor, the applied voltage must be reduced. Otherwise, the current to the motor becomes excessive, as implied by Fig. 7.33, and causes magnetic saturation. This follows from Faraday's law

$$\underline{\mathbf{V}} = j\omega_e n\underline{\Phi} \tag{7.48}$$

where ω_e is the electrical frequency; clearly, the voltage must decrease as the frequency is reduced if the peak flux is to be kept constant. Thus, the square-wave inverter requires decreasing dc voltage as motor speed is reduced below rated speed. For this reason, Fig. 7.36 shows control signals to the ac–dc converter.

IDEA 7 **The Frequency Domain**

Harmonics. The square-wave inverter has two types of problems with harmonics. At the input, the controlled rectifier that produces the variable dc voltage creates harmonics that constitute noise in the power system. These can be filtered, but the added complexity degrades the efficiency and power factor, which are already somewhat poor for a controlled rectifier.

The output waveforms also have problems with harmonics. The stair-step-output waveforms in Fig. 7.37(a) contain only odd harmonics. The third, ninth, and so on, cause no problems because they are in phase and thus self-cancel at the input to the motor. But the other harmonics, principally the fifth and seventh, cause currents that increase losses in the motor and produce no torque. Although these harmonics are filtered somewhat by the inductance of the motor, the combination of these problems has caused designers to look for alternatives. The pulse-width modulation (PWM) system discussed in what follows conquers these problems with a more complicated switching sequence.

pulse-width modulation (PWM)

Pulse-width modulation (PWM). Figure 7.38(b) shows 180 electrical degrees of a waveform that has been generated through pulse-width modulation. The basic idea in PWM is to chop pieces out of the wave to control the fundamental in the output, while in the same operation shifting the harmonics to high frequencies that are easily filtered by the inductance of the motor.

Sine/triangle modulation. Figure 7.38(a) shows one means for controlling the switches to create a PWM waveform. The triangle wave is a carrier, and its frequency is much higher than the electrical frequency applied to the motor. The sinusoidal waveform, of which only one-half of the period is shown, is the modulating waveform that controls the amplitude and frequency of the fundamental applied to the motor. Specifically, the electrical frequency of the ac applied to the motor is equal to that of the mod-

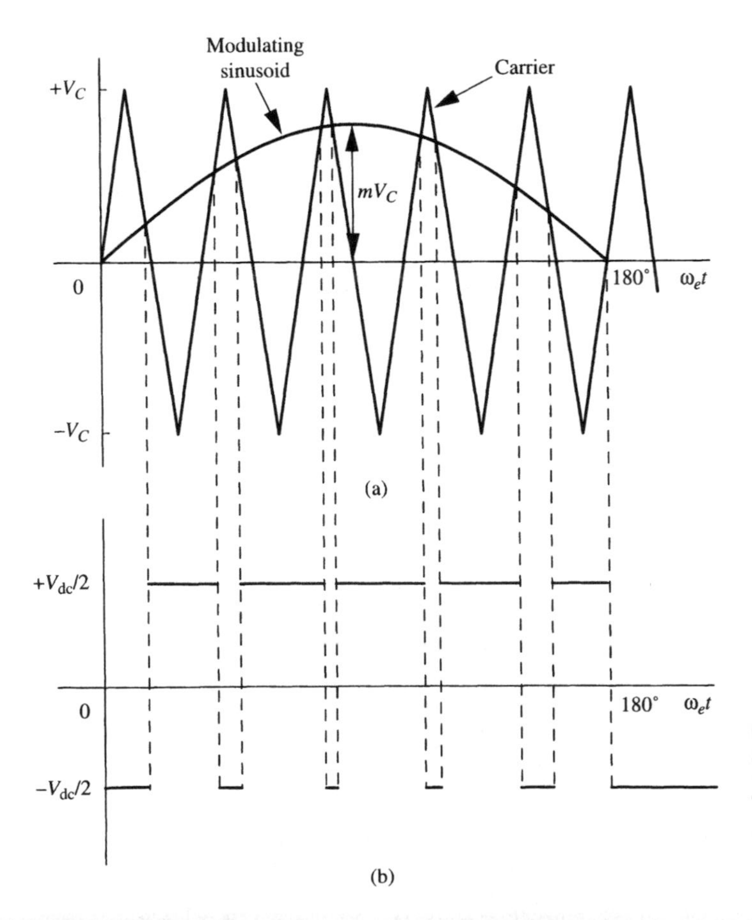

Figure 7.38 (a) The switching sequence is generated by comparing a triangular wave with a sinusoidal wave. The modulation index is m. (b) The resulting waveform has a sinusoidal component proportional to m at the frequency of the sinusoid in part (a).

ulating sinusoid, and the amplitude of the ac is proportional to the modulation index, m, defined in Fig. 7.38(a).

The relationship between the carrier, modulating sinusoid, and switched waveform is evident: Whenever the modulating sinusoid exceeds the carrier, the switches connect the motor to the positive dc voltage, and whenever the modulating sinusoid falls below the carrier, the switches connect the motor to the negative dc voltage. Note that if m were small, the output would have roughly equal amounts of positive and negative voltage for this half-cycle and thus produce a small component of the fundamental. But as m increases, the positive voltage dominates. The opposite would be true for the second half of the cycle. In this way, the modulating sinusoid controls the amplitude and frequency of the output.

The Frequency Domain

Harmonics in PWM. With PWM, the harmonics are shifted to frequencies comparable to the carrier frequency and thus are much higher than the harmonics of the square-wave inverter. The higher harmonics are reduced by the inductance of the motor, and the resulting current is sinusoidal with a small ripple.

We showed the modulated waveform for only one phase of the three-phase inverter. For the other two phases, the modulating sinusoids are phase shifted by 120° and 240° in the usual fashion. In this way, the inverter produces three-phase voltage of controlled frequency and amplitude.

Hysteresis-current control. Figure 7.39 illustrates yet another method for deriving the PWM switching sequence. The output current in each output phase is compared with a reference signal of the desired amplitude and frequency. When the output current falls below the reference signal by a prescribed error, the output is connected to the positive dc bus, and when the output current rises above the reference signal by the specified error, the output is connected to the negative dc bus. Thus, the current bounces between the two error bounds, following the reference signal. This method controls current, and thus torque, directly and is not greatly affected by the variation of the motor impedance with drive frequency.

Benefits of PWM drive. Pulse-width modulation controls motor voltage or current, depending on the modulation scheme, by connecting the motor to the positive or negative buses of a fixed-voltage dc supply. Such a supply has good efficiency and power factor and does not generate excessive harmonics in the power system. The PWM sys-

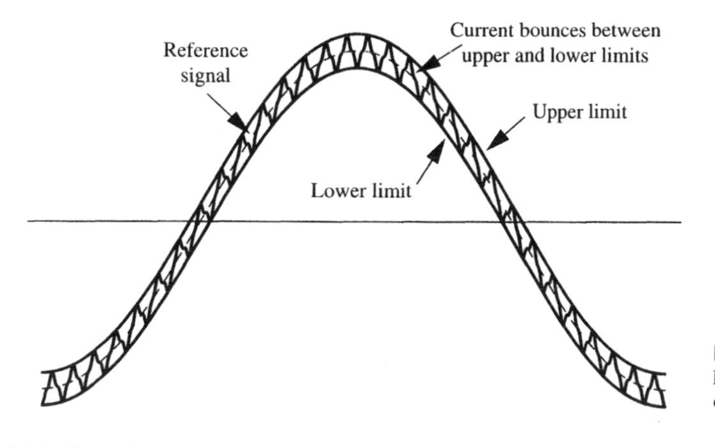

Figure 7.39 Hysteresis-current control keeps the current within a prescribed range of a reference waveform.

tems shift the harmonics to high frequencies that are filtered by the inductance of the motor.

Check Your Understanding

1. The speed of an induction motor can be controlled by varying the drive voltage or frequency, but not both. True or false?
2. Figure 7.38 shows a sine/triangle modulation scheme. Sketch the voltage waveform if the modulation index, m, equals 1.5.
3. The speed of a synchronous motor cannot be controlled by a power electronics system because the slip is always zero. True or false?

Answers. **(1)** False; **(2)** almost a square wave; **(3)** false; the rotor position, and hence its speed, can be controlled precisely.

CHAPTER SUMMARY

This introduction to power electronics began with a presentation of the properties and limitations of various semiconductor switches. We then showed how such switches are used in the simple ac power controller used in the common dimmer-switch circuit.

We analyze in detail a controlled rectifier circuit with a dc motor for a load. The various modes of operation lead to analytical complexities, but the behavior of the system is straightforward. We end by describing several motor controllers for ac induction motors, without detailed analysis.

Objective 1: To understand the variety and limitations of semiconductor switches. The solid-state revolution has produced a variety of semiconductor switches that can handle large amounts of power. We focus primarily on the switching losses and protection needs of such switches.

Objective 2: To understand in full the operation of the common light-dimmer circuit. This circuit in its simplest form uses one four-level diode as a voltage-controlled switch to control the firing of an SCR that in turn controls power to a low-impedance load. The thorough analysis introduces the time-domain analysis of such nonlinear circuits.

Objective 3: To understand the operation of a controlled single-phase, full-wave rectifier driving a dc motor. The unfiltered full-wave rectifier is analyzed with a dc motor armature as a load. We show that the motor characteristics are changed markedly in this application, and that lightly loaded motors can overspeed.

Objective 4: To understand the methods and complexities of ac motor controllers. The characteristics of the three-phase induction motor are examined when ac drive voltage and frequency are varied. The need for constant-volts/hertz drive is shown. Several types of variable-voltage and variable-frequency drives are described qualitatively.

Chopper, p. 306, a converter that takes fixed-voltage dc power and produces variable-voltage dc power.

Commutation, p. 294, the switching of a conducting SCR from the ON state to the OFF state.

Conduction angle, p. 316, the time, measured in electrical phase angle, during which a semiconductor switch is conducting.

Critical delay angle, p. 313, the phase angle in the ac cycle when a semiconductor becomes forward biased.

Delay angle, p. 302, the position in the ac cycle when a semiconductor switch conducts; also called firing angle.

Diac, p. 304, a four-level diode that fires in both directions.

Extinction angle, p. 315, the phase angle in the ac cycle when a semiconductor becomes reverse-biased and hence is line commutated.

Forced commutation, p. 294, when an auxiliary circuit is used to commutate an SCR independent of the line voltage.

Free-wheeling diode, p. 297, a diode placed in parallel with an inductor to give a current path and thus prevent the occurrence of excessive voltage.

Gate-turnoff thyristor, p. 295, a semiconductor switch in which conduction in the forward-bias region is turned ON and OFF by signals at its gate.

Harmonics, p. 327, power converted to frequencies that are multiples of the line frequency due to periodic switching.

Inverter, p. 306, a circuit that converts dc power to ac power.

Line commutation, p. 294, when commutation is accomplished by a voltage reversal occurring in the ac cycle.

Power electronics, p. 292, applying electronic techniques to the control of electrical power.

Pulse-width modulation (PWM), p. 328, a means for modulation of dc power to produce an ac frequency with controlled frequency and amplitude. The hallmark of PWM is that harmonics are created at relatively high frequencies where their effects are lessened.

RC snubber, p. 294, a circuit that protects a semiconductor switch circuit from excessive voltage, current, energy, or power.

Rectifier, p. 294, a converter that converts ac power to dc power.

Regenerative braking, p. 306, using the electronic drive circuit as a load to slow down a motor, also called plugging.

Rise time, p. 295, the time required for the signal to rise from 10% to 90% of its final value.

Switch, p. 292, a device that, when ON, has an impedance much smaller than its load and, when OFF, has an impedance much larger than its load.

Thyristor (SCR), p. 293, a semiconductor switch in which conduction in the forward-bias direction is initiated by a signal at its gate.

Triac, p. 304, an SCR that conducts in both directions.

PROBLEMS

Section 7.1: Introduction to Power Electronics

7.1. A snubber circuit, shown in Fig. P7.1, is placed in parallel with an SCR to limit the rate of voltage rise across the OFF device. If $120\sqrt{2}$ V were suddenly applied to the SCR and snubber circuit with the load shown, what would be the required value of R to keep the dv/dt of the SCR below 100 V/μs? Treat the capacitor as a short circuit for the worst-case transient.

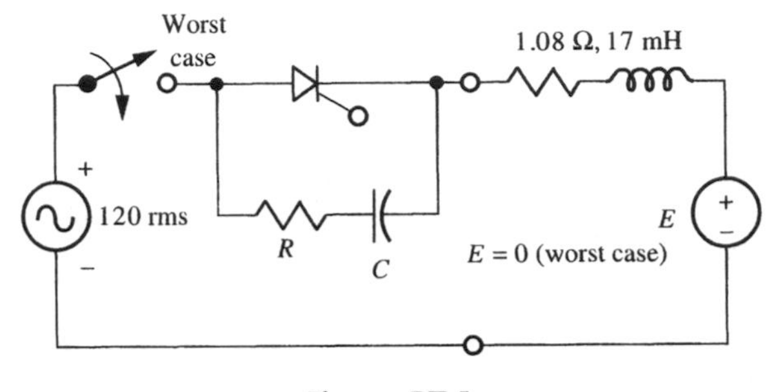

Figure P7.1

7.2. Verify the result of the integration in Eq. (7.3).

7.3. **(a)** Calculate the energy loss in the transistor for the inductive switching trajectory shown in Fig. 7.6(b). The voltage and current for the turn-ON switching operation is shown in Fig. P7.4(b). In both transitions, the collector voltage is at the power-supply voltage while the current makes its transition in $t_r/0.8$.

(b) What is the power in the transistor if it is switching with a frequency, f_{SW}?

7.4. A transistor switching an inductive load, Fig. P7.4(a) has the waveform shown in Fig. P7.4(b) for current and voltage. The switching cycle is short compared to the time constant; hence, the current in the inductor is essentially constant at I_{Cp}.

(a) Explain briefly why the diode is in the circuit.

(b) Find the energy dissipated in the transistor in an OFF → ON transition.

(c) If the same energy is used in the ON → OFF transition and the transistor is driven ON → OFF → ON at a rate f_r, find the time-average power dissipated by the transistor.

7.5. On page 296, we work out the switching losses for a transistor switching a resistive load. The model used is linear; that is, the current increases linearly and the voltage across the transistor decreases linearly. Let us assume an *exponential* model and rework the problem for comparison. Let

$$i_C(t) = I_{Cp}(1 - e^{-2.2t/t_r}) \text{ and } v_{CE}(t) = V_{CC}e^{-2.2t/t_r}$$

In these expressions, t is counted from the time the current starts to increase (same as t' in the book.)

(a) Show that the rise time satisfies the definition on page 295.

(b) Calculate the energy delivered to the transistor during the switching.

7.6. The SCR in Fig. P7.6 has infinite resistance when OFF and a voltage of 0.3 V when ON. A pulse of current occurs at the gate at $t = 0$. Find the following:

(a) The current in the circuit before the gate pulse.

(b) The maximum current that can flow in the circuit.

(c) The time constant of the current.

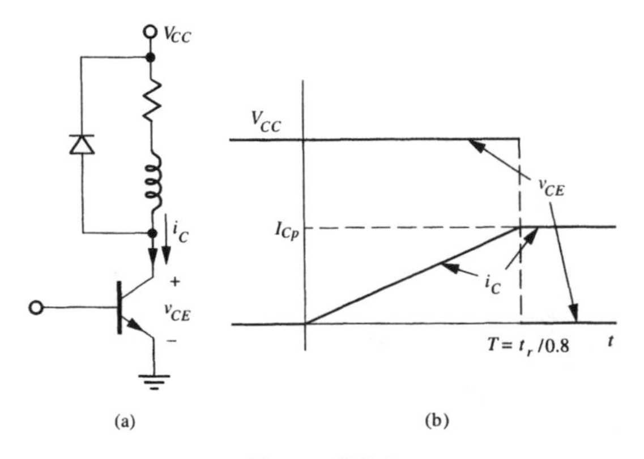

Figure P7.4

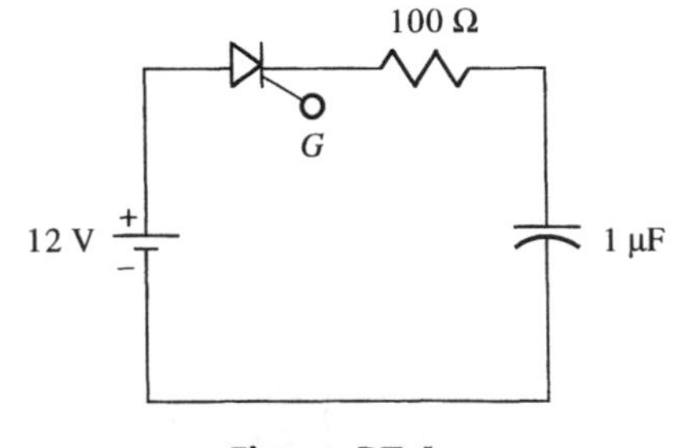

Figure P7.6

7.7. Commercial airplanes carry flashlights that have a light-emitting diode (LED) that flashes occasionally to show that the batteries are still good (and also presumably to allow one to find the flashlights in the dark). A circuit that accomplishes this flashing is shown in Fig. P7.7. Assume that the *pnpn* diode fires at 2.2 V and has 0.7 V when ON. The LED requires at least 10 mA of current at 0.7 V to glow. Find R_1 and R_2 so that the LED glows for at least 4 ms every 10 s.

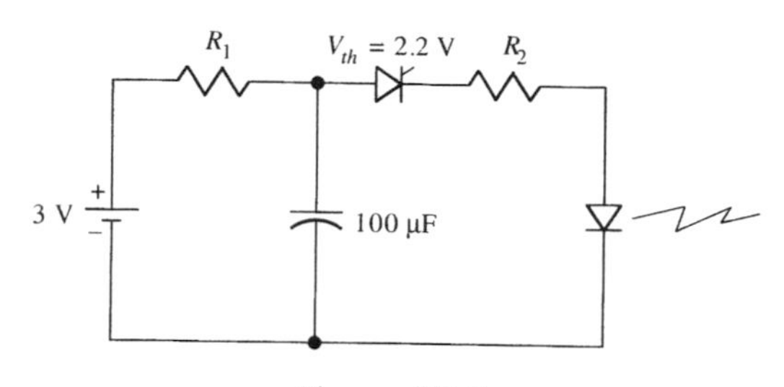

Figure P7.7

7.8. For the circuit shown in Fig. P7.8, the capacitor is initially uncharged and the switch is closed at $t = 0$. The *pnpn* diode has a threshold voltage of 32 V and has an ON voltage of 0.3 V. Find the maximum current in the 100-Ω resistor and the time when the maximum current occurs.

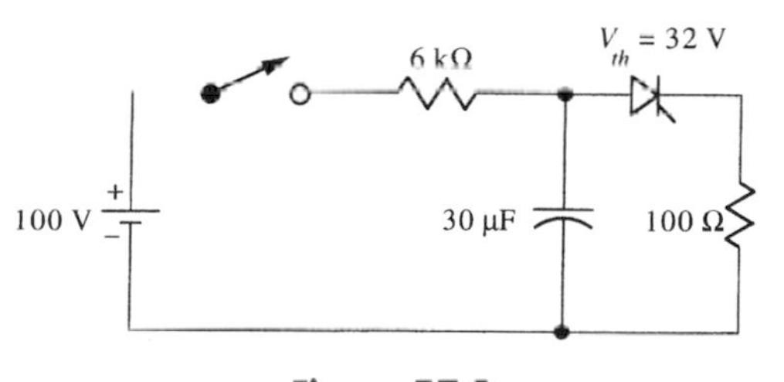

Figure P7.8

7.9. If the gate of the SCR in Fig. 7.12 is limited to 10 mA of current, what value of resistance should be placed in series with the *pnpn* diode to limit the gate current to this value. Assume that the *pnpn* diode fires at 30 V and that the equivalent resistance of the gate is 15 Ω.

7.10. For the circuit in Fig. 7.9 and $V_p = 120\sqrt{2}$ V, determine the exact minimum and maximum delay time if the *pnpn* diode has a threshold voltage of 26 V.

7.11. The circuit shown in Fig. P7.11(b) contains a *pnpn* diode, whose characteristics are shown in Fig. P7.11(a), and an ideal SCR. The switch closes at $t = 0$.

(a) At what time does the SCR conduct?

(b) What is the maximum current in the 5-kΩ resistor?

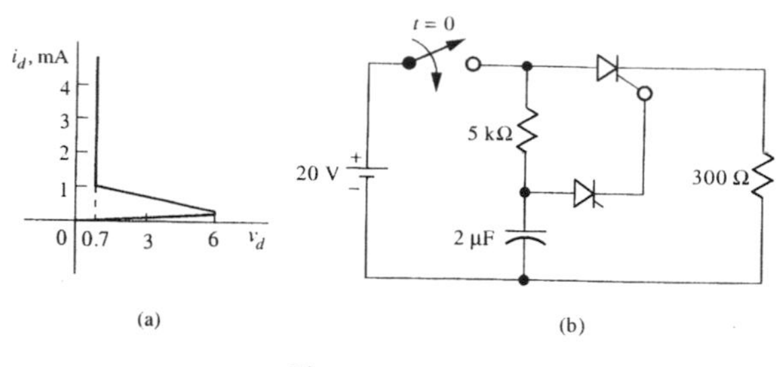

Figure P7.11

(c) What is the maximum current in the 300-Ω resistor?

7.12. For the circuit shown in Fig. P7.12, the switch is closed at $t = 0$. Before the switch is closed, the capacitor is uncharged. The four-level diode turns ON at 10 V, has a voltage of 0.5 V when ON, and turns OFF when the current through it falls below 10 mA. Sketch the voltage across the 10-Ω resistor as a function of time for $t > 0$.

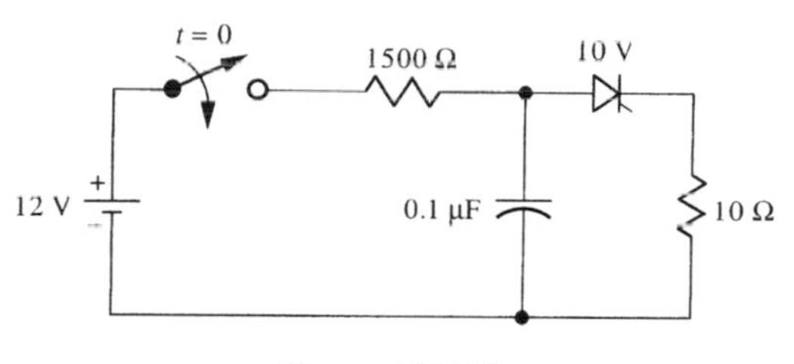

Figure P7.12

7.13. For the circuit in Fig. 7.12, what would be the maximum $R_V C$ time constant for the *pnpn* diode to reach 30 V and fire the *pnpn* diode? *Hint*: The exact solution is difficult due to the nonlinear character of the equations. A good numerical approach would be to ignore the exponential term in Eq. (7.11) for a first approximation and then include it for subsequent approximations.

7.14. The circuit in Fig. 7.14 represents a dimmer switch, and the load represents a 150-watt light bulb. The voltage is turned ON and the circuit is adjusted to cause the bulb to glow slightly. Circle the change (brighter, dimmer, or no change) for the following changes:

(a) The amplitude of the input voltage is increased.

(b) The phase of the input voltage is increased.

(c) The frequency of the input voltage is increased.

(d) The resistance, R_V, is increased.

(e) The capacitance, C, is increased.

7.15. For $V_p = 120\sqrt{2}$, $f = 60$ Hz, $V_{th} = 30$ V, $R_V = 30$ kΩ, and $C = 0.3$ μF, determine the firing time, t_α, from Eq. (7.11). This requires a graphical or numerical solution unless you have a calculator that solves nonlinear equations.

7.16. Normally, the firing time in SCR operation, t_α, shown in Fig. 7.13, is described in terms of a "conduction angle," where $\gamma = 180° - 360° \times t_\alpha/T$, where T is the period of the frequency, usually 1/60 s. Clearly, γ is the amount of phase angle during which the SCR fires each cycle.

(a) For a conduction angle of 90°, $V = 120$ V (rms), and $R_L = 100$ Ω, what is the dc current and the total power in the load? Note that the total power is a measure of the total heating effect on the load and is more than the dc power.

(b) Repeat part (a) for a conduction angle of 120°. Part (a) can be solved without evaluating any integrals, but part (b) requires two integrations.

Section 7.2: DC Motor Controllers

7.17. Consider a separately excited 180-V dc motor with armature resistance of 1.05 Ω and inductance of 22 mH. Nameplate ratings are 1750 rpm (2050 rpm max.), 1.5 hp, and 7.3-A armature current. The motor is driven by an uncontrolled 60-Hz full-wave rectifier with $V_p = 230\sqrt{2}$ V. The emf of the motor is 201.9V. The current is discontinuous and flows for 91.6% of the time. What is the extinction angle β in radians?

7.18. Consider a separately excited 180-V dc motor with an armature resistance of 1.05 Ω and an inductance of 22 mH. Nameplate ratings are 1750 rpm (2050 rpm max.), 1.5 hp, and 7.3-A armature current. The dc motor is driven by an uncontrolled 60-Hz full-wave rectifier.

(a) Find the required input voltage to the rectifier for nameplate operation, assuming that the current to the motor flows continuously.

(b) Find the required emf of the motor for continuous current, considering only the second-harmonic current.

(c) Is the current for this emf below its nameplate value?

7.19. Consider a separately excited 180-V dc motor with an armature resistance of 1.05 Ω and an inductance of 22 mH. Nameplate ratings are 1750 rpm (2050 rpm max.), 1.5 hp, and 7.3-A armature current. The motor is supplied by an uncontrolled single-phase, 60-Hz full-wave rectifier.

(a) Determine the minimum input voltage (rms) such that the armature current flows continuously at 90% of nameplate speed. Consider only the second harmonic.

(b) Find the current under these conditions.

7.20. A series-excited dc motor has the following nameplate information: 24 V, 5 A, 0.8-Ω armature plus field resistance, 0.05 H-armature plus field inductance, and 800 rpm. The motor is driven by a full-wave rectifier connected to the 120-V, 60-Hz power with a center-tapped transformer.

(a) Find the turns ratio of the transformer, n_s/n_p so that the motor is driven at nameplate conditions, assuming that current is continuous.

(b) Confirm that current is continuous.

(c) At what speed does the current in the motor current become discontinuous?

7.21. Consider a separately excited 180-V dc motor with armature resistance of 1.05 Ω and inductance of 22 mH. Nameplate ratings are 1750 rpm (2050 rpm max.), 1.5 hp, and 7.3-A armature current. This motor is driven by a controlled rectifier. The ac source is 230 V rms.

(a) If the firing pulses begin at $\alpha = 0°$, estimate the no-load speed of the motor? Ignore losses. *Hint*: The motor accelerates to a speed where very little current flows; that is, the critical delay angle is approximately 90°.

(b) With no load, what range of α has no effect on the speed?

(c) With $\alpha = 90°$ and $E = 180$ V, the extinction angle, β, is 101.8°.

(1) What is the conduction angle, γ, in radians?

(2) What is the dc armature voltage?

(3) Find the developed torque in the motor.

7.22. Consider a separately excited 180-V dc motor with armature resistance of 1.05 Ω and inductance of 22 mH. Nameplate ratings are 1750 rpm (2050 rpm max.), 1.5 hp, and 7.3-A armature current. The dc motor is driven by a single-phase controlled rectifier with $V_p = 120\sqrt{2}$ V, and 60 Hz.

(a) If the firing pulses begin at $\alpha = 0°$, what is the no-load speed of the motor? Ignore losses.

(b) With $E = 150$ V, what range of α has no effect on the speed?

(c) With the firing pulses beginning at $\alpha = 0°$, what range of emfs gives continuous conduction?

(d) With the firing pulses beginning at $\alpha = 0°$ and $E = 150$ V, the extinction angle, β, is 145.2°.
  (1) What is the conduction angle, γ, in radians?
  (2) What is the dc load voltage?
  (3) Find the armature current.

7.23. Consider a separately excited 180-V dc motor with an armature resistance of 1.05 Ω and inductance of 22 mH. Nameplate ratings are 1750 rpm (2050 rpm max.), 1.5 hp, and 7.3-A armature current. This motor is driven by a single-phase, controlled rectifier drive. The input voltage is 220 V rms, 60 Hz, for each half of the transformer supplying the rectifier. Consider that the motor is running at nameplate speed.

(a) Determine the critical α for the circuit, α_c, in degrees and radians.

(b) Assuming the firing pulses begin at $\alpha = \alpha_c$, write the equation for the current in the period of time $\alpha_c < \omega t < 180°$. Assume zero current at $\omega t = \alpha_c$ and solve for all constants in the equation.

(c) Calculate $i_a(180°)$ to see if we have Case A.

(d) If not Case A, is it Case B?

7.24. A full-wave gated SCR rectifier provides adjustable dc power to an RL load, as shown in Fig. P7.24(a). The resistance is 10 Ω and the inductance is unknown. For a circuit delay angle the voltage applied to the RL load is that shown in Fig. P7.24(b). No free-wheeling diodes are used, so the voltage can go negative due to the inductance.

(a) What is the delay angle?

(b) Is the current continuous or discontinuous? Explain.

(c) Find the minimum value of inductance to give the voltage shown and 60-Hz operation.

(d) Determine the dc current in the RL load.

7.25. The circuit in Fig. P7.25 shows a single-phase, full-wave uncontrolled rectifier battery charger. The load is a dc battery with an emf that varies from 11.8 V (discharged) to 12.6 V (charged), and an equivalent resistance of 0.03 Ω. The peak value of the transformer secondary voltage is 13.3 V, such that no current flows into the battery when fully charged because the diodes require 0.7 V to turn ON.

(a) For a discharged battery, 11.8 V, what is the critical delay angle, α_c, of the diodes?

(b) What is the peak current in the discharged battery?

(c) What is the average current in the discharged battery?

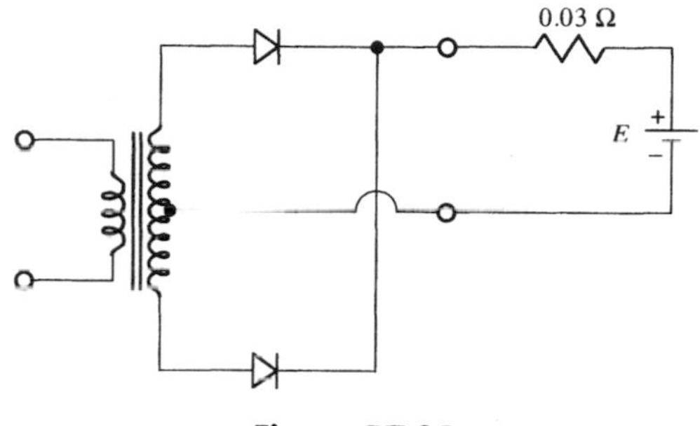

Figure P7.25

7.26. The circuit in Fig. P7.26 shows a single-phase, full-wave controlled rectifier. The top switch is gated at α and the bottom at $\alpha + \pi$. The load is a dc motor with an emf of 160 V, an armature resistance of 1.1 Ω, and negligible armature inductance. The secondary voltage of the transformer is 120 V rms. With negligible inductance, the current cannot go negative, so no free-wheeling diode is required.

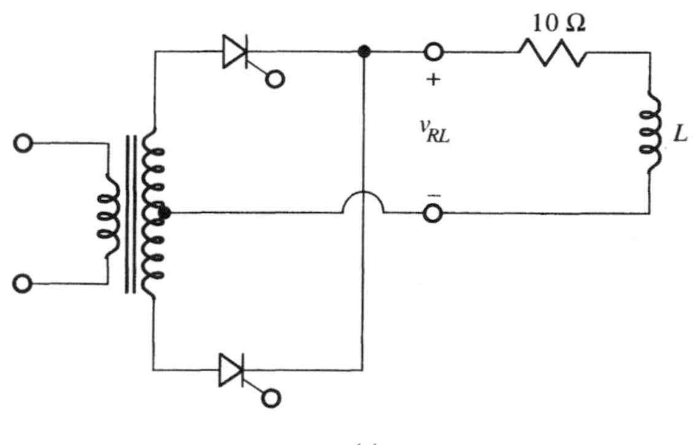

(a)

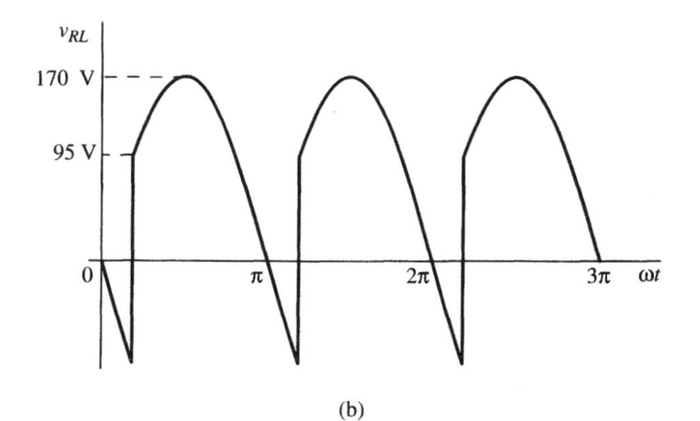

(b)

Figure P7.24

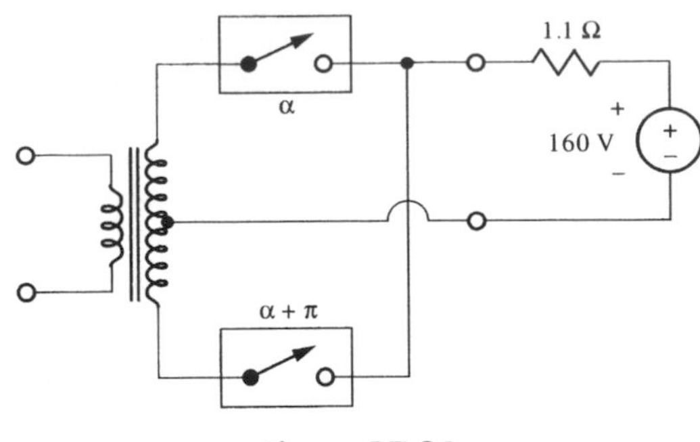

Figure P7.26

(a) What values of α actually affect motor voltage and current?
(b) For $\alpha = 90°$, what is the peak instantaneous current in the motor?
(c) For $\alpha = 90°$, what is the average current in the motor?

7.27. A dc motor has the following nameplate information: 230 V dc, 1750 rpm, 20 hp, 74.0 A dc, 0.180-Ω armature resistance, and 2.93-mH armature inductance. The field current is separately excited for nameplate operation. For the motor turning at nameplate speed, with 230 V rms driving it via a controlled full-wave rectifier, with a prescribed delay angle, the current is as shown in the shaded area in Fig. P7.27. The current begins at a phase angle of 70°, reaches a maximum value of 33 A, and ceases at an angle of 182°. There is a free-wheeling diode.
(a) What is α?
(b) What is β?
(c) Sketch the voltage applied to the motor. (The full-wave rectified sinusoid is given for reference.)
(d) What is the developed power in the motor?

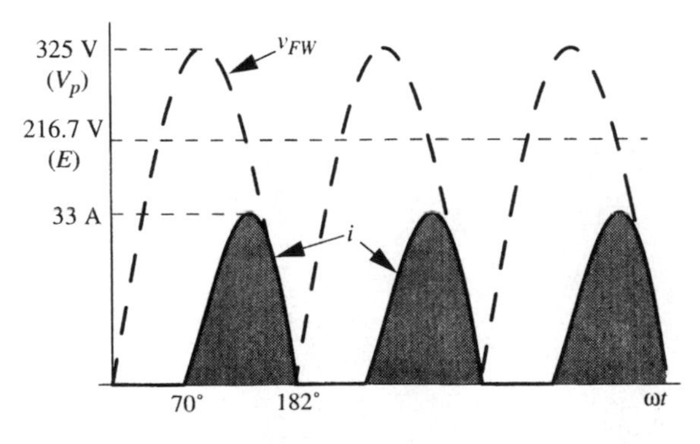

Figure P7.27

7.28. A dc motor is driven by an ungated full-wave rectifier circuit. The motor nameplate information is 1 hp, 1150 rpm, 180 V, 5.0 A, 2.43-Ω armature resistance, and 49-mH armature inductance. The field current is that required for nameplate operation.
(a) Find the rms voltage to the rectifier input such that the motor draws continuous current at nameplate conditions.
(b) Determine approximately the speed of the motor if allowed to run at no load.
(c) Estimate the dc current into the motor if allowed to run at no load.

7.29. A model train locomotive has a permanent-magnet dc motor. For full speed of 1 ft/s, the motor requires 6 V dc, and 2 A dc. The motor armature has 2 Ω in series with 10 mH. The motor is to be driven by a full-wave rectifier using pn-junction diodes (assume ideal), connected to a 120-V, 60-Hz power source via a transformer whose secondary is center-tapped for the rectifier. A variable resistor (rheostat) is connected in series with the motor for speed control.
(a) Draw the circuit of the rectifier, rheostat, and motor model.
(b) Find the turns ratio of the transformer (total primary turns/total secondary turns) to drive the train at full speed with the rheostat set to zero resistance. Assume that current flows continuously.
(c) Confirm that the current flows continuously.
(d) Find the resistance of the rheostat to give a train speed of 0.5 ft/s if the current required is 1.6 A.

7.30. A dc motor has the following nameplate information: armature: 1.5 hp, 2500 rpm (2750 max. with reduced field), 180 V, 7.5 A, 0.563 Ω, and 12 mH; field: 0.56 A (0.43 A for max. speed), 282 Ω, and 86 H. A full-wave rectifier is installed to give the required 180-V input, as you can confirm from the circuit diagram in Fig. P7.30 using the usual equations for the dc from a full-wave rectifier. However, the engineer who designed the power supply (transformer plus the full-wave rectifier) found that, with the motor running with nameplate speed and torque, the voltage at the output is not 180 volts at nameplate operation, but is higher.
(a) Explain why the voltage is high. Support your explanation with numerical calculations.
(b) The engineer then found that adding an external inductor in series with the motor armature, as

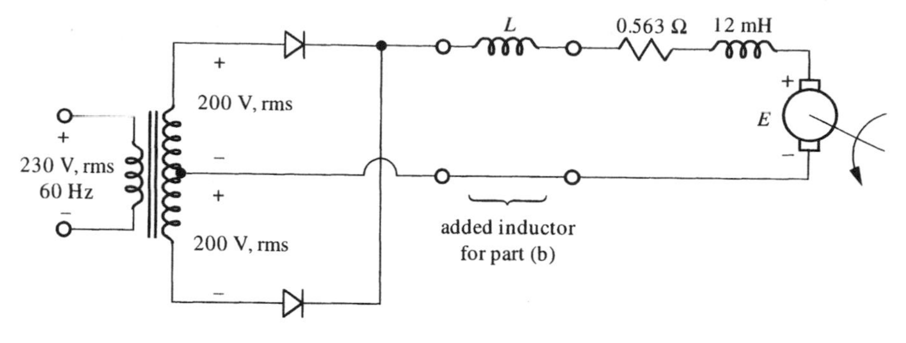

Figure P7.30

shown in Fig. P7.30, reduces the voltage. Calculate the minimum inductance to give the required 180 V at nameplate operation.

7.31. A permanent-magnet-field dc motor is driven by a 60-Hz, uncontrolled, full-wave single-phase rectifier with a free-wheeling diode. The no load speed and current of the motor are 1925 rpm and 0.8 A, respectively. The current becomes continuous at 930 rpm, and the full-load speed and output power are 900 rpm and 2 hp, respectively. Assume rotational losses are proportional to the mechanical speed. The armature resistance is 1 ohm.

(a) Find the emf at full load.

(b) Estimate the armature inductance.

7.32. An engineer is operating a dc motor with an electronic box, but does not know what is in the box. The engineer knows the frequency, 60 Hz, and the armature resistance of the motor, 1.05 Ω. To learn what the electronics was doing, she looked at the output voltage of the box with an oscilloscope. What she saw is shown in Fig. P7.32: The voltage was sinusoidal for part of the cycle, was zero for about 0.5 ms, jumped to 110 V and was constant for a while, then jumped to 130 V, and becomes sinusoidal again. From this, the engineer learned a lot about the system:

(a) Continuous or discontinuous current?

(b) EMF of the motor?

(c) Delay angle of the drive?

(d) Extinction angle?

(e) Time-average motor current?

(f) Developed power?

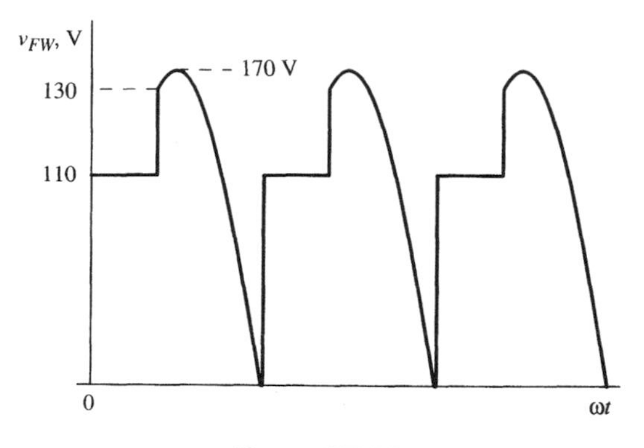

Figure P7.32

7.33. A dc motor has the following nameplate information: 1.5 hp, 2500 (2850 max.) rpm, 180 V, 7.5 A, 0.563-Ω armature resistance, and 12 mH armature inductance. The motor is separately excited at rated field current for operation at nameplate conditions. The motor armature is connected to a 460-V secondary of a transformer, center-tapped for 230 V(rms), 60 Hz each half, and diodes are used to rectify the ac for the armature circuit, as shown in Fig. P7.33. Find the motor speed at which the current becomes continuous. Is the motor operating within its allowed limits?

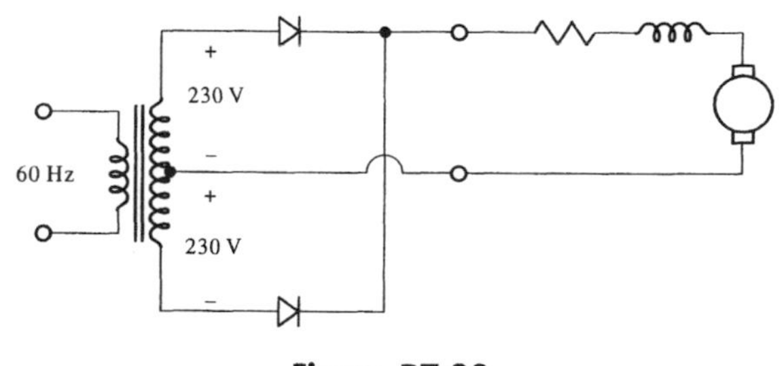

Figure P7.33

7.34. A 230-V, 60-Hz single-phase rectifier-type motor controller drives a dc motor. The motor nameplate information is 180 V, 8.5 A, 1150 rpm, 1.03-Ω armature resistance, 28-mH armature inductance, 2-hp output power. The motor is separately excited for nameplate conditions. A free-wheeling diode is used to reduce harmonics and increase armature current. The motor is turning 1150 rpm. The delay angle is set to 40°, and the first pulse of current is observed as shown in Fig. P7.34. From this information, determine the time-average torque developed by the motor.

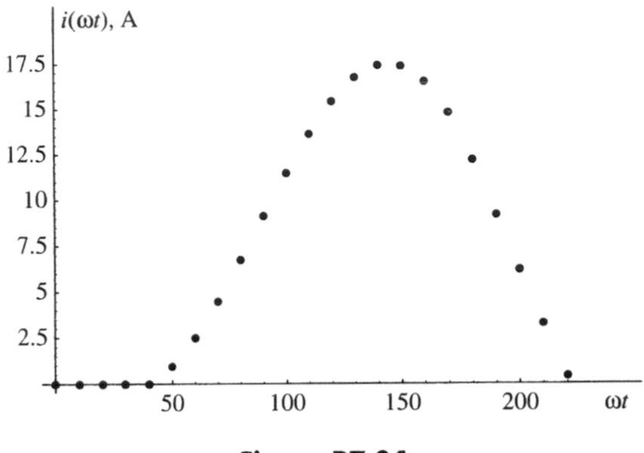

Figure P7.34

7.35. A dc motor has the following specifications: armature: 1 hp, 1150/1380 max. rpm (with reduced field), 5.0 A, 180 V, 2.43 Ω, and 0.049 H; field: base-speed current = 0.54 A; top-speed current = 0.33 A, 200 V, 282 Ω, and 86 H. This motor is to be used in an application where it is to run at 1000 rpm with an output torque of 3.5 N-m, driven by an ac source and a full wave rectifier.

(a) Because 230-V, 60-Hz, single-phase power is available, a transformer must be used to convert the voltage for the application. Determine the transformer voltage required to drive the motor at this speed and load. Assume the nameplate current of 0.54 A for the motor field. Assume rotational power losses proportional to speed.

(b) Draw the entire circuit, including the field circuit and the ON/OFF switch. You are supposed to excite the field from the same rectifier, adding a rheostat if required.

7.36. A dc motor with a free-wheeling diode has the following nameplate information: 5 hp, 1750 rpm, 220 V, 20 A, 0.5-Ω armature resistance, and 2.5-mH armature inductance. The motor is driven by a 60-Hz, single-phase controlled full-wave rectifier with a voltage of 460 V rms, center-tapped for the rectifier ($V_p = 230\sqrt{2}$). For an emf (E) of 210 V (nameplate speed), the conduction angles are calculated from Eq. (7.32), as follows:

α	30°	40°	50°	60°	80°	100°	120°	140°
γ°	148.8	138.8	128.5	117.6	93.8	66.8	36.0	0.000

(a) Confirm one of the data points except the last by direct calculation.

(b) What portion of the calculated data is relevant to system performance?

(c) Estimate the value of α to drive the motor at nameplate conditions.

7.37. A dc motor with a free-wheeling diode is driven by a gated full-wave, single-phase bridge rectifier. The armature voltage is shown in Fig. P7.37, along with critical voltages and angles in electrical degrees. The dc motor nameplate information is as follows: 2500 rpm, 3 hp, 180 V, 14.5 A, 0.313-Ω armature resistance, and 7-mH armature inductance. Assume ideal diodes. Find the following:

(a) The rms voltage of the ac system.

(b) The delay angle, α.

(c) The critical delay angle, α_c.

(d) The extinction angle, β.

(e) The motor speed.

(f) The developed power.

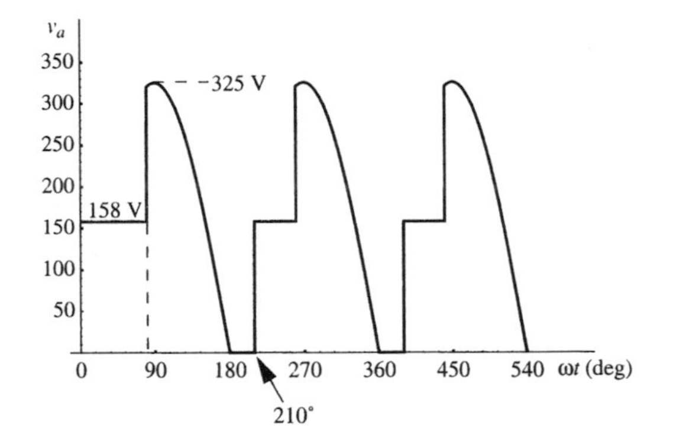

Figure P7.37

7.38. A 240-V, 60-Hz, gated, full-wave single-phase rectifier drives a dc motor with an armature resistance and inductance of 3 Ω and 30 mH, respectively. At a speed of 1350 rpm, the critical delay angle is 20°. A free-wheeling diode is used.

(a) Determine the delay angle required such that the operation is borderline between continuous and discontinuous conduction. This requires a trial-and-error solution.

(b) For the delay angle determined in part (a), find the developed power in the motor in hp.

7.39. A dc motor has the following nameplate information: 1 hp, 1150 rpm (1380 with reduced field), 180 V, 5.0 A, 2.43 Ω, 0.049 H, and 0.076 kg-m² moment (armature only, no load). The field is 200 V, 0.54 A for 1150 rpm and 0.33 A for 1380 rpm. The machine is operated with separately excited field, with nameplate field voltage (200 V) and nameplate current (0.54 A). The armature is connected to a 230-V(rms), 60-Hz, single-phase source via a full-wave rectifier (assume ideal diodes). A series resistor is connected between the power source and motor to limit the voltage to the motor to 180 V under nameplate conditions.

(a) Find the emf under nameplate conditions.

(b) Find the rotational losses under nameplate conditions.

(c) Find the value of the series resistor to ensure nameplate conditions from the rectifier.

(d) Confirm that the current is continuous at nameplate conditions.

(e) Determine the overall efficiency at nameplate conditions, including field losses and losses in the series resistor. Consider only the dc power out of the rectifier; that is, neglect any power in the higher harmonics.

(f) Find the maximum value of the series resistor to give continuous current at $n = 1150$ rpm.

7.40. An electromagnet has a resistance of 120 Ω and an inductance of 0.5 H. The magnet is activated by a 60-Hz, 240-V, single-phase ac power source. Two schemes are considered. One uses ac current only to activate the magnet, and the other uses a full-wave rectifier to convert the ac to dc. In both cases, the lifting power of the magnet is proportional to the square of the rms value of the current.

(a) Draw the circuit diagram for the rectifier scheme, showing the ON/OFF switch and a free-wheeling diode if required. Use a bridge rectifier circuit.

(b) Determine the rms current in the magnet with the rectifier. If you assume a continuous current, show that this assumption is justified.

(c) Determine the rms current with the ac source connected directly. Compare the performance of the two schemes.

7.41. An electromagnet is designed to work in a junkyard lifting junk cars, as shown in Fig. P7.41(a). The geometry is cylindrical, as shown in Fig. P7.41(b), and the coil has 500 turns and a 500-Ω resistance. The OD of the inner pole is 24 cm, and the ID and OD of the outer pole are 30 cm and 38 cm, respectively, as shown. Due to the irregular nature of the loads, assume a nominal air gap of 1 cm. Two methods for exciting the electromagnet are being considered: a 240-V, 60-Hz ac source, and a dc source produced by full-wave rectifying the 240-V source with a bridge rectifier. Determine the current due to each and make your recommendation as to the best approach. Assume lifting power is proportional to the square of the rms current. *Hint*: In the reluctance calculation, consider the two air gaps in series and use the average area.

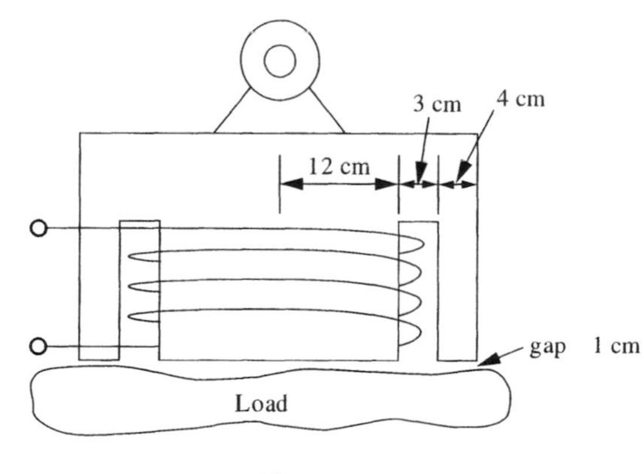

(a)

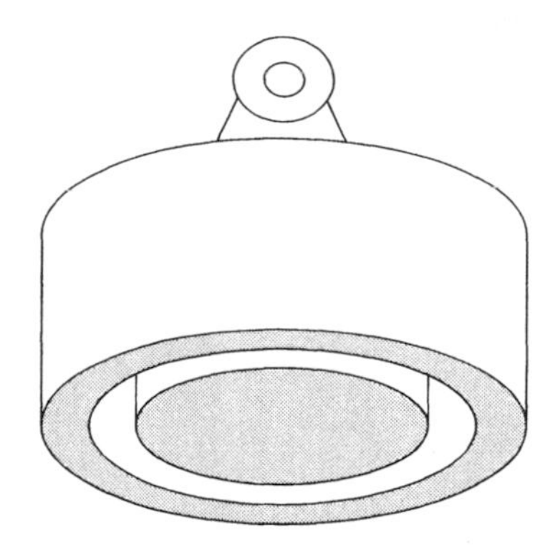
(b)

Figure P7.41

Section 7.3: AC Motor Controllers

7.42. Determine the speed versus percent full-load voltage curve for the motor described in Example 7.9 if the full-load slip were 11% and the rotor resistance were 0.5 Ω. The stator resistance and the total reactance are unchanged. *Hint*: Assume values of s and then calculate the voltage required.

7.43. Figure P7.43 shows an approximate per-phase circuit for a three-phase, two-pole, 60-Hz induction motor at nameplate frequency. At a slip of 3%, the developed torque is 8.18 N-m for nameplate voltage.

(a) Find the torque at 80% of nameplate voltage if the slip is constant.

(b) Find the torque at 80% of nameplate voltage and 80% of nameplate frequency if the slip is changed to 5%.

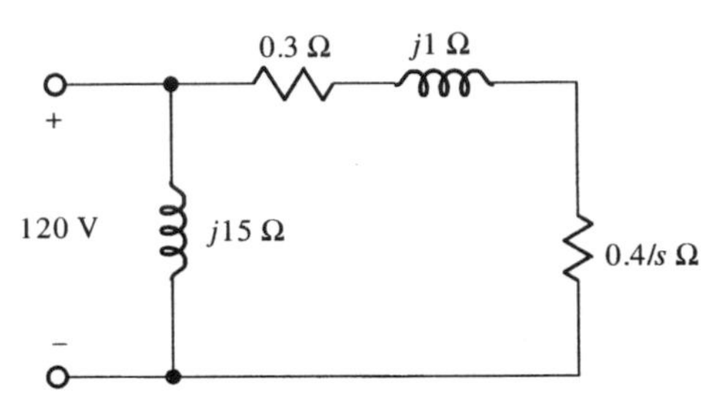

Figure P7.43

7.44. A six-pole, 60-Hz, three-phase induction motor has the simplified per-phase equivalent circuit shown in Fig. P7.44. The motor load requires a torque given by the equation: $T_{dev} = 1.4 \times 10^{-5}\omega^3$ N-m. Find the per-phase voltage required to drive the motor at 1155 rpm.

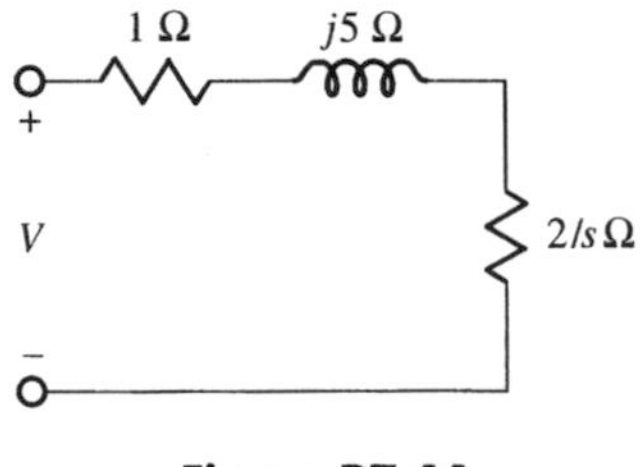

Figure P7.44

7.45. The equivalent circuit in Fig. P7.45 is a simplified model for a 230-V, four-pole, 60-Hz induction motor that is operated from a variable-frequency, variable-voltage power electronic drive. At 60 Hz and nameplate voltage, the motor operates with a 5% slip with a certain power to the load. By keeping the input-current magnitude constant and the developed power at 60% of the previous power, the driver frequency is reduced to 45 Hz. Find the new speed and voltage.

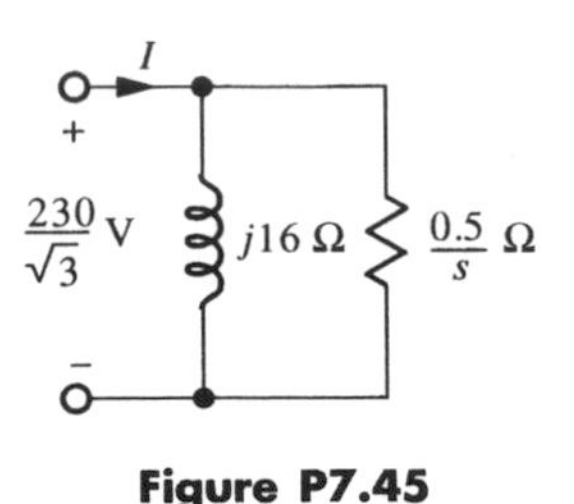

Figure P7.45

7.46. The simplified per-phase equivalent circuit in Fig. P7.46 describes a 230-V, 60-Hz, four-pole, three-phase induction motor in the small-slip region. The motor generates the required torque in a certain application at a slip of 5% when operated at 60 Hz. The motor is driven by an ac controller that can vary the frequency to control speed and the voltage to control the motor current. What voltage should be applied to the machine if operated at 30 Hz if the torque and input-current magnitude are to remain the same as for the nameplate conditions? Ignore mechanical losses.

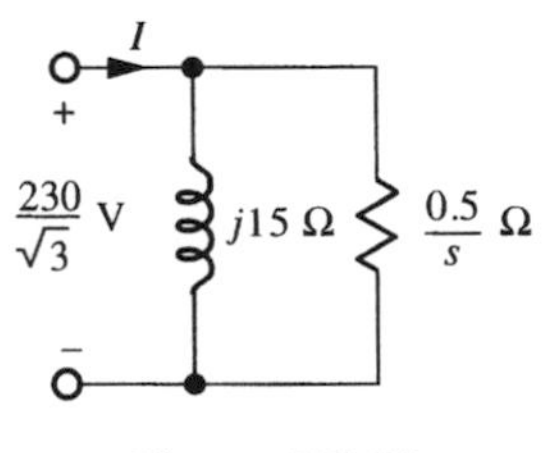

Figure P7.46

7.47. An induction motor runs with a 4% slip at nameplate frequency and voltage. The frequency is reduced to 80% of nameplate frequency and the motor voltage is also reduced to 75% of nameplate voltage. Determine the slip at the reduced speed if the developed torque is the same in both cases. Assume that the motor operates in the small-slip region in both cases.

7.48. The speed of an ac induction motor is controlled by varying the applied voltage and frequency of the ac input voltage. The motor runs at 1164 rpm at full voltage and 60 Hz. Estimate the speed at 60% of nameplate voltage and 70% of nameplate frequency.

Assume small-slip operation and constant load torque.

7.49. The load for a three-phase induction motor requires constant power. The motor is driven by a power electronic circuit to control speed. The motor operates with a slip of 3% at nameplate voltage and frequency. Estimate the slip at 80% of nameplate frequency and 80% of nameplate voltage. Ignore mechanical losses. Assume the motor is operating in the small-slip region.

7.50. The circuit of Fig. P7.50 is a simplified per-phase equivalent circuit for a three-phase induction motor. At the nameplate frequency of 60 Hz and the nameplate voltage of 230-V rms line voltage, the motor draws the nameplate current of 22.1 A, has the nameplate slip of 4%, and develops the nameplate torque of 7.7 N-m. If we allow the frequency and slip to vary, but not the voltage, what is the maximum torque the motor can produce without exceeding the nameplate current of 22.1 A?

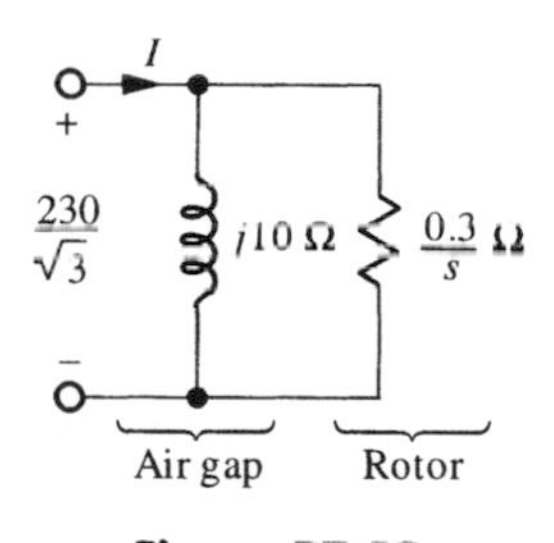

Figure P7.50

7.51. A 60-Hz, four-pole, three-phase induction motor is represented by the per-phase circuit shown in Fig. P7.51 at nameplate voltage and frequency. The developed torque is 10.3 N-m at the nameplate slip of 5%, which you may check if you wish. The motor is driven by a variable-frequency, constant-volts/hertz power electronic drive.

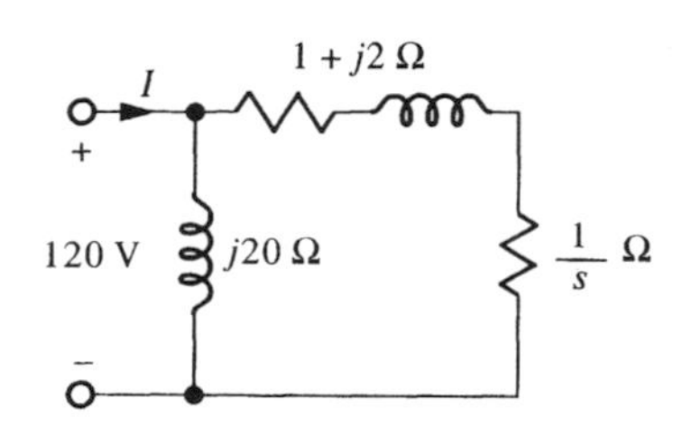

Figure P7.51

(a) Find the drive frequency to produce 3 N-m of developed torque at 875 rpm. Assume operation in the small-slip region.

(b) Find the motor input current at the new frequency and compare (find the ratio) with the nameplate current.

7.52. A two-pole, 60-Hz, three-phase induction motor has the simplified equivalent circuit shown in Fig. P7.52 at 60 Hz. The motor is driven by a constant-volts/hertz variable-frequency drive and drives a load with a torque requirement $T_L = 3 + \omega_m/100$ N-m. Neglect motor mechanical losses.

(a) Find the mechanical speed of the system as a function of the electrical drive frequency. Use the small-slip approximation.

(b) Considering the small-slip approximation valid for $s \leq 0.15$, find the lowest electrical frequency at which the expression in part (a) is valid.

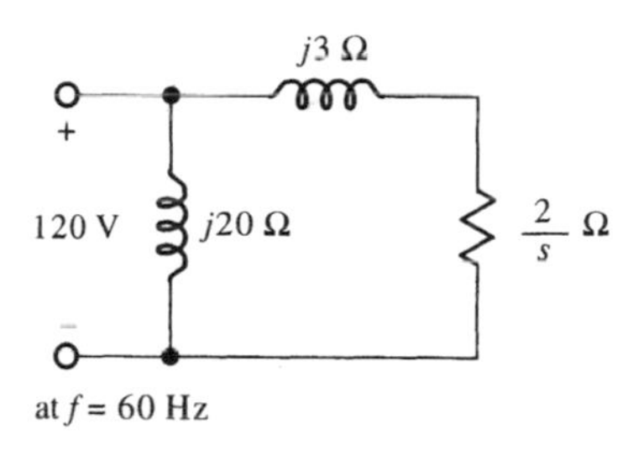

Figure P7.52

7.53. A synchronous motor is to be controlled by a power electronic drive, as shown in Fig. P7.53. The "phase" input corresponds to the $\omega_e t$-axis in Fig. 7.37(a). In effect, the "phase" input consists of triggering signals to the six electronic switches in Fig. 7.36.

(a) If the motor is to turn at 356 rpm, how often is a single diode turned ON and OFF?

(b) If the switches are held in a single state, the rotor takes certain fixed positions. In effect, the motor becomes a stepper motor with certain output angles. What is the smallest increment of angle that can be commanded by this system?

7.54. To speed up an induction motor by 20% and keep the output torque roughly constant, one should (choose one answer): (a) increase the voltage, (b) increase the frequency, (c) increase the voltage and frequency.

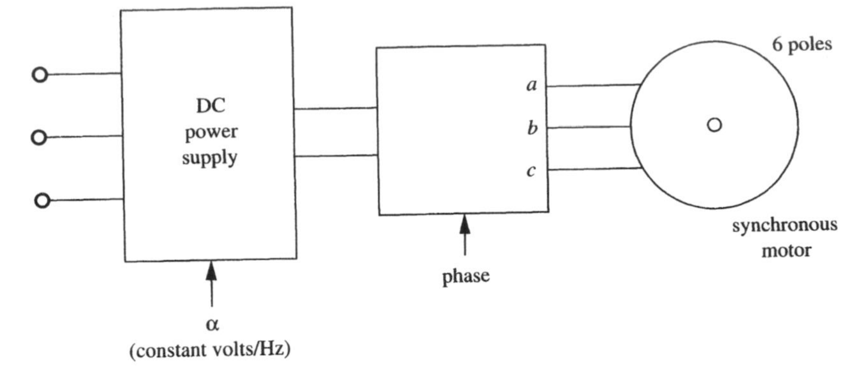

Figure P7.53

Answers to Odd-Numbered Problems

7.1. Less that 10 kΩ.

7.3. **(a)** $V_{CC}I_{Cp}\int_0^{t_r/0.8}\left(\frac{0.8t'}{t_r}\right)dt' = \frac{V_{CC}I_{Cp}t_r}{1.6}$;

(b) $\frac{V_{CC}I_{Cp}}{1.6} \times 2f_{SW}$.

7.5. **(a)** $0.999t_r$; **(b)** $V_{CC}I_{Cp}t_r/4.4$.

7.7. $R_1 = 39$ kΩ; $20\ \Omega < R_2 < 100\ \Omega$.

7.9. 2915 Ω.

7.11. **(a)** 3.57 ms; **(b)** 4 mA; **(c)** 64.3 mA.

7.13. About 26 ms.

7.15. About 3.3 ms.

7.17. 203°.

7.19. **(a)** 254 V peak; **(b)** 6.49 A.

7.21. **(a)** 3303 rpm (should be avoided); **(b)** $0° < \alpha < 90°$; **(c)** **(1)** $11.8° = 0.206$ rad, **(2)** 189 V, **(3)** 8.39 N-m.

7.23. **(a)** 33.6°; **(b)** $i_a(\omega t) = 37.2[\sin(120\pi t - 82.8°) - e^{(0.587 - 120\pi t)/7.90}\sin(33.6° - 82.8°)] - 164[1 - e^{(0.587 - 120\pi t)/7.90}]$; **(c)** $i_a(180°) = 12.0$ A; **(d)** $\beta = 212°$, so Case B.

7.25. **(a)** 70.0°; **(b)** 26.7 A; **(c)** 3.82 A.

7.27. **(a)** 70.0°; **(b)** 182°;

(c)

$v_a(t)$ 325 V, $E = 217$ V, 70°, ωt

(d) 4970 W.

7.29. **(a)**

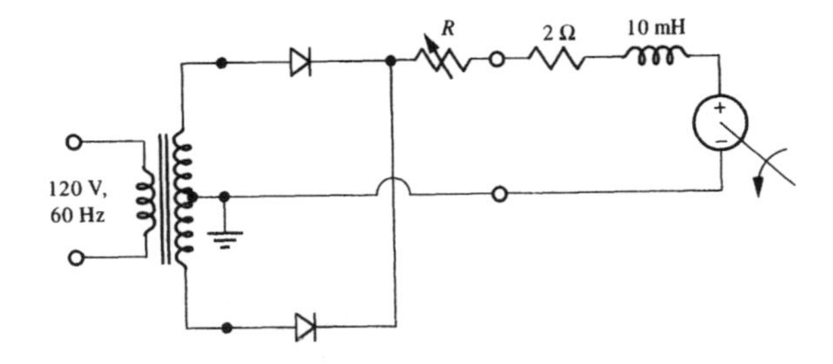

(b) 12.7; **(c)** $E < 4.97$ V; **(d)** 2.13 Ω.

7.31. **(a)** 46.5 V; **(b)** 4.83 mH.

7.33. 2823 rpm, 15.2 A is too much.

7.35. **(a)** 170 V, rms;
(b)

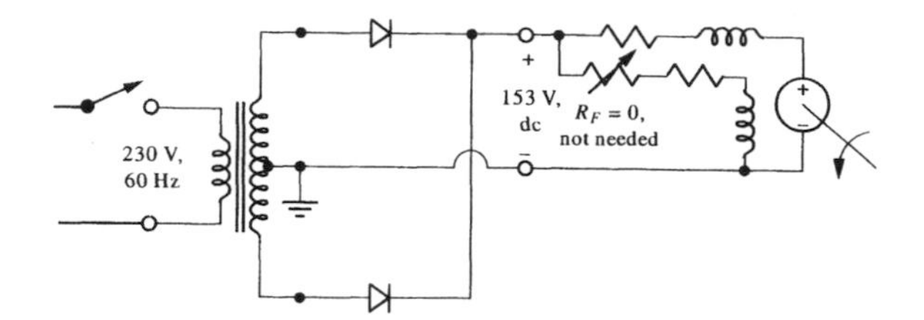

7.37. **(a)** 230 V, rms; **(b)** 80°; **(c)** 29.1°; **(d)** 210°; **(e)** 2251 rpm; **(f)** 3740 W.

7.39. **(a)** 168 V; **(b)** 93.3 W; **(c)** 5.41 Ω; **(d)** $168 < 178$ V, so OK; **(e)** 65.2%; **(f)** 8.52 Ω.

7.41. $I_{dc} = 0.432$ A, $I_{ac} = 0.426$ A, so a rectifier is better.

7.43. **(a)** 5.23 N-m; **(b)** 10.5 N-m.

7.45. 44.2 Hz, 174 V.

7.47. 5.69%.

7.49. 4.77%.

7.51. **(a)** 30.0 Hz (exact = 29.95); **(b)** 0.720.

7.53. **(a)** 56.2 ms; **(b)** 20°.

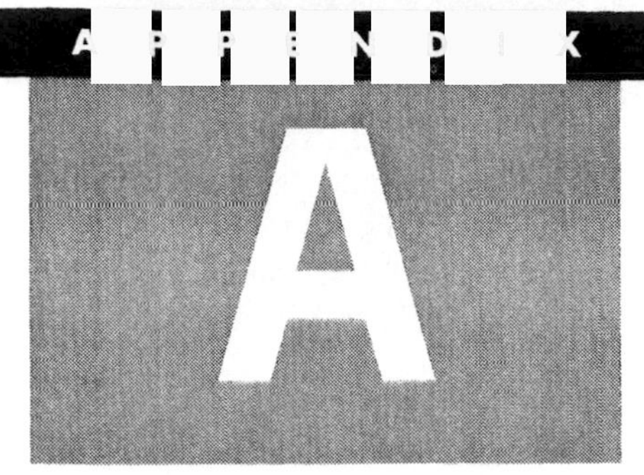

APPENDIX A

Linear Systems

Introduction to Linear Systems

system

What is a system? A *system* consists of several components that together accomplish some purpose. For example, an automobile has a motor, a steering mechanism, lights, padded seats, an entertainment system, and more, operating together to give safe and pleasant transportation. Likewise, a stand-alone ac generator requires a control system to regulate the frequency and voltage of its output. Often, systems are modeled with linear equations. The analysis of such linear systems has furnished a powerful language for system description. This appendix introduces system models and explores basic techniques of linear system description and analysis. Such methods are used in the text in the analysis of motors as system components.

Contents. We begin by generalizing the concept of frequency. We then introduce the language of system notation by defining the generalized impedance and transfer function of electrical circuits. System concepts can be applied broadly, but electric circuits provide a convenient vehicle for developing and illustrating system notation and techniques. From this generalized impedance we determine the natural frequencies and natural response of electrical circuits, and we then investigate the transient response of first- and second-order circuits. Finally we examine the transient response and dynamic stability of a feedback system.

A.1 COMPLEX FREQUENCY

IDEA 7 The Frequency Domain

Definition of complex frequency. A time-domain variable, say, a voltage, is said to have a *complex frequency* $\underline{s}$ when it can be expressed in the form given in Eq. (A.1)

$$v(t) = \text{Re}\{\underline{V}e^{\underline{s}t}\} \tag{A.1}$$

where $\underline{V}$ is a complex number, a phasor, and

$$\underline{s} = \sigma + j\omega \quad \text{s}^{-1} \tag{A.2}$$

where $\underline{s}$, σ, and ω all have units of inverse seconds, s^{-1}. Equation (A.1) is a slightly modified version of the equation used in solving ac circuits using phasor and impedance concepts.

Functions that can be represented by complex frequencies. Functions of the form of Eq. (A.1) are important in linear systems because input excitations of this form will produce outputs or responses of the same form and frequency. Table A.1 gives the responses that may be represented by functions of the form of Eq. (A.1).

One aspect of Table A.1 that requires elaboration is that we may represent a sinusoidal function by two complex frequencies. This is implied by the "real part" operation in Eq. (A.1), because an alternate way to express the real part is through the identity in Eq. (A.3)

$$\text{Re}\{\underline{z}\} = \tfrac{1}{2}(\underline{z} + \underline{z}^*) \tag{A.3}$$

TABLE A.1 Functions That Can Be Represented by Complex Frequencies

Complex frequency, $\underline{s}$	$v(t) = \text{Re}\{\underline{V}e^{\underline{s}t}\}$	*Interpretation*
$\underline{s} = 0$	$v(t) = \text{constant}$	Dc, direct current
Real and negative, $\sigma < 0$	$v(t) = Ve^{-t/\tau}$	First-order transient
Real and positive, $\sigma > 0$	$v(t) = Ve^{+\sigma\tau}$	Unstable growth
Imaginary, $\pm j\omega$	$v(t) = V_p \cos(\omega t + \theta)$	Undamped oscillations, sinusoidal steady state
Complex with real part negative	$v(t) = V_p e^{-t/\tau} \cos(\omega t + \theta)$	Damped, stable oscillations
Complex with real part positive	$v(t) = V_p e^{+\sigma\tau} \cos(\omega t + \theta)$	Growing, unstable oscillations

where $\underline{z}^*$ is the complex conjugate of $\underline{z}$. Thus, Eq. (A.1) can be expressed in the form

$$V_p \cos(\omega t + \theta) = \text{Re}\{\underline{V}e^{j\omega t}\} = \frac{\underline{V}}{2} e^{j\omega t} + \frac{\underline{V}^*}{2} e^{-j\omega t} \tag{A.4}$$

When we compare Eq. (A.4) with Eq. (A.1), we see that two complex frequencies, $\underline{s} = j\omega$ and $\underline{s} = -j\omega$, express a sinusoidal function. In our consideration of complex frequency, we often use this second point of view.

The $\underline{s}$-plane. A complex frequency, $\underline{s}$, can be represented by a point in the complex plane. In Fig. A.1, we show such an $\underline{s}$-plane and identify the regions corresponding to possible time responses.

- The origin corresponds to a constant or dc function.
- Complex frequencies on the real axis correspond to growing and decaying exponential functions.
- Complex frequencies on the imaginary axis correspond to sinusoidal functions.
- The region to the right of the vertical axis, the right-half plane, corresponds to sinusoids that are growing exponentially.
- The left-half plane corresponds to sinusoids that are decreasing exponentially.

IDEA 8 **The Time Domain**

Time domain and frequency domain. The introduction of complex frequency expands the frequency domain to cover a wide class of time-domain behavior. This expansion will allow frequency-domain techniques to be applied to transient problems of linear systems.

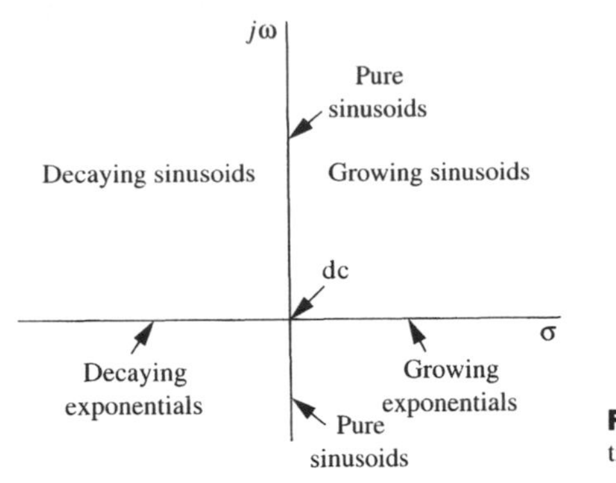

Figure A.1 The $\underline{s}$-plane and associated time functions.

The Frequency Domain

eigenfunction

Importance of $e^{\underline{s}t}$. The function $e^{\underline{s}t}$ is a mathematical probe[1] we use to investigate the properties of linear systems. All linear systems are characterized by certain complex frequencies. Once we determine these frequencies for a given system, we can determine from them the system response in the time domain or the frequency domain. We may thereby examine transient and steady state responses, and we can determine if the system is stable or unstable. In the following section, we show how to determine these characteristic frequencies and how to derive from them the time-domain response of the system.

A.2 IMPEDANCE AND THE TRANSIENT BEHAVIOR OF LINEAR SYSTEMS

Generalized Impedance

generalized impedance

Impedance of *R*, *L*, and *C*. We may use complex frequency to generalize the concept of impedance. To determine the impedance, we excite a circuit with a voltage $\text{Re}\{\underline{\mathbf{V}}(\underline{s})e^{\underline{s}t}\}$ and calculate a response, $\text{Re}\{\underline{\mathbf{I}}(\underline{s})e^{\underline{s}t}\}$. The *generalized impedance* is defined as

$$\underline{\mathbf{Z}}(\underline{s}) = \frac{\underline{\mathbf{V}}(\underline{s})e^{\underline{s}t}}{\underline{\mathbf{I}}(\underline{s})e^{\underline{s}t}} \ \Omega \tag{A.5}$$

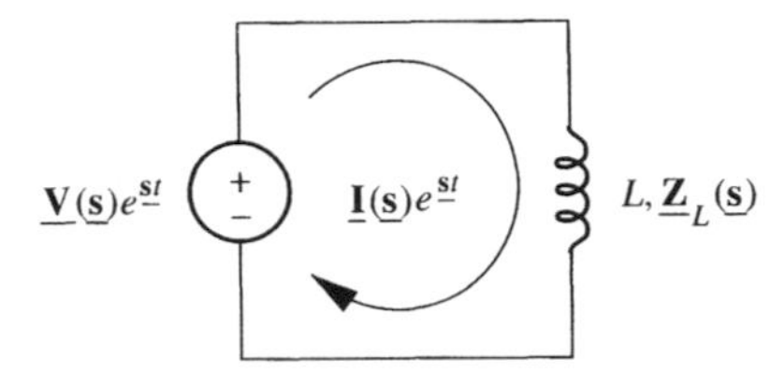

Figure A.2 An inductor. excited by $\underline{\mathbf{V}}e^{\underline{s}t}$.

We illustrate with an inductor, as shown in Fig. A.2. The equations are

$$v(t) = L\frac{d}{dt}i(t) \Rightarrow \underline{\mathbf{V}}(\underline{s})e^{\underline{s}t} = L\frac{d}{dt}\underline{\mathbf{I}}(\underline{s})e^{\underline{s}t} = \underline{s}L\underline{\mathbf{I}}(\underline{s})e^{\underline{s}t} \tag{A.6}$$

[1]The mathematical term for $e^{\underline{s}t}$ is the *eigenfunction* for a linear DE.

where we have omitted the "real part of" for simplicity. The differentiation is performed only on the $e^{\underline{s}t}$ function because this is the only function of time. Thus, the generalized impedance of an inductor is

$$\underline{\mathbf{Z}}_L(\underline{s}) = \frac{\underline{\mathbf{V}}(\underline{s})e^{\underline{s}t}}{\underline{\mathbf{I}}(\underline{s})e^{\underline{s}t}} = \underline{s}L \ \Omega \tag{A.7}$$

In like manner, we can establish the generalized impedances of resistors and capacitors:

$$\underline{\mathbf{Z}}_R(\underline{s}) = R \ \Omega \qquad \text{and} \qquad \underline{\mathbf{Z}}_C(\underline{s}) = \frac{1}{\underline{s}C} \ \Omega \tag{A.8}$$

These are familiar formulas with $j\omega$ replaced by $\underline{s}$.

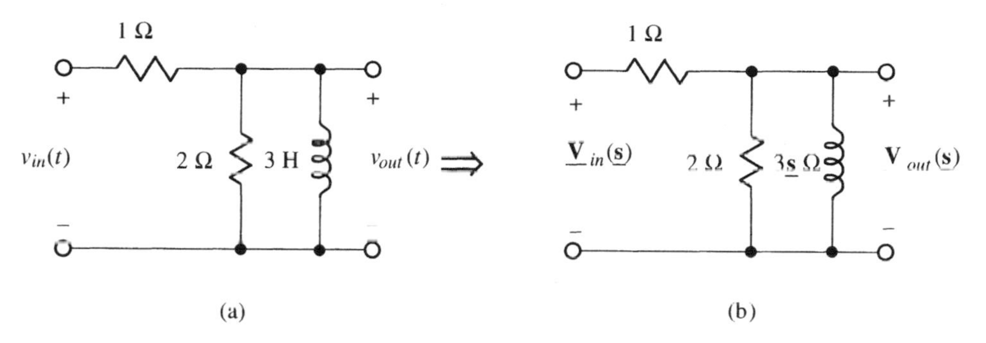

Figure A.3 An *RL* circuit in (a) the time domain; (b) the frequency domain.

Application of generalized impedance. We will analyze the *RL* circuit shown in Fig. A.3(a). In the frequency domain, Fig. A.3(b), the transfer function is determined by combining elements in parallel and then using a voltage divider:

$$\underline{\mathbf{T}}(\underline{s}) = \frac{\underline{\mathbf{V}}_{out}(\underline{s})}{\underline{\mathbf{V}}_{in}(\underline{s})} = \frac{2\|3\underline{s}}{1 + 2\|3\underline{s}} = \frac{6\underline{s}}{9\underline{s} + 2} \tag{A.9}$$

We will use the transfer function in Eq. (A.9) to explore the transient response of the circuit.

Transient Analysis

Forced response from impedance. When we excite a circuit with a voltage or current source, we force a certain output response from the circuit. For example, if we apply a dc source, we expect a dc response. However, part of the transient response of the circuit takes a form that is natural to, and determined by, the circuit itself. Thus the response is always of the form

$$v_{out}(t) = v_{out(f)}(t) + v_{out(n)}(t) \tag{A.10}$$

where $v_{out}(t)$ is the forced response and $v_{out}(t)$ is the natural response. We now explore the role of complex frequency and the transfer function in determining the forced and natural responses of a linear circuit.

Forced response from impedance. The forced response of the circuit may be determined directly from the transfer function when the input voltage is of the form $e^{\underline{s}t}$. We evaluate the transfer function at the value of $\underline{s}$ of the forcing function. For example, if the input has a known complex frequency $\underline{s} = \underline{s}_f$, the output must be of the same form and can be determined from Eq. (A.9)

$$\underline{\mathbf{V}}_{out(f)}(\underline{s}_f) = \underline{\mathbf{V}}_{in}(\underline{s}_f) \times \underline{\mathbf{T}}(\underline{s}_f) \tag{A.11}$$

where $\underline{\mathbf{V}}_{in}$ and $\underline{\mathbf{V}}_{out(f)}(\underline{s}_f)$ are phasors describing the source and force response, respectively, and $\underline{\mathbf{T}}(\underline{s}_f)$ is the transfer function evaluated at the complex frequency of the input. The response in the time domain is thus

$$v_{out}(t) = \text{Re}\{\underline{\mathbf{V}}_{out(f)}(\underline{s}_f)e^{\underline{s}_f t}\} = \text{Re}\{\underline{\mathbf{V}}_{in}(\underline{s}_f) \times \underline{\mathbf{T}}(\underline{s}_f)e^{\underline{s}_f t}\} \tag{A.12}$$

When the exciting function is not of the form $e^{\underline{s}t}$, the forced response must be determined by other methods, such as Laplace transform theory.

Example. For the circuit in Fig. A.3, a voltage source is connected to the input for all time. The complex frequency of this input is $\underline{s}_f = -0.250\ \text{s}^{-1}$. From Eqs. (A.12) and (A.9)

$$\underline{\mathbf{V}}_{out}(-0.250) = 2e^{-0.250} \times \frac{6(-0.250)}{9(-0.250) + 2} = 12e^{-0.250t} \tag{A.13}$$

Thus, $V_{out(f)}(t) = 12e^{-0.250t}$. This is the forced response of the circuit to the input exponential voltage source.

Finding the natural response. The transfer function can also assist us in finding the natural response of the circuit. Consider the case of a circuit excited by a voltage source; we wish to determine the output. The natural response of a linear circuit or system must be of the form

$$v_{out(n)}(t) = \text{Re}\{\underline{\mathbf{V}}_{out(n)}e^{\underline{s}_n t}\} \tag{A.14}$$

where $\underline{\mathbf{V}}_{out(n)}$ is an unknown phasor and $\underline{s}_n$ is the natural frequency of the system. We can determine the natural frequency from the transfer function. The natural frequency corresponds to the case where the input voltage is zero but the output is nonzero, being established solely by the circuit. The transfer function thus becomes

$$\underline{\mathbf{T}}(\underline{s}) = \frac{\underline{\mathbf{V}}_{out(n)}(\underline{s})e^{\underline{s}t}}{\underline{\mathbf{V}}_{out(in)}(\underline{s})e^{\underline{s}t}} = \frac{\neq 0}{0} \tag{A.15}$$

Equation (A.15) can be valid only if the natural frequency makes the transfer function infinite:

$$\underline{\mathbf{T}}(\underline{s}_n) = \infty \tag{A.16}$$

We illustrate with the circuit in Fig. A.3, with the impedance function derived in Eq. (A.9). The circuit has only one natural frequency, which may be determined by setting the transfer function to infinity:

$$\frac{6(\underline{s}_n)}{(9\underline{s}_n + 2)} = \infty \Rightarrow \underline{s}_n = -0.222 \tag{A.17}$$

We will continue the previous example and find the output voltage of the circuit in Fig. A.3 if the voltage source input of Fig. A.4 is applied as an input.

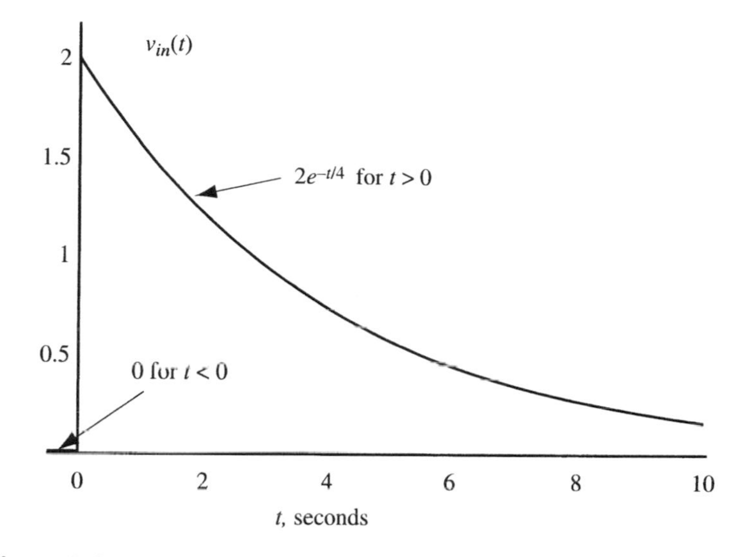

Figure A.4 The input voltage is zero for negative time and a decreasing exponential for positive time.

The output voltage will consist of a forced and a natural component, as in Eq. (A.10). The forced component is derived in Eq. (A.13) and the natural response will be of the form of Eq. (A.14) with $\underline{s}_n = -0.222$, as shown in Eq. (A.17). Thus, the voltage produced at the output by the input voltage in Fig. A.4 must be of the form

$$v_{out}(t) = 12e^{-0.250t} + Ae^{-0.222t} \tag{A.18}$$

where A is to be determined from the initial conditions. In this case, the initial output voltage must be

$$v_{out}(0) = 2 \times \frac{2}{2+1} = 1.333 \text{ V} \tag{A.19}$$

since the initial voltage is 2 V and the inductor acts as an open circuit for sudden changes. Thus, A comes from Eq. (A.18) at $t = 0$

$$1.333 = 12e^0 + Ae^0 \Rightarrow A = -10.666 \tag{A.20}$$

The output voltage is therefore

$$v_{out}(t) = 12e^{-0.250t} - 10.667^{-0.222t} \text{ V} \tag{A.21}$$

which is shown in Fig. A.5.

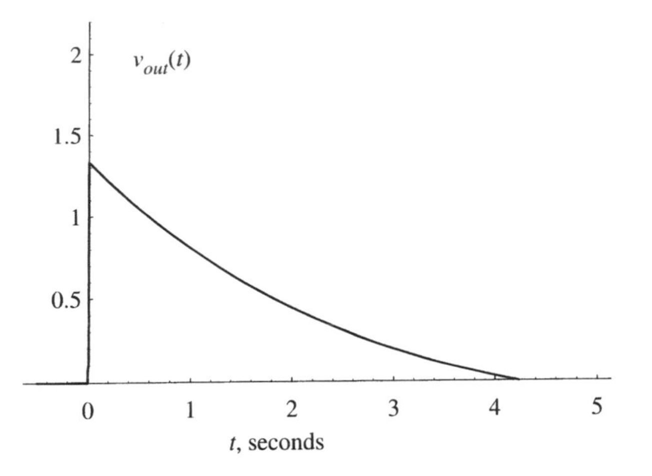

Figure A.5 Response of the circuit of Fig. A.3 to the voltage source of Fig. A.4.

Poles of the system. We now generalize the technique illustrated in the previous example. The impedance or transfer function of a system generally takes the form of Eq. (A.9), with the numerator and denominator polynomials in $\underline{\mathbf{s}}$. Thus, the function can always be factored into the form

$$\underline{\mathbf{T}}(\underline{\mathbf{s}}) = K\frac{(\underline{\mathbf{s}} - \underline{\mathbf{z}}_1)(\underline{\mathbf{s}} - \underline{\mathbf{z}}_2)}{(\underline{\mathbf{s}} - \underline{\mathbf{p}}_1)(\underline{\mathbf{s}} - \underline{\mathbf{p}}_2)(\underline{\mathbf{s}} - \underline{\mathbf{p}}_3)} \tag{A.22}$$

where K is a constant, and we assumed a second-order polynomial in the numerator and a third-order polynomial in the denominator.[2] In Eq. (A.22), $\underline{\mathbf{z}}_1$ and $\underline{\mathbf{z}}_2$ are called the zeros of the function because these are the values of $\underline{\mathbf{s}}$ at which the function goes to zero. Likewise, $\underline{\mathbf{p}}_1$, $\underline{\mathbf{p}}_2$, and $\underline{\mathbf{p}}_3$ are called the poles of the function because these are the values of $\underline{\mathbf{s}}$ at which the function goes to infinity.

Natural frequencies. As we have shown, the poles correspond to the natural frequencies of the system. Except for a multiplicative constant, the poles and zeros of the transfer function fully establish the behavior of the system. They give the natural frequencies, as we have illustrated, and they also directly give the forced response to excitations of the form $e^{\underline{\mathbf{s}}t}$, such as dc ($\underline{\mathbf{s}} = 0$) or ac ($\underline{\mathbf{s}} = j\omega$).

A.3 TRANSIENT RESPONSE OF SECOND-ORDER SYSTEMS

Types of Natural Responses in *RLC* Circuits

Transfer function of an *RLC* circuit. The circuit in Fig. A.6 will be used to investigate the transient response of a second-order system. The *RLC* circuit can exhibit an oscillatory response because the circuit contains two independent energy storage elements (an inductor and a capacitor.) This circuit can be analyzed as a voltage divider:

[2] This indicates a circuit with three independent energy-storage elements.

$$\underline{T}(\underline{s}) = \frac{\underline{V}_{out}(\underline{s})e^{\underline{s}t}}{\underline{V}_{in}(\underline{s})e^{\underline{s}t}} = \frac{R\|1/\underline{s}C}{\underline{s}L + R\|1/\underline{s}C} = \frac{1}{LC}\frac{1}{\underline{s}^2 + (\underline{s}/RC) + 1/(LC)} \quad \text{(A.23)}$$

where $\underline{T}(\underline{s})$ is the transfer function.

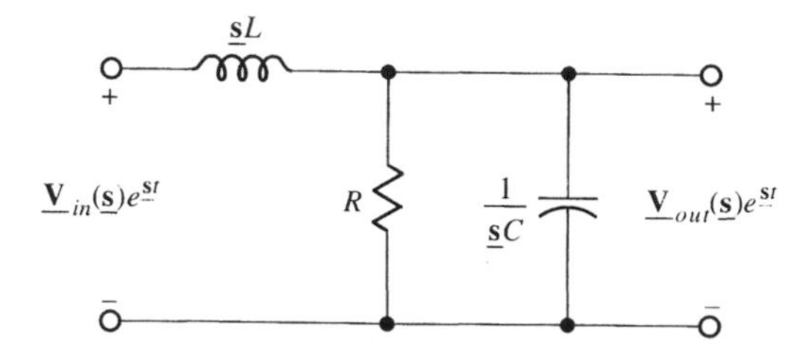

Figure A.6 An *RLC* circuit in the frequency domain. The transfer function is the ratio of output to input voltage.

Natural frequencies of the circuit. The poles of the transfer function correspond to an output voltage with no input voltage, and hence are the natural frequencies of the circuit with the input short-circuited. These poles are investigated in detail because they reveal the various types of behavior that can result with an *RLC* circuit. We may determine the poles with the quadratic formula

$$\underline{s}^2 + \frac{\underline{s}}{RC} + \frac{1}{LC} = 0 \Rightarrow \underline{s}_n = -\frac{1}{2RC} \pm \sqrt{\left(\frac{1}{2RC}\right)^2 - \frac{1}{LC}} \quad s^{-1} \quad \text{(A.24)}$$

There are four possibilities, depending on the relative values of R, L, and C. The possibilities are undamped, underdamped, critically damped, and overdamped responses, and the numerical values of the natural frequencies come from Eq. (A.24). Table A.2 summarizes the results.

TABLE A.2 Transient Responses of a Second-Order System

Condition	*Natural frequency*	*Response*	*Description*
$R = \infty$	$\underline{s} = \pm j\omega 0$	$A\cos(\omega_0 t + \theta)$	Undamped sinusoid, pure oscillation
$R > \frac{1}{2}\sqrt{\frac{L}{C}}$	$\underline{s} = -\alpha \pm j\omega$	$Ae^{-\alpha t}\cos(\omega t + \theta)$	Damped sinusoid, dying oscillation
$R = \frac{1}{2}\sqrt{\frac{L}{C}}$	$\underline{s} = -\alpha, -\alpha$	$Ae^{-\alpha t} + Bte^{-\alpha t}$	Critically damped response
$R < \frac{1}{2}\sqrt{\frac{L}{C}}$	$\underline{s} = -\alpha_1, -\alpha_2$	$Ae^{-\alpha_1 t} + Be^{-\alpha_2 t}$	Overdamped response

Influence of resistancy. The nature of the response is determined by the magnitude of the resistance in the circuit in Fig. A.3. The first column gives R relative to, $\frac{1}{2}\sqrt{L/C}$,which is a critical parameter for the circuit. We begin with $R = \infty$, which is equivalent to removing the resistor from the circuit, and each row shows the effect of increasingly smaller values of resistance. The effect of reducing the resistance is to dampen the response since smaller resistances suck energy out of the circuit at higher rates.

A.4 SYSTEM ANALYSIS

Introduction. In this section we will further explore the notation and techniques commonly used to describe linear systems. Our example will be the feedback control system shown in Fig. A.7. We show that first- and second-order systems are unconditionally stable, although a second-order system response may be unacceptable in practice. We then assert that a third-order system can become unstable for large loop gain.

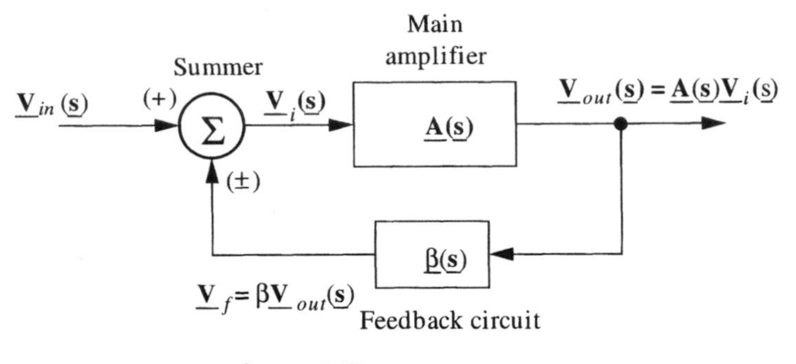

Figure A.7 A feedback system.

System functions A *system function,* $\underline{\mathbf{F}}(\underline{s})$, relates the transforms of input and output in the equation

$$\frac{\underline{\mathbf{X}}_{out}(\underline{\mathbf{s}})}{\underline{\mathbf{X}}_{in}(\underline{\mathbf{s}})} = \underline{\mathbf{F}}(\underline{s}) \tag{A.25}$$

where the $\underline{\mathbf{X}}$'s are frequency-domain transforms of the time-domain input and output variables. The system function is similar to the transfer function, except that the input and output variables may be any of a wide class of variables. Our example uses only voltages as variables, but mechanical and thermal variables are common in systems.

Main amplifier system function. We will assume the main amplifier has a single time constant, τ, and a gain A_0 at dc. To satisfy both properties, the output voltage of the amplifier must satisfy the first-order differential equation

$$\tau\frac{dv_{out}(t)}{dt} + v_{out}(t) = A_0 v_i(t) \tag{A.26}$$

where $v_i(t)$ is the input voltage, τ is the time constant of the amplifier, and $v_{out}(t)$ is the output voltage. We may transform Eq. (A.26) into the frequency domain by assuming input and output voltage to be of the form $e^{\underline{s}t}$, with the results

$$(\tau \underline{s} + 1)\underline{\mathbf{V}}_{out}(\underline{s}) = A_0 \underline{\mathbf{V}}_i(\underline{s}) \Rightarrow \underline{\mathbf{V}}_{out}(\underline{s}) = \underbrace{\frac{A_0/\tau}{\underline{s} + (1/\tau)}}_{\substack{\text{system function} \\ \text{for amplifier}}} \times \underline{\mathbf{V}}_i(\underline{s}) \tag{A.27}$$

where $\underline{\mathbf{V}}_{out}(\underline{s})$ is the transform of the output voltage and $\underline{\mathbf{V}}_i(\underline{s})$ is the transform of the input voltage. Thus, we may represent the main amplifier in the frequency domain by the system component shown in Fig. A.8.

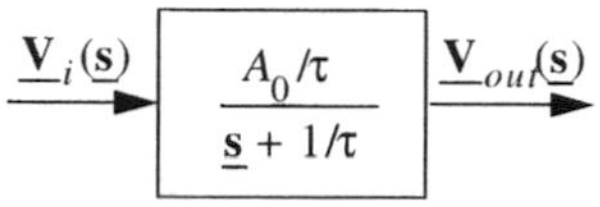

Figure A.8 System function for the main amplifier with a single time constant.

System function with feedback. The transfer function of the system with feedback in Fig. A.7, $\underline{\mathbf{F}}(\underline{s})$, is easily shown to be

$$\frac{\underline{\mathbf{V}}_{out}(\underline{s})}{\underline{\mathbf{V}}_{in}(\underline{s})} = \underline{\mathbf{F}}(\underline{s}) = \frac{A(\underline{s})}{1 - L(\underline{s})} \tag{A.28}$$

The natural frequencies, which show the dynamic properties of the system, occur where

$$1 - L(\underline{s}) = 0 \tag{A.29}$$

First-order systems. As an example of a first-order system, let

$$L(\underline{s}) = \frac{\beta A_0}{\underline{s} + s_n} \tag{A.30}$$

where s_n is the negative of the natural frequency of the main amplifier, a real and positive number. This is the case pictured in Fig. A.7, with $\underline{\beta}(\underline{s}) = \beta$, a constant. In Eq. (A.30), we consider that βA_0 is a positive number and the natural frequency, s_n, is also a positive number. In this case, $1 - L(\underline{s}) = 0$ yields

$$\underline{s}_n + (s_n + \beta A_0) = 0 \Rightarrow \underline{s} = -(s_n + \beta A_0) \tag{A.31}$$

with the accompanying natural response in the time domain of

$$v_{out}(t) = Ae^{-(s_n + \beta A_0)t} \tag{A.32}$$

This response is unconditionally stable. The effect of increasing the magnitude of the feedback is to increase the magnitude of the natural frequency and hence shorten the time constant of the system. This is equivalent to increasing the bandwidth of the system.

Second-order systems. We now consider that the main amplifier has two time constants, and hence we have a second-order system. Let

$$L(\underline{s}) = -\frac{\beta A_0}{(\underline{s} + s_1)(\underline{s} + s_2)} \tag{A.33}$$

In Eq. (A.33), we assume the βA_0 to be positive and the natural frequencies, s_1 and s_2, are also positive. In this case, $1 - L(\underline{s}) = 0$ yields

$$\underline{s}^2 + (s_1 + s_2)\underline{s} + s_1 s_2 + (\beta A_0) = 0 \tag{A.34}$$

The quadratic equation in Eq. (A.34) has the solution

$$\begin{aligned} \underline{s} &= -\frac{s_1 + s_2}{2} \pm \sqrt{\left(\frac{s_1 + s_2}{2}\right) - (s_1 s_2 + \beta A_0)} \\ &= -\frac{s_1 + s_2}{2} \pm \sqrt{\left(\frac{s_1 - s_2}{2}\right)^2 - \beta A_0} \end{aligned} \tag{A.35}$$

The character of the natural response depends on the magnitude of the dc loop gain. We summarize the three types of possible responses in Table A.3.

TABLE A.3 Transient Responses of a Second-Order System

Condition	*Natural frequency*	*Response*
$\beta A_0 < \left(\frac{s_1 - s_2}{2}\right)^2$	Real, negative, and distinct	overdamped $v(t) = Ae^{-t/\tau_1} + Be^{-t/\tau_2} + C$
$\beta A_0 = \left(\frac{s_1 - s_2}{2}\right)^2$	Real, negative, and equal	critical damping $v(t) = Ae^{-t/\tau} + Be^{-t/\tau} + C$
$\beta A_0 > \left(\frac{s_1 - s_2}{2}\right)^2$	Complex conjugates, with negative real part	underdamped oscillation $v(t) = Ae^{-t/\tau_1}\cos(\omega t + B) + C$

These responses are unconditionally stable, meaning that the system never enters a region of uncontrolled oscillations. However, with large loop gain, the system response may be strongly oscillatory. Figure A.9 shows responses of a system with time constants of 30 and 300 seconds, as the dc loop gain takes values of -1, -10, and -50. The responses are normalized to give the same final value of 10. Note that the response with a gain of -50, though stable, has large oscillations and might be unacceptable in practice.

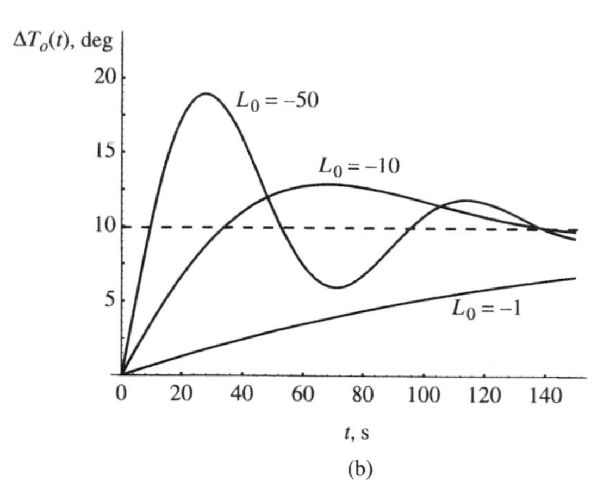

Figure A.9 Responses of a second-order system as the magnitude of the loop gain is increased.

Third- and higher-order systems. Third- and higher-order systems are capable of unstable behavior, with growing oscillations, if the loop gain is sufficiently high. In such systems, as indeed in second-order systems, response of the system is frequently improved through the adding of electrical filters in the feedback loop. Control of the response of feedback systems is an important topic in control theory, but is beyond the scope of this appendix.

Summary. In this appendix we have introduced the notation and techniques of system theory. We used frequency-domain techniques with complex frequency to explore the transient response and stability of systems with and without feedback.

Index

F

G

H

I

L

M

N